21. $\displaystyle\int \sqrt{a^2 + x^2}\, dx = \frac{x}{2}\sqrt{a^2 + x^2} + \frac{a^2}{2}\sinh^{-1}\frac{x}{a} + C$

22. $\displaystyle\int x^2\sqrt{a^2 + x^2}\, dx = \frac{x(a^2 + 2x^2)\sqrt{a^2 + x^2}}{8} - \frac{a^4}{8}\sinh^{-1}\frac{x}{a} + C$

23. $\displaystyle\int \frac{\sqrt{a^2 + x^2}}{x}\, dx = \sqrt{a^2 + x^2} - a\sinh^{-1}\left|\frac{a}{x}\right| + C$

24. $\displaystyle\int \frac{\sqrt{a^2 + x^2}}{x^2}\, dx = \sinh^{-1}\frac{x}{a} - \frac{\sqrt{a^2 + x^2}}{x} + C$

25. $\displaystyle\int \frac{x^2}{\sqrt{a^2 + x^2}}\, dx = -\frac{a^2}{2}\sinh^{-1}\frac{x}{a} + \frac{x\sqrt{a^2 + x^2}}{2} + C$

26. $\displaystyle\int \frac{dx}{x\sqrt{a^2 + x^2}} = -\frac{1}{a}\ln\left|\frac{a + \sqrt{a^2 + x^2}}{x}\right| + C$

27. $\displaystyle\int \frac{dx}{x^2\sqrt{a^2 + x^2}} = -\frac{\sqrt{a^2 + x^2}}{a^2 x} + C$

28. $\displaystyle\int \frac{dx}{\sqrt{a^2 - x^2}} = \sin^{-1}\frac{x}{a} + C$

29. $\displaystyle\int \sqrt{a^2 - x^2}\, dx = \frac{x}{2}\sqrt{a^2 - x^2} + \frac{a^2}{2}\sin^{-1}\frac{x}{a} + C$

30. $\displaystyle\int x^2\sqrt{a^2 - x^2}\, dx = \frac{a^4}{8}\sin^{-1}\frac{x}{a} - \frac{1}{8}x\sqrt{a^2 - x^2}\,(a^2 - 2x^2) + C$

31. $\displaystyle\int \frac{\sqrt{a^2 - x^2}}{x}\, dx = \sqrt{a^2 - x^2} - a\ln\left|\frac{a + \sqrt{a^2 - x^2}}{x}\right| + C$

32. $\displaystyle\int \frac{\sqrt{a^2 - x^2}}{x^2}\, dx = -\sin^{-1}\frac{x}{a} - \frac{\sqrt{a^2 - x^2}}{x} + C$

33. $\displaystyle\int \frac{x^2}{\sqrt{a^2 - x^2}}\, dx = \frac{a^2}{2}\sin^{-1}\frac{x}{a} - \frac{1}{2}x\sqrt{a^2 - x^2} + C$

34. $\displaystyle\int \frac{dx}{x\sqrt{a^2 - x^2}} = -\frac{1}{a}\ln\left|\frac{a + \sqrt{a^2 - x^2}}{x}\right| + C$

35. $\displaystyle\int \frac{dx}{x^2\sqrt{a^2 - x^2}} = -\frac{\sqrt{a^2 - x^2}}{a^2 x} + C$

36. $\displaystyle\int \frac{dx}{\sqrt{x^2 - a^2}} = \cosh^{-1}\frac{x}{a} + C = \ln\left|x + \sqrt{x^2 - a^2}\right| + C$

37. $\displaystyle\int \sqrt{x^2 - a^2}\, dx = \frac{x}{2}\sqrt{x^2 - a^2} - \frac{a^2}{2}\cosh^{-1}\frac{x}{a} + C$

38. $\displaystyle\int \left(\sqrt{x^2 - a^2}\right)^n dx = \frac{x\left(\sqrt{x^2 - a^2}\right)^n}{n + 1} - \frac{na^2}{n + 1}\int \left(\sqrt{x^2 - a^2}\right)^{n-2} dx, \quad n \neq -1$

39. $\displaystyle\int \frac{dx}{\left(\sqrt{x^2 - a^2}\right)^n} = \frac{x\left(\sqrt{x^2 - a^2}\right)^{2-n}}{(2 - n)a^2} - \frac{n - 3}{(n - 2)a^2}\int \frac{dx}{\left(\sqrt{x^2 - a^2}\right)^{n-2}}, \quad n \neq 2$

40. $\displaystyle\int x\left(\sqrt{x^2 - a^2}\right)^n dx = \frac{\left(\sqrt{x^2 - a^2}\right)^{n+2}}{n + 2} + C, \quad n \neq -2$

41. $\displaystyle\int x^2\sqrt{x^2 - a^2}\, dx = \frac{x}{8}(2x^2 - a^2)\sqrt{x^2 - a^2} - \frac{a^4}{8}\cosh^{-1}\frac{x}{a} + C$

42. $\displaystyle\int \frac{\sqrt{x^2 - a^2}}{x}\, dx = \sqrt{x^2 - a^2} - a\sec^{-1}\left|\frac{x}{a}\right| + C$

43. $\displaystyle\int \frac{\sqrt{x^2 - a^2}}{x^2}\, dx = \cosh^{-1}\frac{x}{a} - \frac{\sqrt{x^2 - a^2}}{x} + C$

Continued overleaf.

44. $\int \dfrac{x^2}{\sqrt{x^2 - a^2}}\, dx = \dfrac{a^2}{2} \cosh^{-1} \dfrac{x}{a} + \dfrac{x}{2} \sqrt{x^2 - a^2} + C$

45. $\int \dfrac{dx}{x\sqrt{x^2 - a^2}} = \dfrac{1}{a} \sec^{-1} \left|\dfrac{x}{a}\right| + C = \dfrac{1}{a} \cos^{-1} \left|\dfrac{a}{x}\right| + C$

46. $\int \dfrac{dx}{x^2 \sqrt{x^2 - a^2}} = \dfrac{\sqrt{x^2 - a^2}}{a^2 x} + C$ 47. $\int \dfrac{dx}{\sqrt{2ax - x^2}} = \sin^{-1}\left(\dfrac{x - a}{a}\right) + C$

48. $\int \sqrt{2ax - x^2}\, dx = \dfrac{x - a}{2} \sqrt{2ax - x^2} + \dfrac{a^2}{2} \sin^{-1}\left(\dfrac{x - a}{a}\right) + C$

49. $\int (\sqrt{2ax - x^2})^n\, dx = \dfrac{(x - a)(\sqrt{2ax - x^2})^n}{n + 1} + \dfrac{na^2}{n + 1} \int (\sqrt{2ax - x^2})^{n-2}\, dx,$

50. $\int \dfrac{dx}{(\sqrt{2ax - x^2})^n} = \dfrac{(x - a)(\sqrt{2ax - x^2})^{2-n}}{(n - 2)a^2} + \dfrac{(n - 3)}{(n - 2)a^2} \int \dfrac{dx}{(\sqrt{2ax - x^2})^{n-2}}$

51. $\int x\sqrt{2ax - x^2}\, dx = \dfrac{(x + a)(2x - 3a)\sqrt{2ax - x^2}}{6} + \dfrac{a^3}{2} \sin^{-1} \dfrac{x - a}{a} + C$

52. $\int \dfrac{\sqrt{2ax - x^2}}{x}\, dx = \sqrt{2ax - x^2} + a \sin^{-1} \dfrac{x - a}{a} + C$

53. $\int \dfrac{\sqrt{2ax - x^2}}{x^2}\, dx = -2\sqrt{\dfrac{2a - x}{x}} - \sin^{-1}\left(\dfrac{x - a}{a}\right) + C$

54. $\int \dfrac{x\, dx}{\sqrt{2ax - x^2}} = a \sin^{-1} \dfrac{x - a}{a} - \sqrt{2ax - x^2} + C$

55. $\int \dfrac{dx}{x\sqrt{2ax - x^2}} = -\dfrac{1}{a}\sqrt{\dfrac{2a - x}{x}} + C$

56. $\int \sin ax\, dx = -\dfrac{1}{a} \cos ax + C$ 57. $\int \cos ax\, dx = \dfrac{1}{a} \sin ax + C$

58. $\int \sin^2 ax\, dx = \dfrac{x}{2} - \dfrac{\sin 2ax}{4a} + C$ 59. $\int \cos^2 ax\, dx = \dfrac{x}{2} + \dfrac{\sin 2ax}{4a} + C$

60. $\int \sin^n ax\, dx = \dfrac{-\sin^{n-1} ax \cos ax}{na} + \dfrac{n - 1}{n} \int \sin^{n-2} ax\, dx$

61. $\int \cos^n ax\, dx = \dfrac{\cos^{n-1} ax \sin ax}{na} + \dfrac{n - 1}{n} \int \cos^{n-2} ax\, dx$

62. (a) $\int \sin ax \cos bx\, dx = -\dfrac{\cos (a + b)x}{2(a + b)} - \dfrac{\cos (a - b)x}{2(a - b)} + C, \quad a^2 \neq b^2$

(b) $\int \sin ax \sin bx\, dx = \dfrac{\sin (a - b)x}{2(a - b)} - \dfrac{\sin (a + b)x}{2(a + b)}, \quad a^2 \neq b^2$

(c) $\int \cos ax \cos bx\, dx = \dfrac{\sin (a - b)x}{2(a - b)} + \dfrac{\sin (a + b)x}{2(a + b)}, \quad a^2 \neq b^2$

63. $\int \sin ax \cos ax\, dx = -\dfrac{\cos 2ax}{4a} + C$

64. $\int \sin^n ax \cos ax\, dx = \dfrac{\sin^{n+1} ax}{(n + 1)a} + C, \quad n \neq -1$

This table is continued on the endpapers at the back.

Undergraduate Texts in Mathematics

Editors

J. H. Ewing
F. W. Gehring
P. R. Halmos

Some Springer Books on]

MATH! Encounters with High School Students
1985. ISBN 96129-1

The Beauty of Doing Mathematics
1985. ISBN 96149-6

Geometry. A High School Course (with G. Murrow)
1991. ISBN 96654-4

Basic Mathematics
1988. ISBN 96787-7

A First Course in Calculus
1991. ISBN 96201-8

Calculus of Several Variables
1988. ISBN 96405-3

Introduction to Linear Algebra
1988. ISBN 96205-0

Linear Algebra
1989. ISBN 96412-6

Undergraduate Algebra
1990. ISBN 97279-x

Undergraduate Analysis
1989. ISBN 90800-5

Complex Analysis
1993. ISBN 97886-0

Real and Functional Analysis
1993. ISBN 94001-4

Serge Lang

A First Course in Calculus

Fifth Edition

With 367 Illustrati

Springer-Verlag
New York Berlin Heidelberg London Paris
Tokyo Hong Kong Barcelona Budapest

Serge Lang
Department of Mathematics
Yale University
New Haven, CT 06520
USA

Mathematics Subjects Classifications (1991): 26-01, 26A06

Library of Congress Cataloging in Publication Data
Lang, Serge, 1927–
    A first course in calculus.
    (Undergraduate texts in mathematics)
    Includes index.
    1. Calculus. I. Title. II. Series.
QA303.L26 1986   515   85-17181

Printed on acid-free paper.

Previous editions of this book were published in 1978, 1973, 1968, 1964 by Addison-Wesley Publishing Company, Inc.

Typeset by Composition House Ltd., Salisbury, England.
Printed and bound by R. R. Donnelley & Sons, Harrisonburg, Virginia.
Printed in the United States of America.

9 8 7 6 5 4 (Fourth Corrected Printing, 1993)

ISBN 0-387-96201-8 Springer-Verlag New York Berlin Heidelberg Tokyo
ISBN 3-540-96201-8 Springer-Verlag Berlin Heidelberg New York Tokyo

Foreword

The purpose of a first course in calculus is to teach the student the basic notions of derivative and integral, and the basic techniques and applications which accompany them. The very talented students, with an obvious aptitude for mathematics, will rapidly require a course in functions of one real variable, more or less as it is understood by professional mathematicians. This book is not primarily addressed to them (although I hope they will be able to acquire from it a good introduction at an early age).

I have not written this course in the style I would use for an advanced monograph, on sophisticated topics. One writes an advanced monograph for oneself, because one wants to give permanent form to one's vision of some beautiful part of mathematics, not otherwise accessible, somewhat in the manner of a composer setting down his symphony in musical notation.

This book is written for the students to give them an immediate, and pleasant, access to the subject. I hope that I have struck a proper compromise, between dwelling too much on special details and not giving enough technical exercises, necessary to acquire the desired familiarity with the subject. In any case, certain routine habits of sophisticated mathematicians are unsuitable for a first course.

Rigor. This does not mean that so-called rigor has to be abandoned. The logical development of the mathematics of this course from the most basic axioms proceeds through the following stages:

Set theory	Numbers (i.e. real numbers)
Integers (whole numbers)	Limits
Rational numbers (fractions)	Derivatives and forward.

No one in his right mind suggests that one should begin a course with set theory. It happens that the most satisfactory place to jump into the subject is between limits and derivatives. In other words, any student is ready to accept as intuitively obvious the notions of numbers and limits and their basic properties. Experience shows that the students do *not* have the proper psychological background to accept a theoretical study of limits, and resist it tremendously.

In fact, it turns out that one can have the best of both ideas. The arguments which show how the properties of limits can be reduced to those of numbers form a self-contained whole. Logically, it belongs *before* the subject matter of our course. Nevertheless, we have inserted it as an appendix. If some students feel the need for it, they need but read it and visualize it as Chapter 0. In that case, everything that follows is as rigorous as any mathematician would wish it (so far as objects which receive an analytic definition are concerned). Not one word need be changed in any proof. I hope this takes care once and for all of possible controversies concerning so-called rigor.

Most students will not feel any need for it. My opinion is that epsilon-delta should be entirely left out of ordinary calculus classes.

Language and logic. It is not generally recognized that some of the major difficulties in teaching mathematics are analogous to those in teaching a foreign language. (The secondary schools are responsible for this. Proper training in the secondary schools could entirely eliminate this difficulty.) Consequently, I have made great efforts to carry the student verbally, so to say, in using proper mathematical language. It seems to me essential that students be required to write their mathematics papers in full and coherent sentences. A large portion of their difficulties with mathematics stems from their slapping down mathematical symbols and formulas isolated from a meaningful sentence and appropriate quantifiers. Papers should also be required to be neat and legible. They should not look as if a stoned fly had just crawled out of an inkwell. Insisting on reasonable standards of expression will result in drastic improvements of mathematical performance. The systematic use of words like "let," "there exists," "for all," "if...then," "therefore" should be taught, as in sentences like:

Let $f(x)$ be the function such that....
There exists a number such that....
For all numbers x with $0 < x < 1$, we have....
If f is a differentiable function and K a constant such that $f'(x) = Kf(x)$, then $f(x) = Ce^{Kx}$ for some constant C.

Plugging in. I believe that it is unsound to view "theory" as adversary to applications or "computations." The present book treats both as complementary to each other. Almost always a theorem gives a tool for more efficient computations (e.g. Taylor's formula, for computing values of functions). Different classes will of course put different emphasis on them, omitting some proofs, but I have found that if no excessive pedantry is introduced, students are willing, and even eager, to understand the reasons for the truth of a result, i.e. its proof.

It is a disservice to students to teach calculus (or other mathematics, for that matter) in an exclusive framework of "plugging in" ready-made formulas. Proper teaching consists in making the student adept at handling a large number of techniques in a routine manner (in particular, knowing how to plug in), but it also consists in training students in knowing some general principles which will allow them to deal with new situations for which there are no known formulas to plug in.

It is impossible in one semester, or one year, to find the time to deal with all desirable applications (economics, statistics, biology, chemistry, physics, etc.). On the other hand, covering the proper balance between selected applications and selected general principles will equip students to deal with other applications or situations by themselves.

Worked-out problems and exercises. For the convenience of both students and instructors, a large number of worked-out problems has been added in the present edition. Many of these have been put in the answer section, to be referred to as needed. I did this for at least two reasons. First, in the text, they might obscure the main ideas of the course. Second, it is a good idea to make students think about a problem before they see it worked out. They are then much more receptive, and will retain the methods better for having encountered the difficulties (whatever they are, depending on individual students) by themselves. Both the inclusion of worked-out examples and their placement in the answer section was requested by students. Unfortunately, the requirements for good teaching, testing, and academic pressures are in conflict here. The *de facto* tendency is for students to object to being asked to think (even if they fail), because they are afraid of being penalized with bad grades for homework. Instructors may either make too strong requirements on students, or may take the path of least resistance and never require anything beyond plugging in new numbers in a type of exercise which has already been worked out (in class or in the book). I believe that testing conditions (limited time, pressures of other courses and examinations) make it difficult (if not unreasonable) to *test* students other than with basic, routine problems. I do not conclude that the course should consist only of this type of material. Some students often take the attitude that if something is not on tests, then why should it be covered in the course? I object very much to this attitude. I have no global solution to these conflicting pressures.

General organization. I have made no great innovations in the exposition of calculus. Since the subject was discovered some 300 years ago, such innovations were out of the question.

I have cut down the amount of analytic geometry to what is both necessary and sufficient for a general first course in this type of mathematics. For some applications, more is required, but these applications are fairly specialized. For instance, if one needs the special properties concerning the focus of a parabola in a course on optics, then that is the place to present them, not in a general course which is to serve mathematicians, physicists, chemists, biologists, and engineers, to mention but a few. I regard the tremendous emphasis on the analytic geometry of conics which has been the fashion for many years as an unfortunate historical accident. What is important is that the basic idea of representing a graph by a figure in the plane should be thoroughly understood, together with basic examples. The more abstruse properties of ellipses, parabolas, and hyperbolas should be skipped.

Differentiation and the elementary functions are covered first. Integration is covered second. Each makes up a coherent whole. For instance, in the part on differentiation, rate problems occur three times, illustrating the same general principle but in the contexts of several elementary functions (polynomials at first, then trigonometric functions, then inverse functions). This repetition at brief intervals is pedagogically sound, and contributes to the coherence of the subject. It is also natural to slide from integration into Taylor's formula, proved with remainder term by integrating by parts. It would be slightly disagreeable to break this sequence.

Experience has shown that Chapters III through VIII make up an appropriate curriculum for **one term (differentiation and elementary functions)** while Chapters IX through XIII make up an appropriate curriculum for a **second term (integration and Taylor's formula)**. The first two chapters may be used for a quick review by classes which are not especially well prepared.

I find that all these factors more than offset the possible disadvantage that for other courses (physics, chemistry perhaps) integration is needed early. This may be true, but so are the other topics, and unfortunately the course has to be projected in a totally ordered way on the time axis.

In addition to this, studying the log and exponential before integration has the advantage that we meet in a special concrete case the situation where we find an antiderivative by means of area: $\log x$ is the area under $1/x$ between 1 and x. We also see in this concrete case how $dA(x)/dx = f(x)$, where $A(x)$ is the area. This is then done again in full generality when studying the integral. Furthermore, inequalities involving lower sums and upper sums, having already been used in this concrete case, become more easily understandable in the general case. Classes which start the term on integration without having gone through the part on differentiation might well start with the last section of the chapter on logarithms, i.e. the last section of Chapter VIII.

Taylor's formula is proved with the integral form of the remainder, which is then properly estimated. The proof with integration by parts is more natural than the other (differentiating some complicated expression pulled out of nowhere), and is the one which generalizes to the higher dimensional case. I have placed integration after differentiation, because otherwise one has no technique available to evaluate integrals.

I personally think that the computations which arise naturally from Taylor's formula (computations of values of elementary functions, computation of e, π, $\log 2$, computations of definite integrals to a few decimals, traditionally slighted in calculus courses) are important. This was clear already many years ago, and is even clearer today in the light of the pocket computer proliferation. Designs of such computers rely precisely on effective means of computation by means of the Taylor polynomials. Learning how to estimate effectively the remainder term in Taylor's formula gives a very good feeling for the elementary functions, not obtainable otherwise.

The computation of integrals like

$$\int_0^1 e^{-x^2}\,dx \qquad \text{or} \qquad \int_0^{0.1} e^{-x^2}\,dx$$

which can easily be carried out numerically, without the use of a simple form for the indefinite integral, should also be emphasized. Again it gives a good feeling for an aspect of the integral not obtainable otherwise. Many texts slight these applications in favor of expanded treatment of applications of integration to various engineering situations, like fluid pressure on a dam, mainly by historical accident. I have nothing against fluid pressure, but one should keep in mind that too much time spent on some topics prevents adequate time being spent on others. For instance, Ron Infante tells me that numerical computations of integrals like

$$\int_0^1 \frac{\sin x}{x}\,dx,$$

which we carry out in Chapter XIII, occur frequently in the study of communication networks, in connection with square waves. Each instructor has to exercise some judgment as to what should be emphasized at the expense of something else.

The chapters on functions of several variables are included for classes which can proceed at a faster rate, and therefore have time for additional material during the first year. **Under ordinary circumstances, these chapters will not be covered during a first-year course. For instance, they are not covered during the first-year course at Yale.**

Induction. I think the first course in calculus is a good time to learn induction. However, an attempt to teach induction without having met natural examples first meets with very great psychological difficulties. Hence throughout the part on differentiation, I have not mentioned induction formally. Whenever a situation arises where induction may be used, I carry out stepwise procedures illustrating the inductive procedure. After enough repetitions of these, the student is then ready to see a pattern which can be summarized by the formal "induction," which just becomes a name given to a notion which has already been understood.

Review material. The present edition also emphasizes more review material. Deficient high school training is responsible for many of the difficulties experienced at the college level. These difficulties are not so much due to the problem of understanding calculus as to the inability to handle elementary algebra. A large group of students cannot automatically give the expansion for expressions like

$$(a + b)^2, \qquad (a - b)^2, \qquad \text{or} \qquad (a + b)(a - b).$$

The answers should be memorized like the multiplication table. To memorize by rote such basic formulas is not incompatible with learning general principles. It is complementary.

To avoid any misunderstandings, I wish to state explicitly that the poor preparation of so many high school students cannot be attributed to the "new math" versus the "old math." When I started teaching calculus as a graduate student in 1950, I found the quasi-totality of college freshmen badly prepared. Today, I find only a substantial number of them (it is hard to measure how many). On the other hand, a sizable group at the top has had the opportunity to learn some calculus, even as much as one year, which would have been inconceivable in former times. As bad as the situation is, it is nevertheless an improvement.

I wish to thank my colleagues at Yale and others in the past who have suggested improvements in the book: Edward Bierstone (University of Toronto), Folke Eriksson (University of Gothenburg), R. W. Gatterdam (University of Wisconsin, Parkside), and George Metakides (University of Rochester). I thank Ron Infante for assisting with the proofreading.

I am also much indebted to Anthony Petrello for checking worked-out examples and answers in past editions. I thank Allen Altman and Gimli Khazad for lists of corrections.

S. Lang

Contents

CHAPTER XVIII

Part One

Review of
Basic Material

If you are already at ease with the elementary properties of numbers and if you know about coordinates and the graphs of the standard equations (linear equations, parabolas, ellipses), then you should start immediately with Chapter III on derivatives.

CHAPTER I

Numbers and Functions

In starting the study of any sort of mathematics, we cannot prove everything. Every time that we introduce a new concept, we must define it in terms of a concept whose meaning is already known to us, and it is impossible to keep going backwards defining forever. Thus we must choose our starting place, what we assume to be known, and what we are willing to explain and prove in terms of these assumptions.

At the beginning of this chapter, we shall describe most of the things which we assume known for this course. Actually, this involves very little. Roughly speaking, we assume that you know about numbers, addition, subtraction, multiplication, and division (by numbers other than 0). We shall recall the properties of inequalities (when a number is greater than another). On a few occasions we shall take for granted certain properties of numbers which might not have occurred to you before and which will always be made precise. Proofs of these properties will be supplied in the Appendix for those of you who are interested.

I, §1. INTEGERS, RATIONAL NUMBERS, AND REAL NUMBERS

The most common numbers are the numbers 1, 2, 3,... which are called **positive integers**.

The numbers -1, -2, -3,... are called **negative integers**. When we want to speak of the positive integers together with the negative integers and 0, we call them simply **integers**. Thus the integers are 0, 1, -1, 2, -2, 3, -3,....

The sum and product of two integers are again integers.

In addition to the integers we have **fractions**, like $\frac{3}{4}$, $\frac{5}{7}$, $-\frac{1}{8}$, $-\frac{101}{27}$, $\frac{8}{16}$,..., which may be positive or negative, and which can be written as quotients m/n, where m, n are integers and n is not equal to 0. Such fractions are called **rational numbers**. Every integer m is a rational number, because it can be written as $m/1$, but of course it is not true that every rational number is an integer. We observe that the sum and product of two rational numbers are again rational numbers. If a/b and m/n are two rational numbers (a, b, m, n being integers and b, n unequal to 0), then their sum and product are given by the following formulas, which you know from elementary school:

$$\frac{a}{b}\,\frac{m}{n} = \frac{am}{bn},$$

$$\frac{a}{b} + \frac{m}{n} = \frac{an + bm}{bn}.$$

In this second formula, we have simply put the two fractions over the common denominator bn.

We can represent the integers and rational numbers geometrically on a straight line. We first select a unit length. The integers are multiples of this unit, and the rational numbers are fractional parts of this unit. We have drawn a few rational numbers on the line below.

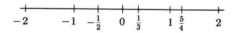

Observe that the negative integers and rational numbers occur to the left of zero.

Finally, we have the numbers which can be represented by infinite decimals, like $\sqrt{2} = 1.414\ldots$ or $\pi = 3.14159\ldots$, and which will be called **real numbers** or simply **numbers**.

The integers and rational numbers are special cases of these infinite decimals. For instance,

$$3 = 3.000000\ldots,$$

and

$$\tfrac{3}{4} = 0.7500000\ldots,$$

$$\tfrac{1}{3} = 0.3333333\ldots.$$

We see that there may be several ways of denoting the same number, for instance as the fraction $\frac{1}{3}$ or as the infinite decimal $0.33333\ldots$. We have written the decimals with dots at the end. If we stop the decimal expansion at any given place, we obtain an approximation to the number. The further off we stop the decimal, the better approximation we obtain.

Finding the decimal expansion for a fraction is easy by the process of long division which you should know from high school.

Later in the course we shall learn how to find decimal expansions for other numbers which you may have heard about, like π. You were probably told that $\pi = 3.14 \ldots$ but were not told why. You will learn how to compute arbitrarily many decimals for π in Chapter XIII.

Geometrically, the numbers are represented as the collection of all points on the above straight line, not only those which are a rational part of the unit length or a multiple of it.

We note that the sum and product of two numbers are numbers. If a is a number unequal to zero, then there is a unique number b such that $ab = ba = 1$, and we write

$$b = \frac{1}{a} \qquad \text{or} \qquad b = a^{-1}.$$

We say that b is the **inverse** of a, or "a inverse." We emphasize that the expression

$$1/0 \qquad \text{or} \qquad 0^{-1} \quad \text{is not defined.}$$

In other words, we cannot divide by zero, and we do not attribute any meaning to the symbols $1/0$ or 0^{-1}.

However, if a is a number, then the prodct $0 \cdot a$ is defined and is equal to 0. The product of any number and 0 is 0. Furthermore, if b is any number unequal to 0, then $0/b$ is defined and equal to 0. It can also be written $0 \cdot (1/b)$.

If a is a rational number $\neq 0$, then $1/a$ is also a rational number. Indeed, if we can write $a = m/n$, with integers m, n both different from 0, then

$$\frac{1}{a} = \frac{n}{m}$$

is also a rational number.

I, §2. INEQUALITIES

Aside from addition, multiplication, subtraction, and division (by numbers other than 0), we shall now discuss another important feature of the real numbers.

We have the **positive numbers**, represented geometrically on the straight line by those numbers unequal to 0 and lying to the right of 0. If a is a positive number, we write $a > 0$. You have no doubt already worked with positive numbers, and with inequalities. The next two properties are the most basic ones, concerning positivity.

POS 1. *If a, b are positive, so is the product ab and the sum a + b.*

POS 2. *If a is a number, then either a is positive, or a = 0, or −a is positive, and these possibilities are mutually exclusive.*

If a number is not positive and not 0, then we say that this number is **negative**. By **POS 2**, if a is negative, then $-a$ is positive.

Although you know already that the number 1 is positive, it can in fact be **proved** from our two properties. It may interest you to see the proof, which runs as follows and is very simple. By **POS 2**, we know that either 1 or −1 is positive. If 1 is not positive, then −1 is positive. By **POS 1**, it must then follow that $(-1)(-1)$ is positive. But this product is equal to 1. Consequently, it must be 1 which is positive, and not −1. Using property **POS 1**, we could now conclude that $1 + 1 = 2$ is positive, that $2 + 1 = 3$ is positive, and so forth.

If $a > 0$, we shall say that a is **greater than** 0. If we wish to say that a is positive or equal to 0, we write

$$a \geqq 0$$

and read this "a greater than or equal to zero."

Given two numbers a, b we shall say that a is **greater than** b and write $a > b$ if $a - b > 0$. We write $a < 0$ (a is **less than** 0) if $-a > 0$ and $a < b$ if $b > a$. Thus $3 > 2$ because $3 - 2 > 0$.

We shall write $a \geqq b$ when we want to say that a is **greater than or equal to** b. Thus $3 \geqq 2$ and $3 \geqq 3$ are both true inequalities.

Other rules concerning inequalities are valid.

In what follows, let a, b, c be numbers.

Rule 1. *If $a > b$ and $b > c$, then $a > c$.*

Rule 2. *If $a > b$ and $c > 0$, then $ac > bc$.*

Rule 3. *If $a > b$ and $c < 0$, then $ac < bc$.*

Rule 2 expresses the fact that an inequality which is multiplied by a positive number is **preserved**. Rule 3 tells us that if we multiply both sides of an inequality by a negative number, then the inequality gets **reversed**. For instance, we have the inequality

$$1 < 3$$

Since $2 > 0$ we also have $2 \cdot 1 < 2 \cdot 3$. But −2 is negative, and if we multiply both sides by −2 we get

$$-2 > -6.$$

In the geometric representation of the real numbers on the line, -2 lies to the right of -6. This gives us the geometric representation of the fact that -2 is greater than -6.

If you wish, you may assume these three rules just as you assume **POS 1** and **POS 2**. All of these are used in practice. It turns out that the three rules can be proved in terms of **POS 1** and **POS 2**. We cannot assume all the inequalities which you will ever meet in practice. Hence just to show you some techniques which might recur for other applications, we show how we can deduce the three rules from **POS 1** and **POS 2**. You may omit these (short) proofs if you wish.

To prove Rule 1, suppose that $a > b$ and $b > c$. By definition, this means that $(a - b) > 0$ and $(b - c) > 0$. Using property **POS 1**, we conclude that

$$a - b + b - c > 0,$$

and canceling b gives us $(a - c) > 0$. By definition, this means $a > c$, as was to be shown.

To prove Rule 2, suppose that $a > b$ and $c > 0$. By definition,

$$a - b > 0.$$

Hence using the property of **POS 1** concerning the product of positive numbers, we conclude that

$$(a - b)c > 0.$$

The left-hand side of this inequality is none other than $ac - bc$, which is therefore > 0. Again by definition, this gives us

$$ac > bc.$$

We leave the proof of Rule 3 as an exercise.

We give an example showing how to use the three rules.

Example. Let a, b, c, d be numbers with $c, d > 0$. Suppose that

$$\frac{a}{c} < \frac{b}{d}.$$

We wish to prove the "cross-multiplication" rule that

$$ad < bc.$$

Using Rule 2, multiplying each side of the original inequality by c, we obtain

$$a < bc/d.$$

Using Rule 2 again, and multiplying each side by d, we obtain

$$ad < bc,$$

as desired.

Let a be a number > 0. Then there exists a number whose square is a. If $b^2 = a$ then we observe that

$$(-b)^2 = b^2$$

is also to a. Thus either b or $-b$ is positive. We agree to denote by $\sqrt{a}$ the **positive** square root and call it simply **the square root of** a. Thus $\sqrt{4}$ is equal to 2 and not -2, even though $(-2)^2 = 4$. This is the most practical convention about the use of the $\sqrt{}$ sign that we can make. Of course, the square root of 0 is 0 itself. A negative number does *not* have a square root in the real numbers.

There are thus two solutions to an equation

$$x^2 = a$$

with $a > 0$. These two solutions are $x = \sqrt{a}$ and $x = -\sqrt{a}$. For instance, the equation $x^2 = 3$ has the two solutions

$$x = \sqrt{3} = 1.732\ldots \qquad \text{and} \qquad x = -\sqrt{3} = -1.732\ldots.$$

The equation $x^2 = 0$ has exactly one solution, namely $x = 0$. The equation $x^2 = a$ with $a < 0$ has no solution in the real numbers.

Definition. Let a be a number. We define the **absolute value** of a to be

$$\boxed{|a| = \sqrt{a^2}.}$$

In particular,

$$\boxed{|a|^2 = a^2.}$$

Thus the absolute value of a number is always $\geqq 0$. The absolute value of a positive number is always positive.

Example. We have

$$|3| = \sqrt{3^2} = \sqrt{9} = 3,$$

but

$$|-3| = \sqrt{(-3)^2} = \sqrt{9} = 3.$$

Also for any number a we get

$$|-a| = \sqrt{(-a)^2} = \sqrt{a^2} = |a|.$$

Theorem 2.1. *If a is any number, then*

$$|a| = \begin{cases} a & \text{if } a \geqq 0, \\ -a & \text{if } a < 0. \end{cases}$$

Proof. If $a \geqq 0$ then a is the unique number $\geqq 0$ whose square is a^2, so $|a| = \sqrt{a^2} = a$. If $a < 0$ then $-a > 0$ and

$$(-a)^2 = a^2,$$

so this time $-a$ is the unique number > 0 whose square is a^2, whence $|a| = -a$. This proves the theorem.

Theorem 2.2. *If a, b are numbers, then*

$$|ab| = |a| \, |b|.$$

Proof. We have:

$$|ab| = \sqrt{(ab)^2} = \sqrt{a^2 b^2} = \sqrt{a^2} \sqrt{b^2} = |a| \, |b|.$$

As an example, we see that

$$|-6| = |(-3) \cdot 2| = |-3| \, |2| = 3 \cdot 2 = 6.$$

There is one final inequality which is extremely important.

Theorem 2.3. *If a, b are two numbers, then*

$$|a + b| \leqq |a| + |b|.$$

Proof. We first observe that either ab is positive, or it is negative, or it is 0. In any case, we have

$$ab \leqq |ab| = |a|\,|b|.$$

Hence, multiplying both sides by 2, we obtain the inequality

$$2ab \leqq 2|a|\,|b|.$$

Using this inequality we find:

$$(a + b)^2 = a^2 + 2ab + b^2$$
$$\leqq a^2 + 2|a|\,|b| + b^2$$
$$= (|a| + |b|)^2.$$

We can take the square root of both sides and use Theorem 2.1 to conclude that

$$|a + b| \leqq |a| + |b|,$$

thereby proving our theorem.

You will find plenty of exercises below to give you practice with inequalities. We shall work out some numerical examples to show you the way.

Example 1. Determine the numbers satisfying the equality

$$|x + 1| = 2.$$

This equality means that either $x + 1 = 2$ or $-(x + 1) = 2$, because the absolute value of $x + 1$ is either $(x + 1)$ itself or $-(x + 1)$. In the first case, solving for x gives us $x = 1$, and in the second case, we get $-x - 1 = 2$ or $x = -3$. Thus the answer is $x = 1$ or $x = -3$.

Let a, b be numbers. We may interpret

$$|a - b| = \sqrt{(a - b)^2}$$

as the distance between a and b.

For instance, if $a > b$ then this is geometrically clear from the figure.

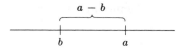

On the other hand, if $a < b$ we have

$$|a - b| = |-(b - a)| = |b - a|,$$

and $b > a$, so again we see that $|a - b| = |b - a|$ is the distance between a and b.

In the preceding example, the set of numbers x such that

$$|x + 1| = 2$$

is the set of numbers whose distance from -1 is 2, because we can write

$$x + 1 = x - (-1).$$

Hence we see again geometrically that this set of numbers consists of 1 and -3, as shown on the figure.

We shall also give an example showing how to determine numbers satisfying certain inequalities. For this we need some terminology. Let a, b be numbers, and assume $a < b$.

The collection of numbers x such that $a < x < b$ is called the **open interval** between a and b, and is sometimes denoted by (a, b).

The collection of numbers x such that $a \leq x \leq b$ is called the **closed interval** between a and b, and is sometimes denoted by $[a, b]$. A single point will also be called a closed interval.

In both above cases, the numbers a, b are called the **end points** of the intervals. Sometimes we wish to include only one of them in an interval, and so we define the collection of numbers x such that $a \leq x < b$ to be **half-closed interval**, and similarly for those numbers x such that $a < x \leq b$.

Finally, if a is a number, we call the collection of numbers $x > a$, or $x \geq a$, or $x < a$, or $x \leq a$ an **infinite interval**. Pictures of intervals are shown below.

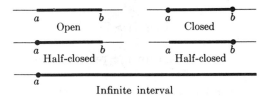

Example 2. Determine all intervals of numbers satisfying

$$|x| \leq 4.$$

We distinguish two cases. The first case is $x \geq 0$. Then $|x| = x$, and in this case, our inequality amounts to

$$0 \leq x \leq 4.$$

The second case is $x < 0$. In this case, $|x| = -x$, and our inequality amounts to $-x \leq 4$, or in other words, $-4 \leq x$. Thus in the second case, the numbers satisfying our inequality are precisely those in the interval.

$$-4 \leq x < 0.$$

Considering now both cases together, we see that the interval of numbers satisfying our inequality $|x| \leq 4$ is the interval

$$-4 \leq x \leq 4.$$

We can also phrase the answer in terms of distance. The numbers x such that $|x| \leq 4$ are precisely those numbers whose distance from the origin is ≤ 4. Thus they constitute the closed interval between -4 and 4 as shown on the figure.

$$-4 \ \ -3 \ -2 \ -1 \quad 0 \quad 1 \quad 2 \quad 3 \quad 4$$

More generally, let a be a positive number. A number x satisfies the inequality $|x| < a$ if and only if

$$-a < x < a.$$

The argument to prove this is the same as in the special case $a = 4$ worked out above.

Example 3. Determine all intervals of numbers satisfying the inequality

$$|x + 1| > 2.$$

This inequality is equivalent with the two inequalities

$$x + 1 > 2 \quad \text{or} \quad -(x + 1) > 2.$$

From the first we get the condition $x > 1$, and from the second, we get the condition $-x - 1 > 2$, or in other words, $x < -3$. Thus there are two (infinite) intervals, namely

$$x > 1 \quad \text{and} \quad x < -3.$$

Example 4. On the other hand, we wish to determine the interval of numbers x such that

$$|x + 1| < 2.$$

These are the numbers x whose distance from -1 is < 2, because we can write

$$x + 1 = x - (-1).$$

Hence it is the interval of numbers satisfying

$$-3 < x < 1$$

as shown on the figure.

I, §2. EXERCISES

Determine all intervals of numbers x satisfying the following inequalities.

1. $|x| < 3$

2. $|2x + 1| \leq 1$

3. $|x^2 - 2| \leq 1$

4. $|x - 5| > 2$

5. $(x + 1)(x - 2) < 0$

6. $(x - 1)(x + 1) > 0$

7. $(x - 5)(x + 5) < 0$

8. $x(x + 1) \leq 0$

9. $x^2(x - 1) \geq 0$

10. $(x - 5)^2(x + 10) \leq 0$

11. $(x - 5)^4(x + 10) \leq 0$

12. $(2x + 1)^6(x - 1) \geq 0$

13. $(4x + 7)^{20}(2x + 8) < 0$

14. $|x + 4| < 1$

15. $0 < |x + 2| < 1$

16. $|x| < 2$

17. $|x - 3| < 5$

18. $|x - 3| < 1$

19. $|x - 3| < 7$

20. $|x - 3| > 7$

21. $|x + 3| > 7$

Prove the following inequalities for all numbers x, y.

22. $|x + y| \geq |x| - |y|$ [*Hint*: Write $x = x + y - y$, and apply Theorem 2.3, together with the fact that $|-y| = |y|$.]

23. $|x - y| \geq |x| - |y|$ 24. $|x - y| \leq |x| + |y|$

25. Let a, b be positive numbers such that $a < b$. Show that $a^2 < b^2$.

26. Let a, b, c, d be numbers > 0, such that $a/b < c/d$. Show that

$$\frac{a}{b} < \frac{a + c}{b + d} \quad \text{and} \quad \frac{a + c}{b + d} < \frac{c}{d}.$$

27. Let a, b be numbers > 0. Show that

$$\sqrt{ab} \leq \frac{a + b}{2}.$$

28. Let $0 < a < b$ and $0 < c < d$. Prove that

$$ac < bd.$$

I, §3. FUNCTIONS

A function, defined for all numbers, is an association which to any given number associates another number.

It is customary to denote a function by some letter, just as a letter "x" denotes a number. Thus if we denote a given function by f, and x is a number, then we denote by $f(x)$ the number associated with x by the function. This of course does not mean "f times x." There is no multiplication involved here. The symbols $f(x)$ are read "f of x." The association of the number $f(x)$ to the number x is sometimes denoted by a special arrow, namely

$$x \mapsto f(x).$$

For example, consider the function which associates to each number x the number x^2. If f denotes this function, then we have $f(x) = x^2$. In particular, the square of 2 is 4 and hence $f(2) = 4$. The square of 7 is 49 and thus $f(7) = 49$. The square of $\sqrt{2}$ is 2, and hence $f(\sqrt{2}) = 2$. The square of $(x + 1)$ is

$$x^2 + 2x + 1$$

and thus $f(x + 1) = x^2 + 2x + 1$. If h is any number,

$$f(x + h) = x^2 + 2xh + h^2.$$

To take another example, let g be the function which to each number x associates the number $x + 1$. Then we may describe g by the symbols

$$x \mapsto x + 1$$

and write $g(x) = x + 1$. Therefore, $g(1) = 2$. Also $g(2) = 3$, $g(3) = 4$, $g(\sqrt{2}) = \sqrt{2} + 1$, and $g(x + 1) = x + 2$ for any number x.

We can view the **absolute value** as a function,

$$x \mapsto |x|$$

defined by the rule: Given any number a, we associate the number a itself if $a \geq 0$, and we associate the number $-a$ if $a < 0$. Let F denote the absolute value function. Then $F(x) = |x|$ for any number x. We have in particular $F(2) = 2$, and $F(-2) = 2$ also. The absolute value is not defined by means of a formula like x^2 or $x + 1$. We give you another example of such a function which is not defined by a formula.

We consider the function G described by the following rule:

$$G(x) = 0 \quad \text{if } x \text{ is a rational number.}$$

$$G(x) = 1 \quad \text{if } x \text{ is not a rational number.}$$

Then in particular, $G(2) = G(\frac{2}{3}) = G(-\frac{3}{4}) = 0$ but

$$G(\sqrt{2}) = 1.$$

You must be aware that you can construct a function just by prescribing arbitrarily the rule associating a number to a given one.

If f is a function and x a number, then $f(x)$ is called the **value** of the function at x. Thus if f is the function

$$x \mapsto x^2,$$

the value of f at 2 is 4 and the value of f at $\frac{1}{2}$ is $\frac{1}{4}$.

In order to describe a function, we need simply to give its value at any number x. That is the reason why we use the notation $x \mapsto f(x)$. Sometimes, for brevity, we speak of a function $f(x)$, meaning by that the function f whose value at x is $f(x)$. For instance, we would say "Let $f(x)$ be the function $x^3 + 5$" instead of saying "Let f be the function which to each number x associates $x^3 + 5$." Using the special arrow $\mapsto$, we could say also "Let f be the function $x \mapsto x^3 + 5$."

We would also like to be able to define a function for some numbers and leave it undefined for others. For instance we would like to say that $\sqrt{x}$ is a function (the square root function, whose value at a number x is the square root of that number), but we observe that a negative number does not have a square root. Hence it is desirable to make the notion of function somewhat more general by stating explicitly for what numbers it is defined. For instance, the square root function is defined only for numbers ≥ 0. This function is denoted by $\sqrt{x}$. The value $\sqrt{x}$ is the unique number ≥ 0 whose square is x.

Thus in general, let S be a collection of numbers. By a **function, defined on** S, we mean an association, which to each number x in S associates a number. We call S the **domain of definition** of the function. For example, the domain of definition of the square root function is the collection of all numbers ≥ 0.

Let us give another example of a function which is not defined for all numbers. Let S be the collection of all numbers $\neq 0$. The function

$$f(x) = \frac{1}{x}$$

is defined for numbers $x \neq 0$, and is thus defined on the domain S. For this particular function, we have $f(1) = 1$, $f(2) = \frac{1}{2}$, $f(\frac{1}{2}) = 2$, and

$$f(\sqrt{2}) = \frac{1}{\sqrt{2}}.$$

In practice, functions are used to denote the dependence of one quantity with respect to another.

Example. The area inside a circle of radius r is given by the formula

$$A = \pi r^2.$$

Thus the area is a function of the radius r, and we can also write

$$A(r) = \pi r^2.$$

If the radius is 2, then the area inside a circle of radius 2 is given by

$$A(2) = \pi 2^2 = 4\pi.$$

Example. A car moves at a constant speed of 50 km/hr. If time is measured in hours, the distance traveled is a function of time, namely if we denote distance by s, then

$$s(t) = 50t.$$

The distance is the product of the speed by the time traveled. Thus after two hours, the distance is

$$s(2) = 50 \cdot 2 = 100 \text{ km.}$$

One final word before we pass to the exercises: There is no magic reason why we should always use the letter x to describe a function $f(x)$. Thus instead of speaking of the function $f(x) = 1/x$ we could just as well say $f(y) = 1/y$ or $f(q) = 1/q$. Unfortunately, the most neutral way of writing would be $f(\text{blank}) = 1/\text{blank}$, and this is really not convenient.

I, §3. EXERCISES

1. Let $f(x) = 1/x$. What is $f(-\frac{2}{3})$?

2. Let $f(x) = 1/x$ again. What is $f(2x + 1)$ (for any number x such that $x \neq -\frac{1}{2}$)?

3. Let $g(x) = |x| - x$. What is $g(1)$, $g(-1)$, $g(-54)$?

4. Let $f(y) = 2y - y^2$. What is $f(z)$, $f(w)$?

5. For what numbers could you define a function $f(x)$ by the formula

$$f(x) = \frac{1}{x^2 - 2}?$$

What is the value of this function for $x = 5$?

6. For what numbers could you define a function $f(x)$ by the formula $f(x) = \sqrt[3]{x}$ (cube root of x)? What is $f(27)$?

7. Let $f(x) = x/|x|$, defined for $x \neq 0$. What is:
   (a) $f(1)$ (b) $f(2)$ (c) $f(-3)$ (d) $f(-\frac{4}{3})$

8. Let $f(x) = x + |x|$. What is:
   (a) $f(\frac{1}{2})$ (b) $f(2)$ (c) $f(-4)$ (d) $f(-5)$

9. Let $f(x) = 2x + x^2 - 5$. What is:
   (a) $f(1)$ (b) $f(-1)$ (c) $f(x + 1)$

10. For what numbers could you define a function $f(x)$ by the formula $f(x) = \sqrt[4]{x}$ (fourth root x)? What is $f(16)$?

11. A function (defined for all numbers) is said to be an **even** function if $f(x) = f(-x)$ for all x. It is said to be an **odd** function if $f(x) = -f(-x)$ for all x. Determine which of the following functions are odd or even.
   (a) $f(x) = x$ (b) $f(x) = x^2$ (c) $f(x) = x^3$
   (d) $f(x) = 1/x$ if $x \neq 0$, and $f(0) = 0$.

12. Let f be any function defined for all numbers. Show that the function $g(x) = f(x) + f(-x)$ is even. What about the function

$$h(x) = f(x) - f(-x),$$

is it even, odd, or neither?

I, §4. POWERS

In this section we just summarize some elementary arithmetic.

Let n be an integer ≥ 1 and let a be any number. Then a^n is the product of a with itself n times. For example, let $a = 3$. If $n = 2$, then $a^2 = 9$. If $n = 3$, then $a^3 = 27$. Thus we obtain a function which is called the n-th **power**. If f denotes this function, then $f(x) = x^n$.

We recall the rule

$$x^{m+n} = x^m x^n$$

for any number x and integers $m, n \geq 1$.

Again, let n be an integer ≥ 1, and let a be a positive number. We define $a^{1/n}$ to be the unique positive number b such that $b^n = a$. (That there exists such a unique number b is taken for granted as part of the properties of numbers.) We get a function called the n-th **root**. Thus if f is the 4th root, then $f(16) = 2$ and $f(81) = 3$.

The n-th root function can also be defined at 0, the n-th root of 0 being 0 itself.

If a, b are two numbers ≥ 0 and n is an integer ≥ 1, then

$$(ab)^{1/n} = a^{1/n} b^{1/n}.$$

There is another useful and elementary rule. Let m, n be integers ≥ 1 and let a be a number ≥ 0. We define $a^{m/n}$ to be $(a^{1/n})^m$ which is also equal to $(a^m)^{1/n}$. This allows us to define fractional powers, and gives us a function

$$f(x) = x^{m/n}$$

defined for $x \geq 0$.

We now come to powers with negative numbers or 0. We want to define x^a when a is a negative rational number or 0 and $x > 0$. We want the fundamental rule

$$x^{a+b} = x^a x^b$$

to be true. This means that we must define x^0 to be 1. For instance, since

$$2^3 = 2^{3+0} = 2^3 2^0,$$

we see from this example that the only way in which this equation holds is if $2^0 = 1$. Similarly, in general, if the relation

$$x^a = x^{a+0} = x^a x^0$$

is true, then x^0 must be equal to 1.

Suppose finally that a is a positive rational number, and let x be a number > 0. We **define**

$$\boxed{x^{-a} = \frac{1}{x^a}.}$$

Thus

$$2^{-3} = \frac{1}{2^3} = \frac{1}{8}, \quad \text{and} \quad 4^{-2/3} = \frac{1}{4^{2/3}}.$$

We observe that in this special case,

$$(4^{-2/3})(4^{2/3}) = 4^0 = 1.$$

In general,

$$x^a x^{-a} = x^0 = 1.$$

We are tempted to define x^a even when a is not a rational number. This is more subtle. For instance, it is absolutely meaningless to say that $2^{\sqrt{2}}$ is the product of 2 square root of 2 times itself. The problem of defining 2^a (or x^a) when a is not rational will be postponed to a later chapter. Until that chapter, when we deal with such a power, we shall assume that there is a function, written x^a, described as we have done above for rational numbers, and satisfying the fundamental relation

$$x^{a+b} = x^a x^b, \qquad x^0 = 1.$$

Example. We have a function $f(x) = x^{\sqrt{2}}$ defined for all $x \geq 0$. It is actually hard to describe its values for special numbers, like $2^{\sqrt{2}}$. It was unknown for a very long time whether $2^{\sqrt{2}}$ is a rational number or not. The solution (*it is not*) was found only in 1927 by the mathematician Gelfond, who became famous for solving a problem that was known to be very hard.

Warning. Do not confuse a function like x^2 and a function like 2^x. Given a number $c > 0$, we can view c^x as a function defined for all x. (It will be discussed in detail in Chapter VIII.) This function is called an **exponential function**. Thus 2^x and 10^x are exponential functions. We shall select a number

$$e = 2.718\ldots$$

and the exponential function e^x as having special properties which make it better than any other exponential function. The meaning of our use of the word "better" will be explained in Chapter VIII.

I, §4. EXERCISES

Find a^x and x^a for the following values of x and a.

1. $a = 2$ and $x = 3$ 2. $a = 5$ and $x = -1$

3. $a = \frac{1}{2}$ and $x = 4$ 4. $a = \frac{1}{3}$ and $x = 2$

5. $a = -\frac{1}{2}$ and $x = 4$ 6. $a = 3$ and $x = 2$

7. $a = -3$ and $x = -1$ 8. $a = -2$ and $x = -2$

9. $a = -1$ and $x = -4$ 10. $a = -\frac{1}{2}$ and $x = 9$

11. If n is an odd integer like $1, 3, 5, 7,\ldots$, can you define an n-th root function for all numbers?

CHAPTER II

Graphs and Curves

The ideas contained in this chapter allow us to translate certain statements backwards and forwards between the language of numbers and the language of geometry.

It is extremely basic for what follows, because we can use our geometric intuition to help us solve problems concerning numbers and functions, and conversely, we can use theorems concerning numbers and functions to yield results about geometry.

II, §1. COORDINATES

Once a unit length is selected, we can represent numbers as points on a line. We shall now extend this procedure to the plane, and to pairs of numbers.

We visualize a horizontal line and a vertical line intersecting at an origin O.

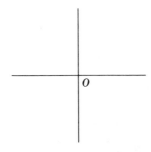

These lines will be called **coordinate axes** or simply **axes**.

We select a unit length and cut the horizontal line into segments of lengths 1, 2, 3,... to the left and to the right, and do the same to the vertical line, but up and down, as indicated in the next figure.

On the vertical line we visualize the points going below 0 as corresponding to the negative integers, just as we visualized points on the left of the horizontal line as corresponding to negative integers. We follow the same idea as that used in grading a thermometer, where the numbers below zero are regarded as negative. See figure.

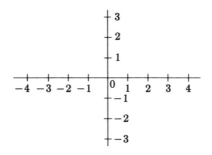

We can now cut the plane into squares whose sides have length 1.

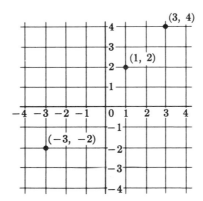

We can describe each point where two lines intersect by a pair of integers. Suppose that we are given a pair of integers like (1, 2). We go to the right of the origin 1 unit and vertically up 2 units to get the point (1, 2) which has been indicated above. We have also indicated the point (3, 4). The diagram is just like a map.

Furthermore, we could also use negative numbers. For instance, to describe the point (−3, −2) we go to the left of the origin 3 units and vertically downwards 2 units.

There is actually no reason why we should limit ourselves to points which are described by integers. For instance we can also have the point $(\frac{1}{2}, -1)$ and the point $(-\sqrt{2}, 3)$ as on the figure below.

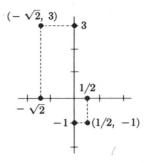

We have not drawn all the squares on the plane. We have drawn only the relevant lines to find our two points.

In general, if we take any point P in the plane and draw the perpendicular lines to the horizontal axis and to the vertical axis, we obtain two numbers x, y as in the figure below.

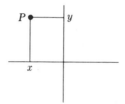

The perpendicular line from P to the horizontal axis determines a number x which is negative in the figure because it lies to the left of the origin. The number y determined by the perpendicular from P to the vertical axis is positive because it lies above the origin. The two numbers x, y are called the **coordinates** of the point P, and we can write $P = (x, y)$.

Every pair of numbers (x, y) determines a point of the plane. We find the point by going a distance x from the origin O in the horizontal direction and then a distance y in the vertical direction. If x is positive we go to the right of O. If x is negative, we go to the left of O. If y is positive we go vertically upwards, and if y is negative we go vertically downwards. The coordinates of the origin are $(0, 0)$. We usually call the horizontal axis the **x-axis** and the vertical axis the **y-axis**. If a point P is described by two numbers, say $(5, -10)$, it is customary to call the first number its x-coordinate and the second number its y-coordinate. Thus 5 is the x-coordinate, and -10 the y-coordinate of our point. Of course, we could use other letters besides x and y, for instance t and s, or u and v.

Our two axes separate the plane into four **quadrants** which are numbered as indicated in the figure:

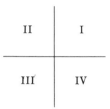

If (x, y) is a point in the first quadrant, then both x and y are > 0. If (x, y) is a point in the fourth quadrant, then $x > 0$ but $y < 0$.

II, §1. EXERCISES

1. Plot the following points: $(-1, 1)$, $(0, 5)$, $(-5, -2)$, $(1, 0)$.

2. Plot the following points: $(\frac{1}{2}, 3)$, $(-\frac{1}{3}, -\frac{1}{2})$, $(\frac{4}{3}, -2)$, $(-\frac{1}{4}, \frac{1}{2})$.

3. Let (x, y) be the coordinates of a point in the second quadrant. Is x positive or negative? Is y positive or negative?

4. Let (x, y) be the coordinates of a point in the third quadrant. Is x positive or negative? Is y positive or negative?

5. Plot the following points: $(1.2, -2.3)$, $(1.7, 3)$.

6. Plot the following points: $(-2.5, \frac{1}{3})$, $(-3.5, \frac{5}{4})$.

7. Plot the following points: $(1.5, -1)$, $(-1.5, -1)$.

II, §2. GRAPHS

Let f be a function. We define the **graph** of f to be the collection of all pairs of numbers $(x, f(x))$ whose first coordinate is any number for which f is defined and whose second coordinate is the value of the function at the first coordinate.

For example, the graph of the function $f(x) = x^2$ consists of all pairs (x, y) such that $y = x^2$. In other words, it is the collection of all pairs (x, x^2), like $(1, 1)$, $(2, 4)$, $(-1, 1)$, $(-3, 9)$, etc.

Since each pair of numbers corresponds to a point on the plane (once a system of axes and a unit length have been selected), we can view the graph of f as a collection of points in the plane. The graph of the function $f(x) = x^2$ has been drawn in the following figure, together with the points which we gave above as examples.

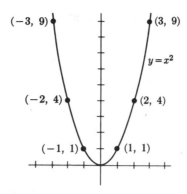

To determine the graph, we plot a lot of points making a table giving the x- and y-coordinates.

x	$f(x)$		x	$f(x)$
1	1		-1	1
2	4		-2	4
3	9		-3	9
$\frac{1}{2}$	$\frac{1}{4}$		$-\frac{1}{2}$	$\frac{1}{4}$

At this stage of the game there is no other way for you to determine the graph of a function other than this trial and error method. Later, we shall develop techniques which give you greater efficiency in doing it.

We shall now give several examples of graphs of functions which occur very frequently in the sequel.

Example 1. Consider the function $f(x) = x$. The points on its graph are of type (x, x). The first coordinate must be equal to the second. Thus $f(1) = 1$, $f(-\sqrt{2}) = -\sqrt{2}$, etc. The graph looks like this:

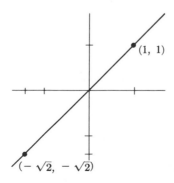

Example 2. Let $f(x) = -x$. Its graph looks this this:

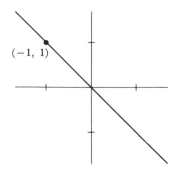

Observe that the graphs of the preceding two functions are straight lines. We shall study the general case of a straight line later.

Example 3. Let $f(x) = |x|$. When $x \geq 0$, we know that $f(x) = x$. When $x \leq 0$, we know that $f(x) = -x$. Hence the graph of $|x|$ is obtained by combining the preceding two, and looks like this:

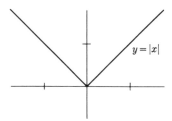

All values of $f(x)$ are ≥ 0, whether x is positive or negative.

Example 4. There is an even simpler type of function than the ones we have just looked at, namely the constant functions. For instance, we can define a function f such that $f(x) = 2$ for all numbers x. In other words, we associate the number 2 to any number x. It is a very simple association, and the graph of this function is a horizontal line, intersecting the vertical axis at the point $(0, 2)$.

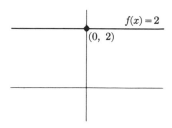

If we took the function $f(x) = -1$, then the graph would be a horizontal line intersecting the vertical axis at the point $(0, -1)$.

In general, let c be a fixed number. The graph of any function $f(x) = c$ is the horizontal line intersecting the vertical axis at the point $(0, c)$. The function $f(x) = c$ is called a **constant** function.

Example 5. The last of our examples is the function $f(x) = 1/x$ (defined for $x \neq 0$). By plotting a few points of the graph, you will see that it looks like this:

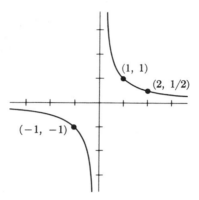

For instance, you can plot the following points:

x	$1/x$		x	$1/x$
1	1		-1	-1
2	$\frac{1}{2}$		-2	$-\frac{1}{2}$
3	$\frac{1}{3}$		-3	$-\frac{1}{3}$
$\frac{1}{2}$	2		$-\frac{1}{2}$	-2
$\frac{1}{3}$	3		$-\frac{1}{3}$	-3

As x becomes very large positive, $1/x$ becomes very small. As x approaches 0 from the right, $1/x$ becomes very large. A similar phenomenon occurs when x approaches 0 from the left; then x is negative and $1/x$ is negative. Hence in that case, $1/x$ is very large negative.

In trying to determine how the graph of a function looks, you can already watch for the following:

The points at which the graph intersects the two coordinate axes.
What happens when x becomes very large positive and very large negative.

On the whole, however, in working out the exercises, your main technique is just to plot a lot of points until it becomes clear to you what the graph looks like.

II, §2. EXERCISES

Sketch the graphs of the following functions and plot at least three points on each graph. In each case we give the value of the function at x.

1. $x + 1$ 2. $2x$ 3. $3x$

4. $4x$ 5. $2x + 1$ 6. $5x + \frac{1}{2}$

7. $\dfrac{x}{2} + 3$ 8. $-3x + 2$ 9. $2x^2 - 1$

10. $-3x^2 + 1$ 11. x^3 12. x^4

13. $\sqrt{x}$ 14. $x^{-1/2}$ 15. $2x + 1$

16. $x + 3$ 17. $|x| + x$ 18. $|x| + 2x$

19. $-|x|$ 20. $-|x| + x$ 21. $\dfrac{1}{x + 2}$

22. $\dfrac{1}{x - 2}$ 23. $\dfrac{1}{x + 3}$ 24. $\dfrac{1}{x - 3}$

25. $\dfrac{2}{x - 2}$ 26. $\dfrac{2}{x + 2}$ 27. $\dfrac{2}{x}$

28. $\dfrac{-2}{x + 5}$ 29. $\dfrac{3}{x + 1}$ 30. $\dfrac{x}{|x|}$

(In Exercises 13, 14, and 21 through 30, the functions are not defined for all values of x.)

31. Sketch the graph of the function $f(x)$ such that:
$f(x) = 0$ if $x \leq 0$. $f(x) = 1$ if $x > 0$.

32. Sketch the graph of the function $f(x)$ such that:
$f(x) = x$ if $x < 0$. $f(0) = 2$. $f(x) = x$ if $x > 0$.

33. Sketch the graph of the function $f(x)$ such that:
$f(x) = x^2$ if $x < 0$. $f(x) = x$ if $x \geq 0$.

34. Sketch the graph of the function $f(x)$ such that:
$f(x) = |x| + x$ if $-1 \leq x \leq 1$.
$f(x) = 3$ if $x > 1$. $[f(x)$ is not defined for other values of x.]

35. Sketch the graph of the function $f(x)$ such that:
$f(x) = x^3$ if $x \leq 0$. $f(x) = 1$ if $0 < x < 2$.
$f(x) = x^2$ if $x \geq 2$.

36. Sketch the graph of the function $f(x)$ such that:
$f(x) = x$ if $0 < x \leq 1$. $f(x) = x - 1$ if $1 < x \leq 2$.
$f(x) = x - 2$ if $2 < x \leq 3$. $f(x) = x - 3$ if $3 < x \leq 4$.

[We leave $f(x)$ undefined for other values of x, but try to define it yourself in such a way as to preserve the symmetry of the graph.]

II, §3. THE STRAIGHT LINE

One of the most basic types of functions is the type whose graph represents a straight line. We have already seen that the graph of the function $f(x) = x$ is a straight line. If we take $f(x) = 2x$, then the line slants up much more steeply, and even more so for $f(x) = 3x$. The graph of the function $f(x) = 10,000x$ would look almost vertical. In general, let a be a positive number $\neq 0$. Then the graph of the function

$$f(x) = ax$$

represents a straight line. The point $(2, 2a)$ lies on the line because $f(2) = 2a$. The point $(\sqrt{2}, \sqrt{2}a)$ also lies on the line, and if c is any number, the point (c, ca) lies on the line. The (x, y) coordinates of these points are obtained by making a similarity transformation, starting with the coordinates $(1, a)$ and multiplying them by some number c.

We can visualize this procedure by means of similar triangles. In the figure below, we have a straight line. If we select a point (x, y) on the line and drop the perpendicular from this point to the x-axis, we obtain a right triangle.

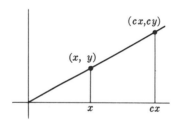

If x is the length of the base of the smaller triangle in the figure, and y its height, and if cx is the length of the base of the bigger triangle, then cy is the height of the bigger triangle: The smaller triangle is similar to the bigger one.

If a is a number < 0, then the graph of the function $f(x) = ax$ is also a straight line, which slants up to the left. For instance, the graphs of

$$f(x) = -x \qquad \text{or} \qquad f(x) = -2x.$$

We now give examples of more general lines, not passing through the origin.

Example 1. Let $g(x) = 2x + 1$. When $x = 0$, then $g(x) = 1$. When $g(x) = 0$, then $x = -\frac{1}{2}$. The graph looks as on the following figure.

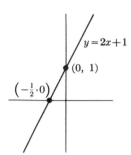

Example 2. Let $g(x) = -2x - 5$. When $x = 0$, then $g(x) = -5$. When $g(x) = 0$, then $x = -\frac{5}{2}$. The graph looks like this:

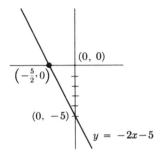

We shall frequently speak of a function $f(x) = ax + b$ as a straight line (although of course, it is its graph which is a straight line).

The number a which is the coefficient of x is called the **slope** of the line. It determines how much the line is slanted. As we have already seen in examples, when the slope is positive, the line is slanted to the right, and when the slope is negative, the line is slanted to the left. The relationship $y = ax + b$ is also called the **equation** of the line. It gives us the relation between the x- and y-coordinates of a point on the line.

Let $f(x) = ax + b$ be a straight line, and let (x_1, y_1) and (x_2, y_2) be two points of the line. It is easy to find the slope of the line in terms of the coordinates of these two points. By definition, we know that

$$y_1 = ax_1 + b$$

and

$$y_2 = ax_2 + b.$$

Subtracting, we get

$$y_2 - y_1 = ax_2 - ax_1 = a(x_2 - x_1).$$

Consequently, if the two points are distinct, $x_2 \neq x_1$, then we can divide by $x_2 - x_1$ and obtain

$$\text{slope of line} = a = \frac{y_2 - y_1}{x_2 - x_1}.$$

This formula gives us the slope in terms of the coordinates of two distinct points on the line.

Geometrically, our quotient

$$\frac{y_2 - y_1}{x_2 - x_1}$$

is simply the ratio of the vertical side and horizontal side of the triangle in the next diagram:

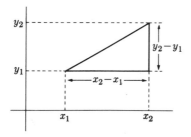

In general, let a be a number, and (x_1, y_1) some point.

We wish to find the equation of the line having slope equal to a, and passing through the point (x_1, y_1).

The condition that a point (x, y) with $x \neq x_1$ be on this line is equivalent with the condition that

$$\frac{y - y_1}{x - x_1} = a.$$

Thus the equation of the desired line is

$$y - y_1 = a(x - x_1).$$

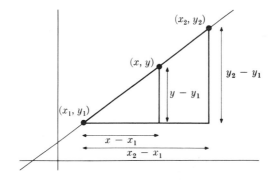

Example 3. Let $(1, 2)$ and $(2, -1)$ be the two points. What is the slope of the line between them? What is the equation of the line?

We first find the slope. We have:

$$\text{slope} = \frac{y_2 - y_1}{x_2 - x_1} = \frac{-1 - 2}{2 - 1} = -3.$$

The line must pass through the given point $(1, 2)$. Hence its equation is

$$y - 2 = -3(x - 1).$$

This is a correct answer. Sometimes it may be useful to put the equation in the form

$$y = -3x + 5,$$

but it is equally valid to leave it in the first form.

Observe that it does not matter which point we call (x_1, y_1) and which we call (x_2, y_2). We would get the same answer for the slope.

We can also determine the equation of a line provided we know the slope and one point.

Example 4. Find the equation of the line having slope -7 and passing through the point $(-1, 2)$.

The equation is

$$y - 2 = -7(x + 1).$$

Example 5. In general, let (x_1, y_1) and (x_2, y_2) be two distinct points with $x_1 \neq x_2$. We wish to find the equation of the line passing through these two points. Its slope must then be equal to

$$\frac{y_2 - y_1}{x_2 - x_1}.$$

Therefore the equation of the line can be expressed by the formula

$$\boxed{\frac{y - y_1}{x - x_1} = \frac{y_2 - y_1}{x_2 - x_1}}$$

for all points (x, y) such that $x \neq x_1$, or for all points by

$$\boxed{y - y_1 = \left(\frac{y_2 - y_1}{x_2 - x_1}\right)(x - x_1).}$$

Finally, we should mention vertical lines. These cannot be represented by equations of type $y = ax + b$. Suppose that we have a vertical line intersecting the x-axis at the point $(2, 0)$. The y-coordinate of any point on the line can be arbitrary. Thus the equation of the line is simply $x = 2$. In general, the equation of the vertical line intersecting the x-axis at the point $(c, 0)$ is $x = c$.

We can find the point of intersection of two lines by solving simultaneously two linear equations.

Example 6. Find the point of intersection of the two lines

$$y = 3x - 5 \quad \text{and} \quad y = -4x + 1.$$

We solve

$$3x - 5 = -4x + 1$$

or equivalently, $7x = 6$. This gives $x = \frac{6}{7}$, whence

$$y = 3 \cdot \frac{6}{7} - 5 = \frac{18}{7} - 5.$$

Hence the common point is

$$\left(\frac{6}{7}, \frac{18}{7} - 5 \right).$$

II, §3. EXERCISES

Sketch the graphs of the following lines:

1. $y = -2x + 5$ 2. $y = 5x - 3$

3. $y = \dfrac{x}{2} + 7$ 4. $y = -\dfrac{x}{3} + 1$

What is the equation of the line passing through the following points?

5. $(-1, 1)$ and $(2, -7)$ 6. $(3, \frac{1}{2})$ and $(4, -1)$

7. $(\sqrt{2}, -1)$ and $(\sqrt{2}, 1)$ 8. $(-3, -5)$ and $(\sqrt{3}, 4)$

What is the equation of the line having the given slope and passing through the given point?

9. slope 4 and point $(1, 1)$ 10. slope -2 and point $(\frac{1}{2}, 1)$

11. slope $-\frac{1}{2}$ and point $(\sqrt{2}, 3)$ 12. slope $\sqrt{3}$ and point $(-1, 5)$

Sketch the graphs of the following lines:

13. $x = 5$ 14. $x = -1$ 15. $x = -3$

16. $y = -4$ 17. $y = 2$ 18. $y = 0$

What is the slope of the line passing through the following points?

19. $(1, \frac{1}{2})$ and $(-1, 1)$ 20. $(\frac{1}{4}, 1)$ and $(\frac{1}{2}, -1)$

21. $(2, 3)$ and $(\sqrt{2}, 1)$ 22. $(\sqrt{3}, 1)$ and $(3, 2)$

What is the equation of the line passing through the following points?

23. $(\pi, 1)$ and $(\sqrt{2}, 3)$ 24. $(\sqrt{2}, 2)$ and $(1, \pi)$

25. $(-1, 2)$ and $(\sqrt{2}, -1)$ 26. $(-1, \sqrt{2})$ and $(-2, -3)$

27. Sketch the graphs of the following lines:
   (a) $y = 2x$ (b) $y = 2x + 1$ (c) $y = 2x + 5$
   (d) $y = 2x - 1$ (e) $y = 2x - 5$

28. Two straight lines are said to be **parallel** if they have the same slope. Let $y = ax + b$ and $y = cx + d$ be the equations of two straight lines with $b \neq d$.
   (a) If they are parallel, show that they have no point in common.
   (b) If they are not parallel, show that they have exactly one point in common.

29. Find the common point of the following pairs of lines:
   (a) $y = 3x + 5$ and $y = 2x + 1$ (b) $y = 3x - 2$ and $y = -x + 4$
   (c) $y = 2x$ and $y = -x + 2$ (d) $y = x + 1$ and $y = 2x + 7$

II, §4. DISTANCE BETWEEN TWO POINTS

Let (x_1, y_1) and (x_2, y_2) be two points in the plane, for instance as in the following diagrams.

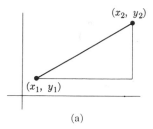

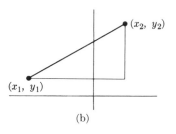

(a)   (b)

We can then make up a right triangle. By the Pythagoras theorem, the length of the line segment joining our two points can be determined from the lengths of the two sides. The square of the bottom side is $(x_2 - x_1)^2$, which is equal to $(x_1 - x_2)^2$.

The square of the length of the vertical side is $(y_2 - y_1)^2$, which is equal to $(y_1 - y_2)^2$. If L denotes the length of the line segment, then by Pythagoras,

$$L^2 = (x_1 - x_2)^2 + (y_1 - y_2)^2$$

and consequently,

$$L = \sqrt{(x_2 - x_1)^2 + (y_2 - y_1)^2}.$$

Example 1. Let the two points be $(1, 2)$ and $(1, 3)$. Then the length of the line segment between them is

$$\sqrt{(1 - 1)^2 + (3 - 2)^2} = 1.$$

The length L is also called the **distance** between the two points.

Example 2. Find the distance between the points $(-1, 5)$ and $(4, -3)$.

The distance is

$$\sqrt{(4 - (-1))^2 + (-3 - 5)^2} = \sqrt{89}.$$

II, §4. EXERCISES

Find the distance between the following points:

1. The points $(-3, -5)$ and $(1, 4)$

2. The points $(1, 1)$ and $(0, 2)$

3. The points $(-1, 4)$ and $(3, -2)$

4. The points $(1, -1)$ and $(-1, 2)$

5. The points $(\frac{1}{2}, 2)$ and $(1, 1)$

6. Find the coordinates of the fourth corner of a rectangle, three of whose corners are $(-1, 2)$, $(4, 2)$, $(-1, -3)$.

7. What are the lengths of the sides of the rectangle in Exercise 6?

8. Find the coordinates of the fourth corner of a rectangle, three of whose corners are $(-2, -2)$, $(3, -2)$, $(3, 5)$.

9. What are the lengths of the sides of the rectangle in Exercise 8?

10. If x, y are numbers, define the distance between these two numbers to be $|x - y|$. Show that this is the same as the distance between the points $(x, 0)$ and $(y, 0)$ in the plane.

II, §5. CURVES AND EQUATIONS

Let $F(x, y)$ be an expression involving a pair of numbers (x, y). Let c be a number. We consider the equation

$$F(x, \ y) = c.$$

Definition. The **graph** of the equation is the collection of points (a, b) in the plane satisfying the equation, that is such that

$$F(a, b) = c$$

This graph is also known as a **curve**, and we will usually not make a distinction between the equation

$$F(x, y) = c$$

and the curve which represents the equation.

For example,

$$x + y = 2$$

is the equation of a straight line, and its graph is the straight line. We shall study below important examples of equations which arise frequently.

If f is a function, then we can form the expression $y - f(x)$, and the graph of the **equation**

$$y - f(x) = 0$$

is none other than the graph of the **function** f as we discussed it in §2.

You should observe that there are equations of type

$$F(x, y) = c$$

which are not obtained from a function $y = f(x)$, i.e. from an equation

$$y - f(x) = 0.$$

For instance, the equation $x^2 + y^2 = 1$ is such an equation.

We shall now study important examples of graphs of equations

$$F(x, y) = 0 \qquad \text{or} \qquad F(x, y) = c.$$

II, §6. THE CIRCLE

The expression $F(x, y) = x^2 + y^2$ has a simple geometric interpretation. By Pythagoras' theorem, it is the square of the distance of the point (x, y) from the origin $(0, 0)$. Thus the points (x, y) satisfying the equation

$$x^2 + y^2 = 1^2 = 1$$

are simply those points whose distance from the origin is 1. They form the circle of radius 1, with center at the origin.

Similarly, the points (x, y) satisfying the equation

$$x^2 + y^2 = 4$$

are those points whose distance from the origin is 2. They constitute the circle of radius 2. In general, if c is any number > 0, then the graph of the equation

$$x^2 + y^2 = c^2$$

is the circle of radius c, with center at the origin.

We have already remarked that the **equation**

$$x^2 + y^2 = 1$$

or $x^2 + y^2 - 1 = 0$ is not of the type $y - f(x) = 0$. However, we can write our equation in the form

$$y^2 = 1 - x^2.$$

For any value of x between -1 and $+1$, we can solve for y and get

$$y = \sqrt{1 - x^2} \quad \text{or} \quad y = -\sqrt{1 - x^2}.$$

If $x \neq 1$ and $x \neq -1$, then we get two values of y for each value of x. Geometrically, these two values correspond to the points indicated on the following diagram.

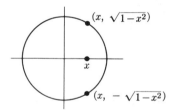

There is a function, defined for $-1 \leqq x \leqq 1$, such that

$$f(x) = \sqrt{1 - x^2},$$

and the graph of this function is the upper half of our circle. Similarly, there is another function

$$g(x) = -\sqrt{1 - x^2},$$

also defined for $-1 \leqq x \leqq 1$, whose graph is the lower half of the circle. Neither of these functions is defined for other values of x.

We now ask for the equation of the circle whose center is $(1, 2)$ and whose radius has length 3. It consists of the points (x, y) whose distance from $(1, 2)$ is 3. These are the points satisfying the equation

$$(x - 1)^2 + (y - 2)^2 = 9.$$

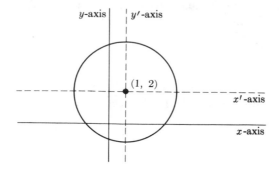

The graph of this equation has been drawn above. We may also put

$$x' = x - 1 \qquad \text{and} \qquad y' = y - 2.$$

In the new coordinate system (x', y') the equation of the circle is then

$$x'^2 + y'^2 = 9.$$

We have drawn the (x', y')-axes as dotted lines in the figure.

To pick another example, we wish to determine those points at a distance 2 from the point $(-1, -3)$. They are the points (x, y) satisfying the equation

$$(x - (-1))^2 + (y - (-3))^2 = 4$$

or, in other words,

$$(x + 1)^2 + (y + 3)^2 = 4.$$

(Observe carefully the cancellation of minus signs!) Thus the graph of this equation is the circle of radius 2 and center $(-1, -3)$.

In general, let a, b be two numbers and r a number > 0. Then the circle of radius r and center (a, b) is the graph of the equation

$$(x - a)^2 + (y - b)^2 = r^2.$$

We may put

$$x' = x - a \qquad \text{and} \qquad y' = y - b.$$

Then in the new coordinates x', y' the equation of the circle is

$$x'^2 + y'^2 = r^2.$$

Completing the square

Example. Suppose we are given an equation

$$x^2 + y^2 + 2x - 3y - 5 = 0,$$

where x^2 and y^2 occur with the same coefficient 1. We wish to see that this is the equation of a circle, and we use the method of **completing the square**, which we now review.

We want the equation to be of the form

$$(x - a)^2 + (y - b)^2 = r^2.$$

because then we know immediately that it represents a circle centered at (a, b) and of radius r. Thus we need $x^2 + 2x$ to be the first two terms of the expansion

$$(x - a)^2 = x^2 - 2ax + a^2.$$

Similarly, we need $y^2 - 3y$ to be the first two terms of the expansion

$$(y - b)^2 = y^2 - 2by + b^2.$$

This means that $a = -1$ and $b = 3/2$. Then

$$x^2 + 2x + y^2 - 3y - 5 = (x + 1)^2 - 1 + \left(y - \frac{3}{2}\right)^2 - \frac{9}{4} - 5.$$

Thus $x^2 + y^2 + 2x - 3y - 5 = 0$ is equivalent with

$$(x + 1)^2 + \left(y - \frac{3}{2}\right)^2 = 5 + 1 + \frac{9}{4} = \frac{33}{4}.$$

Consequently our given equation is the equation of a circle of radius $\sqrt{33/4}$, with center at $(-1, 3/2)$.

II, §6. EXERCISES

Sketch the graph of the following equations:

1. (a) $(x - 2)^2 + (y + 1)^2 = 25$ (b) $(x - 2)^2 + (y + 1)^2 = 4$
   (c) $(x - 2)^2 + (y + 1)^2 = 1$ (d) $(x - 2)^2 + (y + 1)^2 = 9$

2. (a) $x^2 + (y - 1)^2 = 9$ (b) $x^2 + (y - 1)^2 = 4$
   (c) $x^2 + (y - 1)^2 = 25$ (d) $x^2 + (y - 1)^2 = 1$

3. (a) $(x + 1)^2 + y^2 = 1$ (b) $(x + 1)^2 + y^2 = 4$
  (c) $(x + 1)^2 + y^2 = 9$ (d) $(x + 1)^2 + y^2 = 25$

4. $x^2 + y^2 - 2x + 3y - 10 = 0$

5. $x^2 + y^2 + 2x - 3y - 15 = 0$

6. $x^2 + y^2 + x - 2y = 16$

7. $x^2 + y^2 - x + 2y = 25$

II, §7. DILATIONS AND THE ELLIPSE

Dilations

Before studying the ellipse, we have to make some remarks on "stretching," or to use a more standard word, dilations.

Let (x, y) be a point in the plane. Then $(2x, 2y)$ is the point obtained by stretching both its coordinates by a factor of 2, as illustrated on Fig. 1, where we have also drawn $(3x, 3y)$ and $(\frac{1}{2}x, \frac{1}{2}y)$.

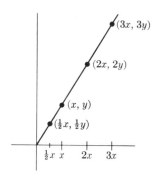

Figure 1

Definition. In general, if $c > 0$ is a positive number, we call (cx, cy) the **dilation** of (x, y) by a factor c.

Example. Let

$$u^2 + v^2 = 1$$

be the equation of the circle of radius 1. Put

$$x = cu \quad \text{and} \quad y = cv.$$

Then

$$u = x/c \quad \text{and} \quad v = y/c.$$

Hence x and y satisfy the equation

$$\frac{x^2}{c^2} + \frac{y^2}{c^2} = 1,$$

or equivalently,

$$x^2 + y^2 = c^2.$$

The set of points (x, y) satisfying this equation is the circle of radius c. Thus we may say:

The dilation of the circle of radius 1 by a factor of $c > 0$ is the circle of radius c.

This is illustrated on Fig. 2, with $c = 3$.

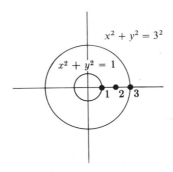

Figure 2

The Ellipse

There is no reason why we should dilate the first and second coordinates by the same factor. We may use different factors. For instance, if we put

$$x = 2u \qquad \text{and} \qquad y = 3v$$

we are dilating the first coordinate by a factor of 2, and we are dilating the second coordinate by a factor of 3. In that case, suppose that (u, v) is a point on the circle of radius 1, in other words suppose we have

$$u^2 + v^2 = 1.$$

Then (x, y) satisfies the equation

$$\frac{x^2}{4} + \frac{y^2}{9} = 1.$$

We interpret this as the equation of a "stretched out circle," as shown on Fig. 3.

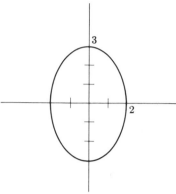

Figure 3

More generally, let a, b be numbers > 0. Let us put

$$x = au \quad \text{and} \quad y = bv.$$

If (u, v) satisfies

$$(*) \qquad\qquad u^2 + v^2 = 1,$$

then (x, y) satisfies

$$(**) \qquad\qquad \frac{x^2}{a^2} + \frac{y^2}{b^2} = 1.$$

Conversely, we may put $u = x/a$ and $v = y/b$ to see that the points satisfying equation $(*)$ correspond to the points of $(**)$ under this transformation, and vice versa.

Definition. An **ellipse** is the set of points satisfying an equation $(**)$ in some coordinate system of the plane. We have just seen that an ellipse is a dilated circle, by means of a dilation by factors a, $b > 0$ in the first and second coordinates respectively.

Example. Sketch the graph of the ellipse

$$\frac{x^2}{4} + \frac{y^2}{25} = 1.$$

This ellipse is a dilated circle by the factors 2 and 5, respectively. Note that

when $x = 0$ we have $\dfrac{y^2}{25} = 1$, so $y^2 = 25$ and $y = \pm 5$.

Also

when $y = 0$, we have $\dfrac{x^2}{4} = 1$, so $x^2 = 4$ and $x = \pm 2$.

Hence the graph of the ellipse looks like Fig. 4.

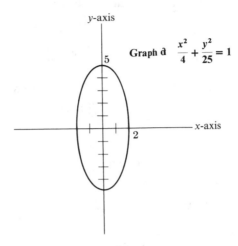

Figure 4

Example. Sketch the graph of the ellipse

$$\frac{(x-1)^2}{25} + \frac{(y+2)^2}{4} = 1.$$

In this case, let us put

$$x' = x - 1 \quad \text{and} \quad y' = y + 2.$$

We know that in (u, v) coordinates

$$u^2 + v^2 = 1$$

is the equation of a circle with center $(1, -2)$ and radius 1. Next we put

$$u = \frac{x'}{5} \quad \text{and} \quad v = \frac{y'}{2}.$$

The original equation is of the form

$$\frac{x'^2}{5^2} + \frac{y'^2}{2^2} = 1,$$

which in terms of u and v can be written

$$u^2 + v^2 = 1.$$

Thus our ellipse is obtained from the circle $u^2 + v^2 = 1$ by the dilation

$$u = x'/5 \qquad \text{and} \qquad v = y'/2,$$

or equivalently,

$$x' = 5u \qquad \text{and} \qquad y' = 2v.$$

The easiest way to sketch its graph is to draw the new coordinate system with coordinates x', y'. To find the intercepts of the ellipse with these new axes, we see that when $y' = 0$, then

$$\frac{x'^2}{5^2} = 1, \qquad \text{so that} \qquad x' = \pm 5.$$

Similarly, when $x' = 0$, then

$$\frac{y'^2}{2^2} = 1, \qquad \text{so that} \qquad y' = \pm 2.$$

Thus the graph looks like:

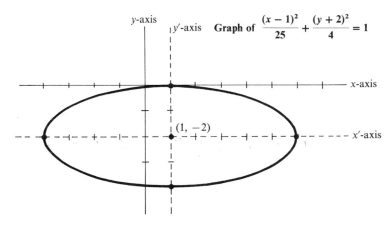

y-axis $\quad$ y'-axis $\quad$ **Graph of** $\dfrac{(x-1)^2}{25} + \dfrac{(y+2)^2}{4} = 1$

x-axis

$(1, -2)$

x'-axis

II, §7. EXERCISES

Sketch the graphs of the following curves.

1. $\dfrac{x^2}{9} + \dfrac{y^2}{16} = 1$ $\qquad\qquad\qquad$ 2. $\dfrac{x^2}{4} + \dfrac{y^2}{9} = 1$

3. $\dfrac{x^2}{5} + \dfrac{y^2}{16} = 1$ 4. $\dfrac{x^2}{4} + \dfrac{y^2}{25} = 1$

5. $\dfrac{(x-1)^2}{9} + \dfrac{(y+2)^2}{16} = 1$ 6. $4x^2 + 25y^2 = 100$

7. $\dfrac{(x+1)^2}{3} + \dfrac{(y+2)^2}{4} = 1$ 8. $25x^2 + 16y^2 = 400$

9. $(x-1)^2 + \dfrac{(y+3)^2}{4} = 1$

II, §8. THE PARABOLA

A **parabola** is a curve which is the graph of a function

$$y = ax^2$$

in some coordinate system, with $a \neq 0$.

Example. We have already seen what the graph of the function $y = x^2$ looks like. Consider now

$$y = -x^2.$$

Then by symmetry, you can easily see that the graph looks as on the figure.

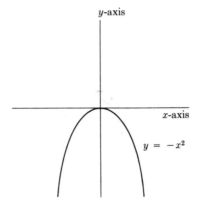

Suppose that we graph the equation $y = (x - 1)^2$. We shall find that it looks exactly the same, but as if the origin were placed at the point $(1, 0)$.

Similarly, the curve $y - 2 = (x - 4)^2$ looks again like $y = x^2$ except that the whole curve has been moved as if the origin were the point $(4, 2)$. The graphs of these equations have been drawn on the next diagram.

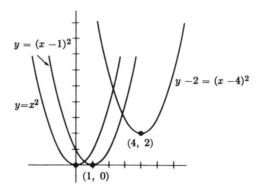

We can formalize these remarks as follows. Suppose that in our given coordinate system we pick a point (a, b) as a new origin. We let new coordinates be $x' = x - a$ and $y' = y - b$. Thus when $x = a$ we have $x' = 0$ and when $y = b$ we have $y' = 0$. If we have a curve

$$y' = x'^2$$

in the new coordinate system whose origin is at the point (a, b), then it gives rise to the equation

$$(y - b) = (x - a)^2$$

in terms of the old coordinate system. This type of curve is known as a **parabola**.

We can apply the same technique of completing the square that we did for the circle.

Example. What is the graph of the equation

$$2y - x^2 - 4x + 6 = 0?$$

Completing the square, we can write

$$x^2 + 4x = (x + 2)^2 - 4.$$

Thus our equation can be rewritten

$$2y = (x + 2)^2 - 10$$

or

$$2(y + 5) = (x + 2)^2.$$

We choose a new coordinate system

$$x' = x + 2 \qquad \text{and} \qquad y' = y + 5$$

so that our equation becomes

$$2y' = x'^2 \quad \text{or} \quad y' = \tfrac{1}{2}x'^2.$$

This is a function whose graph you already know, and whose sketch we leave to you.

We remark that if we have an equation

$$x - y^2 = 0$$

or

$$x = y^2,$$

then we get a parabola which is tilted horizontally.

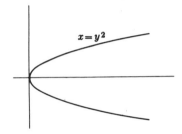

We can then apply the technique of changing the coordinate system to see what the graph of a more general equation is like.

Example. Sketch the graph of

$$x - y^2 + 2y + 5 = 0.$$

We can write this equation in the form

$$(x + 6) = (y - 1)^2$$

and hence its graph looks like this:

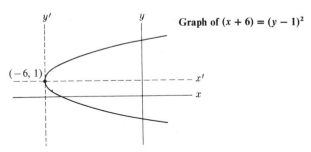

Graph of $(x + 6) = (y - 1)^2$

Suppose we are given the equation of a parabola

$$y = f(x) = ax^2 + bx + c,$$

with $a \neq 0$. We wish to determine where this parabola intersects the x-axis. These are the values for which $f(x) = 0$ and are called the **roots** of f. It is shown in high school that the roots of f are given by the **quadratic formula**:

$$x = \frac{-b \pm \sqrt{b^2 - 4ac}}{2a}.$$

You should read this formula out loud enough times so that it is memorized, just like the multiplication table. *It should be used automatically, without further thinking, to find the roots of a quadratic equation.*

Example. We want to find the roots of the equation

$$-2x^2 + 5x - 1 = 0.$$

The roots are

$$x = \frac{-5 \pm \sqrt{25 - 8}}{2(-2)} = \frac{-5 \pm \sqrt{17}}{-4} = \frac{5 \pm \sqrt{17}}{4}.$$

Thus the two roots are

$$\frac{5 + \sqrt{17}}{4} \quad \text{and} \quad \frac{5 - \sqrt{17}}{4}.$$

These are the two points where the parabola $y = -2x^2 + 5x - 1$ crosses the x-axis, and its graph is shown on the figure.

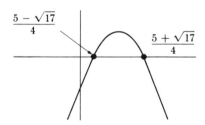

Proof of the quadratic formula. We shall now give the proof of the quadratic formula, to convince you that it is true. So we want to solve

(∗)

$$ax^2 + bx + c = 0.$$

Since we assumed $a \neq 0$ this amounts to solving the equation

$$(**) \qquad x^2 + \frac{b}{a}x + \frac{c}{a} = 0$$

obtained by dividing by a. Recall the formula

$$(x + t)^2 = x^2 + 2tx + t^2.$$

We want to find t such that $x^2 + \left(\frac{b}{a}\right)x$ has the form $x^2 + 2tx$. This means that we let

$$\frac{b}{a} = 2t, \qquad \text{that is} \quad t = \frac{b}{2a}.$$

We now add $\left(\frac{b}{2a}\right)^2$ to both sides of equation $(**)$, and obtain

$$x^2 + \frac{b}{a}x + \left(\frac{b}{2a}\right)^2 + \frac{c}{a} = \left(\frac{b}{2a}\right)^2.$$

This can be rewritten in the form

$$\left(x + \frac{b}{2a}\right)^2 + \frac{c}{a} = \left(\frac{b}{2a}\right)^2,$$

or equivalently,

$$\left(x + \frac{b}{2a}\right)^2 = \left(\frac{b}{2a}\right)^2 - \frac{c}{a} = \frac{b^2 - 4ac}{4a^2}.$$

Taking square roots yields

$$x + \frac{b}{2a} = \pm \frac{\sqrt{b^2 - 4ac}}{2a}$$

whence

$$x = \frac{-b \pm \sqrt{b^2 - 4ac}}{2a},$$

thus proving the quadratic formula.

Remark. It may happen that $b^2 - 4ac < 0$, in which case the quadratic equation does not have a solution in the real numbers.

Example. Find the roots of the equation

$$3x^2 - 2x + 1 = 0.$$

The roots are

$$x = \frac{-(-2) \pm \sqrt{4 - 12}}{6}.$$

Since $4 - 12 = -8 < 0$, the equation does not have roots in the real numbers. The equation

$$y = 3x^2 - 2x + 1$$

is the equation of a parabola, whose graph looks like that on the figure. The graph does not cross the x-axis.

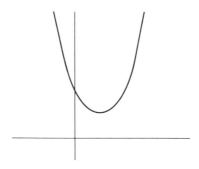

The discussion of the quadratic formula in this section illustrates some general pedagogical principles concerning the relationship of rote memorization and the role of proofs in learning mathematics.

1. You should memorize the quadratic formula aurally:

x equals minus b plus or minus square root of b square minus $4ac$ over $2a$

as you would memorize a poem, by repeating it out loud. Such memorizing is necessary for a few important items in basic math, to induce getting the right answer as a conditioned reflex, without wasting any time.

2. Independently of using the formula as a conditioned reflex, you should see the proof by completing the square. Learning to handle logic and the English language in establishing theorems is also part of mathematics. In addition, the technique of completing the square arises often in the context of graphing circles, ellipses, parabolas, in addition to the quadratic formula. Knowing the formula as a conditioned reflex and knowing the proof serve two different complementary functions in mathematical training, neither of which eliminates the other.

II, §8. EXERCISES

Sketch the graph of the following equations:

1. $y = -x + 2$ 　　　　　　　　　2. $y = 2x^2 + x - 3$

3. $x - 4y^2 = 0$ 　　　　　　　　4. $x - y^2 + y + 1 = 0$

Complete the square in the following equations and change the coordinate system to put them into the form

$$x'^2 + y'^2 = r^2 \qquad \text{or} \qquad y' = cx'^2 \qquad \text{or} \qquad x' = cy'^2$$

with a suitable constant c.

5. $x^2 + y^2 - 4x + 2y - 20 = 0$ 　　　　6. $x^2 + y^2 - 2y - 8 = 0$

7. $x^2 + y^2 + 2x - 2 = 0$ 　　　　　　8. $y - 2x^2 - x + 3 = 0$

9. $y - x^2 - 4x - 5 = 0$ 　　　　　　　10. $y - x^2 + 2x + 3 = 0$

11. $x^2 + y^2 + 2x - 4y = -3$ 　　　　　12. $x^2 + y^2 - 4x - 2y = -3$

13. $x - 2y^2 - y + 3 = 0$ 　　　　　　14. $x - y^2 - 4y = 5$

II, §9. THE HYPERBOLA

We have already seen what the graph of the equation

$$xy = 1 \qquad \text{or} \qquad y = 1/x$$

looks like. It is of course the same as the graph of the function

$$f(x) = 1/x$$

(defined for $x \neq 0$). If we pick a coordinate system whose origin is at the point (a, b), the equation

$$y - b = \frac{1}{x - a}$$

is known as a **hyperbola**. In terms of the new coordinate system

$$x' = x - a$$

and $y' = y - b$, our hyperbola has the old type of equation

$$x'y' = 1.$$

If we are given an equation like

$$xy - 2x + 3y = 1$$

we want to put this equation in the form

$$(x - a)(y - b) = c,$$

or expanding out,

$$xy - ay - bx + ab = c.$$

This tells us what a and b must be. Thus

$$xy - 2x + 3y = (x + 3)(y - 2) + 6.$$

Hence $xy - 2x + 3y = 1$ is equivalent with

$$(x + 3)(y - 2) + 6 = 1$$

or in other words

$$(x + 3)(y - 2) = -5.$$

The graph of this equation has been drawn on the following diagram.

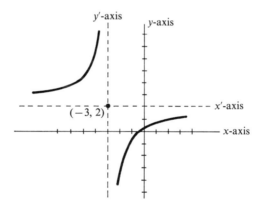

There is another form for the hyperbola. Let us try to graph the equation

$$x^2 - y^2 = 1.$$

If we solve for y we obtain

$$y^2 = x^2 - 1$$

so

$$y = \pm \sqrt{x^2 - 1}.$$

The graph has symmetry, because if (x, y) is a point on the graph then $(-x, y)$, $(x, -y)$ and $(-x, -y)$ is also a point on the graph. So let us look at the graph in the first quadrant when $x \geq 0$ and $y \geq 0$.

Since $x^2 - 1 = y^2$, it follows that $x^2 - 1 \geq 0$ so $x^2 \geq 1$. Hence the graph exists only for $x \geq 1$. We claim that it looks like this in the first quadrant.

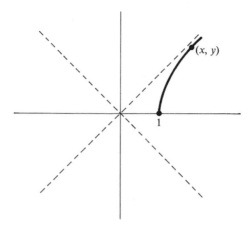

To see this, we could of course make a table of a few values first, to see experimentally what the graph is like. Do this yourself. Here we describe it theoretically.

As x increases, the expression $x^2 - 1$ increases, so $\sqrt{x^2 - 1}$ increases. Thus y also increases.

Also, since $y^2 = x^2 - 1$ it follows that $y^2 < x^2$ so $y < x$ for x, y in the first quadrant. We have drawn the line $y = x$. The set of points with $y < x$ lies below this line in the first quadrant.

Let us divide the equation $y^2 = x^2 - 1$ by x^2. We obtain

$$\frac{y^2}{x^2} = 1 - \frac{1}{x^2}.$$

When x becomes large, then

$$1 - \frac{1}{x^2} \quad \text{approaches 1.}$$

The ratio y/x is the slope of the line from the origin to the point (x, y). Hence this slope approaches 1 when x becomes large. Also from the expression

$$y = \sqrt{x^2 - 1},$$

we see that when x is large, $x^2 - 1$ is nearly equal to x^2, and so its square root is nearly equal to x. Thus the graph of the hyperbola comes closer and closer to the graph of the line $y = x$. This justifies drawing this graph as we have done.

Finally, by symmetry, the full graph of the hyperbola looks like this.

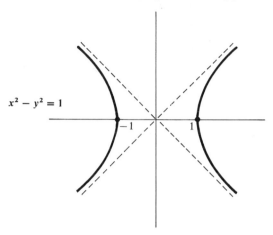

It is obtained by reflecting the graph in the first quadrant over the x-axis and over the y-axis, and also reflecting the graph through the origin.

II, §9. EXERCISES

Sketch the graphs of the following curves:

1. $(x - 1)(y - 2) = 2$

2. $x(y + 1) = 3$

3. $xy - 4 = 0$

4. $y = \dfrac{2}{1 - x}$

5. $y = \dfrac{1}{x + 1}$

6. $(x + 2)(y - 1) = 1$

7. $(x - 1)(y - 1) = 1$

8. $(x - 1)(y - 1) = 1$

9. $y = \dfrac{1}{x - 2} + 4$

10. $y = \dfrac{1}{x + 1} - 2$

11. $y = \dfrac{4x - 7}{x - 2}$

12. $y = \dfrac{-2x - 1}{x + 1}$

13. $y = \dfrac{x + 1}{x - 1}$

14. $y = \dfrac{x - 1}{x + 1}$

15. Graph the equation $y^2 - x^2 = 1$.

16. Graph the equation $(y - 1)^2 - (x - 2)^2 = 1$.

17. Graph the equation $(y + 1)^2 - (x - 2)^2 = 1$.

Part Two

Differentiation and Elementary Functions

In this part, we learn how to differentiate. Geometrically speaking, this amounts to finding the slope of a curve, or its rate of change. We analyze systematically the techniques for doing this, and how they apply to the elementary functions: polynomials, trigonometric functions, exponential and logarithmic functions, and inverse functions.

One of the reasons why we postpone integration till after this section is that the techniques of integration, up to a point, depend on our knowing the derivatives of certain functions, because one of the properties of integration is that it is the inverse operation to differentiation.

You will find applied rate problems similar to each other, but with different types of functions in Chapter III, §9, Chapter IV, §4, and Chapter VII, §4. This is an example of seeing the same idea threaded in a coherent manner through the part on differentiation.

CHAPTER III

The Derivative

The two fundamental notions of this course are those of the derivative and the integral. We take up the first one in this chapter.

The derivative will give us the slope of a curve at a point. It has also applications to physics, where it can be interpreted as the rate of change.

We shall develop some basic techniques which will allow you to compute the derivative in all the standard situations which you are likely to encounter in practice.

III, §1. THE SLOPE OF A CURVE

Consider a curve, and take a point P on the curve. We wish to define the notions of slope of the curve at that point, and tangent line to the curve at that point. Sometimes the statement is made that the tangent to the curve at the point is the line which touches the curve only at that point. This is pure nonsense, as the subsequent pictures will convince you.

Consider a straight line:

Don't you want to say that the line is tangent to itself? If yes, then this certainly contradicts that the tangent is the line which touches the curve at only one point, since the line touches itself at all of its points.

In Figs, 1, 2, and 3, we look at the tangent line to the curve at the point P. In Fig. 1 the line cuts the curve at the other point Q. In Fig. 2 the line is also tangent to the curve at the point Q. In Fig. 3, the

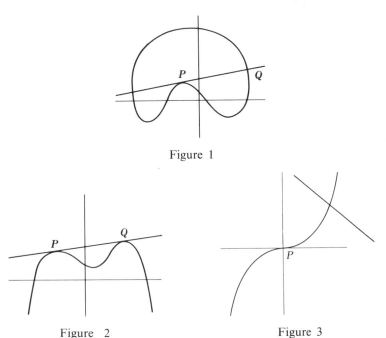

Figure 1

Figure 2   Figure 3

horizontal line is tangent to the curve and "cuts" the curve. The vertical line and the slanted line intersect the curve in only one point, but are not tangent.

Observe also that you cannot get out of the difficulties by trying to distinguish a line "cutting", the curve, or "touching the curve," or by saying that the line should lie on one side of the curve (cf. Fig. 1).

We therefore have to give up the idea of touching the curve only at one point, and look for another idea.

We have to face two problems. One of them is to give the correct geometric idea which allows us to define the tangent to the curve, and the other is to test whether this idea allows us to compute effectively this tangent line when the curve is given by a simple equation with numerical coefficients. It is a remarkable thing that our solution of the first problem will in fact give us a solution to the second.

In Chapter II, we have seen that knowing the slope of a straight line and one point on the straight line allows us to determine the equation of the line. We shall therefore define the slope of a curve at a point and then get its tangent afterward by using the method of Chapter II.

Our examples show us that to define the slope of the curve at P, we should not consider what happens at a point Q which is far removed from P. Rather, it is what happens near P which is important.

Let us therefore take any point Q on the given curve $y = f(x)$, and assume that $Q \neq P$. Then the two points P, Q determine a straight line with a certain slope which depends on P, Q and which we shall write as $S(P, Q)$. Suppose that the point Q approaches the point P on the curve (but stays distinct from P). Then, as Q comes near P, the slope $S(P, Q)$ of the line passing through P and Q should approach the (unkown) slope of the (unknown) tangent line to the curve at P. In the following diagram, we have drawn the tangent line to the curve at P and two lines between P and another point on the curve close to P (Fig. 4). The point

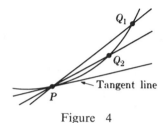

Figure   4

Q_2 is closer to P on the curve and so the slope of the line between P and Q_2 is closer to the slope of the tangent line than is the slope of the line between P and Q_1.

If the limit of the slope $S(P, Q)$ exists as Q approaches P, then it should be regarded as the slope of the curve itself at P. This is the basic idea behind our definition of the slope of the curve at P. We take it as a definition, perhaps the most important definition in this book. To repeat:

Definition. Given a curve $y = f(x)$, let P be a point on the curve. The **slope** of the curve at P is the limit of the slopes of lines through P and another point Q on the curve, as Q approaches P.

The idea of defining the slope in this manner was discovered in the seventeenth century by Newton and Leibnitz. We shall see that this definition allows us to determine the slope effectively in practice.

First we observe that when $y = ax + b$ is a straight line, then the slope of the line between any two distinct points on the curve is always the same, and is the slope of the line as we defined it in the preceding chapter.

Example. Let us now look at the next simplest example,

$$y = f(x) = x^2.$$

We wish to determine the slope of this curve at the point $(1, 1)$.

We look at a point near $(1, 1)$, for instance a point whose x-coordinate is 1.1. Then $f(1.1) = (1.1)^2 = 1.21$. Thus the point $(1.1, 1.21)$ lies on the curve. The slope of the line between two points (x_1, y_1) and (x_2, y_2) is

$$\frac{y_2 - y_1}{x_2 - x_1}.$$

Therefore the slope of the line between $(1, 1)$ and $(1.1, 1.21)$ is

$$\frac{1.21 - 1}{1.1 - 1} = \frac{0.21}{0.1} = 2.1.$$

In general, the x-coordinate of a point near $(1, 1)$ can be written $1 + h$, where h is some small number, positive or negative, but $h \neq 0$. We have

$$f(1 + h) = (1 + h)^2 = 1 + 2h + h^2.$$

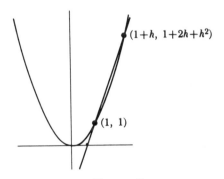

Figure 5

Thus the point $(1 + h, 1 + 2h + h^2)$ lies on the curve. When h is positive, the line between our two points would look like that in Fig. 5. When h is negative, then $1 + h$ is smaller than 1 and the line would look like this:

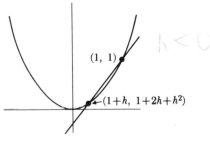

Figure 6

For instance, h could be -0.1 and $1 + h = 0.9$.

The slope of the line between our two points is therefore the quotient

$$\frac{(1 + 2h + h^2) - 1}{(1 + h) - 1},$$

which is equal to

$$\frac{2h + h^2}{h} = 2 + h.$$

As the point whose x-coordinate is $1 + h$ approaches our point $(1, 1)$, the number h approaches 0. As h approaches 0, the slope of the line btween our two points approaches 2, which is therefore the slope of the curve at the point $(1, 1)$ by definition.

You will appreciate how simple the computation turns out to be, and how easy it was to get this slope!

Let us take another example. We wish to find the slope of the same curve $f(x) = x^2$ at the point $(-2, 4)$. Again we take a nearby point whose x-coordinate is $-2 + h$ for small $h \neq 0$. The y-coordinate of this nearby point is

$$f(-2 + h) = (-2 + h)^2 = 4 - 4h + h^2.$$

The slope of the line between the two points is therefore

$$\frac{4 - 4h + h^2 - 4}{-2 + h - (-2)} = \frac{-4h + h^2}{h} = -4 + h.$$

As h approaches 0, the nearby point approaches the point $(-2, 4)$ and we see that the slope approaches -4.

III, §1. EXERCISES

Find the slopes of the following curves at the indicated points:

1. $y = 2x^2$ at the point $(1, 2)$ 2. $y = x^2 + 1$ at the point $(-1, 2)$

3. $y = 2x - 7$ at the point $(2, -3)$ 4. $y = x^3$ at the point $(\frac{1}{2}, \frac{1}{8})$

5. $y = 1/x$ at the point $(2, \frac{1}{2})$ 6. $y = x^2 + 2x$ at the point $(-1, -1)$

7. $y = x^2$ at the point $(2, 4)$ 8. $y = x^2$ at the point $(3, 9)$

9. $y = x^3$ at the point $(1, 1)$ 10. $y = x^3$ at the point $(2, 8)$

11. $y = 2x + 3$ at the point whose x-coordinate is 2.

12. $y = 3x - 5$ at the point whose x-coordinate is 1.

13. $y = ax + b$ at an arbitrary point.

(In Exercises 11, 12, 13 use the h-method, and verify that this method gives the same answer for the slope as that stated in Chapter II, §3.)

III, §2. THE DERIVATIVE

We continue to consider the function $y = x^2$. Instead of picking a definite numerical value for the x-coordinate of a point, we could work at an arbitrary point on the curve. Its coordinates are then (x, x^2). We write the x-coordinate of a point nearby as $x + h$ for some small h, positive or negative, but $h \neq 0$. The y-coordinate of this nearby point is

$$(x + h)^2 = x^2 + 2xh + h^2.$$

Hence the slope of the line between them is

$$\frac{(x + h)^2 - x^2}{(x + h) - x} = \frac{x^2 + 2xh + h^2 - x^2}{x + h - x}$$

$$= \frac{2xh + h^2}{h}$$

$$= 2x + h.$$

As h approaches 0, $2x + h$ approaches $2x$. Consequently, the slope of the curve $y = x^2$ at an arbitrary point (x, y) is $2x$. In particular, when $x = 1$ the slope is 2 and when $x = -2$ the slope is -4, as we found out before by the explicit computation using the special x-coordinates 1 and -2.

This time, however, we have found out a general formula giving us the slope for any point on the curve. Thus when $x = 3$ the slope is 6 and when $x = -10$ the slope is -20.

The example we have just worked out gives us the procedure for treating more general functions.

Given a function $f(x)$, we let the **Newton quotient** be

$$\boxed{\frac{f(x + h) - f(x)}{x + h - x} = \frac{f(x + h) - f(x)}{h}.}$$

This quotient is the slope of the line between the points

$$(x, f(x)) \quad \text{and} \quad (x + h, f(x + h)).$$

It is illustrated on the figure.

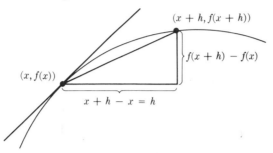

Definition. If the Newton quotient approaches a limit as h approaches 0, then we define the **derivative of f at** x to be this limit, that is

$$\text{derivative of } f \text{ at } x = \lim_{h \to 0} \frac{f(x + h) - f(x)}{h}.$$

The derivative of f at x will be denoted briefly by any one of the notations $f'(x)$, or df/dx, or $df(x)/dx$. Thus by definition

$$f'(x) = \frac{df}{dx} = \frac{df(x)}{dx} = \lim_{h \to 0} \frac{f(x + h) - f(x)}{h}.$$

Thus the two expressions $f'(x)$ and df/dx mean the same thing. We emphasize however that in the expression df/dx we do not multiply f or x by d or divide df by dx. The expression is to be read *as a whole*. We shall find out later that the expression, under certain circumstances, behaves *as if* we were dividing, and it is for this reason that we adopt this classical way of writing the derivative.

The derivative may thus be viewed as a function f', which is defined at all numbers x such that the Newton quotient approaches a limit as h tends to 0. Observe that when taking the limit, both the numerator $f(x + h) - f(x)$ and the denominator h approaches 0. However, their quotient

$$\frac{f(x + h) - f(x)}{h}$$

approaches the slope of the curve at the point $(x, f(x))$.

Definition. The function f is **differentiable** if it has a derivative at all the points for which it is defined.

Example 1. The function $f(x) = x^2$ is differentiable and its derivative is $2x$. Thus we have in this case

$$f'(x) = \frac{df}{dx} = 2x.$$

We have been using systematically the letter x. One can use any other letter: **The truth of mathematical statements is invariant under permutations of the alphabet.** Thus, for instance, if $f(u) = u^2$ then

$$f'(u) = \frac{df}{du} = 2u.$$

What is important here is that the *same letter* u should occur in each one of these places, and that the letter u should be different from the letter f.

We work out some examples before giving you exercises on this section.

Example 2. Let $f(x) = 2x + 1$. Find the derivative $f'(x)$.

We form the Newton quotient. We have $f(x + h) = 2(x + h) + 1$. Thus

$$\frac{f(x + h) - f(x)}{h} = \frac{2x + 2h + 1 - (2x + 1)}{h} = \frac{2h}{h} = 2.$$

As h approaches 0 (which we write also $h \to 0$), this Newton quotient is equal to 2 and hence the limit is 2. Thus

$$\frac{df}{dx} = f'(x) = 2$$

for all values of x. The derivative is constant.

Example 3. Find the slope of the graph of the function $f(x) = 2x^2$ at the point whose x-coordinate is 3, and find the equation of the tangent line at that point.

We may just as well find the slope at an arbitrary point on the graph. It is the derivative $f'(x)$. We have

$$f(x + h) = 2(x + h)^2 = 2(x^2 + 2xh + h^2).$$

The Newton quotient is

$$\frac{f(x+h) - f(x)}{h} = \frac{2(x^2 + 2xh + h^2) - 2x^2}{h}$$

$$= \frac{4xh + 2h^2}{h}$$

$$= 4x + 2h.$$

Hence by definition

$$f'(x) = \lim_{h \to 0}(4x + 2h) = 4x.$$

Thus $f'(x) = 4x$. At the point $x = 3$ we get

$$f'(3) = 12,$$

which is the desired slope.

As for the equation of the tangent line, when $x = 3$ we have $f(3) = 18$. Hence we must find the equation of the line passing through the point $(3, 18)$, having slope 12. This is easy, namely the equation is

$$y - 18 = 12(x - 3).$$

Remark on notation. In the preceding example, we have

$$\frac{df}{dx} = 4x = f'(x).$$

We wanted the derivative when $x = 3$. Here we see the advantage of the $f'(x)$ notation instead of the df/dx notation. We can substitute 3 for x in $f'(x)$ to write

$$f'(3) = 12.$$

We cannot substitute 3 for x in the df/dx notation, because writing

$$\frac{df}{d3}$$

would be very confusing. If we want to use the df/dx notation in such a context, we would have to use some device like writing

$$\frac{df}{dx} \quad \text{at} \quad x = 3 \quad \text{is equal to} \quad 12$$

or sometimes

$$\frac{df}{dx}\bigg|_{x=3} = 12.$$

Still, it is clearly better to use the $f'(x)$ notation in such a context.

Example 4. Find the equation of the tangent line to the curve $y = 2x^2$ at the point whose x-coordinate is -2.

In the preceding example we have computed the general formula for the slope of the tangent line. It is

$$f'(x) = 4x.$$

At the point $x = -2$ the slope is therefore

$$f'(-2) = -8.$$

On the other hand, $f(-2) = 8$. Hence the equation of the tangent line is

$$y - 8 = -8(x + 2).$$

Example 5. Find the derivative of $f(x) = x^3$. We use the Newton quotient, and first write down its numerator:

$$
\begin{aligned}
f(x + h) - f(x) &= (x + h)^3 - x^3 \\
&= x^3 + 3x^2h + 3xh^2 + h^3 - x^3 \\
&= 3x^2h + 3xh^2 + h^3.
\end{aligned}
$$

Then

$$\frac{f(x + h) - f(x)}{h} = 3x^2 + 3xh + h^2.$$

As h approaches 0 the right-hand side approaches $3x^2$, so

$$\frac{d(x^3)}{dx} = 3x^2.$$

In defining the Newton quotient, we can take h positive or negative. It is sometimes convenient when taking the limit to look only at values of h which are positive. We are then looking only at points on the curve which approach the given point from the right. In this manner we get what is called the **right derivative**. If in taking the limit of the Newton quotient we took only negative values for h, we would get the **left derivative**.

Example 6. Let $f(x) = |x|$. Find its right derivative and its left derivative when $x = 0$.

The right derivative is the limit

$$\lim_{\substack{h \to 0 \\ h > 0}} \frac{f(0 + h) - f(0)}{h}.$$

When $h > 0$, we have

$$f(0 + h) = f(h) = h,$$

and $f(0) = 0$. Thus

$$\frac{f(0 + h) - f(0)}{h} = \frac{h}{h} = 1.$$

The limit as $h \to 0$ and $h > 0$ is therefore 1.

The left derivative is the limit

$$\lim_{\substack{h \to 0 \\ h < 0}} \frac{f(0 + h) - f(0)}{h}.$$

When $h < 0$ we have

$$f(0 + h) = f(h) = -h.$$

Hence

$$\frac{f(0 + h) - f(0)}{h} = \frac{-h}{h} = -1.$$

The limit as $h \to 0$ and $h < 0$ is therefore -1.

We see that the right derivative at 0 is 1 and the left derivative is -1. They are not equal. This is illustrated by the graph of our function

$$f(x) = |x|,$$

which looks like that in Fig. 7.

Both the right derivative of f and the left derivative of f exist but they are not equal.

We can rephrase our definition of the derivative and say that the derivative of a function $f(x)$ is defined when the right derivative and the left derivative exist and they are equal, in which case this common value is simply called the **derivative**.

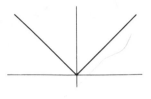

Figure 7

Thus the derivative of $f(x) = |x|$ is not defined at $x = 0$.

Example 7. Let $f(x)$ be equal to x if $0 < x \leq 1$ and $x - 1$ if $1 < x \leq 2$. We do not define f for other values of x. Then the graph of f looks like this:

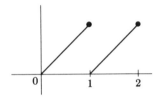

Figure 8

The left derivative of f at 1 exists and is equal to 1, but the right derivative of f at 1 does not exist. We leave the verification of the first assertion to you. To verify the second assertion, we must see whether the limit

$$\lim_{\substack{h \to 0 \\ h > 0}} \frac{f(1 + h) - f(1)}{h}$$

exists. Since $1 + h > 1$ we have

$$f(1 + h) = 1 + h - 1 = h.$$

Also $f(1) = 1$. Thus the Newton quotient is

$$\frac{f(1 + h) - f(1)}{h} = \frac{h - 1}{h} = 1 - \frac{1}{h}.$$

As h approaches 0 the quotient $1/h$ has no limit since it becomes arbitrarily large. Thus the Newton quotient has no limit for $h > 0$ and the function does not have a right derivative when $x = 1$.

III, §2. EXERCISES

Find (a) the derivatives of the following functions, (b) the slope of the graph at the point whose x-coordinate is 2, and find the equation of the tangent line at that point.

1. $x^2 + 1$

2. x^3

3. $2x^3$

4. $3x^2$

5. $x^2 - 5$

6. $2x^2 + x$

7. $2x^2 - 3x$

8. $\frac{1}{2}x^3 + 2x$

9. $\dfrac{1}{x + 1}$

10. $\dfrac{2}{x + 1}$

III, §3. LIMITS

In defining the slope of a curve at a point, or the derivative, we used the notion of limit, which we regarded as intuitively clear. It is indeed. You can see in the Appendix at the end of Part Four how one may define limits using only properties of numbers, but we do not worry about this here. However, we shall make a list of the properties of limits which will be used in the sequel, just to be sure of what we assume about them, and also to give you a technique for computing limits.

We consider functions $F(h)$ defined for all sufficiently small values of h, except that $h \neq 0$. We write

$$\lim_{h \to 0} F(h) = L$$

to mean that $F(h)$ **approaches** L as h **approaches** 0.

First, we note that if F is a constant function, $F(x) = c$ for all x, then

$$\lim_{h \to 0} F(h) = c$$

is the constant itself.
If $F(h) = h$, then

$$\lim_{h \to 0} F(h) = 0.$$

The next properties relate limits with addition, subtraction, multiplication, division, and inequalities.

Suppose that we have two functions $F(x)$ and $G(x)$ which are defined for the same numbers. Then we can form the sum of the two functions

$F + G$, whose value at a point x is $F(x) + G(x)$. Thus when $F(x) = x^4$ and $G(x) = 5x^{3/2}$ we have

$$F(x) + G(x) = x^4 + 5x^{3/2}.$$

The value $F(x) + G(x)$ is also written $(F + G)(x)$. The first property of limits concerns the sum of two functions.

Property 1. *Suppose that we have two functions F and G defined for small values of h, and assume that the limits*

$$\lim_{h \to 0} F(h) \quad \text{and} \quad \lim_{h \to 0} G(h)$$

exist. Then

$$\lim_{h \to 0} [F(h) + G(h)]$$

exists and

$$\lim_{h \to 0} (F + G)(h) = \lim_{h \to 0} F(h) + \lim_{h \to 0} G(h).$$

In other words the limit of a sum is equal to the sum of the limits.

A similar statement holds for the difference $F - G$, namely

$$\lim_{h \to 0} (F(h) - G(h)) = \lim_{h \to 0} F(h) - \lim_{h \to 0} G(h).$$

After the sum we discuss the product. Suppose we have two functions F and G defined for the same numbers. Then we can form their product FG whose value at a number x is

$$(FG)(x) = F(x)G(x).$$

For instance if $F(x) = 2x^2 - 2^x$ and $G(x) = x^2 + 5x$, then the product is

$$(FG)(x) = (2x^2 - 2^x)(x^2 + 5x).$$

Property 2. *Let F, G be two functions for small values of h, and assume that*

$$\lim_{h \to 0} F(h) \quad \text{and} \quad \lim_{h \to 0} G(h)$$

exist. Then the limit of the product exists and we have

$$\lim_{h \to 0} (FG)(h) = \lim_{h \to 0} [F(h)G(h)]$$

$$= \lim_{h \to 0} F(h) \cdot \lim_{h \to 0} G(h).$$

In words, we can say that the product of the limits is equal to the limit of the product.

As a special case, suppose that $F(x)$ is the constant function $F(x) = c$. Then we can form the function cG, product of the constant by G, and we have

$$\lim_{h \to 0} cG(h) = c \cdot \lim_{h \to 0} G(h).$$

Example. Let $F(h) = 3h + 5$. Then $\lim_{h \to 0} F(h) = 5$.

Example. Let $F(h) = 4h^3 - 5h + 1$. Then $\lim_{h \to 0} F(h) = 1$. We can see this by considering the limits

$$\lim_{h \to 0} 4h^3 = 0, \qquad \lim_{h \to 0} 5h = 0, \qquad \lim_{h \to 0} 1 = 1,$$

and taking the appropriate sum.

Example. We have $\lim_{h \to 0} 3xh = 0$, and

$$\lim_{h \to 0}(3xh - 7y) = -7y.$$

Thirdly, we come to quotients. Let F, G be as before, but assume that $G(x) \neq 0$ for any x. Then we can form the quotient function F/G whose value at x is

$$\frac{F}{G}(x) = \frac{F(x)}{G(x)}.$$

Example. Let $F(x) = 2x^3 - 4x$ and $G(x) = x^4 + x^{1/3}$. Then

$$\frac{F}{G}(x) = \frac{F(x)}{G(x)} = \frac{2x^3 - 4x}{x^4 + x^{1/3}}.$$

Property 3. *Assume that the limits*

$$\lim_{h \to 0} F(h) \qquad and \qquad \lim_{h \to 0} G(h)$$

exist, and that

$$\lim_{h \to 0} G(h) \neq 0.$$

Then the limit of the quotient exists and we have

$$\lim_{h \to 0} \frac{F(h)}{G(h)} = \frac{\lim F(h)}{\lim G(h)}.$$

In words, the quotient of the limits is equal to the limit of the quotient.

As we have done above, we shall sometimes omit writing $h \to 0$ for the sake of simplicity.

The next property is stated here for completeness. It will not be used until we find the derivative of sine and cosine, and consequently should be skipped until then.

Property 4. *Let F, G be two functions defined for small values of h, and assume that $G(h) \leqq F(h)$. Assume also that*

$$\lim_{h \to 0} F(h) \qquad and \qquad \lim_{h \to 0} G(h)$$

exist. Then

$$\lim_{h \to 0} G(h) \leqq \lim_{h \to 0} F(h).$$

Property 5. *Let the assumptions be as in Property 4, and in addition, assume that*

$$\lim_{h \to 0} G(h) = \lim_{h \to 0} F(h).$$

Let E be another function defined for same numbers as F, G such that

$$G(h) \leqq E(h) \leqq F(h)$$

for all small values of h. Then

$$\lim_{h \to 0} E(h)$$

exists and is equal to the limits of F and G.

Property 5 is known as the **squeezing process**. You will find many applications of it in the sequel.

Example. Find the limit

$$\lim_{h \to 0} \frac{2xh + 3}{x^2 - 4h}$$

when $x \neq 0$.

The numerator of our quotient approaches 3 when $h \to 0$ and the denominator approaches x^2. Thus the quotient approaches $3/x^2$. We can justify these steps more formally by applying our three properties. For instance:

$$\lim_{h \to 0}(2xh + 3) = \lim_{h \to 0}(2xh) + \lim_{h \to 0} 3$$

$$= \lim_{h \to 0}(2x) \lim(h) + \lim 3$$

$$= 2x \cdot 0 + 3$$

$$= 3.$$

For the denominator, we have

$$\lim(x^2 - 4h) = \lim x^2 + \lim(-4h)$$

$$= x^2 + \lim(-4) \lim(h)$$

$$= x^2 + (-4) \cdot 0$$

$$= x^2.$$

Using the rule for the quotient, we see that the desired limit is equal to $3/x^2$.

Example. In the previous examples, it turns out that we could substitute the value 0 for h and find the appropriate limit. This cannot be done in general. For instance, suppose that we want to find the limit

$$\lim_{h \to 0} \frac{h^2 - h}{h^3 + 2h}.$$

If we substitute $h = 0$ we get a meaningless expression $0/0$, and hence we do not get information on the limit. However, for $h \neq 0$ we can cancel h from the quotient, and we see that

$$\frac{h^2 - h}{h^3 + 2h} = \frac{h(h - 1)}{h(h^2 + 2)} = \frac{h - 1}{h^2 + 2}.$$

From the expression on the right we can determine the limit. Indeed, $h - 1$ approaches -1 as h approaches 0. Also, $h^2 + 2$ approaches 2 as h approaches 0. Hence by the rule for quotients of limits, we conclude that

$$\lim_{h \to 0} \frac{h^2 - h}{h^3 + 2h} = \lim_{h \to 0} \frac{h - 1}{h^2 + 2} = \frac{-1}{2}.$$

Observe that in this example, both the numerator and the denominator approach 0 as h approaches 0. However, the limit exists, and we see that it is $-\frac{1}{2}$.

Example. Find the limit

$$\lim_{h \to 0} \frac{x^2 h^3 - h^2}{3xh - h}.$$

In this case, we can factor h from the numerator and denominator, so that the quotient is equal to

$$\frac{x^2 h^2 - h}{3x - 1}.$$

The numerator is now seen to approach 0 and the denominator approaches $3x - 1$ as h approaches 0. Hence the quotient approaches 0. This is the desired limit.

The properties of limits which we have stated above will allow you to compute limits in determining derivatives. We illustrate this by an example.

Example. Let $f(x) = 1/x$ (defined for $x \neq 0$). Find the derivative df/dx.

The Newton quotient is

$$\frac{f(x + h) - f(x)}{h} = \frac{\dfrac{1}{x + h} - \dfrac{1}{x}}{h}$$

$$= \frac{x - (x + h)}{(x + h)xh}$$

$$= \frac{-h}{(x + h)xh} = \frac{-1}{(x + h)x}.$$

Then we take the limit:

$$\lim_{h \to 0} \frac{f(x+h) - f(x)}{h} = \lim_{h \to 0} \frac{-1}{(x+h)x}$$

$$= \frac{-1}{\lim(x+h)x}$$

(by the rule for the limit of a quotient)

$$= \frac{-1}{x^2}.$$

Thus we have proved:

$$\boxed{\frac{d\left(\dfrac{1}{x}\right)}{dx} = -\frac{1}{x^2}.}$$

III, §3. EXERCISES

Find the derivatives of the following functions, justifying the steps in taking limits by means of the first three properties:

1. $f(x) = 2x^2 + 3x$

2. $f(x) = \dfrac{1}{2x+1}$

3. $f(x) = \dfrac{x}{x+1}$

4. $f(x) = x(x+1)$

5. $f(x) = \dfrac{x}{2x-1}$

6. $f(x) = 3x^3$

7. $f(x) = x^4$

8. $f(x) = x^5$

(It is especially important that you should work out Exercises 7 and 8 to see a developing pattern, to be followed in the next section.)

9. $f(x) = 2x^3$

10. $f(x) = \frac{1}{2}x^3 + x$

11. $2/x$

12. $3/x$

13. $\dfrac{1}{2x-3}$

14. $\dfrac{1}{3x+1}$

15. $\dfrac{1}{x+5}$

16. $\dfrac{1}{x-2}$

17. $1/x^2$

18. $1/(x+1)^2$

III, §4. POWERS

We have seen that the derivative of the function x^2 is $2x$.

Let us consider the function $f(x) = x^3$ and find its derivative. We have

$$f(x + h) = (x + h)^3 = x^3 + 3x^2h + 3xh^2 + h^3.$$

Hence the Newton quotient is

$$\frac{f(x + h) - f(x)}{h} = \frac{x^3 + 3x^2h + 3xh^2 + h^3 - x^3}{h}$$

$$= \frac{3x^2h + h^2(3x + h)}{h} \qquad \text{(after cancellations)}$$

$$= 3x^2 + 3xh + h^2 \qquad \text{(after cancelling } h\text{)}.$$

Using the properties of limits of sums and products, we see that $3x^2$ remains equal to itself as h approaches 0, that $3xh$ and h^3 both approach 0. Hence

$$f'(x) = \lim_{h \to 0} \frac{f(x + h) - f(x)}{h} = 3x^2.$$

This suggests that in general, whenever $f(x) = x^n$ for some positive integer n, the derivative $f'(x)$ should be nx^{n-1}. This is indeed the case, and will be proved in the next theorem. The proof will follow the same pattern as in the case $f(x) = x^3$ above. You should work out the case $f(x) = x^4$ and probably also the case $f(x) = x^5$ in detail (these were exercises for the preceding section) in order to confirm in these special cases that the pattern remains the same, before seeing the general case. Note that when we treated $f(x) = x^3$ we obtained an expression for the Newton quotient which had a numerator

$$3x^2h + h^2(3x + h),$$

containing the term $3x^2h$, and another term containing h^2 as a factor. When we divide by h, $3x^2$ yields the value of the derivative, and the remaining term $h(3x + h)$ still contains h as a factor, and consequently approaches 0 as h approaches 0. You will find explicitly a similar phenomenon for x^4 and x^5.

Theorem 4.1. *Let n be an integer ≥ 1 and let $f(x) = x^n$. Then*

$$\frac{df}{dx} = nx^{n-1}.$$

Proof. We have

$$f(x + h) = (x + h)^n = (x + h)(x + h)\cdots(x + h),$$

the product being taken n times. Selecting x from each factor gives us a term x^n. If we take x from all but one factor and h from the remaining factors, we get hx^{n-1} taken n times. This gives us a term $nx^{n-1}h$. All other terms will involve selecting h from at least two factors, and the corresponding terms will be divisible by h^2. Thus we get

$$f(x + h) = (x + h)^n = x^n + nx^{n-1}h + h^2g(x, h),$$

where $g(x, h)$ is simply some expression involving powers of x and h with numerical coefficients which, as we shall see later in the proof, it is unnecessary for us to determine. However, using the rules for limits of sums and products we can conclude that

$$\lim_{h \to 0} g(x, h)$$

will be some number which it is unnecessary for us to determine.

The Newton quotient is therefore

$$\frac{f(x + h) - f(x)}{h} = \frac{x^n + nx^{n-1}h + h^2g(x, h) - x^n}{h}$$

$$= \frac{nx^{n-1}h + h^2g(x, h)}{h} \qquad \text{(because } x^n \text{ cancels)}$$

$$= nx^{n-1} + hg(x, h) \qquad \text{(divide numerator and denominator by } h\text{)}.$$

As h approaches 0, the term nx^{n-1} remains unchanged. The limit of h as h tends to 0 is 0, and hence by the product rule, the term $hg(x, h)$ approaches 0 when h approaches 0. Thus finally

$$\lim_{h \to 0} \frac{f(x + h) - f(x)}{h} = nx^{n-1},$$

which proves our theorem.

For another proof, see the end of the next section.

Theorem 4.2. *Let a be any number and let $f(x) = x^a$ (defined for $x > 0$). Then $f(x)$ has a derivative, which is*

$$f'(x) = ax^{a-1}.$$

It would not be difficult to prove Theorem 4.2 when a is a negative integer. It is best, however, to wait until we have a rule giving us the derivative of a quotient before doing it. We could also give a proof when a is a rational number. However, we shall prove the general result in a later chapter, and thus we prefer to wait until then, when we have more techniques available.

Examples. If $f(x) = x^{10}$, then $f'(x) = 10x^9$.
If $f(x) = x^{3/2}$ (for $x > 0$), then $f'(x) = \frac{3}{2}x^{1/2}$.
If $f(x) = x^{-5/4}$, then $f'(x) = -\frac{5}{4}x^{-9/4}$.
If $f(x) = x^{\sqrt{2}}$, then $f'(x) = \sqrt{2}x^{\sqrt{2}-1}$.

Note especially the special case when $f(x) = x$. Then $f'(x) = 1$.

Example. We can now find the equations of tangent lines to certain curves which we could not do before. Consider the curve

$$y = x^5$$

We wish to find the equation of its tangent line at the point $(2, 32)$. By Theorem 4.1, if $f(x) = x^5$, then $f'(x) = 5x^4$. Hence the slope of the tangent line at $x = 2$ is

$$f'(2) = 5 \cdot 2^4 = 80.$$

On the other hand, $f(2) = 2^5 = 32$. Hence the equation of the tangent line is

$$y - 32 = 80(x - 2).$$

III, §4. EXERCISES

1. Write out the expression of $(x + h)^4$ in terms of powers of x and h.

2. Find the derivative of the function x^4 directly, using the Newton quotient.

3. What are the derivatives of the following functions?
   (a) $x^{2/3}$ (b) $x^{-3/2}$ (c) $x^{7/6}$

4. What is the equation of the tangent line to the curve $y = x^9$ at the point $(1, 1)$?

5. What is the slope of the curve $y = x^{2/3}$ at the point $(8, 4)$? What is the equation of the tangent line at that point?

6. Give the slope and equation of the tangent line to the curve $y = x^{-3/4}$ at the point whose x-coordinate is 16.

7. Give the slope and equation of the tangent line to the curve $y = \sqrt{x}$ at the point whose x-coordinates is 3.

8. Give the derivatives of the following functions at the indicated points:
   (a) $f(x) = x^{1/4}$ at $x = 5$ (b) $f(x) = x^{-1/4}$ at $x = 7$
   (c) $f(x) = x^{\sqrt{2}}$ at $x = 10$ (d) $f(x) = x^{\pi}$ at $x = 7$

III, §5. SUMS, PRODUCTS, AND QUOTIENTS

In this section we shall derive several rules which allow you to find the derivatives for sums, products, and quotients of functions when you know the derivative of each factor.

We begin with a definition of continuous functions and the reason why a differentiable function is continuous.

Definition. A function is said to be **continuous at a point** x if and only if

$$\lim_{h \to 0} f(x + h) = f(x).$$

A function is said to be **continuous** if it is continuous at every point of its domain of definition.

Let f be a function having a derivative $f'(x)$ at x. Then f is continuous at x.

Proof. The quotient

$$\frac{f(x + h) - f(x)}{h}$$

approaches the limit $f'(x)$ as h approaches 0. We have

$$h \, \frac{f(x + h) - f(x)}{h} = f(x + h) - f(x).$$

Therefore using the rule for the limit of a product, and noting that h approaches 0, we find

$$\lim_{h \to 0} f(x + h) - f(x) = 0 f'(x) = 0.$$

This is another way of stating that

$$\lim_{h \to 0} f(x + h) = f(x).$$

In other words, f is continuous.

Of course, we can never substitute $h = 0$ in our quotient, because then it becomes $0/0$, which is meaningless. Geometrically, letting $h = 0$ amounts to taking the two points on the curve equal to each other. It is then impossible to have a unique straight line through one point. Our procedure of taking the limit of the Newton quotient is meaningful only if $h \neq 0$.

Observe that in the Newton quotient, both the numerator and the denominator approach 0. The quotient itself, however, need not approach 0.

Example. Let $f(x) = |x|$. Then the absolute value function f is continuous at 0, even though it is not differentiable at 0. It is still true that

$$f(0 + h) = f(h) = |h|$$

approaches 0 as h approaches 0, even though the function is not differentiable at 0. As we saw in §2, the function $f(x) = |x|$ is right differentiable at 0 and left differentiable at 0, but not differentiable at 0.

Example. Let $f(x) = 0$ if $x < 0$, and $f(x) = 1$ if $x \geq 0$. The graph of f is shown on Fig. 9.

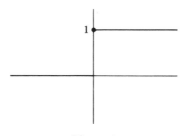

Figure 9

The function f is not continuous at 0. Roughly speaking, the function is not continuous at 0 because its graph has a "break" at 0. In Fig. 10, we show other examples of graphs with discontinuities.

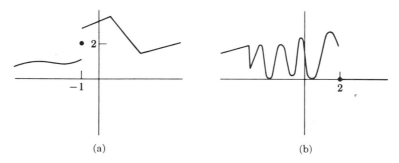

(a)   (b)

Figure 10

In Fig. 10(a), the function is continuous except at $x = -1$. In Fig. 10(b), the function is continuous except at $x = 2$.

The remark at the beginning of this section shows that if a function is differentiable, then it is continuous. Since we are concerned principally about differentiable functions at present, we do not go any deeper into continuous functions, and wait till later, until the notion becomes more relevant to us.

Let c be a number and $f(x)$ a function which has a derivative $f'(x)$ for all values of x for which it is defined. We can multiply f by the constant c to get another function cf whose value at x is $cf(x)$.

Constant times a function. *The derivative of cf is then given by the formula*

$$(cf)'(x) = c \cdot f'(x).$$

In other words, the derivative of a constant times a function is the constant times the derivative of the function.

In the other notation, this reads

$$\boxed{\frac{d(cf)}{dx} = c\,\frac{df}{dx}.}$$

To prove this rule, we use the definition of derivative. The Newton quotient for the function cf is

$$\frac{(cf)(x+h) - (cf)(x)}{h} = \frac{cf(x+h) - cf(x)}{h} = c\,\frac{f(x+h) - f(x)}{h}.$$

Let us take the limit as h approaches 0. Then c remains fixed, and

$$\frac{f(x+h) - f(x)}{h}$$

approaches $f'(x)$. According to the rule for the product of limits, we see that our Newton quotient approaches $cf'(x)$, as was to be proved.

Example. Let $f(x) = 3x^2$. Then $f'(x) = 6x$. If $f(x) = 17x^{1/2}$, then $f'(x) = \frac{17}{2}x^{-1/2}$. If $f(x) = 10x^a$, then $f'(x) = 10ax^{a-1}$.

Next we look at the sum of two functions.

Sum. *Let $f(x)$ and $g(x)$ be two functions which have derivatives $f'(x)$ and $g'(x)$, respectively. Then the sum $f(x) + g(x)$ has a derivative, and*

$$(f + g)'(x) = f'(x) + g'(x).$$

The derivative of a sum is equal to the sum of the derivatives.

In the other notation, this reads:

$$\boxed{\frac{d(f+g)}{dx} = \frac{df}{dx} + \frac{dg}{dx}.}$$

To prove this, we have by definition

$$(f+g)(x+h) = f(x+h) + g(x+h)$$

and

$$(f+g)(x) = f(x) + g(x).$$

Therefore the Newton quotient for $f+g$ is

$$\frac{(f+g)(x+h) - (f+g)(x)}{h} = \frac{f(x+h) + g(x+h) - f(x) - g(x)}{h}.$$

Collecting terms and separating the fraction, we see that this expression is equal to

$$\frac{f(x+h) - f(x) + g(x+h) - g(x)}{h}$$

$$= \frac{f(x+h) - f(x)}{h} + \frac{g(x+h) - g(x)}{h}.$$

Taking the limit as h approaches 0 and using the rule for the limit of a sum, we see that this last sum approaches $f'(x) + g'(x)$ as h approaches 0. This proves what we wanted.

Example.

$$\frac{d}{dx}(x^3 + x^2) = 3x^2 + 2x,$$

$$\frac{d}{dx}(4x^{1/2} + 5x^{-10}) = 2x^{-1/2} - 50x^{-11}.$$

Carried away by our enthusiasm at determining so easily the derivative of functions built up from others by means of constants and sums, we might now be tempted to state the rule that the derivative of a product is the product of the derivatives. Unfortunately, this is false. To see that the rule is false, we look at an example.

Let $f(x) = x$ and $g(x) = x^2$. Then $f'(x) = 1$ and $g'(x) = 2x$. Therefore $f'(x)g'(x) = 2x$. However, the derivative of the product $(fg)(x) = x^3$ is

$3x^2$, which is certainly not equal to $2x$. Thus the product of the derivatives is not equal to the derivative of the product.

Through trial and error the correct rule was discovered. It can be stated as follows:

Product. *Let $f(x)$ and $g(x)$ be two functions having derivatives $f'(x)$ and $g'(x)$. Then the product function $f(x)g(x)$ has a derivative, which is given by the formula*

$$(fg)'(x) = f(x)g'(x) + f'(x)g(x).$$

In words, the derivative of the product is equal to the first times the derivative of the second, plus the derivative of the first times the second.

In the other notation, this reads

$$\frac{d(fg)}{dx} = f(x)\frac{dg}{dx} + \frac{df}{dx}g(x).$$

The proof is not very much more difficult than the proofs we have already encountered. By definition, we have

$$(fg)(x + h) = f(x + h)g(x + h)$$

and

$$(fg)(x) = f(x)g(x).$$

Consequently the Newton quotient for the product function fg is

$$\frac{(fg)(x + h) - (fg)(x)}{h} = \frac{f(x + h)g(x + h) - f(x)g(x)}{h}.$$

At this point, it looks a little hopeless to transform this quotient in such a way that we see easily what limit it approaches as h approaches 0. But we use a trick, and rewrite our quotient by inserting

$$-f(x + h)g(x) + f(x + h)g(x)$$

in the numerator. This certainly does not change the value of our quotient, which now looks like

$$\frac{f(x + h)g(x + h) - f(x + h)g(x) + f(x + h)g(x) - f(x)g(x)}{h}.$$

We can split this fraction into a sum of two fractions:

$$\frac{f(x + h)g(x + h) - f(x + h)g(x)}{h} + \frac{f(x + h)g(x) - f(x)g(x)}{h}.$$

We can factor $f(x + h)$ in the first term, and $g(x)$ in the second term, to obtain

$$f(x + h)\frac{g(x + h) - g(x)}{h} + \frac{f(x + h) - f(x)}{h}g(x).$$

The situation is now well under control. As h approaches 0, $f(x + h)$ approaches $f(x)$, and the two quotients in the expression we have just written approach $g'(x)$ and $f'(x)$ respectively. Thus the Newton quotient for fg approaches

$$f(x)g'(x) + f'(x)g(x),$$

thereby proving our assertion.

Example. Applying the product rule, we find:

$$\frac{d}{dx}(x + 1)(3x^2) = (x + 1)6x + 1 \cdot 3x^2.$$

Similarly,

$$\frac{d}{dx}[(2x^5 + 5x^4)(2x^{1/2} + x^{-1})]$$

$$= (2x^5 + 5x^4)\left(x^{-1/2} - \frac{1}{x^2}\right) + (10x^4 + 20x^3)(2x^{1/2} + x^{-1})$$

which you may and should leave just like that without attempting to simplify the expression.

A special case of the product rule is used all the time, that of the power of a function.

Example.

$$\boxed{\frac{d(f(x)^2)}{dx} = 2f(x)f'(x).}$$

Indeed, differentiating the product $y = f(x)f(x)$, we obtain

$$\frac{dy}{dx} = f(x)f'(x) + f'(x)f(x) = 2f(x)f'(x).$$

The last rule of this section concerns the derivative of a quotient. We begin with a special case.

Let $g(x)$ be a function having a derivative $g'(x)$, and such that $g(x) \neq 0$. Then the derivative of the quotient $1/g(x)$ exists, and is equal to

$$\frac{d}{dx} \frac{1}{g(x)} = \frac{-1}{g(x)^2} g'(x).$$

To prove this, we look at the Newton quotient

$$\frac{\dfrac{1}{g(x+h)} - \dfrac{1}{g(x)}}{h}$$

which is equal to

$$\frac{g(x) - g(x+h)}{g(x+h)g(x)h} = -\frac{1}{g(x+h)g(x)} \frac{g(x+h) - g(x)}{h}.$$

Letting h approach 0 we see immediately that our expression approaches

$$\frac{-1}{g(x)^2} g'(x)$$

as desired.

Example.

$$\frac{d}{dx} \frac{1}{(x^5 - 3x)} = \frac{-1}{(x^5 - 3x)^2} (5x^4 - 3).$$

The general case of the rule for quotients can now be easily stated and proved.

Quotient. *Let $f(x)$ and $g(x)$ be two functions having derivatives $f'(x)$ and $g'(x)$ respectively, and such that $g(x) \neq 0$. Then the derivative of the quotient $f(x)/g(x)$ exists, and is equal to*

$$\frac{g(x)f'(x) - f(x)g'(x)}{g(x)^2}.$$

In words this yields:

The bottom times the derivative of the top, minus the top times the derivative of the bottom, over the bottom squared

(*which you should memorize like a poem*).

In the other notation, this reads:

$$\frac{d(f/g)}{dx} = \frac{g(x)\,df/dx - f(x)\,dg/dx}{g(x)^2}.$$

To prove this rule, we write our quotient in the form

$$\frac{f(x)}{g(x)} = f(x)\,\frac{1}{g(x)}$$

and use the rule for the derivative of a product, together with the special case we have just proved. We obtain its derivative:

$$\frac{d}{dx}\left(\frac{f(x)}{g(x)}\right) = f'(x)\,\frac{1}{g(x)} + f(x)\,\frac{-1}{g(x)^2}\,g'(x)$$

$$= \frac{g(x)f'(x) - f(x)g'(x)}{g(x)^2}.$$

by putting this expression over the common denominator $g(x)^2$. This is the desired derivative.

We work out some examples.

Example.

$$\frac{d}{dx}\left(\frac{x^2 + 1}{3x^4 - 2x}\right) = \frac{(3x^4 - 2x)2x - (x^2 + 1)(12x^3 - 2)}{(3x^4 - 2x)^2}.$$

Example.

$$\frac{d}{dx}\left(\frac{2x}{x + 4}\right) = \frac{(x + 4)\cdot 2 - 2x\cdot 1}{(x + 4)^2}.$$

Example. Find the equation of the tangent line to the curve

$$y = x/(x^2 + 4)$$

at the point $x = -3$.

We let $f(x) = x/(x^2 + 4)$. Then

$$f'(x) = \frac{(x^2 + 4)\cdot 1 - x\cdot 2x}{(x^2 + 4)^2}.$$

Hence

$$f'(-3) = \frac{13 - 18}{13^2} = \frac{-5}{169}.$$

This is the slope at the given point. Furthermore,

$$f(-3) = \frac{-3}{13}.$$

Hence the coordinates of the given point are $(-3, -3/13)$. The equation of the tangent line is therefore

$$y + \frac{3}{13} = \frac{-5}{169} (x + 3).$$

Example. Show that the two curves

$$y = x^2 + 1 \quad \text{and} \quad y = \tfrac{2}{3}x^3 + \tfrac{4}{3}$$

have a common tangent line at the point $(1, 2)$. We let

$$f(x) = x^2 + 1 \quad \text{and} \quad g(x) = \tfrac{2}{3}x^3 + \tfrac{4}{3}.$$

Then

$$f'(x) = 2x \quad \text{and} \quad g'(x) = 2x^2.$$

Since $f'(1) = g'(1) = 2$, the slope of the tangent lines to the graph of f and the graph of g is the same at the point $(1, 2)$. Hence the tangent lines are the same since they pass through the same point, namely

$$y - 2 = 2(x - 1).$$

Appendix. Another proof that $dx^n/dx = nx^{n-1}$

There is another proof that

$$\frac{dx^n}{dx} = nx^{n-1}$$

for any positive integer n, as follows. First we note that when $n = 1$ we have proved directly

$$\frac{dx}{dx} = 1.$$

by using the h-method.

Now we use the rule for the derivative of a product. To get the derivative of x^2, we have:

$$\frac{d(x^2)}{dx} = \frac{d(x \cdot x)}{dx} = x\frac{dx}{dx} + \frac{dx}{dx}x = 2x.$$

Next, we get the derivative of x^3:

$$\frac{d(x^3)}{dx} = \frac{d(x^2 \cdot x)}{dx} = x^2\frac{dx}{dx} + \frac{dx^2}{dx}x$$

$$= x^2 + 2x^2 \qquad \text{(by the preceding step)}$$

$$= 3x^2.$$

Next we get the derivative of x^4:

$$\frac{d(x^4)}{dx} = \frac{d(x^3 \cdot x)}{dx} = x^3\frac{dx}{dx} + \frac{d(x^3)}{dx}x$$

$$= x^3 + 3x^3 \qquad \text{(by the preceding step)}$$

$$= 4x^3.$$

We can proceed like this for any integer n. Suppose we have proved the formula up to some integer n. Then

$$\frac{d(x^{n+1})}{dx} = \frac{d(x^n \cdot x)}{dx} = x^n\frac{dx}{dx} + \frac{d(x^n)}{dx}x$$

$$= x^n + nx^n \qquad \text{(by the preceding step)}$$

$$= (n+1)x^n.$$

This shows how to proceed from one step to the next.

Such a procedure is called **induction**. You will find several instances of such a procedure throughout the course. In each case, work out the steps for $n = 1$, $n = 2$, $n = 3$, $n = 4$, and so forth, in succession. Go as far as you need to recognize the pattern and to be comfortable with it. Then try to formulate and carry out the final step with n instead of a specific number. After you have met enough examples of this type, you will have understood what induction means, and in particular, you will understand the following formal definition.

Suppose we want to prove an assertion $A(n)$ for every positive integer n. Then a **proof by induction** consists in proving:

(1)   The assertion is true when $n = 1$.
(2)   If the assertion is true for a given integer n, then it is true for $n + 1$.

The first step allows us to start the procedure; and the second step allows us to go from one integer to the next, just as we did in the above example.

III, §5. EXERCISES

Find the derivatives of the following functions:

1. (a) $2x^{1/3}$ (b) $3x^{3/4}$ (c) $\frac{1}{2}x^2$ (d) $\frac{3}{4}x^2$

2. (a) $5x^{11}$ (b) $4x^{-2}$ (c) $\frac{1}{3}x^4 - 5x^3 + x^2 - 2$

3. (a) $\frac{1}{2}x^{-3/4}$ (b) $3x - 2x^3$ (c) $4x^5 - 7x^3 + 2x - 1$

4. (a) $7x^3 + 4x^2$ (b) $4x^{2/3} + 5x^4 - x^3 + 3x$

5. (a) $25x^{-1} + 12x^{1/2}$ (b) $2x^3 + 5x^7$ (c) $4x^4 - 7x^3 + x - 12$

6. (a) $\frac{3}{5}x^2 - 2x^8$ (b) $3x^4 - 2x^2 + x - 10$ (c) $\pi x^7 - 8x^5 + x + 1$

7. $(x^3 + x)(x - 1)$ 8. $(2x^2 - 1)(x^4 + 1)$

9. $(x + 1)(x^2 + 5x^{3/2})$ 10. $(2x - 5)(3x^4 + 5x + 2)$

11. $(x^{-2/3} + x^2)\left(x^3 + \dfrac{1}{x}\right)$ 12. $(2x + 3)\left(\dfrac{1}{x^2} + \dfrac{1}{x}\right)$

13. $\dfrac{2x + 1}{x + 5}$ 14. $\dfrac{2x}{x^2 + 3x + 1}$

To break the monotony of the letter x, let us use another.

15. $f(t) = \dfrac{t^2 + 2t - 1}{(t + 1)(t - 1)}$ 16. $\dfrac{t^{-5/4}}{t^2 + t - 1}$

17. What is the slope of the curve

$$y = \frac{t}{t + 5}$$

at the point $t = 2$? What is the equation of the tangent line at this point?

18. What is the slope of the curve?

$$y = \frac{t^2}{t^2 + 1}$$

at $t = 1$? What is the equation of the tangent line?

III, §5. SUPPLEMENTARY EXERCISES

Find the derivatives of the following functions. Do not simplify your answers!

1. $3x^3 - 4x + 5$ 2. $x^2 + 2x + 27$

3. $x^2 + x - 1$ 4. $x^{1/2} - 8x^4 + x^{-1}$

5. $x^{5/2} + x^{-5/2}$

6. $x^7 + 15x^{-1/5}$

7. $(x^2 - 1)(x + 5)$

8. $\left(x^5 + \dfrac{1}{x}\right)(x^5 + 1)$

9. $(x^{3/2} + x^2)(x^4 - 99)$

10. $(x^2 + x + 1)(x^5 - x - 25)$

11. $(2x^2 + 1)\left(\dfrac{1}{x^2} + 4x + 8\right)$

12. $(x^4 - x^2)(x^2 - 1)$

13. $(x + 1)(x + 2)(x + 3)$

14. $5(x - 1)(x - 2)(x^2 + 1)$

15. $x^3(x^2 + 1)(x + 1)$

16. $(x^4 + 1)(x + 5)(2x + 7)$

17. $\dfrac{1}{2x + 3}$

18. $\dfrac{1}{7x + 27}$

19. $\dfrac{-5}{x^3 + 2x^2}$

20. $\dfrac{3}{2x^4 + x^{3/2}}$

21. $\dfrac{-2x}{x + 1}$

22. $\dfrac{x + 1}{x - 5}$

23. $\dfrac{3x^{1/2}}{(x + 1)(x - 1)}$

24. $\dfrac{2x^{1/2} + x^{3/4}}{(x + 1)x^3}$

25. $\dfrac{x^5 + 1}{(x^2 + 1)(x + 7)}$

26. $\dfrac{(x + 1)(x + 5)}{x - 4}$

27. $\dfrac{x^3}{1 - x^2}$

28. $\dfrac{x^5}{x^{3/2} + x}$

29. $\dfrac{x^2 - x}{x^2 + 1}$

30. $\dfrac{x^2 + 2x + 7}{8x}$

31. $\dfrac{2x + 1}{x^2 + x - 4}$

32. $\dfrac{x^5}{x^2 + 3}$

33. $\dfrac{4x - x^3}{x^2 + 2}$

34. $\dfrac{x^3}{x^2 - 5x + 7}$

35. $\dfrac{1 - 5x}{x}$

36. $\dfrac{1 + 6x + x^{3/4}}{7x - 2}$

37. $\dfrac{x^2}{(x + 1)(x - 2)}$

38. $\dfrac{x^{1/2} - x^{-1/2}}{x_{3/4}}$

39. $\dfrac{3x^4 + x^{5/4}}{4x^3 - x^5 + 1}$

40. $\dfrac{x - 1}{(x - 2)(x - 3)}$

Find the equations of the tangent lines to the following curves at the given point.

41. $y = x^{1/4} + 2x^{3/4}$ at $x = 16$

42. $y = 2x^3 + 3$ at $x = \frac{1}{2}$

43. $y = (x - 1)(x - 3)(x - 4)$ at $x = 0$

44. $y = 2x^2 + 5x - 1$ at $x = 2$

45. $y = (x^2 + 1)(2x + 3)$ at $x = 1$

46. $y = \dfrac{x - 1}{x + 5}$ at $x = 2$

47. $y = \dfrac{x^2}{x^3 + 1}$ at $x = 2$

48. $y = \dfrac{x^2 + 1}{x^3 + 1}$ at $x = 2$

49. $y = \dfrac{x^2}{x^2 - 1}$ at $x = 2$

50. $y = \dfrac{x-1}{x^2+1}$ at $x = 1$

51. Show that the line $y = -x$ is tangent to the curve given by the equation

$$y = x^3 - 6x^2 + 8x.$$

Find the point of tangency.

52. Show that the line $y = 9x - 15$ is tangent to the curve

$$y = x^3 - 3x + 1.$$

Find the point of tangency.

53. Show that the graphs of the equations

$$y = 3x^2 \qquad \text{and} \qquad y = 2x^3 + 1$$

have the common tangent line at the point $(1, 3)$. Sketch the graphs.

54. Show that there are exactly two tangent lines to the graph of $y = (x + 1)^2$ which pass through the origin, and find their equations.

55. Find all the points (x_0, y_0) on the curve

$$y = 4x^4 - 8x^2 + 16x + 7$$

such that the tangent line to the curve at (x_0, y_0) is parallel to the line

$$16x - y + 5 = 0.$$

Find the tangent line to the curve at each of these points.

III, §6. THE CHAIN RULE

We know how to build up new functions from old ones by means of sums, products, and quotients. There is one other important way of building up new functions. We shall first give examples of this new way.

Consider the function $(x + 2)^{10}$. We can say that this function is made up of the 10-th power function, and the function $x + 2$. Namely, given a number x, we first add 2 to it, and then take the 10-th power. Let

$$g(x) = x + 2$$

and let f be the 10-th power function. Then we can take the value of f at $x + 2$, namely

$$f(x + 2) = (x + 2)^{10}$$

and we can also write it as

$$f(x + 2) = f(g(x)).$$

Another example: Consider the function $(3x^4 - 1)^{1/2}$. If we let

$$g(x) = 3x^4 - 1$$

and we let f be the square root function, then

$$f(g(x)) = \sqrt{3x^4 - 1} = (3x^4 - 1)^{1/2}.$$

In order not to get confused by the letter x, which cannot serve us any more in all contexts, we use another letter to denote the values of g. Thus we may write $f(u) = u^{1/2}$.

Similarly, let $f(u)$ be the function $u + 5$ and $g(x) = 2x$. Then

$$f(g(x)) = f(2x) = 2x + 5.$$

One more example of the same type: Let

$$f(u) = \frac{1}{u + 2}$$

and

$$g(x) = x^{10}.$$

Then

$$f(g(x)) = \frac{1}{x^{10} + 2}.$$

In order to give you sufficient practice with many types of functions, we now mention several of them whose definitions will be given later. These will be sin and cos (which we read sine and cosine), log (which we read logarithm or simply log), and the exponential function exp. We shall select a special number e (whose value is approximately 2.718...), such that the function exp is given by

$$\exp(x) = e^x.$$

We now see how we make new functions out of these:

Let $f(u) = \sin u$ and $g(x) = x^2$. Then

$$f(g(x)) = \sin(x^2).$$

Let $f(u) = e^u$ and $g(x) = \cos x$. Then

$$f(g(x)) = e^{\cos x}.$$

Let $f(v) = \log v$ and $g(t) = t^3 - 1$. Then

$$f(g(t)) = \log(t^3 - 1).$$

Let $f(w) = w^{10}$ and $g(z) = \log z + \sin z$. Then

$$f(g(z)) = (\log z + \sin z)^{10}.$$

Whenever we have two functions f and g such that f is defined for all numbers which are values of g, then we can build a new function denoted by $f \circ g$ whose value at a number x is

$$(f \circ g)(x) = f(g(x)).$$

The rule defining this new function is: Take the number x, find the number $g(x)$, and then take the value of f at $g(x)$. This is the value of $f \circ g$ at x. The function $f \circ g$ is called the **composite function** of f and g. We say that g is the **inner** function and that f is the **outer** function. For example, in the function $\log \sin x$, we have

$$f \circ g = \log \circ \sin.$$

The outer function is the log, and the inner function is the sine.

It is important to keep in mind that we can compose two functions only when the outer function is defined at all values of the inner function. For instance, let $f(u) = u^{1/2}$ and $g(x) = -x^2$. Then we cannot form the composite function $f \circ g$ because f is defined only for positive numbers (or 0) and the values of g are all negative, or 0. Thus $(-x^2)^{1/2}$ does not make sense.

However, for the moment you are asked to learn the mechanism of composite functions just the way you learned the multiplication table, in order to acquire efficient conditioned reflexes when you meet composite functions. Hence for the drills given by the exercises at the end of the section, you should forget for a while the meaning of the symbols and operate with them formally, just to learn the formal rules properly.

Even though we have not defined the exponential function e^x nor have we dealt formally with $\sin x$ or other of the functions just mentioned, nevertheless, we don't need to know their definitions in order to manipulate them. If we limited ourselves only to those functions which we have explicitly dealt with so far, we would be restricted in seeing how composite functions work, and how the chain rule works below.

We come to the problem of taking the derivative of a composite function.

We start with an example. Suppose we want to find the derivative of the function $(x + 1)^{10}$. The Newton quotient would be a very long expression, which it would be essentially hopeless to disentangle by brute force, the way we have up to now. It is therefore a pleasant surprise that there will be an easy way of finding the derivative. We tell you the answer right away: The derivative of this function is $10(x + 1)^9$. This looks very much related to the derivative of powers.

Before stating and proving the general theorem, we give you other examples.

Example.

$$\frac{d}{dx}(x^2 + 2x)^{3/2} = \frac{3}{2}(x^2 + 2x)^{1/2}(2x + 2).$$

Observe carefully the extra term $2x + 2$, which is the derivative of the expression $x^2 + 2x$. We may describe the answer in the following terms. We let $u = x^2 + 2x$ so that $du/dx = 2x + 2$. Let $f(u) = u^{3/2}$. Then we have

$$\frac{d(f(u(x)))}{dx} = \frac{df}{du}\frac{du}{dx}.$$

Example.

$$\frac{d}{dx}(x^2 + x)^{10} = 10(x^2 + x)^9(2x + 1).$$

Observe again the presence of the term $2x + 1$, which is the derivative of $x^2 + x$. Again, if we let $u = x^2 + x$ and $f(u) = u^{10}$, then

$$\frac{df(u(x))}{dx} = \frac{df}{du}\frac{du}{dx}, \quad \text{where} \quad \frac{df}{du} = 10u^9 \quad \text{and} \quad \frac{du}{dx} = 2x + 1.$$

Can you guess the general rule from the preceding assertions? The general rule was also discovered by trial and error, but we profit from three centuries of experience, and thus we are able to state it and prove it very simply, as follows.

Chain rule. *Let f and g be two functions having derivatives, and such that f is defined at all numbers which are values of g. Then the composite function $f \circ g$ has a derivative, given by the formula*

$$(f \circ g)'(x) = f'(g(x))g'(x).$$

This can be expressed in words by saying that we take the *derivative of the outer function times the derivative of the inner function* (*or the derivative of what's inside*).

The preceding assertion is known as the **chain rule**.

If we put $u = g(x)$, then we may express the chain rule in the form

$$\frac{df(u(x))}{dx} = \frac{df}{du}\frac{du}{dx},$$

or also

$$\frac{d(f \circ g)}{dx} = \frac{df}{du}\frac{du}{dx}.$$

Thus the derivative behaves *as if* we could cancel the du. As long as we have proved this result, there is nothing wrong with working like a machine in computing derivatives of composite functions, and we shall give you several examples before the exercises.

Example. Let $F(x) = (x^2 + 1)^{10}$. Then

$$F(x) = f(g(x)),$$

where

$$u = g(x) = x^2 + 1 \quad \text{and} \quad f(u) = u^{10}.$$

Then

$$f'(u) = 10u^9 = \frac{df}{du} \quad \text{and} \quad g'(x) = 2x = \frac{dg}{dx}.$$

Thus

$$\frac{d(f \circ g)}{dx} = \frac{df}{du}\frac{du}{dx} = 10u^9 \cdot 2x = 10(x^2 + 1)^9 2x.$$

Example. Let $f(u) = 2u^{1/2}$ and $g(x) = 5x + 1$. Then

$$f'(u) = u^{-1/2} = \frac{df}{du} \quad \text{and} \quad g'(x) = 5 = \frac{dg}{dx}.$$

Thus

$$\frac{d}{dx} 2(5x + 1)^{1/2} = 2 \cdot \tfrac{1}{2}(5x + 1)^{-1/2} \cdot 5 = (5x + 1)^{-1/2} \cdot 5.$$

(Pay attention to the constant 5, which is the derivative of $5x + 1$. You are very likely to forget it.)

In order to give you more extensive drilling than would be afforded by the functions we have considered, like powers, we summarize the derivatives of the elementary functions which are to be considered later.

$$\frac{d(\sin x)}{dx} = \cos x, \qquad \frac{d(\cos x)}{dx} = -\sin x.$$

$$\frac{d(e^x)}{dx} = e^x \qquad \text{(yes, } e^x, \text{ the same as the function!)}.$$

$$\frac{d(\log x)}{dx} = \frac{1}{x}.$$

Example.

$$\frac{d}{dx}(\sin x)^7 = 7(\sin x)^6 \cos x.$$

In this example, $f(u) = u^7$ and $df/du = 7u^6$. Also

$$u = \sin x, \qquad \text{and} \qquad \frac{du}{dx} = \cos x.$$

Example.

$$\frac{d}{dx}(\sin 3x)^7 = 7(\sin 3x)^6 \cos 3x \cdot 3.$$

The last factor of 3 occurring on the right-hand side is the derivative $d(3x)/dx$.

Example. Let n be any integer. For any differentiable function $f(x)$,

$$\boxed{\frac{d}{dx} f(x)^n = nf(x)^{n-1} \frac{df}{dx}.}$$

Example.

$$\frac{de^{\sin x}}{dx} = \frac{df}{du}\frac{du}{dx} = e^{\sin x}(\cos x).$$

In this example, $f(u) = e^u$, $df/du = e^u$, and $u = \sin x$.

Example.

$$\frac{d \cos(2x^2)}{dx} = \frac{df}{du}\frac{du}{dx} = -\sin(2x^2) \cdot 4x.$$

If this example, $f(u) = \cos u$ and $df/du = -\sin u$. Also $u = 2x^2$ so that $du/dx = 4x$.

Example.

$$\frac{d \cos 4x}{dx} = -\sin(4x) \cdot 4.$$

In this example, $f(u) = \cos u$ and $u = 4x$ so $du/dx = 4$.

We emphasize what we have already stated. *If we limited ourselves just to polynomials or quotients of polynomials, we would not have enough examples to drill the mechanism of the chain rule. There is nothing wrong in **using** the properties of functions which have not yet been formally defined in the course.* We could in fact create totally imaginary functions to achieve the same end.

Example. Suppose there is a function called schmoo(x), whose derivative is given by

$$\frac{d \text{ schmoo}(x)}{dx} = \frac{1}{x + \sin x}.$$

Then

$$\frac{d}{dx} \text{schmoo}(x^3 + 4x) = \frac{d \text{ schmoo}(u)}{du} \frac{du}{dx}$$

$$= \frac{1}{(x^3 + 4x) + \sin(x^3 + 4x)} (3x^2 + 4).$$

Example. Suppose there is a function cow(x) such that $\text{cow}'(x) = \text{schmoo}(x)$. Then

$$\frac{d \text{ cow}(x^2)}{dx} = \text{schmoo}(x^2) \cdot 2x.$$

Proof of the Chain Rule. We must consider the Newton quotient of the composite function $f \circ g$. By definition, it is

$$\frac{f[g(x + h)] - f[g(x)]}{h}.$$

Put $u = g(x)$, and let

$$k = g(x + h) - g(x).$$

Then k depends on h, and tends to 0 as h approaches 0. Our Newton quotient is equal to

$$\frac{f(u + k) - f(u)}{h}.$$

For the present argument *suppose that k is unequal to* 0 for all small values of h. Then we can multiply and divide this quotient by k, and obtain

$$\frac{f(u + k) - f(u)}{k} \frac{k}{h} = \frac{f(u + k) - f(u)}{k} \frac{g(x + h) - g(x)}{h}.$$

If we let h approach 0 and use the rule for the limit of a product, we see that our Newton quotient approaches

$$f'(u)g'(x),$$

and this would prove our chain rule, under the assumption that k is not 0.

It does not happen very often that $k = 0$ for arbitrarily small values of h, but when it does happen, the preceding argument must be refined. For those of you who are interested, we shall show you how the argument can be slightly modified so as to be valid in all cases. **The uninterested reader can just skip it.**

We distinguish two kinds of numbers h. The first kind, those for which $g(x + h) - g(x) \neq 0$, and the second kind, those for which

$$g(x + h) - g(x) = 0.$$

Let H_1 be the set of h of the first kind, and H_2 the set of h of the second kind. We assume that we have

$$g(x + h) - g(x) = 0$$

for arbitrarily small values of h. Then the Newton quotient

$$\frac{g(x + h) - g(x)}{h}$$

is 0 for such values, that is for h in H_2, and consequently

$$g'(x) = \lim_{h \to 0} \frac{g(x + h) - g(x)}{h} = 0.$$

Furthermore,

$$\lim_{\substack{h \to 0 \\ h \, in \, H_2}} \frac{f(g(x + h)) - f(g(x))}{h} = 0,$$

because h is of second kind, so $g(x + h) - g(x) = 0$, $g(x + h) = g(x)$ and therefore $f(g(x + h)) - f(g(x)) = 0$. The limit here is taken with h approaching 0, but h of the second kind.

On the other hand, if we take the limit with h of the first kind, then the original argument applies, i.e. we can divide and multiply by

$$k = g(x + h) - g(x),$$

and we find

$$\lim_{\substack{h \to 0 \\ h \, in \, H_1}} \frac{f(g(x + h)) - f(g(x))}{h} = f'(g(x))g'(x)$$

as before. But $g'(x) = 0$. Hence the limit is $0 = f'(g(x))g'(x)$ when h approaches 0, whether h is of first kind or second kind. This concludes the proof.

III, §6. EXERCISES

Find the derivatives of the following functions.

1. $(x + 1)^8$ 2. $(2x - 5)^{1/2}$

3. $(\sin x)^3$ 4. $(\log x)^5$

5. $\sin 2x$ 6. $\log(x^2 + 1)$

7. $e^{\cos x}$ 8. $\log(e^x + \sin x)$

9. $\sin\left(\log x + \dfrac{1}{x}\right)$ 10. $\dfrac{x + 1}{\sin 2x}$

11. $(2x^2 + 3)^3$ 12. $\cos(\sin 5x)$

13. $\log(\cos 2x)$ 14. $\sin[(2x + 5)^2]$

15. $\sin[\cos(x + 1)]$ 16. $\sin(e^x)$

17. $\dfrac{1}{(3x - 1)^4}$ 18. $\dfrac{1}{(4x)^3}$

19. $\dfrac{1}{(\sin 2x)^2}$ 20. $\dfrac{1}{(\cos 2x)^2}$

21. $\dfrac{1}{\sin 3x}$ 22. $(\sin x)(\cos x)$

23. $(x^2 + 1)e^x$

24. $(x^3 + 2x)(\sin 3x)$

25. $\dfrac{1}{\sin x + \cos x}$

26. $\dfrac{\sin 2x}{e^x}$

27. $\dfrac{\log x}{x^2 + 3}$

28. $\dfrac{x + 1}{\cos 2x}$

29. $(2x - 3)(e^x + x)$

30. $(x^3 - 1)(e^{3x} + 5x)$

31. $\dfrac{x^3 + 1}{x - 1}$

32. $\dfrac{x^2 - 1}{2x + 3}$

33. $(x^{4/3} - e^x)(2x + 1)$

34. $(\sin 3x)(x^{1/4} - 1)$

35. $\sin(x^2 + 5x)$

36. $e^{3x^2 + 8}$

37. $\dfrac{1}{\log(x^4 + 1)}$

38. $\dfrac{1}{\log(x^{1/2} + 2x)}$

39. $\dfrac{2x}{e^x}$

40. Let f be a function such that $f'(u) = \dfrac{1}{1 + u^3}$. Let $g(x) = f(x^2)$. Find $g'(x)$ and $g'(2)$. Do not attempt to evaluate $f(u)$.

41. Relax.

III, §6. SUPPLEMENTARY EXERCISES

Find the derivatives of the following functions.

1. $(2x + 1)^2$

2. $(2x + 5)^3$

3. $(5x + 3)^7$

4. $(7x - 2)^{81}$

5. $(2x^2 + x - 5)^3$

6. $(2x^3 - 3x)^4$

7. $(3x + 1)^{1/2}$

8. $(2x - 5)^{5/4}$

9. $(x^2 + x - 1)^{-2}$

10. $(x^4 + 5x + 6)^{-1}$

11. $(x + 5)^{-5/3}$

12. $(x^3 + 2x + 1)^3$

13. $(x - 1)(x - 5)^3$

14. $(2x^2 + 1)^2(x^2 + 3x)$

15. $(x^3 + x^2 - 2x - 1)^4$

16. $(x^2 + 1)^3(2x + 5)^2$

17. $\dfrac{(x + 1)^{3/4}}{(x - 1)^{1/2}}$

18. $\dfrac{(2x + 1)^{1/2}}{(x + 5)^5}$

19. $\dfrac{(2x^2 + x - 1)^{5/2}}{(3x + 2)^9}$

20. $\dfrac{(x^2 + 1)(3x - 7)^8}{(x^2 + 5x - 4)^3}$

21. $\sqrt{2x + 1}$

22. $\sqrt{x + 3}$

23. $\sqrt{x^2 + x + 5}$

24. $\sqrt{2x^3 - x + 1}$

In the following exercises, we may assume that there are functions $\sin u$, $\cos u$, $\log u$, and e^u whose derivatives are given by the following formulas:

$$\frac{d \sin u}{du} = \cos u, \qquad \frac{d \cos u}{du} = -\sin u,$$

$$\frac{d(e^u)}{du} = e^u, \qquad \frac{d \log u}{du} = \frac{1}{u}.$$

Find the derivative of each function (with respect to x):

25. $\sin(x^3 + 1)$

26. $\cos(x^3 + 1)$

27. $e^{x^3 + 1}$

28. $\log(x^3 + 1)$

29. $\sin(\cos x)$

30. $\cos(\sin x)$

31. $e^{\sin(x^3 + 1)}$

32. $\log[\sin(x^3 + 1)]$

33. $\sin[(x + 1)(x^2 + 2)]$

34. $\log(2x^2 + 3x + 5)$

35. $e^{(x + 1)(x - 3)}$

36. $e^{2x + 1}$

37. $\sin(2x + 5)$

38. $\cos(7x + 1)$

39. $\log(2x + 1)$

40. $\log \dfrac{2x + 1}{x + 3}$

41. $\sin \dfrac{x - 5}{2x + 4}$

42. $\cos \dfrac{2x - 1}{x + 3}$

43. $e^{2x^2 + 3x + 1}$

44. $\log(4x^3 - 2x)$

45. $\sin[\log(2x + 1)]$

46. $\cos(e^{2x})$

47. $\cos(3x^2 - 2x + 1)$

48. $\sin\left(\dfrac{x^2 - 1}{2x^3 + 1}\right)$

49. $(2x + 1)^{80}$

50. $(\sin x)^{50}$

51. $(\log x)^{49}$

52. $(\sin 2x)^4$

53. $(e^{2x + 1} - x)^5$

54. $(\log x)^{20}$

55. $(3 \log(x^2 + 1) - x^3)^{1/2}$

56. $(\log(2x + 3))^{4/3}$

57. $\dfrac{\sin 2x}{\cos 3x}$

58. $\dfrac{\sin(2x + 5)}{\cos(x^2 - 1)}$

59. $\dfrac{\log 2x^2}{\sin x^3}$

60. $\dfrac{e^{x^3}}{x^2 - 1}$

61. $\dfrac{x^4 + 4}{\cos 2x}$

62. $\dfrac{\sin(x^3 - 2)}{\sin 2x}$

63. $\dfrac{(2x^2 + 1)^4}{(\cos x^3)}$

64. $\dfrac{e^{-x}}{\cos 2x}$

65. e^{-3x}

66. e^{-x^2}

67. $e^{-4x^2 + x}$

68. $\sqrt{e^x + 1}$

69. $\dfrac{\log(x^2 + 2)}{e^{-x}}$

70. $\dfrac{\log(2x + 1)}{\sin(4x + 5)}$

III, §7. HIGHER DERIVATIVES

Given a differentiable function f defined on an interval, its derivative f' is also a function on this interval. If it turns out to be also differentiable (this being usually the case), then its derivative is called the **second derivative** of f and is denoted by $f''(x)$.

Example. Let $f(x) = (x^3 + 1)^2$. Then

$$f'(x) = 2(x^3 + 1)3x^2 = 6x^5 + 6x^2 \quad \text{and} \quad f''(x) = 30x^4 + 12x.$$

There is no reason to stop at the second derivative, and one can of course continue with the third, fourth, etc. provided they exist. Since it is notationally inconvenient to pile up primes after f to denote successive derivatives, one writes

$$f^{(n)}$$

for the n-th derivative of f. Thus f'' is also written $f^{(2)}$. If we wish to refer to the variable x, we also write

$$\boxed{f^{(n)}(x) = \frac{d^n f}{dx^n}.}$$

Example. Let $f(x) = x^3$. Then

$$\frac{df}{dx} = 3x^2, \qquad \frac{d^2 f}{dx^2} = f''(x) = f^{(2)}(x) = 6x,$$

$$\frac{d^3 f}{dx^3} = f^{(3)}(x) = 6, \qquad \frac{d^4 f}{dx^4} = f^{(4)}(x) = 0.$$

Example. Let $f(x) = 5x^3$. Then

$$f'(x) = 15x^2 = \frac{df}{dx},$$

$$f^{(2)}(x) = 30x = \frac{d^2 f}{dx^2},$$

$$f^{(3)}(x) = 30 = \frac{d^3 f}{dx^3},$$

$$f^{(4)}(x) = 0 = \frac{d^4 f}{dx^4}.$$

Example. Let $f(x) = \sin x$. Then:

$$f'(x) = \cos x,$$

$$f^{(2)}(x) = -\sin x.$$

III, §7. EXERCISES

Find the second derivatives of the following functions:

1. $3x^3 + 5x + 1$ 2. $(x^2 + 1)^5$

3. Find the 80-th derivative of $x^7 + 5x - 1$.

4. Find the 7-th derivative of $x^7 + 5x - 1$.

5. Find the third derivative of $x^2 + 1$.

6. Find the third derivative of $x^3 + 2x - 5$.

7. Find the third derivative of the function $g(x) = \sin x$.

8. Find the fourth derivative of the function $g(x) = \cos x$.

9. Find the 10-th derivative of $\sin x$.

10. Find the 10-th derivative of $\cos x$.

11. Find the 100-th derivative of $\sin x$.

12. Find the 100-th derivative of $\cos x$.

13. (a) Find the 5-th derivative of x^5.
    (b) Find the 7-th derivative of x^7.
    (c) Find the 13-th derivative of x^{13}.

In the process of finding these derivatives, you should observe a pattern. Let n be a positive integer. Define $n!$ to be the product of the first n integers. Thus

$$n! = n(n - 1)(n - 2)\cdots 2\cdot 1.$$

This is called n **factorial**. For example:

$$2! = 2, \qquad 3! = 3\cdot 2 = 6, \qquad 4! = 4\cdot 3\cdot 2 = 24.$$

Compute $5!$, $6!$, $7!$. You will find $n!$ used especially in the chapter on Taylor's formula, much later in the course.

14. In general, let k be a positive integer. Let

$$f(x) = x^k.$$

(a) What is $f^{(k)}(x)$?
(b) What is $f^{(k)}(0)$?
(c) Let n be a positive integer $> k$. What is $f^{(n)}(0)$?
(d) Let n be a positive integer $< k$. What is $f^{(n)}(0)$?

III, §8. IMPLICIT DIFFERENTIATION

Suppose that a curve is defined by an equation

$$F(x, y) = 0$$

like a circle, $x^2 + y^2 = 7$, or an ellipse, or more generally an equation
like

$$3x^3y - y^4 + 5x^2 + 5 = 0.$$

It is usually the case that for most values of x, one can solve back for y
as a function of x, that is find a differentiable function

$$y = f(x)$$

satisfying the equation. For example, in the case of the circle,

$$x^2 + y^2 = 7,$$

we have

$$y^2 = 7 - x^2$$

and hence we get two possibilities for y,

$$y = \sqrt{7 - x^2} \qquad \text{or} \qquad y = -\sqrt{7 - x^2}.$$

The graph of the first function is the upper semicircle, and the graph of
the second function is the lower semicircle.

In the example $3x^3y - y^4 + 5x^2 + 5 = 0$, it is a mess to solve for y,
and we don't do it. However, assuming that $y = f(x)$ is a differentiable
function satisfying this equation, we can find an expression for the derivative more easily, and we shall do so in an example below.

Example. Find the derivative dy/dx if $x^2 + y^2 = 7$, in terms of x
and y.

We differentiate both sides of the equation using the chain rule, and
the fact that $dx/dx = 1$. We then obtain:

$$2x \frac{dx}{dx} + 2y \frac{dy}{dx} = 0, \qquad \text{that is} \qquad 2x + 2y \frac{dy}{dx} = 0.$$

Hence,

$$\frac{dy}{dx} = \frac{-2x}{2y}.$$

Example. Find the tangent line to the circle $x^2 + y^2 = 7$ at the point $x = 2$ and $y = \sqrt{3}$.

The slope of the line at this point is given by

$$\frac{dy}{dx}\bigg|_{(2,\sqrt{3})} = \frac{-2 \cdot 2}{2\sqrt{3}} = \frac{-2}{\sqrt{3}}.$$

Hence the equation of the tangent line is

$$y - \sqrt{3} = -\frac{2}{\sqrt{3}}(x - 2).$$

Example. Find the derivative dy/dx in terms of x and y if

$$3x^3 y - y^4 + 5x^2 = -5.$$

Again we assume that y is a function of x. We differentiate both sides using the rule for derivative of a product, and the chain rule. We then obtain:

$$3x^3 \frac{dy}{dx} + 9x^2 y - 4y^3 \frac{dy}{dx} + 10x = 0,$$

or factoring out,

$$\frac{dy}{dx}(3x^3 - 4y^3) = -10x - 9x^2 y.$$

This yields

$$\frac{dy}{dx} = -\frac{10x + 9x^2 y}{3x^3 - 4y^3}.$$

Example. Find the equation of the tangent line in the preceding example at the point $x = 1$, $y = 2$.

We first find the slope at the given point. This is obtained by substituting $x = 1$ and $y = 2$ in the expression for dy/dx obtained in the preceding example. Thus:

$$\frac{dy}{dx}\bigg|_{(1,2)} = -\frac{10 + 18}{3 - 32} = \frac{28}{29}.$$

Then the equation of the tangent line at $(1, 2)$ is

$$y - 2 = \tfrac{28}{29}(x - 1).$$

III, §8. EXERCISES

Find dy/dx in terms of x and y in the following problems.

1. $x^2 + xy = 2$

2. $(x - 3)^2 + (y + 1)^2 = 37$

3. $x^3 - xy + y^3 = 1$

4. $y^3 - 2x^3 + y = 1$

5. $2xy + y^2 = x + y$

6. $\dfrac{1}{x} + \dfrac{1}{y} = 1$

7. $y^2 + 2x^2y + x = 0$

8. $x^2y^2 = x^2 + y^2$

Find the tangent line of the following curves at the indicated point.

9. $x^2y^2 = 9$ at $(-1, 3)$

10. $x^2 + y^3 + 2x - 5y - 19 = 0$ at $(3, -1)$

11. $(y - x)^2 = 2x + 4$ at $(6, 2)$

12. $2x^2 - y^3 + 4xy - 2x = 0$ at $(1, -2)$

13. $x^2 + y^2 = 25$ at $(3, -4)$

14. $x^2 - y^2 + 3xy + 12 = 0$ at $(-4, 2)$

15. $x^2 + xy - y^2 = 1$ at $(2, 3)$

III, §9. RATE OF CHANGE

The derivative has an interesting physical interpretation, which was very closely connected with it in its historical development, and is worth mentioning.

Suppose that a particle moves along some straight line a certain distance depending on time t. Then the distance s is a function of t, which we write $s = f(t)$.

For two values of the time, t_1 and t_2, the quotient

$$\frac{f(t_2) - f(t_1)}{t_2 - t_1}$$

can be regarded as a sort of average speed of the particle since it gives the total distance covered divided by the total time elapsed. At a given time t_0, it is therefore reasonable to regard the limit

$$\lim_{t \to t_0} \frac{f(t) - f(t_0)}{t - t_0}$$

as the rate of change of s with respect to t. This is none other than the derivative $f'(t)$, which is called the **speed**.

Let us denote the **speed** by $v(t)$. Then

$$v(t) = \frac{ds}{dt}.$$

The rate of change of the speed is called the **acceleration**. Thus

$$\frac{dv}{dt} = a(t) = \text{acceleration} = \frac{d^2s}{dt^2}.$$

Example. If the particle is an object dropping under the influence of gravity, then experimental data show that

$$s = \tfrac{1}{2}Gt^2,$$

where G is the gravitational constant. In that case,

$$\frac{ds}{dt} = Gt$$

is its speed. The acceleration is then

$$\frac{d^2s}{dt^2} = G = \text{gravitational constant}.$$

Example. A particle is moving so that at time t the distance traveled is given by the function

$$s(t) = t^2 + 1.$$

The derivative $s'(t)$ is equal to $2t$. Thus the speed of the particle is equal to 0 at time $t = 0$. Its speed is equal to 4 at time $t = 2$.

In general, given a function $y = f(x)$, the derivative $f'(x)$ is interpreted as the **rate of change of** y **with respect to** x. Thus f' is also a function. If y is increasing as x is increasing, this means that the derivative is positive, in other words $f'(x) > 0$. If y is decreasing, this means that the rate of change of y with respect to x is negative, that is $f'(x) < 0$.

Example. Suppose that a particle moves with uniform speed along a straight line, say along the x-axis toward the right, away from the origin. Suppose the speed is 5 cm/sec. We may then write

$$\frac{dx}{dt} = 5.$$

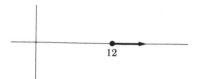

Suppose that at some time the particle is 12 cm to the right of the origin. After each further second of motion, the distance increases by another 5 cm, so after 3 seconds, the distance of the particle from the origin will be

$$12 + 3 \cdot 5 = 12 + 15 = 27 \text{ cm}.$$

One can find the x-coordinate as function of time. Under uniform speed, distance traveled is equal to the product of speed with time. Thus if the particle starts from the origin at time $t = 0$ then

$$x(t) = 5t.$$

If, on the other hand, the particle starts from another point x_0, then

$$x(t) = 5t + x_0.$$

Indeed, if we substitute 0 for t in this equation, we find

$$x(0) = 5 \cdot 0 + x_0 = x_0.$$

Hence x_0 is the value of x when $t = 0$.

Example. Suppose that a particle is moving to the left at a rate of 5 cm/sec. Then we write

$$\frac{dx}{dt} = -5.$$

Again suppose that at some time the particle is 12 cm to the right of the origin. Then after 2 seconds, the distance of the particle from the origin will be

$$12 - 2 \cdot 5 = 12 - 10 = 2 \text{ cm}.$$

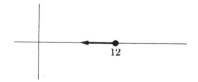

Finally, suppose the particle does not start when $t = 0$ but starts later, say after 25 seconds, but still moves with the same constant speed. We could measure time in terms of a new coordinate t'. In terms of t', the x-coordinate is given by

$$x = -5t'.$$

We can give t' as a function of t, by

$$t' = t - 25.$$

Then

$$x = -5(t - 25)$$

gives x as a function of t.

In many applications, we have to consider related rates of change, which involve the chain rule. Suppose, that y is a function of x, and also that x is given as some function of time, say, $x = g(t)$, then we can determine both the rate of change of y with respect to x, namely dy/dx, but also the rate of change of y with respect to t, namely

$$\frac{dy}{dt} = \frac{dy}{dx} \frac{dx}{dt}$$

by the chain rule.

Example. A square is expanding in such a way that its edge is changing at a rate of 2 cm/sec. When its edge is 6 cm long, find the rate of change of its area.

The area of a square as a function of its side is given by the function

$$A(x) = x^2.$$

If the side x is given as a function of time t, say $x = x(t)$, then the rate of change of the area with respect to time is by definition

$$\frac{d(A(x(t)))}{dt}.$$

Thus we use the chain rule, and if we denote the area by A, we find

$$\frac{dA}{dt} = \frac{dA}{dx} \frac{dx}{dt} = 2x \frac{dx}{dt}.$$

We are told that x increases at a rate of 2 cm/sec. This means that

$$\frac{dx}{dt} = 2.$$

Thus when $x(t) = 6$, we find that

$$\frac{dA}{dt} = 2 \cdot 6 \cdot 2 = 24 \text{ cm/sec.}$$

Example. A point moves along the graph of $y = x^3$ so that its x-coordinate is decreasing at the rate of 2 units per second. What is the rate of change of its y-coordinate when $x = 3$?

We have by the chain rule,

$$\frac{dy}{dt} = \frac{dy}{dx}\frac{dx}{dt} = 3x^2 \frac{dx}{dt}.$$

We are told that x is *decreasing*. This means that

$$\frac{dx}{dt} = -2.$$

Hence the rate of change of y when $x = 3$ is equal to

$$\left.\frac{dy}{dt}\right|_{x=3} = 3(3^2)(-2) = -54 \text{ units per second.}$$

That dy/dt comes out negative means that y is decreasing when $x = 3$.

Example. A light on top of a lamppost shines 25 ft above the ground. A man 6 ft tall is walking away from the light. What is the length of his shadow when he is 40 ft away from the base of the lamppost? If he is walking away at the rate of 5 ft/sec, how fast is his shadow increasing at this point?

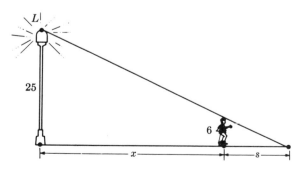

We need to establish a relationship between the length of the shadow and the distance of the man from the post. Let s be the length of the shadow, and let x be the distance between the man and the base of the post. Then using similar triangles, we see that

$$\frac{25}{x+s} = \frac{6}{s}.$$

After cross-multiplying we see that $25s = 6x + 6s$, whence

$$s = \tfrac{6}{19}x.$$

Therefore, $ds/dx = 6/19$, and

$$\frac{ds}{dt} = \frac{6}{19}\frac{dx}{dt}.$$

Since $dx/dt = 5$, we get what we want:

$$\frac{ds}{dt} = \frac{6}{19}\cdot 5 = \frac{30}{19}\ \text{ft/sec.}$$

Also, when $x = 40$, we find that the length of his shadow is given by

$$s = \frac{6\cdot 40}{19} = \frac{240}{19}\ \text{ft.}$$

Remark. If the man is walking toward the post, then the distance x is decreasing, and hence in this case,

$$\frac{dx}{dt} = -5.$$

Consequently, a similar argument shows that

$$\frac{ds}{dt} = -\frac{30}{19}\ \text{ft/sec.}$$

Example. The area of a disc of radius r is given by the formula

$$A = \pi r^2,$$

where r is the radius. Let s be the diameter. Then $s = 2r$ so $r = s/2$ and we can give A as a function of s by

$$A = \pi\left(\frac{s}{2}\right)^2 = \frac{\pi s^2}{4}.$$

Hence the rate of change of A with respect to s is

$$\frac{dA}{ds} = \frac{2\pi s}{4} = \frac{\pi s}{2}.$$

Example. A cylinder is being compressed from the side and stretched, so that the radius of the base is decreasing at a rate of 2 cm/sec and the height is increasing at a rate of 5 cm/sec. Find the rate at which the volume is changing when the radius is 6 cm and the height is 8 cm.

The volume is given by the formula

$$V = \pi r^2 h,$$

where r is the radius of the base and h is the height.

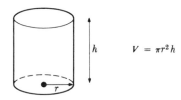

We are given $dr/dt = -2$ (note the negative sign because the radius is decreasing), and $dh/dt = 5$. Hence using the formula for the derivative of a product, we find

$$\frac{dV}{dt} = \pi\left[r^2 \frac{dh}{dt} + h2r \frac{dr}{dt}\right] = \pi[5r^2 - 4hr].$$

When $r = 6$ and $h = 8$ we get

$$\frac{dV}{dt}\bigg|_{\substack{r=6 \\ h=8}} = \pi(5 \cdot 6^2 - 4 \cdot 8 \cdot 6) = -12\pi.$$

The negative sign in the answer means that the volume is decreasing when $r = 6$ and $h = 8$.

Example. Two trains leave a station 3 hours apart. The first one moves north at a speed of 100 km/hr. The second moves east at a speed of 60 km/hr. The second leaves 3 hours after the first. At what rate is the distance between the trains changing 2 hours after the second train has left?

Let y be the distance of the first train and x the distance of the second train from the station. Then

$$\frac{dy}{dt} = 100 \qquad \text{and} \qquad \frac{dx}{dt} = 60.$$

We have $y = 100t$, and since the second train leaves three hours later, we have

$$x = 60(t - 3).$$

Let $f(t)$ be the distance between them. Then

$$f(t) = \sqrt{60^2(t - 3)^2 + 100^2 t^2}.$$

Hence

$$f'(t) = \tfrac{1}{2}[3600(t - 3)^2 + 10000t^2]^{-1/2}[2 \cdot 60^2(t - 3) + 2 \cdot 100^2 t].$$

The time 2 hours after the second train leaves is $t = 2 + 3 = 5$. The desired rate is therefore $f'(5)$, so

$$f'(5) = \tfrac{1}{2}[14400 + 250000]^{-1/2}[14400 + 100000].$$

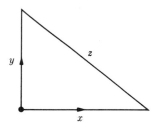

Example. We shall work the preceding example by using another method. Let z be the distance between the trains. We have

$$z^2 = x^2 + y^2.$$

Furthermore, x, y, z are functions of t. Differentiating with respect to t yields:

$$2z \frac{dz}{dt} = 2x \frac{dx}{dt} + 2y \frac{dy}{dt}.$$

We cancel 2 to get

$$z \frac{dz}{dt} = x \frac{dx}{dt} + y \frac{dy}{dt}.$$

Write $x(t)$, $y(t)$, $z(t)$ for x, y, z as functions of t. Then

$$x(5) = 120, \qquad y(5) = 500, \qquad \text{and by Pythagoras,} \quad z(5) = \sqrt{120^2 + 500^2}.$$

Substituting $t = 5$, we find

$$z(5) \left.\frac{dz}{dt}\right|_{t=5} = x(5) \left.\frac{dx}{dt}\right|_{t=5} + y(5) \left.\frac{dy}{dt}\right|_{t=5}.$$

Dividing by $z(5)$ yields:

$$\left.\frac{dz}{dt}\right|_{t=5} = \frac{120 \cdot 60 + 500 \cdot 100}{\sqrt{120^2 + 500^2}}.$$

III, §9. EXERCISES

For some of these exercises, the following formulas can be used.

> Volume of a sphere of radius r is $\frac{4}{3}\pi r^3$.
> Area of a sphere of radius r is $4\pi r^2$.
> Volume of a cone of height h and radius of base r is $\frac{1}{3}\pi r^2 h$.
> Area of circle of radius r is πr^2.
> Circumference of circle of radius r is $2\pi r$.

1. A particle is moving so that at time t, the distance is given by $s(t) = t^3 - 2t$. At what time is the acceleration equal to
   (a) 1   (b) 0   (c) -5?

2. A particle is moving so that at time t, the distance is given by the function $s(t) = 2t^4 + t^2$. At what time is the speed equal to 0?

3. An object travels on a straight line with speed given by the function $v(t) = 4t^5$. Find the acceleration at time $t = 2$.

4. A particle is moving so that at time t, the distance traveled is given by $s(t) = t^3 - 2t + 1$. At what time is the acceleration equal to 0?

5. A cube is expanding in such a way that its edge is changing at a rate of 5 m/sec. When its edge is 4 m long, find the rate of change of its volume.

6. A sphere is increasing so that its radius increases at the rate of 1 cm/sec. How fast is its volume changing when its radius is 3 cm?

7. What is the rate of change of the area of a circle with respect to its radius, diameter, circumference?

8. A point moves along the graph of $y = 1/(x^2 + 4)$ in a way such that $dx/dt = -3$ units per second. What is the rate of change of its y-coordinate when $x = 2$?

9. A light shines on top of a lamppost 20 ft above the ground. A woman 5 ft tall walks away from the light. Find the rate at which her shadow is increasing if she walks at the rate of
   (a) 4 ft/sec   (b) 3 ft/sec.

10. If in Exercise 9 the woman walks toward the light, find the rate at which her shadow is decreasing if she walks at the rate of
    (a) 5 ft/sec   (b) 6 ft/sec.

11. A particle moves differentiably on the parabola $y = x^2$. At what point on the curve are its x- and y-coordinates moving at the same rate? (You may assume dx/dt and $dy/dt \neq 0$ for all t.)

12. One side of a right triangle decreases 1 cm/min and the other side increases 2 cm/min. At some time the first side is 8 cm and the second side is 6 cm long. At what rate is the area increasing 2 min after this time?

13. The length of the side of a square is increasing at the rate of 3 cm per second. Find the rate at which the area is increasing when the side is 15 cm long.

14. A ladder 17 ft long leans against a vertical wall. If the lower end of the ladder is being moved away from the foot of the wall at the rate of 3 ft/sec, how fast is the top descending when the lower end is 8 ft from the wall?

15. A swimming pool is 25 ft wide, 40 ft long, 3 ft deep at one end and 9 ft deep at the other, the bottom being an inclined plane. If water is pumped into the pool at the rate of 10 ft³/min, how fast is the water level rising when it is 4 ft deep at the deep end?

16. A reservoir has the shape of a cone, vertex down, 10 ft high. The radius of the top is 4 ft. Water is poured into the reservoir at the rate of 5 ft³/min. How fast is the water rising when the depth of the water is 5 ft?

17. A particle is moving so that at time t the distance traveled is given by $s(t) = 2t^2 - t$. At what time is the speed equal to 0? What is the acceleration of the particle?

18. A point moves on the parabola with equation $y = x^2 - 6x$. Find the point on the curve at which the rate of change of the y-coordinate is four times the rate of change of the x-coordinate. (Assume $dx/dt \neq 0$ for all x.)

19. Water is flowing into a tank in the form of a hemisphere of radius 10 ft with flat side up at the rate of 4 ft³/min. Let h be the depth of the water, r the radius of the surface of the water, and V the volume of the water in the tank. Assume that $dV/dt = \pi r^2 \, dh/dt$. Find how fast the water level is rising when $h = 5$ ft.

20. A train leaves a station at a certain time and travels north at the rate of 50 mi/hr. A second train leaves the same station 2 hr after the first train leaves, and goes east at the rate of 60 mi/hr. Find the rate at which the two trains are separating 1.5 hr after the second train leaves the station.

21. Sand is falling on a pile, always having the shape of a cone, at the rate of 3 ft³/min. Assume that the diameter at the base of the pile is always three times the altitude. At what rate is the altitude increasing when the altitude is 4 ft?

22. The volume of a sphere is decreasing at the rate of 12π cm³/min. Find the rate at which the radius and the surface area of the sphere are changing when the radius is 20 cm.

23. Water runs into a conical reservoir at the constant rate of 2 m³/min. The vertex is 18 m down and the radius of the top is 24 m. How fast is the water level rising when it is 6 m deep?

24. Sand is falling on a pile, forming a cone. Let V be the volume of sand, so $V = \frac{1}{3}\pi r^2 h$. What is the rate of change of the volume of sand when $r = 10$ ft, if the radius of the base is expanding at a rate of 2 ft/sec, and the height is increasing at a rate of 1 ft/sec? You may assume $r = h = 0$ when $t = 0$.

CHAPTER IV

Sine and Cosine

From the sine of an angle and the cosine of an angle, we shall define functions of numbers, and determine their derivatives.

It is convenient to recall all the facts about trigonometry which we need in the sequel, especially the formula giving us the sine and cosine of the sum of two angles. Thus our treatment of the trigonometric functions is self-contained—you do not need to know anything about sine and cosine before starting to read this chapter. However, most of the proofs of statements in §1 come from plane geometry and will be left to you.

IV, §0. REVIEW OF RADIAN MEASURE

In order to eliminate some confusion of terminology, it is often convenient to use two different words for a circle, and a circle together with the region lying inside. Thus we reserve the word **circle** for the former, and call the circle together with its inside a **disc**. Thus we speak of the length of a circle, but the area of a disc.

We suppose fixed a measure of length. This determines a measure of area. For instance, if length is measured in meters, then area is measured in square meters.

For our immediate purposes, we define π to be the **area of the disc of radius** 1. It is, of course, a problem to find a decimal expansion for π, which you probably have been told is approximately equal to 3.14159.... Later in the course, you will learn to compute π to any degree of accuracy.

The disc of radius r is obtained by a dilation (blow up) of the disc of radius 1, as shown on the figure at the top of the next page.

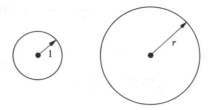

How does area change under dilation? Let us first look at rectangles. Let R be a rectangle whose sides have length a, b. Let r be a positive number. Let rR be the rectangle whose sides have length ra, rb, as shown on the figure. Then the area of rR is $rarb = r^2ab$. If A is the area of R, then the area of rR is r^2A.

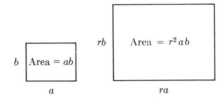

Under dilation by a factor r the area of a rectangle changes by a factor r^2. This applies as a general principle to any region:

> **Let S be any region in the plane. Let A be the area of S. Under dilation by a factor r, the dilated region has area r^2A.**

We shall prove this by approximating S by the squares of a grid, as shown on the figure. If we blow up S by a factor of r, we obtain a region rS. Let A be the area of S. Then the area of rS will again be r^2A, because each small square is blown up by a factor of r, so the area of a small square changes by a factor of r^2. The sum of the areas of the squares gives an approximation to the area of the figure. We want to

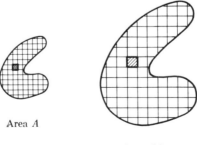

Area A

Area r^2A

estimate how good the approximation is. The difference between the sum of the areas of all the little squares contained in the figure and the area of the figure itself is at most the area of all the small squares which touch the boundary of the figure. We can give an estimate for this as follows.

Suppose we make a grid so that the squares have sides of length c. Then the diagonal of such a square has length $c\sqrt{2}$. If a square intersects the boundary, then any point on the square is at distance at most $c\sqrt{2}$ from the boundary. Look at the figure. This is because the distance between any two points of the square is at most $c\sqrt{2}$. Let us

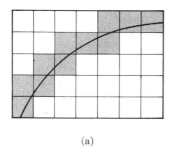

(a)

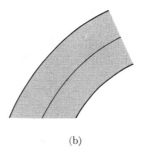

(b)

draw a band of width $c\sqrt{2}$ on each side of the boundary, as shown in (b) of the above figure. Then all the squares which intersect the boundary must lie within that band. It is very plausible that the area of the band is at most equal to

$$2c\sqrt{2} \quad \text{times the length of the boundary.}$$

Thus if we take c to be very small, i.e. if we take the grid to be a very fine grid, then the area of the band is small, and the area of the figure is approximated by the area covered by the squares lying entirely inside the figure. Under dilation, a similar argument applies to the dilated band for the dilated figure, so that the area of the dilated band is at most

$$r^2 \cdot c\sqrt{2} \quad \text{times the length of the boundary.}$$

As c approaches 0, the areas of these bands approach 0. This justifies our assertion that area changes by a factor of r^2 under dilation by a factor of r.

Since we defined π to be the area of the disc of radius 1, we now obtain:

The area of a disc of radius r is πr^2.

We select a unit of measurement for angles such that the flat angle is equal to π times the unit angle. (See the following figure.) The right angle has measure $\pi/2$. The full angle going once around has then measure 2π. This unit of measurement for which the flat angle has measure π is called the **radian**. Thus the right angle has $\pi/2$ radians.

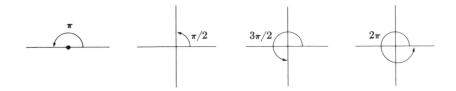

There is another current unit of measurement for which the flat angle is 180. This unit is called the **degree**. Thus the flat angle has 180 degrees, and the right angle has 90 degrees. We also have

$$360 \text{ degrees} = 2\pi \text{ radians,}$$

$$60 \text{ degrees} = \pi/3 \text{ radians,}$$

$$45 \text{ degrees} = \pi/4 \text{ radians,}$$

$$30 \text{ degrees} = \pi/6 \text{ radians.}$$

We shall deal mostly with radian measure, which makes some formulas come out more easily later on. It is easy to convert from one measure to the other.

Example. A wheel is turning at the rate of $50°$ per minute. Find its rate in rad/min and rpm (revolutions per minute).
  We have

$$1° = \frac{2\pi}{360} \text{ radians} = \frac{\pi}{180} \text{ radians.}$$

Hence $50°$ per minute is equal to $50\pi/180 = 5\pi/18$ radians per minute.
  On the other hand, a full revolution is 2π radians, so 1 radian is $1/2\pi$ revolutions. Hence the wheel is turning at the rate of

$$\frac{5\pi}{18} \cdot \frac{1}{2\pi} = \frac{5}{36} \text{ rpm.}$$

A **sector** is the region of the plane lying inside an angle. Often we also speak of a **sector in a disc**, meaning the portion of the sector lying inside the disc, as illustrated on the figure.

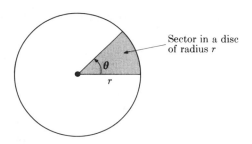

Sector in a disc
of radius r

A sector is measured by its angle. In the figure, we have labeled this angle θ (theta), with $0 \leq \theta \leq 2\pi$. It is measured in radians. The area of the sector is a certain fraction of the total area of the disc. Namely, we know that the total area is πr^2. The fraction is $\theta/2\pi$. Hence if we let S be the sector having angle θ in a disc of radius r, then the area of S is

$$\frac{\theta}{2\pi} \cdot \pi r^2 = \frac{\theta r^2}{2}.$$

We box this for reference:

$$\boxed{\textbf{Area of sector of angle } \theta \textbf{ radians in disc of radius } r = \frac{\theta r^2}{2}.}$$

If the radius is 1, then the area of the sector is $\theta/2$. We shall use this in §4.

Example. The area of a sector of angle $\pi/3$ in a disc of radius 1 is $\pi/6$, because the total area of the disc is π, and the sector represents one-sixth of the total area. This is illustrated on the figure.

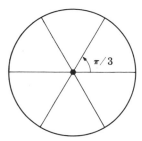

$\pi/3$

The disc in the figure is cut up into six sectors, each angle measuring $\pi/3$ radians, for a total of 2π radians.

Next we come to the length of arc of the circle.

Let c be the circumference of a circle of radius r. Then

$$c = 2\pi r.$$

Proof. The proof will be a beautiful example of the idea of differentiation. We consider a circle of radius r, and a circle of slightly bigger radius, which we write $r + h$. We suppose that these circles have the same center, and so obtain a circular band between them.

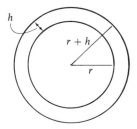

Let c be the length of the small circle. If we had a rectangular band of length c and height h then its area would be ch.

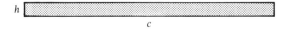

Suppose we wrap this band around the small circle.

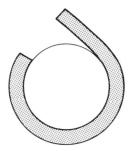

Since the circle curves outward, the rectangular band has to be stretched if it is to cover the band between the circle of radius r and the circle of radius h. Thus we have an inequality for areas:

$$ch < \text{area of circular band.}$$

Similarly, if C is the circumference of the bigger circle, then

$$\text{area of circular band} < Ch.$$

But the area of the circular band is the difference between the areas of the discs, which is

$$\text{area of circular band} = \pi(r + h)^2 - \pi r^2$$
$$= \pi(2rh + h^2).$$

Therefore we obtain the inequalities

$$ch < \pi(2rh + h^2) < Ch.$$

We divide these inequalities by the positive number h to obtain

$$c < \pi(2r + h) < C.$$

Now let h approach 0. Then the circumference of the big circle C approaches the circumference of the small circle c, and $\pi(2r + h)$ approaches $2\pi r$. It follows that

$$c = 2\pi r,$$

thus proving our formula.

Observe that the length of the circumference is just the derivative of the area.

An arc on a circle can be measured by its angle in radians. What is the length L of this arc, as on the following figure?

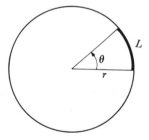

The answer is as follows.

Let L be the length of an arc of θ radians on a circle of radius r. Then

$$\boxed{L = r\theta.}$$

Proof. The total length of the circle is $2\pi r$, and L is the fraction $\theta/2\pi$ of $2\pi r$, which gives precisely $r\theta$.

Example. The length of arc of $\pi/3$ radians in a circle of radius r is $r\pi/3$.

IV, §0. EXERCISE

Suppose you are given that the volume of a sphere of radius r is $\frac{4}{3}\pi r^3$. Can you figure out an argument to obtain the area of the sphere?

IV, §1. THE SINE AND COSINE FUNCTIONS

Suppose that we have given coordinate axes, and a certain angle A, as shown on the figure.

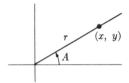

We select a point (x, y) (not the origin) on the ray determining our angle A. We let $r = \sqrt{x^2 + y^2}$. Then r is the distance from $(0,0)$ to the point (x, y). We define

$$\text{sine } A = \frac{y}{r} = \frac{y}{\sqrt{x^2 + y^2}},$$

$$\text{cosine } A = \frac{x}{r} = \frac{x}{\sqrt{x^2 + y^2}}.$$

If we select another point (x_1, y_1) on the ray determining our angle A and use its coordinates to get the sine and cosine, then we shall obtain the same values as with (x, y). Indeed, there is a positive number c such that

$$x_1 = cx \qquad \text{and} \qquad y_1 = cy.$$

Thus we can substitute these values in

$$\frac{y_1}{\sqrt{x_1^2 + y_1^2}}$$

to obtain

$$\frac{y_1}{\sqrt{x_1^2 + y_1^2}} = \frac{cy}{\sqrt{c^2x^2 + c^2y^2}}.$$

We can factor c from the denominator, and then cancel c in both the numerator and denominator to get

$$\frac{y}{\sqrt{x^2 + y^2}}.$$

In this way we see that sine A does not depend on the choice of the point (x, y).

The geometric interpretation of the above argument simply states that the triangles in the following diagram are similar.

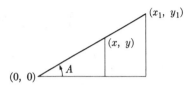

The angle A can go all the way around. For instance, we could have an angle determined by a point (x, y) in the second or third quadrant.

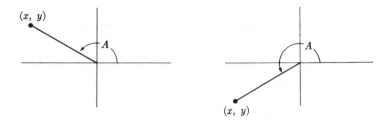

When the angle A is in the first quadrant, then its sine and cosine are positive because both coordinates x, y are positive. When the angle A is in the second quadrant, its sine is positive because y is positive, but its cosine is negative because x is negative.

When A is in the third quadrant, sine A is negative and cosine A is negative also.

As mentioned in the introductory section, we use radian measure for angles. Suppose A is an angle of θ radians, and let (x, y) be a point on the circle of radius 1, also lying on the line determining the angle A as shown on the figure. Then in this case,

$$r = \sqrt{x^2 + y^2} = 1$$

and therefore

$$\boxed{(x, y) = (\cos \theta, \sin \theta).}$$

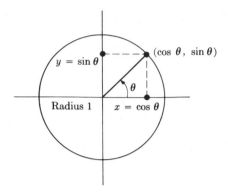

In general for arbitrary radius r, we have the relations:

$$x = r \cos \theta,$$
$$y = r \sin \theta.$$

We can also define the sine and cosine of angles in a right triangle as shown on the figure, by the formulas:

$$\sin \theta = \frac{\text{Opposite side}}{\text{Hypotenuse}}$$

$$\cos \theta = \frac{\text{Adjacent side}}{\text{Hypotenuse}}$$

We make a table of the sines and cosines of some angles.

Angle	Sine	Cosine
$\pi/6$	$1/2$	$\sqrt{3}/2$
$\pi/4$	$1/\sqrt{2}$	$1/\sqrt{2}$
$\pi/3$	$\sqrt{3}/2$	$1/2$
$\pi/2$	1	0
π	0	-1
2π	0	1

Unless otherwise specified, we *always use the radian measure*, and our table is given for this measure.

The values of this table are easily determined, using properties of similar triangles and plane geometry. For instance, we get the sine of the angle $\pi/4$ radians $= 45°$ from a right triangle with two equal sides:

We can determine the sine of $\pi/4$ by means of the point $(1, 1)$. Then $r = \sqrt{2}$ and sine $\pi/4$ radians is $1/\sqrt{2}$. Similarly for the cosine.

We get the sine of an angle of $\pi/6$ radians by considering a triangle such that two angles have $\pi/6$ and $\pi/3$ radians (that is, $30°$ and $60°$ respectively).

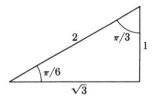

If we let the side opposite the angle of $30°$ have length 1, then the hypotenuse has length 2, and the side adjacent to the angle of $30°$ has length $\sqrt{3}$.

Hence we find that

$$\sin \pi/6 = \frac{1}{2} \quad \text{and} \quad \cos \pi/6 = \frac{\sqrt{3}}{2}.$$

On the other hand, we have

$$\sin 5\pi/6 = \frac{1}{2}, \quad \cos 5\pi/6 = \frac{-\sqrt{3}}{2}$$

as is clear from the figure.

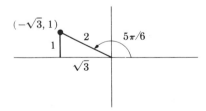

The choice of radian measure allows us to define the **sine of a number** rather than the sine of an angle as follows.

Let x be a number with $0 \leq x < 2\pi$. We define **sin x** to be the sine of x radians.

For an arbitrary number x, we write

$$x = x_0 + 2\pi n$$

where n is an integer, $0 \leq x_0 < 2\pi$, and define

$$\sin x = \sin x_0.$$

From this definition, we see that

$$\boxed{\sin x = \sin(x + 2\pi) = \sin(x + 2\pi n)}$$

for any integer n.

Of course, we call this function the **sine function**. Thus $\sin \pi = 0$, $\sin \pi/2 = 1$, $\sin 2\pi = 0$, $\sin 0 = 0$.

Similarly, we have the **cosine function**, which is defined for all numbers x by the rule:

cos x is the number which is the cosine of the angle x radians.

Thus $\cos 0 = 1$ and $\cos \pi = -1$.

If we had used the measure of angles in degrees we would obtain *another* sine function which is not equal to the sine function which we defined in terms of radians. Suppose we call this other sine function sin*. Then

$$\sin^*(180) = \sin \pi,$$

and in general

$$\sin^*(180x) = \sin \pi x$$

for any number x. Thus

$$\sin^* x = \sin\left(\frac{\pi}{180} x\right)$$

is the formula relating our two sine functions. It will become clear later why we always pick the radian measure instead of any other.

At present we have no means of computing values for the sine and cosine other than the very special cases listed above (and similar ones, based on simple symmetries of right triangles). In Chapter XIII we shall develop a method which will allow us to find $\sin x$ and $\cos x$ for any value of x, up to any degree of accuracy that you wish.

On the next figure we illustrate an angle of θ radians, and by convention the angle of $-\theta$ radians is the reflection around the x-axis.

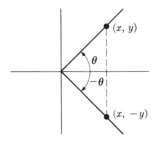

If (x, y) are the coordinates of a point on the ray defining the angle of θ radians, then $(x, -y)$ are the coordinates of a point on the reflected ray. Thus

$$\sin(-\theta) = -\sin \theta,$$
$$\cos(-\theta) = \cos \theta.$$

Finally we recall the definitions of the other trigonometric functions:

$$\tan \theta = \frac{\sin \theta}{\cos \theta}, \qquad \cot \theta = \frac{1}{\tan \theta} = \frac{\cos \theta}{\sin \theta},$$

$$\sec \theta = \frac{1}{\cos \theta}, \qquad \csc \theta = \frac{1}{\sin \theta}.$$

The most important among these are the sine, cosine, and tangent. We make a few additional remarks on the tangent.

The tangent is of course defined for all numbers θ such that

$$\cos \theta \neq 0.$$

These are the numbers unequal to

$$\frac{\pi}{2}, \quad \frac{3\pi}{2}, \quad \frac{5\pi}{2}, \dots$$

in general, $\theta \neq (2n + 1)/2$ for some integer n. We make a table of some values of the tangent.

θ	$\tan \theta$
0	0
$\pi/6$	$1/\sqrt{3}$
$\pi/4$	1
$\pi/3$	$\sqrt{3}$

You should complete this table for all similar values of θ in all four quadrants.

Consider an angle of θ radians, and let (x, y) be a point on the ray determining this angle, with $x \neq 0$. Then

$$\tan \theta = y/x$$

so that

$$y = (\tan \theta)x.$$

Conversely, any point (x, y) satisfying this equation is a point on the line making an angle θ with the x-axis, as shown on the next figure.

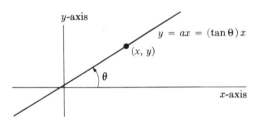

In an equation $y = ax$ where a is the slope, we can say that

$$a = \tan \theta,$$

where θ is the angle which the line makes with the x-axis.

Example. Take $\theta = \pi/6$. Then $\tan \theta = 1/\sqrt{3}$. Hence the line making an angle of θ with the x-axis has the equation

$$y = \left(\tan \frac{\pi}{6}\right) x, \qquad \text{or also} \qquad y = \frac{1}{\sqrt{3}} x.$$

Example. Take $\theta = 1$. There is no easier way to express $\tan 1$ other than writing $\tan 1$. In Chapter XIII you will learn how to compute arbitrarily close decimal approximations. Here we don't care. We simply point out that the equation of the line making an angle of 1 radian with the x-axis is given by

$$y = (\tan 1)x.$$

Similarly, the equation of the line making an angle of 2 radians with the x-axis is given by

$$y = (\tan 2)x.$$

Suppose given that π is approximately equal to 3.14. Then 1 is approximately equal to $\pi/3$. Thus the line making an angle of 1 radian with the x-axis, as shown on Fig. 1(a) is close to the line making an angle of $\pi/3$ radians. Similarly, the line making an angle of 2 radians with the x-axis, as shown on Fig. 1(b), is close to the line making an angle of $2\pi/3$ radians.

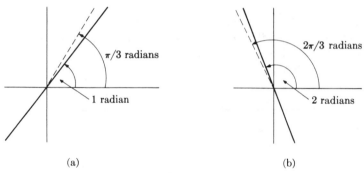

(a)   (b)

Figure 1

IV, §1. EXERCISES

Find the following values of the sin function and cos function:

1. $\sin 3\pi/4$
2. $\sin 2\pi/6$
3. $\sin \dfrac{2\pi}{3}$
4. $\sin\left(\pi - \dfrac{\pi}{6}\right)$

5. $\cos\left(\pi + \dfrac{\pi}{6}\right)$
6. $\cos\left(\pi + \dfrac{2\pi}{6}\right)$
7. $\cos\left(2\pi - \dfrac{\pi}{6}\right)$
8. $\cos \dfrac{5\pi}{4}$

Find the following values:

9. $\tan \dfrac{\pi}{4}$
10. $\tan \dfrac{2\pi}{6}$
11. $\tan \dfrac{5\pi}{4}$
12. $\tan\left(2\pi - \dfrac{\pi}{4}\right)$

13. $\sin \dfrac{7\pi}{6}$
14. $\cos \dfrac{7\pi}{6}$
15. $\cos 2\pi/3$
16. $\cos(-\pi/6)$

17. $\cos(-5\pi/6)$
18. $\cos(-\pi/3)$

19. Complete the following table.

θ	$\sin \theta$	$\cos \theta$	$\tan \theta$
$2\pi/3$			
$3\pi/4$			
$5\pi/6$			
π			
$7\pi/6$			
$5\pi/4$			
$7\pi/4$			

IV, §2. THE GRAPHS

sin x

We wish to sketch the graph of the sine function.

We know that $\sin 0 = 0$. As x goes from 0 to $\pi/2$, the sine of x increases until x reaches $\pi/2$, at which point the sine is equal to 1.

As x ranges from $\pi/2$ to π, the sine decreases until it becomes

$$\sin \pi = 0.$$

As x ranges from π to $3\pi/2$ the sine becomes negative, but otherwise behaves in a similar way to the first quadrant, until it reaches

$$\sin \frac{3\pi}{2} = -1.$$

Finally, as x goes from $3\pi/2$ to 2π, the sine of x goes from -1 to 0, and we are ready to start all over again.

The graph looks like this:

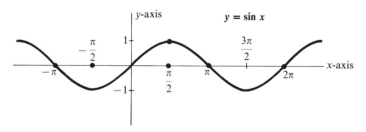

If we go once around by 2π, the sine and cosine each take on the same values they had originally, in other words

$$\sin(x + 2\pi) = \sin x,$$

$$\cos(x + 2\pi) = \cos x$$

for all x. This holds whether x is positive or negative, and the same would be true if we took $x - 2\pi$ instead of $x + 2\pi$.

You might legitimately ask why one arch of the sine (or cosine) curve looks the way we have drawn it, and not the following way:

In the next section, we shall find the slope of the curve $y = \sin x$. It is equal to $\cos x$. Thus when $x = 0$, the slope is $\cos 0 = 1$. Furthermore, when $x = \pi/2$, we have $\cos \pi/2 = 0$ and hence the slope is 0. This means that the curve becomes horizontal, and cannot have a peak the way we have drawn it above.

At present we have no means for computing more values of $\sin x$ and $\cos x$. However, using the few that we know and the derivative, we can convince ourselves that the graph looks as we have drawn it.

cos x

The graph of the cosine will look like that of the sine, but it starts with $\cos 0 = 1$.

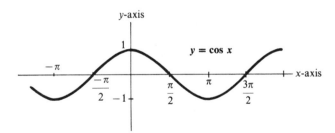

sin $2x$

Next let us graph $y = f(x) = \sin 2x$. Since the sine changes its behavior in intervals of length $\pi/2$, $\sin 2x$ will change behavior in intervals of length $\pi/4$. Thus we make a table with x ranging over intervals of length $\pi/4$.

x	$2x$	$\sin 2x$
inc. 0 to $\pi/4$	inc. 0 to $\pi/2$	inc. 0 to 1
inc. $\pi/4$ to $\pi/2$	inc. $\pi/2$ to π	dec. 1 to 0
inc. $\pi/2$ to $3\pi/4$	inc. π to $3\pi/2$	dec. 0 to -1
inc. $3\pi/4$ to π	inc. $3\pi/2$ to 2π	inc. -1 to 0

Then the graph repeats. Hence the graph of sin 2x looks like this.

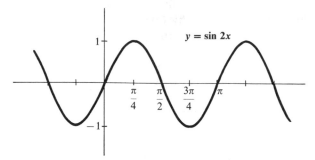

We see that the graph of $y = \sin 2x$ has twice as many wiggles over an interval as the graph of $y = \sin x$.

Similarly, you would see that $y = \sin \frac{1}{2}x$ has half as many wiggles as the graph of $y = \sin x$.

tan x

Finally, let us graph $y = \tan x$. Note that $\tan 0 = 0$. We take the interval

$$-\frac{\pi}{2} < x < \frac{\pi}{2}.$$

If x is near $-\pi/2$ then the tangent is very large negative. Namely

$$\tan x = \frac{\sin x}{\cos x}.$$

When x is near $-\pi/2$ then $\cos x$ is near 0 and $\sin x$ is near -1. You can also see this from a right triangle.

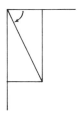

As x increases from $-\pi/2$ to 0, $\sin x$ increases from -1 to 0. On the other hand, as x increases from $-\pi/2$ to 0, $\cos x$ increases from 0 to 1. Hence $1/\cos x$ decreases from very large negative to 1. Hence as x increases from $-\pi/2$ to 0,

$$\tan x = \frac{\sin x}{\cos x} \quad \text{increases from large negative to 0.}$$

Similarly, as x increases from 0 to $\pi/2$, $\sin x$ increases from 0 to 1 and $\cos x$ decreases from 1 to 0. Hence $1/\cos x$ increases from 1 to very large positive, and so

$$\tan x = \frac{\sin x}{\cos x} \quad \text{increases from 0 to large positive.}$$

Thus the graph of $y = \tan x$ looks like this.

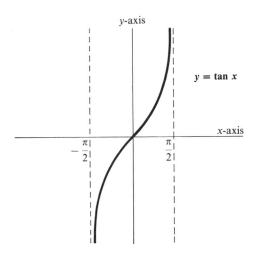

Remark on notation. Do not confuse $\sin 2x$ and $(\sin 2)x$. We usually write

$$\sin(2x) = \sin 2x$$

without parentheses to mean the sine of $2x$. On the other hand, $(\sin 2)x$ is the number $\sin 2$ times x. The graph of

$$y = (\sin 2)x$$

is a straight line, just like $y = cx$ for some fixed number c.

IV, §2. EXERCISES

1. Draw the graph of $\tan x$ for all values of x.

2. Let $\sec x = 1/\cos x$ be defined when $\cos x \neq 0$. Draw the graph of $\sec x$.

3. Let $\cot x = 1/\tan x$. Draw the graph of $\cot x$.
   (Sec and cot are abbreviations for the secant and cotangent.)

4. Sketch the graphs of the following functions:
   (a) $y = \sin 2x$ (b) $y = \sin 3x$
   (c) $y = \cos 2x$ (d) $y = \cos 3x$

5. Sketch the graphs of the following functions:
   (a) $y = \sin \frac{1}{2}x$ (b) $y = \sin \frac{1}{3}x$ (c) $y = \sin \frac{1}{4}x$
   (d) $y = \cos \frac{1}{2}x$ (e) $y = \cos \frac{1}{3}x$ (f) $y = \cos \frac{1}{4}x$

6. Sketch the graphs of:
   (a) $y = \sin \pi x$ (b) $y = \cos \pi x$
   (c) $y = \sin 2\pi x$ (d) $y = \cos 2\pi x$

7. Sketch the graph of the following functions:
   (a) $y = |\sin x|$ (b) $y = |\cos x|$.

8. Let $f(x) = \sin x + \cos x$. Plot approximate values of $f(n\pi/4)$, for

$$n = 0, 1, 2, 3, 4, 5, 6, 7, 8.$$

IV, §3. ADDITION FORMULA

In this section we shall state and prove the most important formulas about sine and cosine.

To begin with, using the Pythagoras theorem, we shall see that

$$\boxed{(\sin x)^2 + (\cos x)^2 = 1}$$

for all x. To show this, we take an angle A and we determine its sine and cosine from the right triangle, as in the following figure.

Then by Pythagoras we have

$$a^2 + b^2 = r^2.$$

Dividing by r^2 yields

$$\left(\frac{a}{r}\right)^2 + \left(\frac{b}{r}\right)^2 = 1.$$

The same argument works when A is greater than $\pi/2$, by means of a triangle like this one:

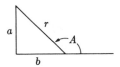

In both cases, we have sine $A = a/r$ and cosine $A = b/r$, so that we have the relation

$$(\text{sine } A)^2 + (\text{cosine } A)^2 = 1.$$

It is customary to write the square of the sine and cosine as $\sin^2 A$ and $\cos^2 A$. In the second case note that b is negative.

Our main result is the **addition formula**, which should be memorized.

Theorem 3.1. *For any angles A and B, we have*

$$
\begin{aligned}
\sin(A + B) &= \sin A \cos B + \cos A \sin B, \\
\cos(A + B) &= \cos A \cos B - \sin A \sin B.
\end{aligned}
$$

Proof. We shall prove the second formula first.
We consider two angles A, B and their sum:

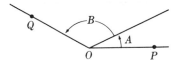

We take two points P, Q as indicated, at a distance 1 from the origin O. We shall now compute the distance from P to Q, using two different coordinate systems.

First, we take a coordinate system as usual:

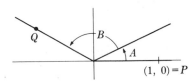

Then $P = (1, 0)$ and

$$Q = (\cos(A + B), \sin(A + B)).$$

The square of the distance between $P = (x_1, y_1)$ and $Q = (x_2, y_2)$ is

$$(y_2 - y_1)^2 + (x_2 - x_1)^2.$$

Hence

$$\text{dist}(P, Q)^2 = \sin^2(A + B) + (\cos(A + B) - 1)^2,$$
$$= -2\cos(A + B) + 2.$$

Next we place the coordinate system as shown in the figure below.

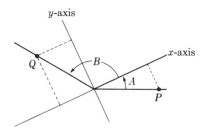

Then the coordinates of P become

$$(\cos(-A),\ \sin(-A)) = (\cos A,\ -\sin A).$$

Those of Q are simply $(\cos B, \sin B)$. Hence

$$\text{dist}(P, Q)^2 = (\sin B + \sin A)^2 + (\cos B - \cos A)^2,$$
$$= \sin^2 B + 2\sin B \sin A + \sin^2 A$$
$$+ \cos^2 B - 2\cos B \cos A + \cos^2 A$$
$$= 2 + 2\sin A \sin B - 2\cos A \cos B.$$

If we set the squares of the two distances equal to each other, we get our formula.

From the addition formula for the cosine, we get some formulas relating sine and cosine.

$$\sin x = \cos\left(x - \frac{\pi}{2}\right),$$

$$\cos x = \sin\left(x + \frac{\pi}{2}\right).$$

To prove the first one, we start with the right-hand side:

$$\cos\left(x - \frac{\pi}{2}\right) = \cos x \cos\left(-\frac{\pi}{2}\right) - \sin x \sin\left(-\frac{\pi}{2}\right)$$

$$= 0 + \sin x = \sin x$$

because $\cos(-\pi/2) = 0$ and $-\sin(-\theta) = \sin\theta$ (with $\theta = \pi/2$).

The second relation follows from the first, namely in the first relation put $x = z + \pi/2$. Then

$$\sin\left(z + \frac{\pi}{2}\right) = \cos\left(z + \frac{\pi}{2} - \frac{\pi}{2}\right) = \cos z.$$

This proves the second relation.

Similarly, you can prove

$$\sin\left(\frac{\pi}{2} - x\right) = \cos x.$$

This will be used in the next argument.

The addition formula for the sine can be obtained from the addition formula for the cosine by the following device:

$$\sin(A + B) = \cos\left(A + B - \frac{\pi}{2}\right)$$

$$= \cos A \cos\left(B - \frac{\pi}{2}\right) - \sin A \sin\left(B - \frac{\pi}{2}\right)$$

$$= \cos A \sin B + \sin A \sin\left(\frac{\pi}{2} - B\right)$$

$$= \cos A \sin B + \sin A \cos B,$$

thereby proving the addition formula for the sine.

Example. Find $\sin(\pi/12)$.

We write

$$\frac{\pi}{12} = \frac{\pi}{3} - \frac{\pi}{4}.$$

Then $\sin(\pi/12) = \sin(\pi/3)\cos(\pi/4) - \cos(\pi/3)\sin(\pi/4)$, and substituting the known values we find

$$\sin(\pi/12) = \frac{\sqrt{3} - 1}{2\sqrt{2}}.$$

Example. We have

$$\cos\left(x + \frac{\pi}{2}\right) = -\sin x,$$

because

$$\cos\left(x + \frac{\pi}{2}\right) = \cos x \cos \frac{\pi}{2} - \sin x \sin \frac{\pi}{2} = -\sin x,$$

since $\cos \pi/2 = 0$ and $\sin \pi/2 = 1$.

In the exercises, you will derive a few more useful formulas for the sine and cosine, notably:

$$\sin 2x = 2 \sin x \cos x,$$

$$\cos 2x = \cos^2 x - \sin^2 x,$$

$$\cos^2 x = \frac{1 + \cos 2x}{2} \quad \text{and} \quad \sin^2 x = \frac{1 - \cos 2x}{2}.$$

You will remember them better for having worked them out, so we don't spoil this in the text.

IV, §3. EXERCISES

1. Find $\sin 7\pi/12$. [*Hint*: Write $7\pi/12 = 4\pi/12 + 3\pi/12$.]

2. Find $\cos 7\pi/12$.

3. Find the following values:
   (a) $\sin \pi/12$
   (b) $\cos \pi/12$
   (c) $\sin 5\pi/12$
   (d) $\cos 5\pi/12$
   (e) $\sin 11\pi/12$
   (f) $\cos 11\pi/12$
   (g) $\sin 10\pi/12$
   (h) $\cos 10\pi/12$

4. Prove the following formulas.
   (a) $\sin 2x = 2 \sin x \cos x$
   (b) $\cos 2x = \cos^2 x - \sin^2 x$
   (c) $\cos^2 x = \dfrac{1 + \cos 2x}{2}$
   (d) $\sin^2 x = \dfrac{1 - \cos 2x}{2}$

   [*Hint*: For (c) and (d), start with the special case of the addition formula for $\cos 2x$. Then use the identity

   $$\sin^2 x + \cos^2 x = 1.]$$

5. Find a formula for $\sin 3x$ in terms of $\sin x$ and $\cos x$. Similarly for $\cos 3x$.

6. Prove that $\sin(\pi/2 - x) = \cos x$, using only the addition formula for the cosine.

7. Prove the formulas

   $$\sin mx \sin nx = \tfrac{1}{2}[\cos(m - n)x - \cos(m + n)x],$$

   $$\sin mx \cos nx = \tfrac{1}{2}[\sin(m + n)x + \sin(m - n)x],$$

   $$\cos mx \cos nx = \tfrac{1}{2}[\cos(m + n)x + \cos(m - n)x].$$

[*Hint*: Expand the right-hand side by the addition formula, and then cancel as much as you can. The left-hand side should be forthcoming. Note that, for instance,

$$\cos(m - n)x = \cos(mx - nx)$$
$$= \cos mx \cos nx + \sin mx \sin nx.]$$

IV, §4. THE DERIVATIVES

We shall prove:

Theorem 4.1. *The functions* $\sin x$ *and* $\cos x$ *have derivatives and*

$$\frac{d(\sin x)}{dx} = \cos x,$$

$$\frac{d(\cos x)}{dx} = -\sin x.$$

Proof. We shall first determine the derivative of $\sin x$. We have to look at the Newton quotient of $\sin x$. It is

$$\frac{\sin(x + h) - \sin x}{h}.$$

Using the addition formula to expand $\sin(x + h)$, we see that the Newton quotient is equal to

$$\frac{\sin x \cos h + \cos x \sin h - \sin x}{h}.$$

We put together the two terms involving $\sin x$:

$$\frac{\cos x \sin h + \sin x(\cos h - 1)}{h}$$

and separate our quotient into a sum of two terms:

$$\cos x \, \frac{\sin h}{h} + \sin x \, \frac{\cos h - 1}{h}.$$

We now face the problem of finding the limit of

$$\frac{\sin h}{h} \quad \text{and} \quad \frac{\cos h - 1}{h} \quad \text{as } h \text{ approaches } 0.$$

This is a somewhat more difficult problem than those we encountered previously. We cannot tell right away what these limits will be. In the next section, we shall prove that

$$\lim_{h \to 0} \frac{\sin h}{h} = 1 \quad \text{and} \quad \lim_{h \to 0} \frac{\cos h - 1}{h} = 0.$$

Once we know these limits, then we see immediately that the first term approaches $\cos x$ and the second term approaches

$$(\sin x) \cdot 0 = 0.$$

Hence,

$$\lim_{h \to 0} \frac{\sin(x + h) - \sin x}{h} = \cos x.$$

This proves that

$$\frac{d(\sin x)}{dx} = \cos x.$$

To find the derivative of $\cos x$, we could proceed in the same way, and we would encounter the same limits. However, there is a trick which avoids this.

We know that $\cos x = \sin\left(x + \dfrac{\pi}{2}\right)$. Let $u = x + \dfrac{\pi}{2}$ and use the chain rule. We get

$$\frac{d(\cos x)}{dx} = \frac{d(\sin u)}{du} \frac{du}{dx}.$$

However, $du/dx = 1$. Hence

$$\frac{d(\cos x)}{dx} = \cos u = \cos\left(x + \frac{\pi}{2}\right) = -\sin x,$$

thereby proving our theorem.

Remark. It is not true that the derivative of the function $\sin^* x$ is $\cos^* x$. Using the chain rule, find out what its derivative is. The reason for using the radian measure of angles is to get a function $\sin x$ whose derivative is $\cos x$.

Example. Find the tangent line of the curve $y = \sin 4x$ at the point $x = \pi/16$. This is easily done. Let $f(x) = \sin 4x$. Then

$$f'(x) = 4 \cos 4x.$$

Hence the slope of the tangent line at $x = \pi/16$ is equal to

$$f'\left(\frac{\pi}{16}\right) = 4\cos\left(\frac{4\pi}{16}\right) = 4\cos\left(\frac{\pi}{4}\right) = \frac{4}{\sqrt{2}}.$$

On the other hand, we have

$$f\left(\frac{\pi}{16}\right) = \sin\left(\frac{4\pi}{16}\right) = \sin\frac{\pi}{4} = \frac{1}{\sqrt{2}}.$$

Hence the equation of the tangent line is

$$y - \frac{1}{\sqrt{2}} = \frac{4}{\sqrt{2}}\left(x - \frac{\pi}{16}\right).$$

Theorem 4.2. *We have*

$$\boxed{\frac{d(\tan x)}{dx} = 1 + \tan^2 x = \sec^2 x.}$$

Proof. We use the rule for the derivative of a quotient. So:

$$\frac{d}{dx}\left(\frac{\sin x}{\cos x}\right) = \frac{(\cos x)(\cos x) - (\sin x)(-\sin x)}{(\cos x)^2}$$

$$= \frac{\cos^2 x + \sin^2 x}{\cos^2 x} = 1$$

$$= \frac{1}{\cos^2 x} = \sec^2 x.$$

But also

$$\frac{\cos^2 x + \sin^2 x}{\cos^2 x} = 1 + \frac{\sin^2 x}{\cos^2 x} = 1 + \tan^2 x.$$

This proves the theorem.

Example. A balloon is going up, starting at a point P. An observer standing 200 ft away looks at the balloon, and the angle θ which the balloon makes increases at the rate of $\frac{1}{20}$ rad/sec. Find the rate at which the distance of the balloon from the ground is increasing when $\theta = \pi/4$.

The picture looks as follows, where y is the distance from the balloon to the ground.

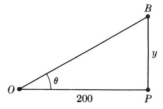

We have $\tan \theta = y/200$, whence

$$y = 200 \cdot \tan \theta.$$

We want to find the rate at which y is increasing, i.e. we want to find dy/dt. Taking the derivative with respect to time t yields

$$\frac{dy}{dt} = 200 \frac{d \tan \theta}{dt}.$$

$$= 200(1 + \tan^2 \theta) \frac{d\theta}{dt}$$

by Theorem 4.2 and the chain rule. Hence

$$\frac{dy}{dt}\bigg|_{\theta = \pi/4} = 200\left(1 + \tan^2 \frac{\pi}{4}\right) \frac{1}{20}$$

$$= 10(1 + 1)$$

$$= 20 \text{ ft/sec.}$$

This is our answer.

Example. In the preceding example, find the rate at which the distance of the balloon from the ground is increasing when $\sin \theta = 0.2$.
We have

$$\frac{dy}{dt} = 200 \frac{d \tan \theta}{d\theta} \frac{d\theta}{dt} = 200(1 + \tan^2 \theta) \frac{d\theta}{dt}.$$

When $\sin \theta = 0.2$, we have $\sin^2 \theta = 0.04$, and

$$\cos^2 \theta = 1 - \sin^2 \theta = 0.96.$$

Hence

$$\tan^2 \theta = \frac{\sin^2 \theta}{\cos^2 \theta} = \frac{4}{96} = \frac{1}{24}.$$

Since we are given $d\theta/dt = 1/20$, we find

$$\frac{dy}{dt}\bigg|_{\sin \theta = 0.2} = 200\left(1 + \frac{1}{24}\right)\frac{1}{20}$$

$$= \frac{10 \cdot 25}{24}$$

$$= \frac{125}{12} \text{ ft/sec.}$$

IV, §4. EXERCISES

1. What is the derivative of cot x?

Find the derivative of the following functions:

2. $\sin(3x)$ 3. $\cos(5x)$

4. $\sin(4x^2 + x)$ 5. $\tan(x^3 - 5)$

6. $\tan(x^4 - x^3)$ 7. $\tan(\sin x)$

8. $\sin(\tan x)$ 9. $\cos(\tan x)$

10. What is the slope of the curve $y = \sin x$ at the point whose x-coordinate is π?

Find the slope of the following curves at the indicated point (we just give the x-coordinate of the point):

11. $y = \cos(3x)$ at $x = \pi/3$

12. $y = \sin x$ at $x = \pi/6$

13. $y = \sin x + \cos x$ at $x = 3\pi/4$

14. $y = \tan x$ at $x = -\pi/4$

15. $y = \dfrac{1}{\sin x}$ at $x = -\pi/6$

16. Give the equation of the tangent line to the following curves at the indicated point.

    (a) $y = \sin x$ at $x = \pi/2$ (b) $y = \cos x$ at $x = \pi/6$

    (c) $y = \sin 2x$ at $x = \pi/4$ (d) $y = \tan 3x$ at $x = \pi/4$

    (e) $y = 1/\sin x$ at $x = \pi/2$ (f) $y = 1/\cos x$ at $x = \pi/4$

    (g) $y = 1/\tan x$ at $x = \pi/4$ (h) $y = \tan \dfrac{x}{2}$ at $x = \pi/2$

(i) $y = \sin \dfrac{x}{2}$ at $x = \pi/3$ (j) $y = \cos \dfrac{\pi x}{3}$ at $x = 1$

(k) $y = \sin \pi x$ at $x = \frac{1}{2}$ (l) $y = \tan \pi x$ at $x = \frac{1}{6}$

17. In the following right triangle, suppose that θ is decreasing at the rate of
$\frac{1}{30}$ rad/sec. Find each one of the indicated derivatives:

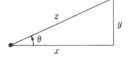

(a) dy/dt, when $\theta = \pi/3$ and x is constant, $x = 12$.
(b) dz/dt, when $\theta = \pi/4$ if y is constant, $y = 10\sqrt{2}$.
(c) dx/dt, when $x = 1$ if x and y are both changing,
   but z is constant, $z = 2$.

Remember that θ is decreasing, so $d\theta/dt = -\frac{1}{30}$.

18. A Ferris wheel 50 ft in diameter makes 1 revolution every 2 min. If the
center of the wheel is 30 ft above the ground, how fast is a passenger in the
wheel moving vertically when he is 42.5 ft above the ground?

19. A balloon is going up, starting at a point P. An observer O, standing 300 ft
away, looks at the balloon, and the angle θ which the balloon makes increases at the rate of 0.3 rad/sec. Find the rate at which the distance of the
balloon from the ground is increasing, when

(a) $\theta = \pi/4$,   (b) $\theta = \pi/3$,   (c) $\cos \theta = 0.2$,
(d) $\sin \theta = 0.3$,   (e) $\tan \theta = 4$.

20. An airplane is flying horizontally on a straight line at a speed of 1,000 km/hr,
at an elevation of 10 km. An automatic camera is photographing a point
directly ahead on the ground. How fast must the camera be turning when
the angle between the path of the plane and the line of sight to the point is
30°?

21. A Beacon light is located 1000 ft from a sea wall, and rotates at the constant
rate of 2 revolutions per minute.
(a) How fast is the lighted spot on the wall moving along the wall at the
   nearest point to the beacon?
(b) How fast is the spot of light moving at a point 500 ft from this nearest
   point?

22. An airplane flying at an altitude of 20,000 ft and on a horizontal course
passes directly over an observer on the ground below. The observer notes
that when the angle between the ground and his line of sight is 60°, the angle
is decreasing at a rate of 2° per second. What is the speed of the airplane?

23. A vertical pole is 30 ft high and is located 30 ft east of a tall building. If the
sun is rising at the rate of 18° per hour, how fast is the shadow of the pole
on the building shortening when the elevation of the sun is 30°? [*Hint*: The
rate of rise is the rate of change of the angle of elevation θ of the sun. First
convert the degrees to radians per hour, namely

$$18 \text{ deg/hr} = 18\,\pi/180 = \pi/10 \text{ rad/hr}.$$

If s is the length of the shadow, you then have $\tan \theta = (30 \quad s)/30$.]

24. A weather balloon is released on the ground 1500 ft from an observer and
rises vertically at the constant rate 250 ft/min. How fast is the angle between

the observer's line of sight and the ground increasing when the balloon is at an altitude of 2000 ft? Give the answer in degrees per minute.

25. A ladder 30 ft long leans against a wall. Suppose the bottom of the ladder slides away from the wall at the rate of 3 ft/sec. How fast is the angle between the ladder and the ground changing when the bottom of the ladder is 15 ft from the wall?

26. A rocket leaves the ground 2000 ft from an observer and rises vertically at the constant rate of 100 ft/sec. How fast is the angle between the observer's line of sight and the ground increasing after 20 sec? Give the answer in degrees per second.

27. A kite at an altitude of 200 ft moves horizontally at the rate of 20 ft/sec. At what rate is the angle between the line and the ground changing when 400 ft of line is out?

28. Two airplanes are flying in the same direction, and at constant altitude. At $t = 0$ airplane A is 1000 ft vertically above airplane B. Airplane A travels at a constant speed of 600 ft/sec, and B at a constant speed of 400 ft/sec. Find the rate of change of the angle of elevation θ of A relative to B at time $t = 10$ sec.

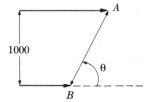

IV, §5. TWO BASIC LIMITS

We shall first prove that

$$\lim_{h \to 0} \frac{\sin h}{h} = 1.$$

Both the numerator and the denominator approach 0 as h approaches 0, and we get no information by trying some cancellation procedure, the way we did it for powers.

Let us assume first that h is positive, and look at the following diagram.

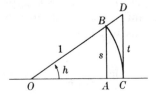

We take a circle of radius 1 and an angle of h radians. Let s be the altitude of the small triangle OAB, and t that of the big triangle OCD. Then, using the small triangle OAB,

$$\sin h = \frac{s}{1} = s$$

and using the large triangle, OCD,

$$\tan h = \frac{t}{1} = t = \frac{\sin h}{\cos h}.$$

We see that:

area of triangle OAB < area of sector OCB < area of triangle OCD.

The base OA of the small triangle is equal to $\cos h$ and its altitude AB is $\sin h$.

The base OC of the big triangle is equal to 1. Its altitude CD is

$$t = \frac{\sin h}{\cos h}.$$

The area of each triangle is $\frac{1}{2}$ the base times the altitude.

The area of the sector is the fraction $h/2\pi$ of the area of the circle, which is π. Hence the area of the sector is $h/2$. Thus we obtain:

$$\frac{1}{2}\cos h \sin h < \frac{1}{2}h < \frac{1}{2}\frac{\sin h}{\cos h}.$$

We multiply everywhere by 2 and get

$$\cos h \sin h < h < \frac{\sin h}{\cos h}.$$

Since we assumed that $h > 0$ it follows that $\sin h > 0$ and we divide both the inequalities by $\sin h$ to obtain

$$\cos h < \frac{h}{\sin h} < \frac{1}{\cos h}.$$

As h approaches 0, both $\cos h$ and $1/\cos h$ approach 1. Thus $h/\sin h$ is squeezed between two quantities which approach 1, and therefore $h/\sin h$ must approach 1 also. Thus we may write

$$\lim_{h \to 0} \frac{h}{\sin h} = 1.$$

Since

$$\frac{\sin h}{h} = \frac{1}{h/\sin h}$$

and the limit of a quotient is the quotient of the limits, it follows that

$$\lim_{h \to 0} \frac{\sin h}{h} = 1,$$

as was to be shown.

We computed our limit when $h > 0$. Suppose that $h < 0$. We can write

$$h = -k$$

with $k > 0$. Then

$$\frac{\sin(-k)}{-k} = \frac{-\sin k}{-k} = \frac{\sin k}{k}.$$

As h tends to 0, so does k. Hence we are reduced to our previous limit because $k > 0$.

We still have to prove the limit

$$\boxed{\lim_{h \to 0} \frac{\cos h - 1}{h} = 0.}$$

We have

$$\frac{\cos h - 1}{h} = \frac{(\cos h - 1)(\cos h + 1)}{h(\cos h + 1)}$$

$$= \frac{\cos^2 h - 1}{h(\cos h + 1)}$$

$$= \frac{-\sin^2 h}{h(\cos h + 1)}$$

$$= -\frac{\sin h}{h} (\sin h) \frac{1}{\cos h + 1}.$$

We shall use the property concerning the product of limits. We have a product of three factors. The first is

$$-\frac{\sin h}{h}$$

and approaches -1 as h approaches 0.

The second is $\sin h$ and approaches 0 as h approaches 0.

The third is

$$\frac{1}{\cos h + 1}$$

and its limit is $\frac{1}{2}$ as h approaches 0.

Therefore the limit of the product is 0, and everything is proved!

IV, §6. POLAR COORDINATES

Instead of describing a point in the plane by its coordinates with respect to two perpendicular axes, we can also describe it as follows. We draw a ray between the point and a given origin. The **angle** θ which this ray makes with the horizontal axis and the **distance** r between the point and the origin determine our point. Thus the point is described by a pair of numbers (r, θ), which are called its **polar coordinates**.

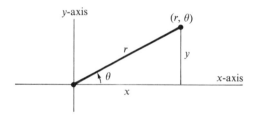

If we have our usual axes and x, y are the ordinary coordinates of our point, then we see that

$$\boxed{\frac{x}{r} = \cos \theta \qquad \text{and} \qquad \frac{y}{r} = \sin \theta,}$$

whence

$$\boxed{x = r \cos \theta \qquad \text{and} \qquad y = r \sin \theta.}$$

This allows us to change from polar coordinates to ordinary coordinates. **It is to be understood that r is always supposed to be ≥ 0.** In terms of the ordinary coordinates, we have

$$\boxed{r = \sqrt{x^2 + y^2}.}$$

By Pythagoras, r is the distance of the point (x, y) from the origin $(0, 0)$. Note that distance is always ≥ 0.

Example 1. Find polar coordinates of the point whose ordinary coordinates are $(1, \sqrt{3})$.

We have $x = 1$ and $y = \sqrt{3}$, so that $r = \sqrt{1 + 3} = 2$. Also

$$\cos \theta = \frac{x}{r} = \frac{1}{2}, \qquad \sin \theta = \frac{y}{r} = \frac{\sqrt{3}}{2}.$$

Hence $\theta = \pi/3$, and the polar coordinates are $(2, \pi/3)$.

We observe that we may have several polar coordinates corresponding to the same point. The point whose polar coordinates are $(r, \theta + 2\pi)$ is the same as the point (r, θ). Thus in our example above, $(2, \pi/3 + 2\pi)$ would also be polar coordinates for our point. In practice, we usually use the value for the angle which lies between 0 and 2π.

Suppose a bug is traveling in the plane. Its position is completely determined if we know the angle θ and the distance of the bug from the origin, that is if we know the polar coordinates. If the distance r from the origin is given as a function of θ, then the bug is traveling along a curve and we can sketch this curve.

Example 2. Sketch the graph of the function $r = \sin \theta$ for $0 \leq \theta \leq \pi$.

If $\pi < \theta < 2\pi$, then $\sin \theta < 0$ and hence for such θ we don't get a point on the curve. Next, we make a table of values. We consider intervals of θ such that $\sin \theta$ is always increasing or always decreasing over these intervals. This tells us whether the point is moving further away from the origin, or coming closer to the origin, since r is the distance of the point from the origin. Intervals of increase and decrease for $\sin \theta$ can be taken to be of length $\pi/2$. Thus we find the following table:

θ	$\sin \theta = r$
inc. 0 to $\pi/2$	inc. 0 to 1
inc. $\pi/2$ to π	dec. 1 to 0
$\pi/6$	1/2
$\pi/4$	$1\sqrt{2}$
$\pi/3$	$\sqrt{3/2}$

Put in words: as θ increases from 0 to $\pi/2$, then $\sin \theta$ and therefore r increases until r reaches 1. As θ increases from $\pi/2$ to π then $\sin \theta$ and thus r decreases from 1 to 0. Hence the graph looks like this.

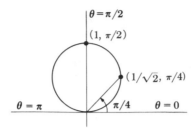

We have drawn the graph like a circle. Actually, we don't know whether it is a circle or not. The graph could be flatter in one direction than in another. In the next example, we shall see that it actually must be a circle.

Example 3. Change the equation

$$r = \sin \theta$$

to rectangular coordinates.

We substitute the expressions

$$r = \sqrt{x^2 + y^2}$$

and

$$\sin \theta = y/r = y/\sqrt{x^2 + y^2}$$

in the polar equation, to obtain

$$\sqrt{x^2 + y^2} = \frac{y}{\sqrt{x^2 + y^2}}.$$

Of course, this substitution is valid only when $r \neq 0$, i.e. $r > 0$. We can then simplify the equation we have just obtained, multiplying both sides by $\sqrt{x^2 + y^2}$. We then obtain

$$x^2 + y^2 = y.$$

You should know from Chapter II that this is the equation of a circle, by completing the square. We recall here how this is done. We write the equation in the form

$$x^2 + y^2 - y = 0.$$

We would like this equation to be of the form

$$x^2 + (y - b)^2 = c^2,$$

because then we know immediately that this is a circle of center $(0, b)$
and radius c. We know that

$$(y - b)^2 = y^2 - 2by + b^2.$$

Therefore we let $2b = 1$ and $b = \frac{1}{2}$. Then

$$x^2 + y^2 - y = x^2 + (y - \tfrac{1}{2})^2 - \tfrac{1}{4}$$

because the $\frac{1}{4}$ cancels. Thus the equation

$$\boxed{x^2 + y^2 - y = 0}$$

is equivalent with

$$\boxed{x^2 + (y - \tfrac{1}{2})^2 = \tfrac{1}{4}.}$$

This is the equation of a circle of center $(0, \frac{1}{2})$ and radius $\frac{1}{2}$. The point
corresponding to the polar coordinate $r = 0$ is the point with rectangular
coordinates $x = 0$ and $y = 0$.

Example 4. The equation of the circle of radius 3 and center at the
origin in polar coordinates is simply

$$r = 3 \qquad \text{or} \qquad \sqrt{x^2 + y^2} = 3 \qquad \text{or} \qquad x^2 + y^2 = 9.$$

This expresses the condition that that distance of the point (x, y) from
the origin is the constant 3. The angle θ can be arbitrary.

Example 5. Consider the equation $\theta = 1$ in polar coordinates. A
point with polar coordinates (r, θ) satisfies this equation if and only if its
angle θ is 1 and there is no restriction on its r-coordinate, i.e. $r \geq 0$.
Thus geometrically, this set of points can be described as a half line, or a
ray, as on the figure (a).

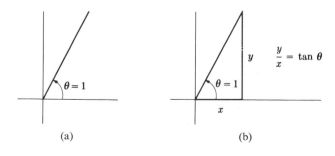

(a)   (b)

By the definition of the tangent, if (x, y) are the ordinary coordinates of a point on this ray, and $y \neq 0$, then

$$y/x = \tan 1 \qquad \text{and} \qquad x > 0,$$

whence

$$y = (\tan 1)x \qquad \text{and} \qquad x > 0.$$

Of course, the point with $x = y = 0$ also lies on the ray. Conversely, any point whose ordinary coordinates (x, y) satisfy

$$y = (\tan 1)x \qquad \text{and} \qquad x \geq 0$$

lies on the ray. Hence the ray defined in polar coordinates by the equation $\theta = 1$ is defined in ordinary coordinates by the pair of conditions

$$y = (\tan 1)x \qquad \text{and} \qquad x \geq 0.$$

Instead of 1 we could take any number. For instance, the ray defined by the equation $\theta = \pi/6$ in polar coordinates is also defined by the pair of conditions

$$y = (\tan \pi/6)x \qquad \text{and} \qquad x \geq 0.$$

Since $\tan \pi/6 = 1/\sqrt{3}$, we may write the equivalent pair of conditions

$$y = \frac{1}{\sqrt{3}} x \qquad \text{and} \qquad x \geq 0.$$

Note that there is no simpler way of expressing $\tan 1$ than just writing $\tan 1$. Only when dealing with fractional multiples of π do we have a way of writing the trigonometric functions in terms of roots, like $\tan \pi/6 = 1/\sqrt{3}$.

Example 6. Let us sketch the curve given in polar coordinates by the equation

$$r = |\sin 2\theta|.$$

The absolute value sign makes the right-hand side always ≥ 0, and so there is a value of r for every value of θ. Regions of increase and decrease for $\sin 2\theta$ will occur when 2θ ranges over intervals of length $\pi/2$. Hence it is natural to look at intervals for θ of length $\pi/4$. We now make a table of the increasing and decreasing behavior of $|\sin 2\theta|$ and r over such intervals.

θ	$r = |\sin 2\theta|$
inc. 0 to $\pi/4$	inc. 0 to 1
inc. $\pi/4$ to $\pi/2$	dec. 1 to 0
inc. $\pi/2$ to $3\pi/4$	inc. 0 to 1
inc. $3\pi/4$ to π	dec. 1 to 0
and so forth	

The graph therefore looks like this:

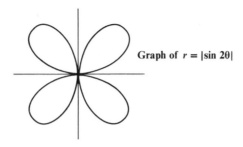

Graph of $r = |\sin 2\theta|$

Because of the absolute value sign, for any value of θ we obtain a value for r which is ≥ 0. According to our convention, if we wanted to graph

$$r = \sin 2\theta$$

without the absolute value sign, then we would have to omit those portions of the above graph for which $\sin 2\theta$ is negative, i.e. those portions of the graph for which

$$\frac{\pi}{2} < \theta < \pi$$

and

$$\frac{3\pi}{2} < \theta < 2\pi.$$

Thus the graph of $r = \sin 2\theta$ would look like that in the next figure.

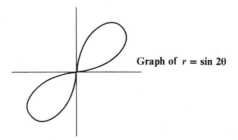

Graph of $r = \sin 2\theta$

Example 7. We want to sketch the curve given in polar coordinates by the equation

$$r = 1 + \sin \theta$$

We look at the behavior of r when θ ranges over the intervals.

$$[0, \pi/2], \qquad [\pi/2, \pi], \qquad [\pi, 3\pi/2], \qquad [3\pi/2, 2\pi].$$

θ	$\sin \theta$	r
inc. from 0 to $\pi/2$	inc. 0 to 1	inc. 1 to 2
inc. from $\pi/2$ to π	dec. 1 to 0	dec. 2 to 1
inc. from π to $3\pi/2$	dec. 0 to -1	dec. 1 to 0.
inc. from $3\pi/2$ to 2π	inc. -1 to 0	inc. 0 to 1

Thus the graph looks roughly like this:

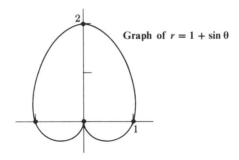

Graph of $r = 1 + \sin \theta$

Example 8. Let us look at the slightly different equation

$$r = 1 + 2 \sin \theta.$$

A similar analysis will work, but we must be careful of the possibility that the expression on the right-hand side is negative. According to our convention, this will not yield a point since we assume that $r \geq 0$ for polar coordinates. Thus when $2 \sin \theta < -1$, we do not get any point. This occurs precisely in the interval

$$\frac{7\pi}{6} < \theta < \frac{11\pi}{6}.$$

The graph will therefore look like the next figure.

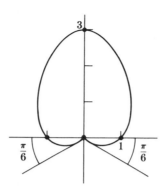

We have also drawn the rays determining angles of $\dfrac{7\pi}{6}$ and $\dfrac{11\pi}{6}$.

IV, §6. EXERCISES

1. Plot the following points in polar coordinates:
   (a) $(2, \pi/4)$ (b) $(3, \pi/6)$ (c) $(1, -\pi/4)$ (d) $(2, -3\pi/6)$

2. Same directions as in Exercise 1.
   (a) $(1, 1)$ (b) $(4, -3)$
   (These are polar coordinates. Just show approximately the angle represented by the given coordinates.)

3. Find polar coordinates for the following points given in the usual x- and y-coordinates:
   (a) $(1, 1)$ (b) $(-1, -1)$ (c) $(3, 3\sqrt{3})$ (d) $(-1, 0)$

4. Sketch the following curves and put the equation in rectangular coordinates.
   (a) $r = 2 \sin \theta$ (b) $r = 3 \cos \theta$

5. Change the following to rectangular coordinates and sketch the curve. We assume $a > 0$.
   (a) $r = a \sin \theta$ (b) $r = a \cos \theta$
   (c) $r = 2a \sin \theta$ (d) $r = 2a \cos \theta$

Sketch the graphs of the following curves given in polar coordinates.

6. $r^2 = \cos \theta$ 7. $r^2 = \sin \theta$

8. (a) $r = \sin^2 \theta$ (b) $r = \cos^2 \theta$ 9. $r = 4 \sin^2 \theta$

10. $r = 5$ 11. $r = 4$

12. (a) $r = \dfrac{1}{\cos \theta}$ (b) $r = \dfrac{1}{\sin \theta}$ 13. $r = 3/\cos \theta$

14. $r = 1 + \cos \theta$ 15. $r = 1 - \sin \theta$ 16. $r = 1 - \cos \theta$

17. $r = 1 - 2 \sin \theta$ 18. $r = \sin 3\theta$ 19. $r = \sin 4\theta$

20. $r = \cos 2\theta$ 21. $r = \cos 3\theta$ 22. $r = |\cos 2\theta|$

23. $r = |\sin 3\theta|$ 24. $r = |\cos 3\theta|$ 25. $r = \theta$

26. $r = 1/\theta$

In the next three problems put the equation in rectangular coordinates and sketch the curve.

27. $r = \dfrac{1}{1 - \cos \theta}$ 28. $r = \dfrac{2}{2 - \cos \theta}$ 29. $r = \dfrac{4}{1 + 2 \cos \theta}$

Sketch the following curves given in polar coordinates.

30. $r = \tan \theta$ 31. $r = 5 + 2 \sin \theta$

32. $r = |1 + 2 \cos \theta|$ 33. (a) $r = 2 + \sin 2\theta$
                                            (b) $r = 2 - \sin 2\theta$

34. $\theta = \pi$ 35. $\theta = \pi/2$

36. $\theta = -\pi/2$ 37. $\theta = 5\pi/4$

38. $\theta = 3\pi/2$ 39. $\theta = 3\pi/4$

CHAPTER V

The Mean Value Theorem

Given a curve, $y = f(x)$, we shall use the derivative to give us information about the curve. For instance, we shall find the maximum and minimum of the graph, and regions where the curve is increasing or decreasing. We shall use the mean value theorem, which is basic in the theory of derivatives.

V, §1. THE MAXIMUM AND MINIMUM THEOREM

Definition. Let f be a differentiable function. A **critical point** of f is a number c such that

$$f'(c) = 0.$$

The derivative being zero means that the slope of the tangent line is 0 and thus that the tangent line itself is horizontal. We have drawn three examples of this phenomenon.

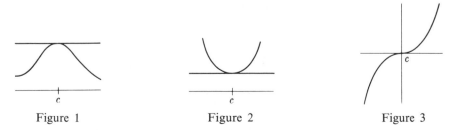

| Figure 1 | Figure 2 | Figure 3 |

The third example is that of a function like $f(x) = x^3$. We have $f'(x) = 3x^2$ and hence when $x = 0$, $f'(0) = 0$.

The other two examples are those of a maximum and a minimum, respectively, if we look at the graph of the function only near our point c. We shall now formalize these notions.

Let a, b be two numbers with $a < b$. We shall repeatedly deal with the interval of numbers between a and b. Sometimes we want to include the end points a and b, and sometimes we do not. We recall the standard terminology.

The collection of numbers x such that $a < x < b$ is called the **open interval** between a and b.

The collection of numbers x such that $a \leq x \leq b$ is called the **closed interval** between a and b. We denote this closed interval by the symbols $[a, b]$. (A single point will also be called a closed interval.)

If we wish to include only one end point, we shall say that the interval is **half-closed**. We have of course two half-closed intervals, namely the one consisting of the numbers x with $a \leq x < b$, and the other one consisting of the numbers x with $a < x \leq b$.

Sometimes, if a is a number, we call the collection of numbers $x > a$ (or $x < a$) an open interval. The context will always make this clear.

Let f be a function, and c a number at which f is defined.

Definition. We shall say that c is a **maximum point** of the function f if and only if

$$f(c) \geq f(x)$$

for all numbers x at which f is defined. If the condition $f(c) \geq f(x)$ holds for all numbers x in some interval, then we say that **the function has a maximum at c in that interval**. We call $f(c)$ a **maximum value**.

Example 1. Let $f(x) = \sin x$. Then f has a maximum at $\pi/2$ because $f(\pi/2) = 1$ and $\sin x \leq 1$ for all values of x. This is illustrated in Fig. 4. Note that $-3\pi/2$ is also a maximum for $\sin x$.

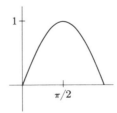

Figure 4

Example 2. Let $f(x) = 2x$, and view f as a function defined only on the interval

$$0 \leq x \leq 2.$$

Then the function has a maximum at 2 in this interval because $f(2) = 4$ and $f(x) \leqq 4$ for all x in the interval. This is illustrated in Fig. 5.

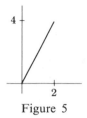

Figure 5

Example 3. Let $f(x) = 1/x$. We know that f is not defined for $x = 0$. This function has no maximum. It becomes arbitrarily large when x comes close to 0 and $x > 0$. This is illustrated in Fig. 6.

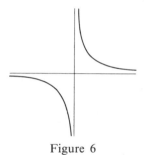

Figure 6

Definition. A **minimum point** for f is a number c such that

$$f(x) \geqq f(c) \qquad \text{for all } x \text{ where } f \text{ is defined.}$$

A **minimum value** for the function is the value $f(c)$, taken at a minimum point.

We illustrate various minima with the graphs of certain functions.

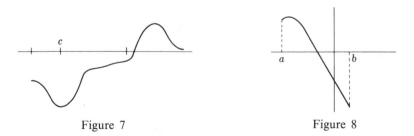

Figure 7   Figure 8

In Fig. 7 the function has a minimum. In Fig. 8 the minimum is at the end point of the interval. In Figs. 3 and 6 the function has no minimum.

In the following picture, the point c_1 looks like a maximum and the point c_2 looks like a minimum, provided we stay close to these points, and don't look at what happens to the curve farther away.

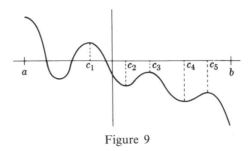

Figure 9

There is a name for such points. We shall say that a point c is a **local minimum** or **relative minimum** of the function f if there exists an interval

$$a_1 < c < b_1$$

such that $f(c) \leqq f(x)$ for all numbers x with $a_1 \leqq x \leqq b_1$.

Similarly, we define the notion of **local maximum** or **relative maximum**. (Do it yourself.) In Fig. 9, the point c_3 is a local maximum, c_4 is a local minimum, and c_5 is a local maximum.

The actual maximum and minimum occur at the end points.

Using basic properties of numbers, one can prove the next theorem which is, however, rather obvious, and so we omit the proof.

Theorem 1.1. *Let f be a continuous function over a closed interval $[a, b]$. Then there exists a point in the interval where f has a maximum, and there exists a point where f has a minimum.*

We wish to have some ideas of the range of values of the given function. The next theorem tells us this informa. on.

Theorem 1.2. Intermediate value theorem. *Let f be a continuous function on the interval $[a, b]$. Let $\alpha = f(a)$ and $\beta = f(b)$. Let γ be a number between α and β. For instance, if $\alpha < \beta$, let $\alpha < \gamma < \beta$, and if $\alpha > \beta$ then let*

$$\alpha > \gamma > \beta.$$

Then there exists a number c such that $a < c < b$ and such that

$$f(c) = \gamma.$$

The theorem is intuitively obvious since a continuous function has no breaks. It is illustrated on the figure.

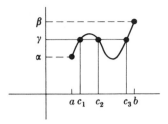

The proof belongs to the range of ideas in the appendix on epsilon-delta and can safely be omitted. Observe that there may be several points c in the interval $[a, b]$ such that $f(c) = \gamma$. In the figure, there are three such points, labeled c_1, c_2, c_3.

As we have mentioned, the point where f has a maximum may occur at the end points of the interval. However, when such a point is not an end point, and the function is differentiable, then we are in a situation similar to that of Fig. 4 or 9, where we see that the tangent to the curve at that point is a horizontal line; in other words the derivative of the function is 0. We can prove this as a theorem.

Theorem 1.3. *Let f be a function which is defined and differentiable in the open interval $a < x < b$. Let c be a number in the interval at which the function has a local maximum or a local minimum. Then*

$$f'(c) = 0.$$

Proof. We give the proof in the case of a local maximum. If we take small values of h (positive or negative), the number $c + h$ will lie in the interval. We are going to find the limit of the Newton quotient as we approach c from the right and from the left, and in that way, determine the value $f'(c)$.

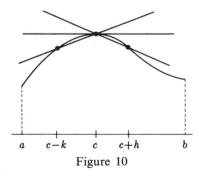

$$a \quad c-k \quad c \quad c+h \quad b$$

Figure 10

Let us first take h positive (see Fig. 10). We must have

$$f(c) \geq f(c + h)$$

no matter what h is (provided h is small) since $f(c)$ is the local maximum. Therefore $f(c + h) - f(c) \leq 0$. Since $h > 0$, the Newton quotient satisfies

$$\frac{f(c + h) - f(c)}{h} \leq 0.$$

Hence the limit is ≤ 0, or in symbols:

$$\lim_{\substack{h \to 0 \\ h > 0}} \frac{f(c + h) - f(c)}{h} \leq 0.$$

Now take h negative, say $h = -k$ with $k > 0$. Then

$$f(c - k) - f(c) \leq 0, \qquad f(c) - f(c - k) \geq 0$$

and the quotient is

$$\frac{f(c - k) - f(c)}{-k} = \frac{f(c) - f(c - k)}{k}.$$

Thus the Newton quotient is ≥ 0. Taking the limit as h (or k) approaches 0, we see that

$$\lim_{\substack{h \to 0 \\ h < 0}} \frac{f(c + h) - f(c)}{h} \geq 0.$$

The only way in which our two limits can be equal is that they should both be 0. Therefore $f'(c) = 0$. This concludes the proof.

We can interpret our arguments geometrically by saying that the line between our two points slants up to the left when we take $h > 0$ and slants up to the right when we take $h < 0$. As h approaches 0, both lines must approach the tangent line to the curve. The only way this is possible is for the tangent line at the point whose x-coordinate is c to be horizontal. This means that its slope is 0, i.e. $f'(c) = 0$.

In practice, a function usually has only a finite number of critical points, and it is easy to find all points c such that $f'(c) = 0$. One can then determine by inspection which of these are maxima, which are minima, and which are neither.

Example 4. Find the critical points of the function $f(x) = x^3 - 1$.

We have $f'(x) = 3x^2$. Hence there is only one critical point, namely $x = 0$, since $3x^2 = 0$ only when $x = 0$.

Example 5. Find the critical points of the function

$$y = x^3 - 2x + 1.$$

The derivative is $3x^2 - 2$. It is equal to 0 precisely when

$$x^2 = \tfrac{2}{3},$$

which means $x = \sqrt{2/3}$ or $-\sqrt{2/3}$. These are the critical points.

We shall find various ways in the next sections to see if the critical point is a local maximum or minimum. In simple cases, sketching the curve, you can often see it by inspection.

V, §1. EXERCISES

Find the critical points of the following functions.

1. $x^2 - 2x + 5$

2. $2x^2 - 3x - 1$

3. $3x^2 - x + 1$

4. $-x^2 + 2x + 2$

5. $-2x^2 + 3x - 1$

6. $x^3 + 2$

7. $x^3 - 3x$

8. $\sin x + \cos x$

9. $\cos x$

10. $\sin x$

V, §2. INCREASING AND DECREASING FUNCTIONS

Let f be a function defined on some interval (which may be open or closed).

Definition. We shall say that f is **increasing** over this interval if

$$f(x_1) \leqq f(x_2)$$

whenever x_1 and x_2 are two points of the interval such that

$$x_1 \leqq x_2.$$

Thus, if a number lies to the right of another, the value of the function at the larger number must be greater than or equal to the value of the function at the smaller number.

In the next figure, we have drawn the graph of an increasing function.

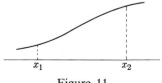

Figure 11

We say that a function defined on some interval is **decreasing** over this interval if

$$f(x_1) \geq f(x_2)$$

whenever x_1 and x_2 are two points of the interval such that $x_1 \leq x_2$.

Observe that a constant function (whose graph is horizontal) is both increasing and decreasing.

If we want to omit the equality sign in our definitions, we shall use the word **strictly** to qualify decreasing or increasing. Thus a function f is **strictly increasing** if

$$x_1 < x_2 \qquad \text{implies} \qquad f(x_1) < f(x_2)$$

and f is **strictly decreasing** if

$$x_1 < x_2 \qquad \text{implies} \qquad f(x_1) > f(x_2)$$

Suppose that a function has a positive derivative throughout an interval, as shown for instance on Fig. 11. Then we can interpret this as meaning that the rate of change of the function is always positive, and therefore that the function is increasing. We state this as a theorem.

Theorem 2.1. *Let f be a function which is continuous in some interval, and differentiable in the interval (excluding the end points).*

If $f'(x) > 0$ in the interval (excluding the end points), then f is strictly increasing.

If $f'(x) < 0$ in the interval (excluding the end points), then f is strictly decreasing.

If $f'(x) = 0$ in the interval (excluding the end points), then f is constant.

In this last statement, the hypothesis that $f'(x) = 0$ in the interval means that the rate of change is 0, and so it is quite plausible that the function is constant. To see how these statements fit into a more formal context, see §3.

Application. Graphs of parabolas

Example. Let us graph the curve

$$y = f(x) = x^2 - 3x + 5,$$

which you should know is a parabola as in Chapter II. We treat the graph here by the method which works in more general cases. First, we have

$$f'(x) = 2x - 3,$$

and

$f'(x) = 0$ if and only if $x = 3/2,$ so $x = 3/2$ is the only critical point.

$f'(x) > 0$ if and only if $2x - 3 > 0$

if and only if $x > 3/2.$

$f'(x) < 0$ if and only if $x < 3/2.$

Thus the function is strictly increasing for $x > 3/2$ and strictly decreasing for $x < 3/2$. Thus by using the derivative, we are able to find the peak of the parabola.

The points where $f(x) = 0$, that is where the graph crosses the x-axis, are given by the quadratic formula:

$$x = \frac{3 \pm \sqrt{9 - 20}}{2} = \frac{3 \pm \sqrt{-11}}{2}.$$

There are no such points. The graph therefore looks like this.

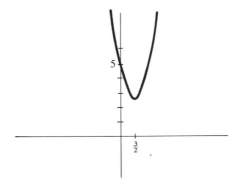

Observe that even if we did not know before the general shape of a parabola, we could deduce it now, and we would know that $x = 3/2$ is a minimum point of the graph. This is because $f(x)$ is strictly decreasing for $x < 3/2$ and strictly increasing for $x > 3/2$. Thus $x = 3/2$ must be a minimum.

Example. Sketch the graph of

$$y = f(x) = x^2 - 5x + 9/4.$$

This time, we have

$$f'(x) = 2x - 5.$$

Hence there is exactly one critical point, namely:

$$f'(x) = 0 \quad \text{if and only if} \quad x = 5/2.$$
$$f'(x) > 0 \quad \text{if and only if} \quad 2x - 5 > 0$$
$$\text{if and only if} \quad x > 5/2.$$
$$f'(x) < 0 \quad \text{if and only if} \quad 2x - 5 < 0$$
$$\text{if and only if} \quad x < 5/2.$$

So f is strictly increasing for $x > 5/2$ and strictly decreasing for $x < 5/2$. Hence f has a minimum at $x = 5/2$.

The x-intercepts of the graph of f are

$$x = \frac{5 \pm \sqrt{25 - 9}}{2} = \frac{9}{2} \quad \text{and} \quad \frac{1}{2}.$$

Definition. The x-intercepts of the graph of f are also called the **roots** of f. In the case of a quadratic polynomial the roots are computed by the quadratic formula.

Therefore the graph of f looks like this.

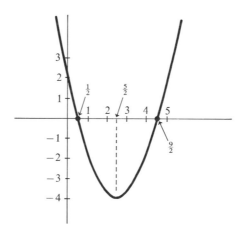

In the above examples, the function was defined for all numbers. In the next example, we look at a function defined only over an interval.

Example. Let $f(x) = x^2 - 5x + 9/4$. Find the minimum and maximum of f for $x \leq 2$.

By the preceding example, we know that f is strictly decreasing for $x \leq 2$. Therefore the minimum is at the end point of the interval, that is $x = 2$, as shown on the figure. Note that at this end point, $f'(2) \neq 0$. Thus the test with critical points is valid only on open intervals.

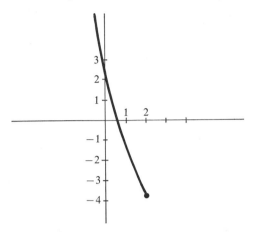

The expression "if and only if" will recur quite frequently. We shall therefore use an abbreviating symbol for it, and we write

$$\Leftrightarrow \qquad \text{to mean,} \qquad \text{"if and only if."}$$

Thus we could write the assertion:

$$x^2 = 3 \quad \Leftrightarrow \quad x = \sqrt{3} \quad \text{or} \quad x = -\sqrt{3}.$$

Similarly,

$$x^2 - 3x + 1 = 0 \quad \Leftrightarrow \quad x = \frac{3 \pm \sqrt{5}}{2}.$$

Example. Show that among all rectangles of given area, the one with least perimeter is a square.

Let a be the given area, and let x be the length of one side of the possible rectangle with area a. We shall express the perimeter as a function $f(x)$. We then differentiate with respect to x, keeping in mind that a is constant, and this will give us a value for x which will show that the rectangle is a square. To carry this out, we take $0 < x$ because the side of an actual rectangle cannot be 0 or have negative length. If y is the length of the other side, then $xy = a$, so that $y = a/x$ is the length of the other side. Hence the perimeter is

$$f(x) = 2\left(x + \frac{a}{x}\right)$$

We have

$$f'(x) = 2\left(1 - \frac{a}{x^2}\right),$$

and:

$$f'(x) = 0 \quad \Leftrightarrow \quad \frac{x^2 - a}{x^2} = 0$$

$$\Leftrightarrow x^2 - a = 0$$

$$\Leftrightarrow x^2 = a$$

$$\Leftrightarrow x = \sqrt{a},$$

because we consider only $x > 0$. Thus the only critical point of f for $x > 0$ is when $x = \sqrt{a}$. Furthermore $x^2 > 0$ for all $x \neq 0$. Hence the fraction $(x^2 - a)/x^2$ is positive if and only if its numerator $x^2 - a$ is positive. Thus:

$$f'(x) > 0 \quad \Leftrightarrow \quad x^2 - a > 0 \quad \Leftrightarrow \quad x^2 > a \quad \Leftrightarrow \quad x > \sqrt{a},$$

$$f'(x) < 0 \quad \Leftrightarrow \quad x^2 - a < 0 \quad \Leftrightarrow \quad x^2 < a \quad \Leftrightarrow \quad x < \sqrt{a},$$

Hence:

$$f \text{ is strictly increasing} \quad \Leftrightarrow \quad x > \sqrt{\alpha},$$

$$f \text{ is strictly decreasing} \quad \Leftrightarrow \quad x < \sqrt{\alpha}.$$

Hence finally $x = \sqrt{a}$ is a minimum for f. When $x = \sqrt{a}$ we have $y = \sqrt{a}$ also, because

$$y = a/x = a/\sqrt{a}.$$

This proves that the rectangle is a square.

Example. Show that among all rectangular fences with given length, the one encompassing the largest area must be a square.

To do this, let c be the fixed length, and let x be one of the sides. If y is the other side, then

$$2x + 2y = c,$$

so that $y = (c - 2x)/2$. Therefore the area encompassed by the fence is equal to

$$xy = \frac{x(c - 2x)}{2} = \frac{xc - 2x^2}{2} = A(x).$$

This area $A(x)$ is a function of x, which has a critical point when $A'(x) = 0$. But

$$A'(x) = \tfrac{1}{2}(c - 4x).$$

Thus $A'(x) = 0$ if and only if $c = 4x$, that is, $x = c/4$ is the only critical point.

We must now see that this is a maximum. The function

$$A(x) = \frac{xc - 2x^2}{2}$$

has a graph which is a parabola. When x becomes large positive or negative, then $A(x)$ becomes large negative. Hence the parabola is shaped as in the following figure.

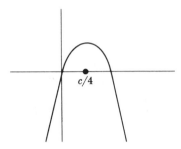

It has only one maximum, and that maximum must therefore be at $x = c/4$. We then find that $y = c/4$ also. In other words, the fence must be a square.

Inequalities

The derivative test for increasing and decreasing functions can also be used to prove inequalities.

Example. Prove that $\sin x < x$ for all $x > 0$.
Let $f(x) = x - \sin x$. Then

$$f'(x) = 1 - \cos x.$$

First take $0 < x < \pi/2$. Then $f'(x) > 0$ because $\cos x < 1$ in this interval. Therefore $f(x)$ is strictly increasing for $0 \le x \le \pi/2$. But

$$f(0) = 0 - \sin 0 = 0.$$

Hence we must have $f(x) > 0$ for $0 < x \le \pi/2$.

If $x \geq \pi/2$, then $x > 1$ (because π is approximately 3.14), and so $\sin x < x$ whenever $x > \pi/2$. Thus the desired inequality holds for simpler reasons when $x > \pi/2$.

The preceding example illustrates a technique which is used for proving certain inequalities between functions. In general:

> *Suppose we have two functions f and g over a certain interval $[a, b]$ and we assume that f, g are differentiable. Suppose that*
>
> $$f(a) \leq g(a),$$
>
> *and that $f'(x) \leq g'(x)$ throughout the interval. Then $f(x) \leq g(x)$ in the interval.*

Proof. We let

$$h(x) = g(x) - f(x).$$

Then by assumption,

$$h'(x) = g'(x) - f'(x) \geq 0,$$

so h is increasing throughout the interval. Since

$$h(a) = g(a) - f(a) \geq 0,$$

it follows that $h(x) \geq 0$ throughout the interval, whence

$$g(x) \geq f(x).$$

The principle just stated can be visualized in the following picture, drawn for the case when $f(a) = g(a)$.

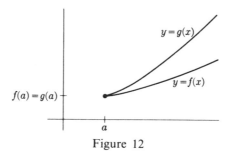

Figure 12

In other words, if g is bigger or equal to f at $x = a$, and if g grows faster than f, then $g(x)$ is bigger than $f(x)$ for all $x > a$.

Example. Show that for any integer $n \geq 1$ and any number $x \geq 1$ one has the inequality

$$x^n - 1 \geq n(x - 1).$$

Let $f(x) = x^n - 1 - n(x - 1)$. Then

$$f'(x) = nx^{n-1} - n.$$

Since $x \geq 1$ it follows that $x^{n-1} \geq 1$ and so $f'(x) \geq 0$. Hence f is increasing for $x \geq 1$. But $f(1) = 0$. Hence $f(x) \geq 0$ for $x \geq 1$. This is equivalent to the desired inequality.

On the other hand, the next theorem tells us what happens if two functions have the same derivative throughout an interval.

Constants

Theorem 2.2. *Let $f(x)$ and $g(x)$ be two functions which are differentiable in some interval and assume that*

$$f'(x) = g'(x)$$

for all x in the interval. Then there is a constant C such that

$$f(x) = g(x) + C$$

for all x in the interval.

Proof. Let $h(x) = f(x) - g(x)$ be the difference of our two functions. Then

$$h'(x) = f'(x) - g'(x) = 0.$$

Hence $h(x)$ is constant by Theorem 2.1, that is $h(x) = C$ for some number C and all x. This proves the theorem.

Remark. The theorem is the *converse* of the statement:

If a function is constant, then its derivative is equal to 0.

We shall use Theorem 2.2 in a fundamental way when we come to the chapter on integration.

For the applications of the theorem, see the beginning of the chapter on logarithms, and also the beginning of Chapter X, §1. We give simpler applications here.

Example. Let f be a function of x such that $f'(x) = 5$. Suppose that $f(0) = 2$. Determine $f(x)$ completely.

We know from past experience that the function

$$g(x) = 5x$$

has the derivative

$$g'(x) = 5.$$

Hence there is constant C such that

$$f(x) = 5x + C.$$

We are also given $f(0) = 2$. Hence

$$2 = f(0) = 0 + C.$$

Therefore $C = 2$. Thus finally

$$f(x) = 5x + 2.$$

Example. A particle moves on the x-axis toward the left at a rate of 5 cm/sec. At time $t = 5$ the particle is at the point 8 cm to the right of the origin. Determine the x-coordinate $x = f(t)$ completely as a function of time.

We are given

$$\frac{dx}{dt} = f'(t) = -5.$$

Let $g(t) = -5t$. Then $g'(t) = -5$ also. Hence there is a constant C such that

$$f(t) = -5t + C.$$

But we are also given $f(5) = 8$. Hence

$$8 = -5 \cdot 5 + C = -25 + C.$$

Therefore $C = 8 + 25 = 33$, so finally

$$f(t) = -5t + 33.$$

V, §2. EXERCISES

Determine the intervals on which the following functions are increasing and decreasing.

1. $f(x) = x^3 + 1$

2. $f(x) = x^2 - x + 5$

3. $f(x) = x^3 + x - 2$

4. $f(x) = -x^3 + 2x + 1$

5. $f(x) = 2x^3 + 5$

6. $f(x) = 5x^2 + 1$

7. $f(x) = -4x^3 - 2x$

8. $f(x) = 5x^3 + 6x$

Sketch the graphs of the following parabolas. Determine the critical point in each case.

9. $f(x) = x^2 - x - 1$

10. $f(x) = x^2 + x + 1$

11. $f(x) = -x^2 + x - 1$

12. $f(x) = -x^2 - x - 1$

13. $f(x) = x^2 + 3x + 1$

14. $f(x) = x^2 - 5x + 1$

15. $f(x) = -2x^2 + 4x - 1$

16. $f(x) = 2x^2 - 4x - 3$

For each of the following functions, find the maximum, minimum for all x in the given interval.

17. $x^2 - 2x - 8$, $[0, 4]$

18. $x^2 - 2x + 1$, $[-1, 4]$

19. $4 - 4x - x^2$, $[-1, 4]$

20. $x - x^2$, $[-1, 2]$

21. $3x - x^3$, $[-2, \sqrt{3}]$

22. $(x - 4)^5$, $[3, 6]$

23. The following steps show how to prove inequalities for the sine and cosine. We start with the inequality proved as an example in the text, namely

(a) $\sin x \leq x$ for all $x \geq 0$.

Let $f_1(x) = x - \sin x$. Then this inequality is equivalent with the inequality

(1) $f_1(x) \geq 0$ for all $x \geq 0$.

Now prove:

(b) $1 - \dfrac{x^2}{2} \leq \cos x$ for $x \geq 0$.

[*Hint*: Let $f_2(x) = \cos x - \left(1 - \dfrac{x^2}{2}\right)$ and use (1), to prove

(2) $f_2(x) \geq 0$ for all $x \geq 0$.]

(c) $x - \dfrac{x^3}{3 \cdot 2} \leq \sin x$. $\left[\textit{Hint}:\ \text{Let } f_3(x) = \sin x - \left(x - \dfrac{x^3}{3 \cdot 2}\right). \right]$

(d) $\cos x \leq 1 - \dfrac{x^2}{2} + \dfrac{x^4}{4 \cdot 3 \cdot 2}$

(e) $\sin x \leq x - \dfrac{x^3}{3 \cdot 2} + \dfrac{x^5}{5 \cdot 4 \cdot 3 \cdot 2}$

24. Prove that $\tan x > x$ if $0 < x < \pi/2$.

25. (a) Prove that

$$t + \frac{1}{t} \geq 2 \quad \text{for} \quad t > 0.$$

   [*Hint*: Let $f(t) = t + 1/t$. Show that f is strictly decreasing for $0 < t \leq 1$ and f is strictly increasing for $1 \leq t$. What is $f(1)$?]
   (b) Let a, b be two positive numbers. Let

$$f(x) = ax + \frac{b}{x} \quad \text{for} \quad x > 0.$$

   Show that the minimum value of f is $2\sqrt{ab}$.

26. A box with open top is to be made with a square base and a constant surface C. Determine the sides of the box if the volume is to be a maximum.

27. A container in the shape of a cylinder with open top is to have a fixed surface area C. Find the radius of its base and its height if it is to have maximum volume.

28. Do the above two problems when the box and the container are closed at the top. (The area of a circle of radius x is πx^2 and its length is $2\pi x$. The volume of a cylinder whose base has radius x and of height y is $\pi x^2 y$.)

29. Assume that there is a function $f(x)$ such that $f(x) \neq 0$ for all x, and $f'(x) = f(x)$. Let $g(x)$ be any function such that $g'(x) = g(x)$. Show that there is a constant C such that $g(x) = Cf(x)$. [*Hint*: Differentiate the quotient g/f.]

30. Suppose that f is a differentiable function of t such that (a) $f'(t) = -3$, (b) $f'(t) = 2$. What can you say about $f(t)$?

31. Suppose that $f'(t) = -3$ and $f(0) = 1$. Determine $f(t)$ completely.

32. Suppose that $f'(t) = 2$ and $f(0) = -5$. Determine $f(t)$ completely.

33. A particle is moving on the x-axis toward the right at a constant speed of 7 ft/sec. If at time $t = 9$ the particle is at a distance 2 ft to the right of the origin, find its x-coordinate as a function of t.

34. Water is dripping out of a vertical tank so that the height of the water is falling at a rate of 2 ft/day. When the tank is full, the height of the water is 30 ft. Find explicitly the height of the water as a function of time.

V, §3. THE MEAN VALUE THEOREM

The theorems in this section are fairly obvious intuitively, and therefore you might omit the proofs of Rolle's theorem and Theorem 3.2 if you wish, **after understanding their statement.**

First suppose we have a function over a closed interval $[a, b]$, whose graph looks like this.

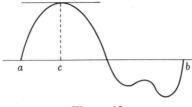

Figure 13

Then we have the following theorem about this function.

Theorem 3.1. Rolle's theorem. *Let a, b be two numbers, $a < b$. Let f be a function which is continuous over the closed interval*

$$a \leqq x \leqq b$$

and differentiable on the open interval $a < x < b$. Assume that

$$f(a) = f(b) = 0.$$

Then there exists a point c such that

$$a < c < b$$

and such that $f'(c) = 0$.

Proof. If the function is constant in the interval, then its derivative is 0 and any point in the open interval $a < x < b$ will do.

If the function is not constant, then there exists some point in the interval where the function is not 0, and this point cannot be one of the end points a or b. Suppose that some value of our function is positive. By Theorem 1.1, the function has a maximum at a point c. Then $f(c)$ must be greater than 0, and c cannot be either one of the end points because $f(a) = f(b) = 0$. Consequently

$$a < c < b.$$

By Theorem 1.3, we must have $f'(c) = 0$. This proves our theorem in case the function is positive somewhere in the interval.

If the function is negative for some number in the interval, then we use Theorem 1.1 to get a minimum, and we argue in a similar way, using Theorem 1.3 (applied to a minimum). (Write out the argument in full as an exercise.)

Let $f(x)$ be a function which is differentiable for $a < x < b$, and continuous in the closed interval

$$a \leqq x \leqq b.$$

We continue to assume throughout that $a < b$. This time we do not assume, as in Theorem 3.1, that $f(a) = f(b) = 0$. We shall prove that there exists a point c between a and b such that the slope of the tangent line at $(c, f(c))$ is the same as the slope of the line between the end points of our graph. In other words, the tangent line is parallel to the line passing through the end points of our graph.

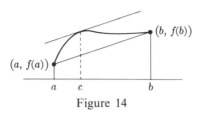

Figure 14

The slope of the line between the two end points is

$$\frac{f(b) - f(a)}{b - a}$$

because the coordinates of the end points are $(a, f(a))$ and $(b, f(b))$ respectively. Thus we have to find a point c such that

$$f'(c) = \frac{f(b) - f(a)}{b - a}.$$

Theorem 3.2. Mean value theorem. *Let $a < b$ as before. Let f be a function which is continuous in the closed interval $a \leqq x \leqq b$, and differentiable in the interval $a < x < b$. Then there exists a point c such that $a < c < b$ and*

$$f'(c) = \frac{f(b) - f(a)}{b - a}.$$

Proof. The equation of the line between the two end points is

$$y = \frac{f(b) - f(a)}{b - a} (x - a) + f(a).$$

Indeed, the slope

$$\frac{f(b) - f(a)}{b - a}$$

is the coefficient of x. When $x = a$, $y = f(a)$. Hence we have written down the equation of the line having the given slope and passing through a given point. When $x = b$, we note that $y = f(b)$.

We now consider geometrically the difference between $f(x)$ and the straight line. This difference becomes 0 at the end points. This geometric idea will allow us to apply Rolle's theorem. In other words, we consider the function

$$g(x) = f(x) - \frac{f(b) - f(a)}{b - a} (x - a) - f(a).$$

Then

$$g(a) = f(a) - f(a) = 0$$

and

$$g(b) = f(b) - f(b) = 0$$

also.

We can therefore apply Theorem 3.1 to the function $g(x)$. We know that there is a point c between a and b, and not equal to a or b, such that

$$g'(c) = 0.$$

But

$$g'(x) = f'(x) - \frac{f(b) - f(a)}{b - a}.$$

Consequently

$$0 = g'(c) = f'(c) - \frac{f(b) - f(a)}{b - a}.$$

This gives us the desired value for $f'(c)$, and concludes the proof.

The point of the mean value theorem is not so much to find explicitly a value c such that

$$f'(c) = \frac{f(b) - f(a)}{b - a},$$

as to use it for theoretical considerations.

Corollary 3.3. *Let f be a function which is differentiable in some interval, and such that $f'(x) = 0$ for all x in the interval. Then f is constant.*

Proof. Let a, b be distinct numbers in the interval. By the mean value theorem, there exists c between a and b such that

$$f'(c) = \frac{f(b) - f(a)}{b - a}.$$

But $f'(c) = 0$ by assumption. Hence $f(b) - f(a) = 0$ and therefore $f(b) = f(a)$. Hence f has the same value at all points of the interval, so f is constant, as was to be shown.

Similarly, we may now give the rest of the proof of Theorem 2.1 from the preceding section:

Corollary 3.4. *If $f'(x) > 0$ for x in an interval, excluding the end points, and f is continuous in the interval, then f is strictly increasing.*

Proof. Let x_1 and x_2 be two points of the interval, and suppose $x_1 < x_2$. By the mean value theorem, there exists a point c such that $x_1 < c < x_2$ and

$$f'(c) = \frac{f(x_2) - f(x_1)}{x_2 - x_1}.$$

The difference $x_2 - x_1$ is positive, and we have

$$f(x_2) - f(x_1) = (x_2 - x_1)f'(c).$$

If the derivative $f'(x)$ is > 0 for all x in the interval, excluding the end points, then $f'(c) > 0$ (because c is in the interval). Hence the product $(x_2 - x_1)f'(c)$ is positive, and $f(x_2) - f(x_1) > 0$, so that

$$f(x_1) < f(x_2).$$

This proves that the function is increasing.

We leave the proof of the assertion concerning decreasing functions as an exercise.

When we study Taylor's formula, we shall use mean value theorems to estimate various functions. Even though we don't know the exact value $f'(c)$, we still may have an estimate for the derivative. For instance the functions $\sin x$ and $\cos x$ can both be estimated in absolute value by 1. In many applications, this is all that matters.

CHAPTER VI

Sketching Curves

We have developed enough techniques to be able to sketch curves and graphs of functions much more efficiently than before. We shall investigate systematically the behavior of a curve, and the mean value theorem will play a fundamental role.

We shall especially look for the following aspects of the curve:

1. Intersections with the coordinate axes.
2. Critical points.
3. Regions of increase.
4. Regions of decrease.
5. Maxima and minima (including the local ones).
6. Behavior as x becomes large positive and large negative.
7. Values of x near which y becomes large positive or large negative.

These seven pieces of information will be quite sufficient to give us a fairly accurate idea of what the graph looks like. We shall devote a section to considering one other aspect, namely:

8. Regions where the curve is bending up or down.

VI, §1. BEHAVIOR AS *x* BECOMES VERY LARGE

Suppose we have a function f defined for all sufficiently large numbers. Then we get substantial information concerning our function by investigating how it behaves as x becomes large.

For instance, $\sin x$ oscillates between -1 and $+1$ no matter how large x is.

However, polynomials don't oscillate. When $f(x) = x^2$, as x becomes large positive, so does x^2. Similarly with the function x^3, or x^4 (etc.). We consider this systematically.

Parabolas

Example 1. Consider a parabola,

$$y = ax^2 + bx + c,$$

with $a \neq 0$. There are two essential cases, when $a > 0$ and $a < 0$. We shall see that the parabola looks like those drawn in the figure.

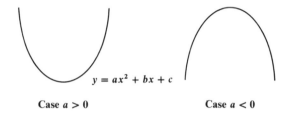

$$y = ax^2 + bx + c$$

Case $a > 0$ **Case $a < 0$**

We look at numerical examples.

Example 2. Sketch the graph of the curve

$$y = f(x) = -3x^2 + 5x - 1.$$

We recognize this as a parabola. Factoring out x^2 shows that

$$f(x) = x^2 \left(-3 + \frac{5}{x} - \frac{1}{x^2} \right).$$

When x is large positive or negative, then x^2 is large positive and the factor on the right is close to -3. Hence $f(x)$ is large negative. This means that the parabola has the shape as shown on the figure.

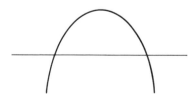

We have $f'(x) = -6x + 5$. Thus $f'(x) = 0$ if and only if $-6x + 5 = 0$, or in other words,

$$x = \tfrac{5}{6}.$$

There is exactly one critical point. We have

$$f(\tfrac{5}{6}) = -3(\tfrac{5}{6})^2 + \tfrac{25}{6} - 1 > 0.$$

The critical point is a maximum, because we have already seen that the parabola bends down.

The curve crosses the x-axis exactly when

$$-3x^2 + 5x - 1 = 0.$$

By the quadratic formula (see Chapter II, §8), this is the case when

$$x = \frac{-5 \pm \sqrt{25 - 12}}{-6} = \frac{5 \pm \sqrt{13}}{6}.$$

Hence the graph of the parabola looks as on the figure.

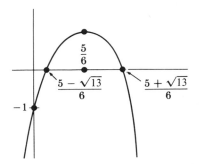

The same principle applies to sketching any parabola.

(i) Looking at what happens when x becomes large positive or negative tells us whether the parabola bends up or down.

(ii) A quadratic function

$$f(x) = ax^2 + bx + c, \qquad \text{with} \quad a \neq 0$$

has only one critical point, when

$$f'(x) = 2ax + b = 0$$

so when

$$x = -b/2a.$$

Knowing whether the parabola bends up or down tells us whether the critical point is a maximum or minimum, and the value $x = -b/2a$ tells us exactly where this critical point lies.

(iii) The points where the parabola crosses the x-axis are determined by the quadratic formula.

Example 3. Cubics. Consider a polynomial

$$f(x) = x^3 + 2x - 1.$$

We can write it in the form

$$x^3\left(1 + \frac{2}{x^2} - \frac{1}{x^3}\right).$$

When x becomes very large, the expression

$$1 + \frac{2}{x^2} - \frac{1}{x^3}$$

approaches 1. In particular, given a small number $\delta > 0$, we have, for all x sufficiently large, the inequality

$$1 - \delta < 1 + \frac{2}{x^2} - \frac{1}{x^3} < 1 + \delta.$$

Therefore $f(x)$ satisfies the inequality

$$x^3(1 - \delta) < f(x) < x^3(1 + \delta).$$

This tells us that $f(x)$ behaves very much like x^3 when x is very large. In particular:

If x becomes large positive, then $f(x)$ becomes large positive.

If x becomes large negative, then $f(x)$ becomes large negative.

A similar argument can be applied to any polynomial.

It is convenient to use an abbreviation for the expression "become large positive." Instead of saying x becomes large positive, we write

$$x \to \infty$$

and also say that x **approaches,** or **goes to infinity. Warning: there is no number called infinity**. The above symbols merely abbreviate the notion of becoming large positive. We have a similar notation for x becoming large negative, when we write

$$x \to -\infty$$

and say that x **approaches minus infinity**. Thus in the case when

$$f(x) = x^3 + 2x - 1,$$

we can assert:

$$\text{If } x \to \infty \quad \text{then } f(x) \to \infty.$$
$$\text{If } x \to -\infty \text{ then } f(x) \to -\infty.$$

Example 4. Consider a quotient of polynomials like

$$Q(x) = \frac{x^3 + 2x - 1}{2x^3 - x + 1}.$$

We factor out the highest power of x from the numerator and denominator, and therefore write $Q(x)$ in the form

$$Q(x) = \frac{x^3(1 + 2/x^2 - 1/x^3)}{x^3(2 - 1/x^2 + 1/x^3)} = \frac{1 + 2/x^2 - 1/x^3}{2 - 1/x^2 + 1/x^3}.$$

As x becomes very large, the numerator approaches 1 and the denominator approaches 2. Thus our fraction approaches $\frac{1}{2}$. We may express this in the form

$$\lim_{x \to \infty} Q(x) = \frac{1}{2}.$$

Or we may write:

$$\text{If } x \to \pm \infty \text{ then } Q(x) \to \tfrac{1}{2}.$$

Example 5. Consider the quotient

$$Q(x) = \frac{x^2 - 1}{x^3 - 2x + 1}.$$

Does it approach a limit as x becomes very large?

We write

$$Q(x) = \frac{x^2(1 - 1/x^2)}{x^3(1 - 2/x^2 + 1/x^3)}$$

$$= \frac{1}{x} \frac{1 - 1/x^2}{1 - 2/x^2 + 1/x^3}.$$

As x becomes large, the term $1/x$ approaches 0, and the other factor approaches 1. Hence $Q(x)$ approaches 0 as x becomes large negative or positive.

We may also write

$$\text{If } x \to \pm\infty \text{ then } Q(x) \to 0,$$

or

$$\lim_{x \to \pm\infty} Q(x) = 0.$$

Example 6. Consider the quotient

$$Q(x) = \frac{x^3 - 1}{x^2 + 5}$$

and determine what happens when x becomes large.

We write

$$Q(x) = \frac{x^3(1 - 1/x^3)}{x^2(1 + 5/x^2)}$$

$$= x \frac{1 - 1/x^3}{1 + 5/x^2}.$$

As x becomes large, positive or negative, the quotient

$$\frac{1 - 1/x^3}{1 + 5/x^2}$$

approaches 1. Hence $Q(x)$ differs from x by a factor near 1. Hence $Q(x)$ becomes large positive when x is large positive, and becomes large negative when x is large negative. We may express this by saying:

$$\text{If } x \to \infty \quad \text{then } Q(x) \to \infty.$$

$$\text{If } x \to -\infty \text{ then } Q(x) \to -\infty.$$

We may also write these assertions in the form of a limit:

$$\lim_{x \to \infty} Q(x) = \infty \quad \text{and} \quad \lim_{x \to -\infty} Q(x) = -\infty.$$

However, even though we use this notation, and may say that the limit of $Q(x)$ is $-\infty$ when x becomes large negative, we emphasize that $-\infty$ is not a number, and so this limit is not quite the same as when the limit is a number. It is correct to say that there is no number which is the limit of $Q(x)$ as $x \to \infty$ or as $x \to -\infty$.

These four examples are typical of what happens when we deal with quotients of polynomials.

Later when we deal with exponents and logarithms, we shall again meet the problem of comparing the quotient of two expressions which become large. There will be a common ground for some of the arguments, summarized by the following table:

> Large positive times large positive is large positive.
> Large positive times large negative is large negative.
> Large negative times large negative is large positive.
> Small positive times large positive: you can't tell without knowing more information.

VI, §1. EXERCISES

Find the limits of the following quotients $Q(x)$ as x becomes large positive or negative. In other words, find

$$\lim_{x \to \infty} Q(x) \qquad \text{and} \qquad \lim_{x \to -\infty} Q(x).$$

1. $\dfrac{2x^3 - x}{x^4 - 1}$

2. $\dfrac{\sin x}{x}$

3. $\dfrac{\cos x}{x}$

4. $\dfrac{x^2 + 1}{\pi x^2 - 1}$

5. $\dfrac{\sin 4x}{x^3}$

6. $\dfrac{5x^4 - x^3 + 3x + 2}{x^3 - 1}$

7. $\dfrac{-x^2 + 1}{x + 5}$

8. $\dfrac{2x^4 - 1}{-4x^4 + x^2}$

9. $\dfrac{2x^4 - 1}{-4x^3 + x^2}$

10. $\dfrac{2x^4 - 1}{-4x^5 + x^2}$

Describe the behavior of the following polynomials as x becomes large positive and large negative.

11. $x^3 - x + 1$

12. $-x^3 - x + 1$

13. $x^4 + 3x^3 + 2$

14. $-x^4 + 3x^3 + 2$

15. $2x^5 + x^2 - 100$

16. $-3x^5 + x + 1000$

17. $10x^6 - x^4$

18. $-3x^6 + x^3 + 1$

19. A function $f(x)$ which can be expressed as follows:

$$f(x) = a_n x^n + a_{n-1} x^{n-1} + \cdots + a_0,$$

where n is a positive integer and the $a_n, a_{n-1}, \ldots, a_0$ are numbers, is called a polynomial. If $a_n \neq 0$, then n is called the **degree** of the polynomial. Describe the behavior of $f(x)$ as x becomes large positive or negative, n is odd or even, and $a_n > 0$ or $a_n < 0$. You will have eight cases to consider. Fill out the following table.

n	a_n	$x \to \infty$	$x \to -\infty$
Odd	> 0	$f(x) \to ?$	$f(x) \to ?$
Odd	< 0	$f(x) \to ?$	$f(x) \to ?$
Even	> 0	$f(x) \to ?$	$f(x) \to ?$
Even	< 0	$f(x) \to ?$	$f(x) \to ?$

20. Using the intermediate value theorem, show that any polynomial of odd degree has a root.

VI, §2. BENDING UP AND DOWN

Let a, b be numbers, $a < b$. Let f be a continuous function defined on the interval $[a, b]$. Assume that f' and f'' exist on the interval $a < x < b$. We view the second derivative f'' as the rate of change of the slope of the curve $y = f(x)$ over the interval. If the second derivative is positive in the interval $a < x < b$, then the slope of the curve is increasing, and we interpret this as meaning that the curve is **bending up**. If the second derivative is negative, we interpret this as meaning that the curve is **bending down**. The following two figures illustrate this.

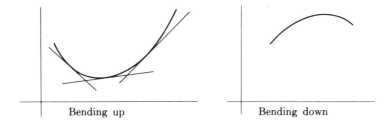

Bending up   Bending down

Example 1. The curve $y = x^2$ is bending up. We can see this using the second derivative. Let $f(x) = x^2$. Then $f''(x) = 2$, and the second derivative is always positive. The present considerations justify drawing the curve as we have always done, i.e. bending up.

Example 2. Let $f(x) = \sin x$. We have $f''(x) = -\sin x$, and thus $f''(x) > 0$ on the interval $\pi < x < 2\pi$. Hence the curve is bending up on this interval. Similarly, $f''(x) < 0$ on the interval $0 < x < \pi$. Hence the curve is bending down on this interval, as shown on the next figures. Of course, this merely justifies the drawings which we have always made for the graph of the sine function.

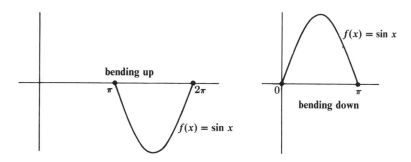

Example 3. Determine the intervals where the curve

$$y = -x^3 + 3x - 5$$

is bending up and bending down.

Let $f(x) = -x^3 + 3x - 5$. Then $f''(x) = -6x$. Thus:

$$f''(x) > 0 \quad \Leftrightarrow \quad x < 0,$$
$$f''(x) < 0 \quad \Leftrightarrow \quad x > 0.$$

Hence f is bending up if and only if $x < 0$; and f is bending down if and only if $x > 0$. The graph of this curve will be discussed fully in the next section when we graph cubics systematically.

A point where a curve changes its behavior from bending up to down (or vice versa) is called an **inflection point**. If the curve is the graph of a function f whose second derivative exists and is continuous, then we must have $f''(x) = 0$ at that point. The following picture illustrates this:

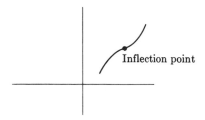

In Example 3 above, the point $(0, -5)$ is an inflection point.

The determination of regions of bending up or down and inflection points gives us worthwhile pieces of information concerning curves. For instance, knowing that a curve in a region of decrease is actually bending down tells us that the decrease occurs essentially as in this example:

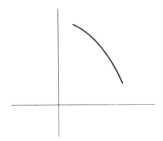

and not as in these examples:

The second derivative can also be used as a test whether a critical point is a **local** maximum or minimum.

Second derivative test. *Let f be twice continuously differentiable on an open interval, and suppose that c is a point where*

$$f'(c) = 0 \qquad and \qquad f''(c) > 0.$$

Then c is a local minimum point of f. On the other hand, if

$$f''(c) < 0$$

then c is a local maximum point of f.

To see this, suppose that $f''(c) > 0$. Then $f''(x) > 0$ for all x close to c because we assumed that the second derivative is continuous. Thus the curve is bending up. Consequently the picture of the graph of f is as on Fig. 1(a) and c is a local minimum.

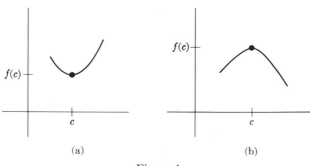

Figure 1

A similar argument shows that if $f''(c) < 0$ then c is a local maximum as on Fig. 1(b).

VI, §2. EXERCISES

1. Determine all inflection points of $\sin x$.

2. Determine all inflection points of $\cos x$.

3. Determine the inflection points of $f(x) = \tan x$ for $-\pi/2 < x < \pi/2$.

4. Sketch the curve $y = \sin^2 x$. Determine the critical points and the inflection points. Compare with the graph of $|\sin x|$.

5. Sketch the curve $y = \cos^2 x$. Determine the critical points and the inflection points. Compare with the graph of $|\cos x|$.

Determine the inflection points and the intervals of bending up and bending down for the following curves.

6. $y = x + \dfrac{1}{x}$

7. $y = \dfrac{x}{x^2 + 1}$

8. $y = \dfrac{x}{x^2 - 1}$

9. Sketch the curve $y = f(x) = \sin x + \cos x$ for $0 \leq x \leq 2\pi$. First plot all values $f(n\pi/4)$ with $n = 0, 1, 2, 3, 4, 5, 6, 7, 8$. Then determine all the critical points. Then determine the regions of increase and decrease. Then determine the inflection points, and the regions where the curve bends up or down.

VI, §3. CUBIC POLYNOMIALS

We can now sketch the graphs of cubic polynomials systematically.

Example 1. Sketch the graph of $f(x) = x^3 - 2x + 1$.

1.

> If $x \to \infty$ then $f(x) \to \infty$ by §1.
>
> If $x \to -\infty$ then $f(x) \to -\infty$ by §1.

2. We have $f'(x) = 3x^2 - 2$. Thus

$$f'(x) = 0 \quad \Leftrightarrow \quad x = \pm\sqrt{2/3}.$$

The critical points of f are $x = \sqrt{2/3}$ and $x = -\sqrt{2/3}$.

3. Let $g(x) = f'(x) = 3x^2 - 2$. Then the graph of g is a parabola, and the x-intercepts of the graph of g are precisely the critical points of f. (Do not confuse the functions f and $f' = g$.) The graph of g is a parabola bending up, as follows.

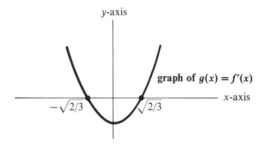

Therefore:

$$f'(x) > 0 \quad \Leftrightarrow \quad x > \sqrt{2/3} \quad \text{and} \quad x < -\sqrt{2/3}, \quad \text{where} \quad g(x) > 0$$

and f is strictly increasing on the intervals

$$x \geqq \sqrt{2/3} \qquad \text{and} \qquad x \leqq -\sqrt{2/3}.$$

Similarly:

$$f'(x) < 0 \quad \Leftrightarrow \quad -\sqrt{2/3} < x < \sqrt{2/3}, \quad \text{where} \quad g(x) < 0,$$

and f is strictly decreasing on this interval. Therefore $-\sqrt{2/3}$ is a local maximum for f, and $\sqrt{2/3}$ is a local minimum.

4. $f''(x) = 6x$, and $f''(x) > 0$ if and only if $x > 0$. Also $f''(x) < 0$ if and only if $x < 0$. Therefore f is bending up for $x > 0$ and bending down for $x < 0$. There is an inflection point at $x = 0$.

Putting all this together, we find that the graph of f looks like this.

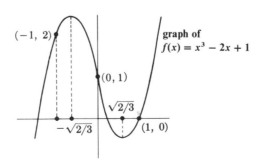

Observe how we used a quadratic polynomial, namely $f'(x) = 3x^2 - 2$, as an intermediate step in the arguments.

Remark 1. Instead of using the quadratic polynomial, we can also argue as follows, after we know that the only critical points of f are $x = \sqrt{2/3}$ and $x = -\sqrt{2/3}$. Consider the interval $x < -\sqrt{2/3}$. Then $f'(x) \neq 0$ for all $x < \sqrt{-2/3}$. Hence $f'(x)$ is either > 0 for all $x < -\sqrt{2/3}$, or $f'(x) < 0$ for all $x < -\sqrt{2/3}$ by the intermediate value theorem. Which is it? We just try one value, say with $x = -10$, to see that $f'(x) > 0$ for $x < -\sqrt{2/3}$, because $f'(-10) = 3 \cdot 10^2 - 2 = 298$. Hence we must have $f'(x) > 0$ for $x < -\sqrt{2/3}$.

Remark 2. For a cubic polynomial it is much more difficult to determine the roots, that is the x-intercepts, and we usually do not do so, unless there is a simple way of doing it, by accident. In the above case when

$$f(x) = x^3 - 2x + 1,$$

there is such an accident, since $f(1) = 0$. Therefore 1 is a root of f. Hence $f(x)$ factors

$$x^3 - 2x + 1 = (x - 1)(x^2 + x - 1).$$

The other roots of f are the roots of $x^2 + x - 1$, which can be found by the quadratic formula:

$$x = \frac{-1 \pm \sqrt{5}}{2}.$$

In the next example, however, there is no such simple way of finding the roots, and we do not find them.

Example 2. Sketch the graph of the curve

$$y = -x^3 + 3x - 5.$$

1.   When $x = 0$, we have $y = -5$. With polynomials of degree ≥ 3 there is in general no simple formula for those x such that $f(x) = 0$, so we do not give explicitly the intersection of the graph with the x-axis.
2.   The derivative is

$$f'(x) = -3x^2 + 3.$$

The graph of $f'(x)$ is a parabola bending down, as you should know from previous experience with parabolas. We have

$$f'(x) = 0 \iff x^2 = 1 \iff x = 1 \text{ and } x = -1.$$

Thus there are two critical points of f, namely $x = 1$ and $x = -1$.

3.   The graph of $f'(x)$ looks like a parabola bending down, as follows.

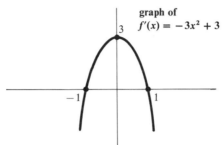

graph of
$f'(x) = -3x^2 + 3$

Then:

f is strictly decreasing $\iff$ $f'(x) < 0$

$\qquad\qquad\qquad\qquad\quad \iff$ $x < -1$ and $x > 1.$

f is strictly increasing $\iff$ $f'(x) > 0$

$\qquad\qquad\qquad\qquad\quad \iff$ $-1 < x < 1.$

Therefore f has a local minimum at $x = -1$, and has a local maximum at $x = 1$.

4.

If $x \to \infty$ then $f(x) \to -\infty$ by §1.

If $x \to -\infty$ then $f(x) \to +\infty$ by §1.

5.   We have $f''(x) = -6x$. Hence $f''(x) > 0$ if and only if $x < 0$ and $f''(x) < 0$ if and only if $x > 0$. There is an inflection point at $x = 0$.

Putting all this information together, we see that the graph of f looks like this:

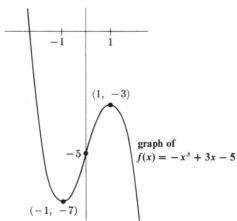

(1, −3)

−5

graph of
$f(x) = -x^3 + 3x - 5$

(−1, −7)

Remark. When f is a polynomial of degree 3, its derivative $f'(x)$ is a polynomial of degree 2, and in general this polynomial has two roots, giving the two critical points of the curve $y = f(x)$. In the preceding example, these critical points are at $(-1, -7)$ and $(1, -3)$.

Again note how we used the graph of a parabola, namely the graph of $f'(x)$, in the process of determining the graph of f itself.

In the last two examples, the cubic polynomial had two bumps, at the two critical points. This is the most general form of cubic polynomials. However, there may be special cases, when there is no critical point, or only one critical point.

Example 3(a). Let $f(x) = 4x^3 + 2$. Sketch the graph of f.

Here we have $f'(x) = 12x^2 > 0$ for all $x \neq 0$. There is only one critical point, when $x = 0$. Hence the function is strictly increasing for all x, and its graph looks like this.

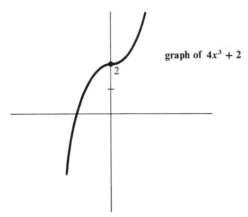

graph of $4x^3 + 2$

Example 3(b). Let $f(x) = 4x^3 + 3x$. Sketch the graph of f.

Here we have $f'(x) = 12x^2 + 3 > 0$ for *all* x. Therefore the graph of f looks like this. There is no critical point.

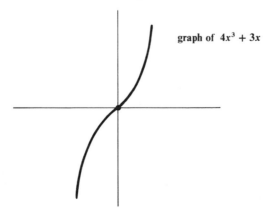

graph of $4x^3 + 3x$

In both examples, we have

$$f''(x) = 24x.$$

Thus in both examples, there is an inflection point at $x = 0$. The graph of f bends down for $x < 0$ and bends up for $x > 0$. The difference between case (a) and case (b) is that in case (a) the inflection point is a critical point, where the derivative of f is equal to 0, so that the curve is flat at the critical point. In case (b), the derivative at the inflection point is

$$f'(0) = 3,$$

so in case (b) the derivative at the inflection point is positive.

VI, §3. EXERCISES

1. Show that a curve

$$y = ax^3 + bx^2 + cx + d$$

with $a \neq 0$ has exactly one inflection point.

Sketch the graphs of the following curves.

2. $x^3 - 2x^2 + 3x$

3. $x^3 + x^2 - 3x$

4. $2x^3 - x^2 - 3x$

5. $\frac{1}{3}x^3 + x^2 - 2x$

6. $x^3 - 3x^2 + 6x - 3$

7. $x^3 + x - 1$

8. $x^3 - x - 1$

9. $-x^3 + 2x + 5$

10. $-2x^3 + x + 2$

11. $x^3 - x^2 + 1$

12. $y = x^4 + 4x$

13. $y = x^5 + x$

14. $y = x^6 + 6x$

15. $y = x^7 + x$

16. $y = x^8 + x$

17. Which of the following polynomials have a minimum (for all x)?
   (a) $x^6 - x + 2$
   (b) $x^5 - x + 2$
   (c) $-x^6 - x + 2$
   (d) $-x^5 - x + 2$
   (e) $x^6 + x + 2$
   (f) $x^5 + x + 2$
   Sketch the graphs of these polynomials.

18. Which of the polynomials in Exercise 17 have a maximum (for all x)?

In the following two problems:
(a) Show that f has exactly two inflection points.
(b) Sketch the graph of f. Determine the critical points explictly. Determine the regions of bending up or down.

19. $f(x) = x^4 + 3x^3 - x^2 + 5$

20. $f(x) = x^4 - 2x^3 + x^2 + 3$

21. Sketch the graph of the function

$$f(x) = x^6 - \tfrac{3}{2}x^4 + \tfrac{9}{16}x^2 - \tfrac{1}{32}.$$

Find the critical points. Find the values of f at these critical points. Sketch the graph of f. It will come out much neater than may be apparent at first.

VI, §4. RATIONAL FUNCTIONS

We shall now consider quotients of polynomials.

Example. Sketch the graph of the curve

$$y = f(x) = \frac{x - 1}{x + 1}$$

and determine the eight properties stated in the introduction.

1. When $x = 0$, we have $f(x) = -1$. When $x = 1$, $f(x) = 0$.
2. The derivative is

$$f'(x) = \frac{2}{(x + 1)^2}.$$

(You can compute it using the quotient rule.) It is never 0, and therefore the function f has no critical points.

3. The denominator is a square and hence is always positive, wherever it is defined, that is for $x \neq -1$. Thus $f'(x) > 0$ for all $x \neq -1$. The function is increasing for all x, Of course, the function is not defined for $x = -1$ and neither is the derivative. Thus it would be more accurate to say that the function is increasing in the region

$$x < -1$$

and is increasing in the region $x > -1$.

4. There is no region of decrease.
5. Since the derivative is never 0, there is no relative maximum or minimum.

6.   The second derivative is

$$f''(x) = \frac{-4}{(x+1)^3}.$$

There is no inflection point since $f''(x) \neq 0$ for all x where the function is defined. If $x < -1$, then the denominator $(x+1)^3$ is negative, and $f''(x) > 0$, so the graph is bending upward. If $x > -1$, then the denominator is positive, and $f''(x) < 0$ so the graph is bending downward.

7.   As x becomes large positive, our function approaches 1 (using the method of §1). As x becomes large negative, our function also approaches 1.

There is one more useful piece of information which we can look into, when $f(x)$ itself becomes large positive or negative. This occurs near points where the denominator of $f(x)$ is 0. On the present instance, $x = -1$.

8.   As x approaches -1, the denominator approaches 0 and the numerator approaches -2. If x approaches -1 from the right so $x > -1$, then the denominator is positive, and the numerator is negative. Hence the fraction

$$\frac{x-1}{x+1}$$

is negative, and is large negative.

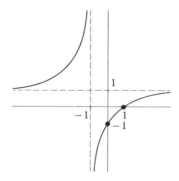

Figure 2

If x approaches -1 from the left so $x < -1$, then $x - 1$ is negative, but $x + 1$ is negative also. Hence $f(x)$ is positive and large, since the denominator is small when x is close to -1.

Putting all this information together, we see that the graph looks like that in the preceding figure.

We have drawn the two lines $x = -1$ and $y = 1$, as these play an important role when x approaches -1 and when x becomes large, positive or negative.

Remark. Again let

$$y = f(x) = \frac{x - 1}{x + 1}.$$

Then we can rewrite this relation to see directly that the graph of f is a hyperbola, as follows. We write the relation in the form

$$y = \frac{x + 1 - 2}{x + 1} = 1 - \frac{2}{x + 1},$$

that is, $y - 1 = -2/(x + 1)$. Clearing denominators, this gives

$$(y - 1)(x + 1) = -2.$$

By Chapter II, you should know that this is a hyperbola. We worked out a sketch by a more general method, because it also works in cases when you cannot reduce the equation to one of the standard curves, like circles, parabolas, or hyperbolas.

Example. Sketch the graph of $f(x) = \dfrac{x^2 + x}{x - 1}$.

Note that f is not defined at $x = 1$. We can rewrite

$$f(x) = \frac{x(x + 1)}{x - 1}.$$

We have $f(x) = 0$ if and only if the numerator $x(x + 1) = 0$. Thus:

$$f(x) = 0 \quad \text{if and only if} \quad x = 0 \text{ or } x = -1.$$

Next we look at the derivative, which is

$$f'(x) = \frac{x^2 - 2x - 1}{(x - 1)^2}.$$

(Compute it using the quotient rule.) Then

$$f'(x) = 0 \quad \Leftrightarrow \quad x^2 - 2x - 1 = 0,$$
$$\Leftrightarrow \quad x = 1 \pm \sqrt{2} \quad \text{(by the quadratic formula)}$$

These are the critical points of f.

The denominator $(x - 1)^2$ in $f'(x)$ is a square and hence is always positive, wherever it is defined, that is for $x \neq 1$. Therefore the sign of $f'(x)$ is the same as the sign of its numerator $x^2 - 2x - 1$. Let

$$g(x) = x^2 - 2x - 1.$$

The graph of g is a parabola, and since the coefficient of x^2 is $1 > 0$, this parabola is bending up as shown on the figure.

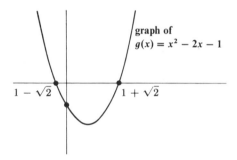

The two roots of $g(x) = 0$ are $x = 1 - \sqrt{2}$ and $1 + \sqrt{2}$. From the graph of $g(x)$ we see that

$$g(x) < 0 \quad \text{when} \quad 1 - \sqrt{2} < x < 1 + \sqrt{2},$$
$$g(x) > 0 \quad \text{when} \quad x < 1 - \sqrt{2} \text{ or } x > 1 + \sqrt{2}.$$

This gives us the regions of increase and decrease for $f(x)$.

For $\quad x \leqq 1 - \sqrt{2}, \qquad f(x)$ is strictly increasing.

For $\quad 1 - \sqrt{2} \leqq x < 1, \quad f(x)$ is strictly decreasing.

For $\quad 1 < x \leqq 1 + \sqrt{2}, \quad f(x)$ is strictly decreasing.

For $\quad 1 + \sqrt{2} \leqq x, \qquad f(x)$ is strictly increasing.

It follows that f has a local maximum at $x = 1 - \sqrt{2}$ and f has a local minimum at $x = 1 + \sqrt{2}$.

As x becomes large positive, $f(x)$ becomes large positive as is seen from the expression

$$f(x) = \frac{x^2 + x}{x - 1} = \frac{x^2(1 + 1/x)}{x(1 - 1/x)} = x \, \frac{1 + 1/x}{1 - 1/x}.$$

As x becomes large negative, $f(x)$ becomes large negative.

As x approaches 1 and $x < 1$, the function $f(x)$ becomes large negative because the denominator $x - 1$ approaches 0, and is negative, while the numerator $x^2 + x$ approaches 2.

As x approaches 1 and $x > 1$, the function $f(x)$ becomes large positive because the denominator $x - 1$ approaches 0 and both numerator and denominator are positive, while the numerator approaches 2.

Hence the graph looks as drawn on Fig. 3.

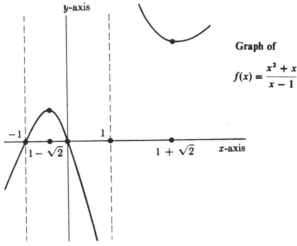

Figure 3

VI, §4. EXERCISES

Sketch the following curves, indicating all the information stated in the introduction. Regard convexity possibly as optional.

1. $y = \dfrac{x^2 + 2}{x - 3}$

2. $y = \dfrac{x - 3}{x^2 + 1}$

3. $y = \dfrac{x + 1}{x^2 + 1}$

4. $y = \dfrac{x^2 - 1}{x} = x - \dfrac{1}{x}$

5. $\dfrac{x}{x^3 - 1}$

6. $y = \dfrac{2x^2 - 1}{x^2 - 2}$

7. $y = \dfrac{2x - 3}{3x + 1}$

8. $\dfrac{4x}{x^2 - 9}$

9. $x + \dfrac{3}{x}$

10. $\dfrac{x^2 - 4}{x^3}$

11. $\dfrac{3x - 2}{2x + 3}$

12. $\dfrac{x}{3x - 5}$

13. $\dfrac{2x}{x+4}$

14. $\dfrac{x^2}{\sqrt{x+1}}$

15. $\dfrac{x+1}{x^2+5}$

16. $\dfrac{x+1}{x^2-5}$

17. $\dfrac{x^2+1}{x^2-1}$

18. $\dfrac{x^2-1}{x^2-4}$

19. Sketch the graph of $f(x) = x + 1/x$.

20. Let a, b be two positive numbers. Let

$$f(x) = ax + \frac{b}{x}.$$

Show that the minimum value of $f(x)$ for $x > 0$ is $2\sqrt{ab}$. Give reasons for your assertions. Deduce that $\sqrt{ab} \leq (a+b)/2$. Sketch the graph of f for $x > 0$.

VI, §5. APPLIED MAXIMA AND MINIMA

This section deals with word problems concerning maxima and minima, and applies the techniques discussed previously. In each case, we want to maximize or minimize a function, which is at first given in terms of perhaps two variables. We proceed as follows.

1. Enough data are given so that one of these variables can be expressed in terms of the other, by some relation. We end up dealing with a function of only one variable.

2. We then find its critical points, setting the derivative equal to 0, and then determine whether the critical points are local maxima or minima.

3. We verify if these local maxima or minima are also maxima or minima for the whole interval of definition of the function. If the function is given only on some finite interval, it may happen that the maximum, say, occurs at an end point, where the derivative test does not apply.

Example 1. Find the point on the graph of the equation $y^2 = 4x$ which is nearest to the point $(2, 3)$.

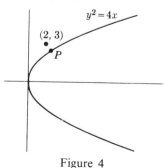

Figure 4

To minimize the distance between a point (x, y) and $(2, 3)$, it suffices to minimize the square of the distance, which has the advantage that no square root occurs in its formula. Indeed, suppose z_0^2 is a minimum value for the square of the distance, with z_0 positive. Then z_0 itself is a minimum value for the distance, because a positive number has a unique positive square root. The square of the distance is equal to

$$z^2 = (2 - x)^2 + (3 - y)^2.$$

Thus z^2 is expressed in terms of the two variables x, y. But we know that the point (x, y) lies on the curve whose equation is $y^2 = 4x$. Hence we can solve for one variable in terms of the other, namely $y = 2\sqrt{x}$. Substituting $y = 2\sqrt{x}$, we find an expression for the square of the distance only in terms of x, namely

$$f(x) = (2 - x)^2 + (3 - 2\sqrt{x})^2$$
$$= 4 - 4x + x^2 + 9 - 12\sqrt{x} + 4x$$
$$= 13 + x^2 - 12\sqrt{x}.$$

We now determine the critical points of f. We have

$$f'(x) = 2x - \frac{6}{\sqrt{x}},$$

so

$$f'(x) = 0 \quad \Leftrightarrow \quad 2x\sqrt{x} = 6$$
$$\Leftrightarrow \quad x = \sqrt[3]{9}.$$

Furthermore, we have

$$f'(x) > 0 \quad \Leftrightarrow \quad 2x\sqrt{x} > 6$$
$$\Leftrightarrow \quad x^3 > 9.$$
$$f'(x) < 0 \quad \Leftrightarrow \quad x^3 < 9.$$

Hence $f(x)$ is strictly increasing when $x > \sqrt[3]{9}$ and strictly decreasing when $x < \sqrt[3]{9}$. Hence $\sqrt[3]{9}$ is a minimum. When $x = \sqrt[3]{9}$ the corresponding value for y is

$$y = 2\sqrt{x} = 2\sqrt[3]{3}.$$

Hence the point on the graph of $y^2 = 4x$ closest to $(2, 3)$ is the point

$$P = (\sqrt[3]{9}, 2\sqrt[3]{3}).$$

Example 2. An oil can is to be made in the form of a cylinder to contain one quart of oil. What dimensions should it have so that the surface area is minimal (in other words, minimize the cost of material to make the can)?

Let r be the radius of the base of the cylinder and let h be its height. Then the volume is

$$V = \pi r^2 h.$$

The total surface area is the sum of the top, bottom, and circular sides, namely

$$A = 2\pi r^2 + 2\pi r h.$$

Thus the area is given in terms of the two variables r and h. However, we are also given that the volume V is constant, $V = 1$. Thus we get a relation between r and h,

$$\pi r^2 h = 1,$$

and we can solve for h in terms of r, namely

$$h = 1/\pi r^2.$$

Hence the area can be expressed entirely in terms of r, that is

$$A = 2\pi r^2 + 2\pi r/\pi r^2 = 2\pi r^2 + 2/r.$$

We want the area to be minimum. We first find the critical points of A. We have:

$$A'(r) = 4\pi r - 2/r^2 = 0 \quad \Leftrightarrow \quad 4\pi r = 2/r^2$$

$$\Leftrightarrow \quad \pi r^3 = \tfrac{1}{2}.$$

Thus we find exactly one critical point

$$r = \left(\frac{1}{2\pi}\right)^{1/3}$$

By physical considerations, we could see that this corresponds to a minimum, but we can also argue as follows. When r becomes large positive, or when r approaches 0, the function $A(r)$ becomes large, and so there has to be a minimum of the function for some value $r > 0$. This minimum is a critical point, and we have found that there is only one critical point. Hence we have found that the minimum occurs when r is the critical point. In this case we can solve back for h, namely

$$h = \frac{1}{\pi r^2} = \frac{(2\pi)^{2/3}}{\pi} = \frac{2^{2/3}}{\pi^{1/3}}.$$

This gives us the required dimensions.

Example 3. A truck is to be driven 200 mi at constant speed x mph. Speed laws require $30 \leq x \leq 60$. Assume that gasoline costs 50 cents/gallon and is consumed at the rate of

$$3 + \frac{x^2}{500} \text{ gal/hr.}$$

If the driver's wages are \$8 per hour, find the most economical speed.

We express the total cost as a sum of the cost of gasoline and the wages. The total time taken for the trip will be

$$\frac{200}{x}$$

because (time)(speed) = (distance) if the speed is constant. The cost of gas is then equal to the product of

(price per gallon)(number of gallons used per hr)(total time)

so that the cost of gasoline is

$$G(x) = \frac{1}{2} \left(3 + \frac{x^2}{500} \right) \frac{200}{x}.$$

(We write 1/2 because 50 cents $= \frac{1}{2}$ dollar.) On the other hand, the wages are given by the product

(wage per hour)(total time),

so that the cost of wages is

$$W(x) = 8 \cdot \frac{200}{x}.$$

Hence the total cost of the trip is

$$f(x) = G(x) + W(x)$$

$$= \frac{1}{2}\left(3 + \frac{x^2}{500}\right)\frac{200}{x} + \frac{8 \cdot 200}{x}$$

$$= 100\left(\frac{3}{x} + \frac{x}{500}\right) + \frac{1600}{x}.$$

We have

$$f'(x) = -\frac{300}{x^2} + \frac{1}{5} - \frac{1600}{x^2}$$

$$= -\frac{1900}{x^2} + \frac{1}{5}.$$

Therefore $f'(x) = 0$ if and only if

$$\frac{1900}{x^2} = \frac{1}{5}$$

or in other words,

$$x^2 = 9500.$$

Thus $x = 10\sqrt{95}$. [We take x positive since this is the solution which has physical significance.]

Now we observe that $10\sqrt{95}$ is approximately equal to 10×10, and in any case is >60, so is beyond the speed limit of 60 which was assigned to begin with. Furthermore, if $0 < x < 10\sqrt{95}$ then

$$f'(x) < 0.$$

Hence the function $f(x)$ is decreasing for $0 \leq x \leq 10\sqrt{95}$. Its graph may be sketched as on the figure.

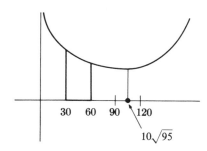

Since to begin with we restricted the possible speed to the interval $30 \leq x \leq 60$, it follows that the minimum of f over this interval must occur when $x = 60$. This is therefore the speed which minimizes the total cost.

Example 4. In the preceding example, suppose there is no speed limit. Then we see that if $x > 10\sqrt{95}$ then $f'(x) > 0$ so that $f(x)$ is increasing for

$$x > 10\sqrt{95}.$$

Therefore $10\sqrt{95}$ is a minimum point for f when no restriction is placed on x. Consequently, in this case, the speed which minimizes the cost is

$$x = 10\sqrt{95}.$$

Example 5. When light from a point source strikes a plane surface, the intensity of illumination is proportional to the cosine of the angle of incidence, and inversely proportional to the square of the distance from the source. How high should a light be located above the center of a circle of radius 12 ft to give the best illumination along the circumference?

The **angle of incidence** is measured from the perpendicular to the plane. The picture is as follows.

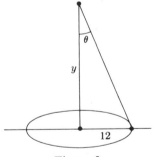

Figure 5

We denote by θ the angle of incidence, and by y the height of the light. Let I be the intensity of illumination. Two quantities are proportional means that there is a constant such that one is equal to the constant times the other. Thus there is a constant c such that

$$I(y) = c \cos \theta \, \frac{1}{12^2 + y^2}$$

$$= c \, \frac{y}{\sqrt{12^2 + y^2}} \, \frac{1}{12^2 + y^2}$$

$$= \frac{cy}{(12^2 + y^2)^{3/2}}.$$

The critical points of $I(y)$ are those points where $I'(y) = 0$. We have:

$$I'(y) = c \left[\frac{(12^2 + y^2)^{3/2} - y \cdot \frac{3}{2}(12^2 + y^2)^{1/2}(2y)}{(12^2 + y^2)^3} \right],$$

and this expression is equal to 0 precisely when the numerator is equal to 0, that is,

$$(12^2 + y^2)^{3/2} = 3y^2(12^2 + y^2)^{1/2}.$$

Canceling $(12^2 + y^2)^{1/2}$, we see that this is equivalent to

$$12^2 + y^2 = 3y^2,$$

or in other words,

$$12^2 = 2y^2.$$

Solving for y yields

$$y = \pm \frac{12}{\sqrt{2}}.$$

Only the positive value of y has physical significance, and thus the height giving maximum intensity is $12/\sqrt{2}$ ft, provided that we know that this critical point is a maximum for the function $I(y)$, for $y > 0$. This can be seen as follows.

If y is very close to 0, then the numerator cy of $I(y)$ is close to 0, and the denominator $(12^2 + y^2)^{3/2}$ is close to $(12^2)^{3/2}$ so $I(y)$ tends to 0 when y is near 0. For large y, the denominator of $I(y)$ is analyzed by factoring out y^2, namely

$$(12^2 + y^2)^{3/2} = \left(\frac{12^2}{y^2} + 1 \right)^{3/2} y^3.$$

Therefore $I(y)$ tends to 0 as y becomes large, because

$$I(y) = \frac{cy}{(\text{term near 1})y^3} = \frac{c}{(\text{term near 1})} \frac{1}{y^2}$$

if y is large positive. Hence we have shown that $I(y)$ tends to 0 when y approaches 0 or y becomes large. It follows that $I(y)$ reaches a maximum for some value of $y > 0$, and this maximum must be a critical point. On the other hand, we have also proved that there is only one critical point. Hence this critical point is the maximum, as desired. Thus $y = 12/\sqrt{2}$ is a maximum for the function. In view of the preceding discussion, the graph can be sketched as on the figure.

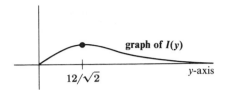

Example 6. A business makes automobile transmissions selling for $400. The total cost of marketing x units is

$$f(x) = 0.02x^2 + 160x + 400,000.$$

How many transmissions should be sold for maximum profit?

Let $P(x)$ be the profit coming from selling x units. Then $P(x)$ is the difference between the total receipts and the cost of marketing. Hence

$$P(x) = 400x - (0.02x^2 + 160x + 400,000)$$
$$= -0.02x^2 + 240x - 400,000.$$

We want to know when $P(x)$ is maximum. We have:

$$P'(x) = -0.04x + 240$$

so the derivative is 0 when

$$0.04x = 240,$$

or in other words,

$$x = \frac{240}{0.04} = 6,000.$$

The equation $y = P(x)$ is a parabola, which bends down because the leading coefficient is -0.02 (negative). Hence the critical point is a maximum, and the answer is therefore 6,000 units.

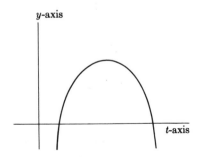

Parabola $y = at^2 + bt + c$ with $a < 0$.

Example 7. A farmer buys a bull weighing 600 lbs, at a cost of $180. It costs 15 cents per day to feed the animal, which gains 1 lb per day. Every day that the bull is kept, the sale price per pound declines according to the formula

$$B(t) = 0.45 - 0.00025t$$

where t is the number of days. How long should the farmer wait to maximize profits?

To do this, note that the total cost after time t is given by

$$f(t) = 180 + 0.15t.$$

The total sales amount to the product of the price $B(t)$ per pound times the weight of the animal, in other words,

$$S(t) = (0.45 - 0.00025t)(600 + t)$$

$$= -0.00025t^2 + 0.30t + 270.$$

Hence the profit is

$$P(t) = S(t) - f(t)$$

$$= -0.00025t^2 + 0.15t + 90.$$

Therefore

$$P'(t) = -0.0005t + 0.15$$

and $P'(t) = 0$ exactly when $0.0005t = 0.15$, or in other words,

$$t = \frac{0.15}{0.0005} = 300.$$

Hence the answer is 300 days for the farmer to wait before selling the bull, provided we can show that this value of t gives a maximum. But the formula for the profit is a quadratic expression in t, of the form

$$P(t) = at^2 + bt + c,$$

and $a < 0$. Hence $P(t)$ is a parabola, and since $a < 0$ this parabola opens downward as on the figure. Hence the critical point must be a maximum, as desired.

VI, §5. EXERCISES

1. Find the length of the sides of a rectangle of largest area which can be inscribed in a semicircle, the lower base being on the diameter.

2. A rectangular box has a square base and no top. The combined area of the sides and bottom is 48 ft^2. Find the dimensions of the box of maximum volume meeting these requirements.

3. Prove that, among all rectangles of given area, the square has the least perimeter.

4. A truck is to be driven 300 km at a constant speed of x km/hr. Speed laws require $30 \leq x \leq 60$. Assume that gasoline costs 30 cents/gallon and is consumed at the rate of $2 + x^2/600$ gal/hr. If the driver's wages are D dollars per hour, find the most economical speed and the cost of the trip if (a) $D = 0$, (b) $D = 1$, (c) $D = 2$, (d) $D = 3$, (e) $D = 4$.

5. A rectangle is to have an area of 64 m^2. Find its dimensions so that the distance from one corner to the mid-point of a non-adjacent side shall be a minimum.

6. Express the number 4 as the sum of two positive numbers in such a way that the sum of the square of the first and the cube of the second is as small as possible.

7. A wire 24 cm long is cut in two, and one part is bent into the shape of a circle, and the other into the shape of a square. How should it be cut if the sum of the areas of the circle and the square is to be (a) minimum, (b) maximum?

8. Find the point on the graph of the equation $y^2 = 4x$ which is nearest to the point $(2, 1)$.

9. Find the points on the hyperbola $x^2 - y^2 = 1$ nearest to the point $(0, 1)$.

10. Show that $(2, 2)$ is the point on the graph of the equation $y = x^3 - 3x$ that is nearest the point $(11, 1)$.

11. Find the coordinates of the points on the curve $x^2 - y^2 = 16$ which are nearest to the point $(0, 6)$.

12. Find the coordinates of the points on the curve $y^2 = x + 1$ which are nearest to the origin.

13. Find the coordinates of the point on the curve $y^2 = \frac{5}{2}(x + 1)$ which is nearest to the origin.

14. Find the coordinates of the points on the curve $y = 2x^2$ which are closest to the point $(9, 0)$.

15. A circular ring of radius b is uniformly charged with electricity, the total charge being Q. The force exerted by this charge on a particle at a distance x from the center of the ring, in a direction perpendicular to the plane of the ring, is given by $F(x) = Qx(x^2 + b^2)^{-3/2}$. Find the maximum of F for all $x \geq 0$.

16. Let F be the rate of flow of water over a certain spillway. Assume that F is proportional to $y(h-y)^{1/2}$, where y is the depth of the flow, and h is the height, and is constant. What value of y makes F a maximum?

17. Find the point on the x-axis the sum of whose distances from (2, 0) and (0, 3) is a minimum.

18. A piece of wire of length L is cut into two parts, one of which is bent into the shape of an equilateral triangle and the other into the shape of a circle. How should the wire be cut so that the sum of the enclosed areas is: (a) a minimum, (b) a maximum?

19. A fence $13\frac{1}{2}$ ft high is 4 ft from the side wall of a house. What is the length of the shortest ladder, one end of which will rest on the level ground outside the fence and the other on the side wall of the house?

20. A tank is to have a given volume V and is to be made in the form of a right circular cylinder with hemispheres attached to each end. The material for the ends costs twice as much per square meter as that for the sides. Find the most economical proportions. [You may assume that the area of a sphere is $4\pi r^2$.]

21. Find the length of the longest rod which can be carried horizontally around a corner from a corridor 8 ft wide into one 4 ft wide.

22. Let P, Q be two points in the plane on the same side of the x-axis. Let R be a point on the x-axis (Fig. 6). Show that the sum of the distances PR and QR is smallest when the angles θ_1 and θ_2 are equal.

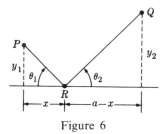

Figure 6

[*Hint*: First use the Pythagoras theorem to give an expression for the distances PR and RQ in terms of x and the fixed quantities y_1, y_2. Let $f(x)$ be the sum of the distances. Show that the condition $f'(x) = 0$ means that $\cos\theta_1 = \cos\theta_2$. Using values of x near 0 and a, show that $f(x)$ is decreasing near $x = 0$ and increasing near $x = a$. Hence the minimum must be in the open interval $0 < x < a$, and is therefore the critical point.]

23. Suppose the velocity of light is v_1 in air and v_2 in water. A ray of light traveling from a point P_1 above the surface of water to a point P_2 below the surface will travel by the path which requires the least time. Show that the ray will cross the surface at the point Q in the vertical plane through P_1 and P_2 so placed that

$$\frac{\sin\theta_1}{v_1} = \frac{\sin\theta_2}{v_2},$$

where θ_1 and θ_2 are the angles shown in the following figure:

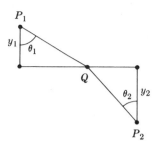

Figure 7

(You may assume that the light will travel in the vertical plane through P_1 and P_2. You may also assume that when the velocity is constant, equal to v, throughout a region, and s is the distance traveled, then the time t is equal to $t = s/v$.)

24. Let p be the probability that a certain event will occur, at any trial. In n trials, suppose that s successes have been observed. The likelihood function L is defined as $L(p) = p^s(1 - p)^{n-s}$. Find the value of p which maximizes the likelihood function. (Take $0 \leq p \leq 1$.) View n, s as constants.

25. Find an equation for the line through the following points making with the coordinate axes a triangle of minimum area in the first quadrant:
    (a) through the point $(3, 1)$.
    (b) through the point $(3, 2)$.

26. Let $a_1, \ldots, a_n$ be numbers. Show that there is a single number x such that

$$(x - a_1)^2 + \cdots + (x - a_n)^2$$

is a minimum, and find this number.

27. When light from a point source strikes a plane surface, the intensity of illumination is proportional to the cosine of the angle of incidence and inversely proportional to the square of the distance from the source. How high should a light be located above the center of a circle of radius 25 cm to give the best illumination along the circumference? (The angle of incidence is measured from the perpendicular to the plane.)

28. A horizontal reservoir has a cross section which is an inverted isosceles triangle, where the length of a leg is 60 ft. Find the angle between the equal legs to give maximum capacity.

29. A reservoir has a horizontal plane bottom and a cross section as shown on the figure. Find the angle of inclination of the sides from the horizontal to give maximum capacity.

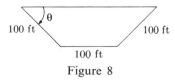

Figure 8

30. Determine the constant a such that the function

$$f(x) = x^2 + \frac{a}{x}$$

has (a) a local minimum at $x = 2$, (b) a local minimum at $x = -3$. (c) Show that the function cannot have a local maximum for any value of a.

31. The intensity of illumination at any point is proportional to the strength of the light source and varies inversely as the square of the distance from the source. If two sources of strengths a and b respectively are a distance c apart, at what point on the line joining them will the intensity be a minimum?

32. A window is in the shape of a rectangle surmounted by a semicircle. Find the dimensions when the perimeter is 12 ft and the area is as large as possible.

33. Find the radius and angle of the circular sector of maximum area if the perimeter is
(a) 20 cm   (b) 16 cm.

34. You are watering the lawn and aiming the hose upward at an angle of inclination θ. Let r be the range of the hose, that is, the distance from the hose to the point of impact of the water. Then r is given by

$$r = \frac{2v^2}{g} \sin \theta \cos \theta,$$

where v, g are constants. For what angle is the range a maximum?

35. A ladder is to reach over a fence 12 ft high to a wall 2 ft behind the fence. What is the length of the shortest ladder that can be used?

36. A cylindrical tank is supposed to have a given volume V. Find the dimension of the radius of the base and the height in terms of V so that the surface area is minimal. The tank should be open on top, but closed at the bottom.

37. A flower bed is to have the shape of a circular sector of radius r and central angle θ. Find r and θ if the area is fixed and the perimeter is a minimum in case:
(a) $0 < \theta \leqq \pi$ and   (b) $0 < \theta \leqq \pi/2$.

Recall that the area of a sector is

$$A = \pi r^2 \cdot \frac{\theta}{2\pi} = \theta r^2 / 2.$$

The length of an arc of a circle of radius r is

$$L = 2\pi r \cdot \frac{\theta}{2\pi} = r\theta.$$

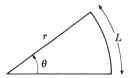

38. A firm sells a product at \$50 per unit. The total cost of marketing x units is given by the function

$$f(x) = 5000 + 650x - 45x^2 + x^3.$$

How many units should be produced per day to maximize profits? What is the daily profit for this number of units?

39. The daily cost of producing x units of a product is given by the formula

$$f(x) = 2002 + 120x - 5x^2 + \tfrac{1}{3}x^3.$$

Each unit sells for \$264. How many units should be produced per day to maximize profits? What is the daily profit for this number of units?

40. A product is marketed at 50 dollars per unit. The total cost of marketing x units of the product is

$$f(x) = 1000 + 150x - 100x^2 + 2x^3.$$

How many units should be produced to maximize the profit? What is the daily profit for this x?

41. A company is the sole producer of a product, whose cost function is

$$f(x) = 100 + 20x + 2x^2.$$

If the company increases the price, then fewer units are sold, and in fact if we express the price $p(x)$ as a function of the number of units x, then

$$p(x) = 620 - 8x.$$

How many units should be produced to maximize profits? What is this maximum profit? [*Hint*: The total revenue is equal to the product $xp(x)$, number of units times the price.]

42. The cost of producing x units of a product is given by the function

$$f(x) = 10x^2 + 200x + 6,000.$$

If p is the price per unit, then the number of units sold at that price is given by

$$x = \frac{1000 - p}{10}.$$

For what value of x will the profit be positive? How many units should be produced to give maximum profits?

CHAPTER VII

Inverse Functions

Suppose that we have a function, for instance

$$y = 3x - 5.$$

Then we can solve for x in terms of y, namely

$$x = \tfrac{1}{3}(y + 5).$$

Thus x can be expressed as a function of y.

Although we are able to solve by means of an explicit formula, there are interesting cases where x can be expressed as a function of y, but without such an explicit formula. In this chapter, we shall investigate such cases.

VII, §1. DEFINITION OF INVERSE FUNCTIONS

Let $y = f(x)$ be a function, defined for all x in some interval. If, for each value y_1 of y, there exists exactly *one* value x_1 of x in the interval such that $f(x_1) = y_1$, then we can define the **inverse function**

$x = g(y) = $ *the unique number x such that $y = f(x)$.*

Our inverse function is defined only at those numbers which are values of f. We have the fundamental relation

$$f(g(y)) = y \qquad \text{and} \qquad g(f(x)) = x.$$

Example 1. Consider the function $y = x^2$, which we view as being defined only for $x \geq 0$. Every positive number (or 0) can be written uniquely as the square of a positive number (or 0). Hence we can define the inverse function, which will also be defined for $y \geq 0$, but not for $y < 0$. It is the square root function, $x = \sqrt{y}$.

Example 2. Suppose that $y = 5x - 7$. Then we can solve for x in terms of y, namely

$$x = \tfrac{1}{5}(y + 7).$$

If $f(x) = 5x - 7$, then its inverse function is the function $g(y)$ such that

$$g(y) = \tfrac{1}{5}(y + 7).$$

In these examples, we could write down the inverse function by explicit formulas. In general, this is not possible, but there are criteria which tell us when the inverse function exists, for instance when the graph of the function f looks like this.

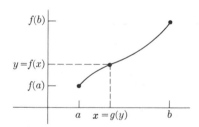

In this case, f is strictly increasing, and is defined on the interval $[a, b]$. To each point x in this interval, there is a value $f(x) = y$, and to each y between $f(a)$ and $f(b)$, there is a unique x between a and b such that $f(x) = y$. We formalize this in the next theorem.

Theorem 1.1. *Let $f(x)$ be a function which is strictly increasing. Then the inverse function exists, and is defined on the set of values of f.*

Proof. This is practically obvious: Given a number y_1 and a number x_1 such that $f(x_1) = y_1$, there cannot be another number x_2 such that $f(x_2) = y_1$ unless $x_2 = x_1$, because if $x_2 \neq x_1$, then

$$\text{either} \quad x_2 > x_1, \text{ in which case } f(x_2) > f(x_1),$$

$$\text{or} \quad x_2 < x_1, \text{ in which case } f(x_2) < f(x_1).$$

Since the positivity of the derivative gives us a good test when a function is strictly increasing, we are able to define inverse functions whenever the function is differentiable and its derivative is positive.

As usual, what we have said above applies as well to functions which are strictly decreasing, and whose derivatives are negative.

The following theorem is intuitively clear and is proved in an appendix. We already recalled it as Theorem 1.2 of Chapter V.

Intermediate value theorem. *Let* f *be a continuous function on a closed interval* $[a, b]$. *Let* v *be a number between* $f(a)$ *and* $f(b)$. *Then there exists a point* c *between* a *and* b *such that* $f(c) = v$.

This theorem says that the function f takes on every intermediate value between the values at the end points of the interval, and is illustrated on the next figure.

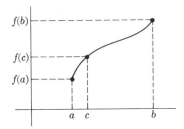

Using the intermediate value theorem, we now conclude:

Theorem 1.2. *Let* f *be a continuous function on the closed interval*

$$a \leq x \leq b$$

and assume that f *is strictly increasing. Let* $f(a) = \alpha$ *and* $f(b) = \beta$. *Then the inverse function is defined on the closed interval* $[\alpha, \beta]$.

Proof. Given any number γ between α and β, there exists a number c between a and b such that $f(c) = \gamma$, by the intermediate value theorem. Our assertion now follows from Theorem 1.1.

If we let g be this inverse function, then $g(\alpha) = a$ and $g(\beta) = b$. Furthermore, the inverse function is characterized by the relation

$$f(x) = y \qquad \text{if and only if} \qquad x = g(y).$$

Note that we can easily visualize the graph of an inverse function. If we want x in terms of y, we just flip the page over a $45°$ angle, reversing

the roles of the x- and y-axes. Thus the graph of $y = f(x)$ gets reflected
across the slanted line at $45°$ to give us the graph of $x = g(y)$.

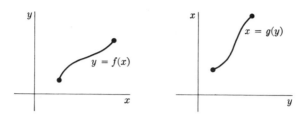

We shall now give some examples of how to define inverse functions
over certain intervals.

Example 3. Take for $f(x)$ a polynomial of degree 3. When the coefficient of x^3 is positive, and when f has local maxima and minima, its
graph looks like this:

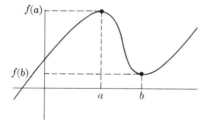

To any given value of y between $f(a)$ and $f(b)$ there correspond three
possible values for x, and hence the inverse function cannot be defined
unless we make other specifications. To do this, suppose first that we
view f as defined only for those numbers $\leqq a$. Then the graph of f looks
like this:

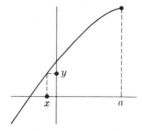

The inverse function is defined in this case. Similarly, we could view f as
defined on the interval $[a, b]$, or on the interval of all $x \geqq b$. In each
one of these cases, illustrated on the next figures, the inverse function
would be defined.

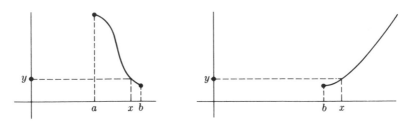

In each case, we have drawn a point y and the corresponding value x of the inverse function. They are different in the three cases.

Example 4. Let us consider a numerical example. Let

$$f(x) = x^3 - 2x + 1,$$

viewed as a function on the interval $x > \sqrt{2/3}$. Can we define the inverse function? For what numbers? If g is the inverse function, what is $g(0)$? What is $g(5)$?

We have $f'(x) = 3x^2 - 2$. The graph of $f'(x)$ is a parabola bending up, which crosses the x-axis at $x = \pm\sqrt{2/3}$.

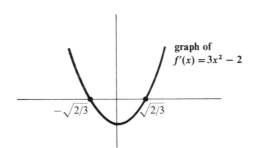

Hence

$$f'(x) > 0 \quad \Leftrightarrow \quad x > \sqrt{2/3} \text{ and } x < -\sqrt{2/3}.$$

Consider the interval $x > \sqrt{2/3}$. Then f is strictly increasing on this interval, and so the inverse function g is defined. Since $f(x) \to \infty$ when $x \to \infty$, it follows that the inverse function $g(y)$ is defined for all $y > f(\sqrt{2/3})$, that is

$$y > f(\sqrt{2/3}) = (2/3)^{3/2} - 2(2/3)^{1/2} + 1.$$

Now $1 > \sqrt{2/3}$, so 1 lies in the interval $x > \sqrt{2/3}$, and $f(1) = 0$. Therefore $g(0) = 1$.

Similarly $f(2) = 5$ and 2 lies in the interval $x > \sqrt{2/3}$, so $g(5) = 2$.

Note that we do not give an explicit formula for our inverse function. When dealing with polynomials of degree ≥ 3, no single formula can be given.

Example 5. On the other hand, take f defined by the same formula,

$$f(x) = x^3 - 2x + 1,$$

but viewed as a function on the interval

$$-\sqrt{\tfrac{2}{3}} \leq x \leq \sqrt{\tfrac{2}{3}}.$$

The derivative of f is given by $f'(x) = 3x^2 - 2$. We have

$$f'(x) < 0 \quad \Leftrightarrow \quad -\sqrt{\tfrac{2}{3}} < x < \sqrt{\tfrac{2}{3}}.$$

Hence f is strictly decreasing on this interval, and the inverse function is defined, but is quite different from that of Example 4. For instance, 0 is in the interval, and $f(0) = 1$, so if h denotes the inverse function, we have $h(1) = 0$.

Example 6. Let $f(x) = x^n$ (n being a positive integer). We view f as defined only for numbers $x > 0$. Since $f'(x)$ is nx^{n-1}, the function is strictly increasing. Hence the inverse function exists. This inverse function g is in fact what we mean by the n-th root.

In all the exercises of the previous chapter you determined intervals over which certain functions increase and decrease. You can now define inverse functions for such intervals. In most cases, you cannot write down a simple explicit formula for such inverse functions.

VII, §1. EXERCISES

For each of the following functions, determine whether there is an inverse function g, and determine those numbers at which g is defined.

1. $f(x) = 3x + 2$,   all x

2. $f(x) = x^2 + 2x - 3$, $0 \leq x$

3. $f(x) = x^3 + 4x - 5$,   all x

4. $f(x) = \dfrac{x}{x + 1}$, $-1 < x$

5. $f(x) = \dfrac{x}{x + 2}$, $-2 < x$

6. $f(x) = \dfrac{x + 1}{x - 1}$, $1 < x$

7. $f(x) = \dfrac{1}{x^2}$, $0 < x \leq 1$

8. $f(x) = \dfrac{x^2}{x^2 + 1}$, $0 \leq x \leq 5$

9. $f(x) = \dfrac{x + 2}{x - 2}, \quad 0 \leqq x < 2$ 10. $f(x) = x + \dfrac{1}{x}, \quad 1 \leqq x \leqq 10$

11. $f(x) = x + \dfrac{1}{x}, \quad 0 < x \leqq 1$ 12. $f(x) = x - \dfrac{1}{x}, \quad 0 < x \leqq 1$

13. $f(x) = \dfrac{2x}{1 + x^2}, \quad -1 \leqq x \leqq 1$ 14. $f(x) = \dfrac{2x}{1 + x^2}, \quad 1 \leqq x$

VII, §2. DERIVATIVE OF INVERSE FUNCTIONS

We shall state a theorem which allows us to determine the derivative of an inverse function when we know the derivative of the given function.

Theorem 2.1. *Let a, b be two numbers, a < b. Let f be a function which is differentiable on the interval a < x < b and such that its derivative f'(x) is > 0 for all x in this open interval. Then the inverse function x = g(y) exists, and we have*

$$g'(y) = \frac{1}{f'(x)} = \frac{1}{f'(g(y))}.$$

Proof. We are supposed to investigate the Newton quotient

$$\frac{g(y + k) - g(y)}{k}.$$

The following picture illustrates the situation:

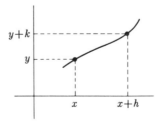

By the intermediate value theorem, every number of the form $y + k$ with small values of k can be written as a value of f. We let $x = g(y)$ and we let $h = g(y + k) - g(y)$. Then

$$x = g(y) \quad \text{and} \quad g(y + k) = x + h.$$

Furthermore, $y + k = f(x + h)$ and hence

$$k = f(x + h) - f(x).$$

The Newton quotient for g can therefore be written

$$\frac{g(y + k) - g(y)}{k} = \frac{x + h - x}{f(x + h) - f(x)} = \frac{h}{f(x + h) - f(x)},$$

and we see that it is the reciprocal of the Newton quotient for f, namely

$$\frac{1}{\dfrac{f(x + h) - f(x)}{h}}.$$

As h approaches 0, we know that k approaches 0 since

$$k = f(x + h) - f(x).$$

Conversely, as k approaches 0, we know that there exists exactly one value of h such that $f(x + h) = y + k$, because the inverse function is defined. Consequently, the corresponding value of h must also approach 0.

If we now take the limit of the reciprocal of the Newton quotient of f, as h (or k) approaches 0, we get

$$\frac{1}{f'(x)}.$$

By definition, this is the derivative $g'(y)$ and our theorem is proved.

Example 1. Let $f(x) = x^3 - 2x + 1$. Find an interval such that the inverse function g of f is defined, and find $g'(0)$, $g'(5)$.

By inspection, we see that

$$f(1) = 0 \quad \text{and} \quad f(2) = 5.$$

We must therefore find an interval containing 1 and 2 such that the inverse function of f is defined for that interval. But

$$f'(x) = 3x^2 - 2$$

and $f'(x) > 0$ if and only if $x > \sqrt{2/3}$ or $x < -\sqrt{2/3}$. (See Example 4 of the preceding section.) We select the interval $x > \sqrt{2/3}$, which contains both 1 and 2. Then we can apply the general theorem on the derivative of the inverse function, which states that if $y = f(x)$ then

$$g'(y) = \frac{1}{f'(x)}.$$

This gives:

$$g'(0) = \frac{1}{f'(1)} = 1 \quad \text{and} \quad g'(5) = \frac{1}{f'(2)} = \frac{1}{10}.$$

Please note that the derivative $g'(y)$ is given in terms of $f'(x)$. We don't have a formula in terms of y.

The theorem giving us the derivative of the inverse function could also be expressed by saying that

$$\frac{dx}{dy} = \frac{1}{dy/dx}.$$

Here also, the derivative behaves *as if* we were taking a quotient. Thus the notation is very suggestive and we can use it from now on without thinking, because we proved a theorem justifying it.

Remark. In Theorem 2.1, we have proved that in fact, the derivative of the inverse function g exists, and is given by $g'(y) = 1/f'(x)$. If one *assumes* that this derivative exists, then one can give a much shorter argument to find its value, using the chain rule. Indeed, we have

$$f(g(y)) = y, \quad \text{since } g(y) = x.$$

Differentiating with respect to y, we find by the chain rule,

$$f'(g(y))g'(y) = 1, \quad \text{because } \frac{dy}{dy} = 1.$$

Hence

$$g'(y) = \frac{1}{f'(g(y))},$$

as was to be shown.

VII, §2. EXERCISES

In each one of the exercises from 1 through 10, restrict f to an inverval so that the inverse function g is defined in an interval containing the indicated point, and find the derivative of the inverse function at the indicated point.

0. $f(x) = -x^3 + 2x + 1$. Find $g'(2)$.

1. $f(x) = x^3 + 1$. Find $g'(2)$.

2. $f(x) = (x - 1)(x - 2)(x - 3)$. Find $g'(6)$.

3. $f(x) = x^2 - x + 5$. Find $g'(7)$.

4. $f(x) = \sin x + \cos x$. Find $g'(-1)$.

5. $f(x) = \sin 2x$ $(0 \leq x \leq 2\pi)$. Find $g'(\sqrt{3}/2)$.

6. $f(x) = x^4 - 3x^2 + 1$. Find $g'(-1)$.

7. $f(x) = x^3 + x - 2$. Find $g'(0)$.

8. $f(x) = -x^3 + 2x + 1$. Find $g'(2)$.

9. $f(x) = 2x^3 + 5$. Find $g'(21)$.

10. $f(x) = 5x^2 + 1$. Find $g'(11)$.

11. Let f be a continuous function on the interval $[a, b]$. Assume that f is twice differentiable on the open interval $a < x < b$, and that $f'(x) > 0$ and $f''(x) > 0$ on this interval. Let g be the inverse function of f.
    (a) Find an expression for the second derivative of g.
    (b) Show that $g''(y) < 0$ on its interval of definition. Thus g bends in the opposite direction to f.

VII, §3. THE ARCSINE

It is impossible to define an inverse function for the function $y = \sin x$ because to each value of y there correspond infinitely many values of x because $\sin(x + 2\pi) = \sin x = \sin(\pi - x)$. However, if we restrict our attention to special intervals, we can define the inverse function.

We restrict the sine function to the interval

$$-\frac{\pi}{2} \leq x \leq \frac{\pi}{2}.$$

The derivative of $\sin x$ is $\cos x$ and in that interval, we have

$$0 < \cos x, \qquad \text{so the derivative is positive,}$$

except when $x = \pi/2$ or $x = -\pi/2$ in which case the cosine is 0.
    Therefore, in the interval

$$-\frac{\pi}{2} \leq x \leq \frac{\pi}{2}.$$

the function is strictly increasing. The inverse function exists, and is called the **arcsine**.

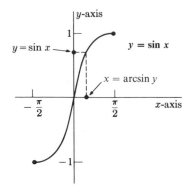

Let $f(x) = \sin x$, and $x = \arcsin y$, the inverse function. Since $f(0) = 0$ we have $\arcsin 0 = 0$. Furthermore, since

$$\sin(-\pi/2) = -1 \quad \text{and} \quad \sin(\pi/2) = 1,$$

we know that the inverse function is defined over the interval going from -1 to $+1$, that is for

$$-1 \leqq y \leqq 1.$$

In words, we can say loosely that $\arcsin x$ *is the angle whose sine is* x. (We throw in the word *loosely* because, strictly speaking, $\arcsin x$ is a number, and not an angle, and also because we mean the angle between $-\pi/2$ and $\pi/2$.)

Example 1. Let $f(x) = \sin x$ and let $g(y) = \arcsin y$. Then

$$\arcsin (1/\sqrt{2}) = \pi/4$$

because

$$\sin \pi/4 = 1/\sqrt{2}.$$

Similarly,

$$\arcsin 1/2 = \pi/6$$

because

$$\sin \pi/6 = 1/2.$$

For any value of x *in the interval* $-\pi/2 \leqq x \leqq \pi/2$ *we have*

$$\boxed{\arcsin \sin x = x,}$$

by definition of the inverse function. However, if x is not in this interval, then we do **not** have

$$\arcsin \sin x = x.$$

Example 2. Let $x = -\pi$. Then

$$\sin(-\pi) = 0,$$

and

$$\arcsin(\sin(-\pi)) = \arcsin 0 = 0 \neq -\pi.$$

Example 3. We have

$$\arcsin \sin(3\pi/4) = \pi/4,$$

because $\sin 3\pi/4 = 1/\sqrt{2}$, and $\arcsin 1/\sqrt{2} = \pi/4$.

We now consider the derivative and the graph of the inverse function. The derivative of $\sin x$ is positive in the interval

$$-\pi/2 < x < \pi/2.$$

Since the derivative of the inverse function $x = g(y)$ is $1/f'(x)$, the derivative of $\arcsin y$ is also positive, in the interval

$$-1 < y < 1.$$

Therefore the inverse function is strictly increasing in that interval. Its graph looks like the figure shown below.

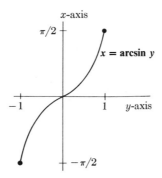

According to the general rule for the derivative of inverse functions, we know that when $y = \sin x$ and $x = \arcsin y$ the derivative is

$$\frac{dx}{dy} = \frac{1}{dy/dx} = \frac{1}{\cos x}.$$

When x is very close to $\pi/2$, we know that cos x is close to 0. Therefore the derivative is very large. Hence the curve is almost vertical. Similarly, when x is close to $-\pi/2$ and y is close to -1, the curve is almost vertical, as drawn.

Finally, it turns out that we can express our derivative explicitly as a function of y. Indeed, we have the relation

$$\sin^2 x + \cos^2 x = 1,$$

whence

$$\cos^2 x = 1 - \sin^2 x.$$

In the interval between $-\pi/2$ and $\pi/2$, the cosine is ≥ 0. Hence we can take the square root, and we get

$$\cos x = \sqrt{1 - \sin^2 x}$$

in that interval. Since $y = \sin x$, we can write our derivative in the form

$$\boxed{\frac{dx}{dy} = \frac{1}{\sqrt{1 - y^2}}}$$

which is expressed entirely in terms of y.

Example 4. Let $x = \arcsin y$. Find dx/dy when $y = 1/\sqrt{2}$. This is easily done, namely if $g(y) = \arcsin y$, then

$$g'(1/\sqrt{2}) = \frac{1}{\sqrt{1 - (1/\sqrt{2})^2}} = \sqrt{2}.$$

Having now obtained all the information we want concerning the arcsine, we shift back our letters to the usual ones. We state the main properties as a theorem.

Theorem 3.1. *View the sine function as defined on the interval*

$$[-\pi/2, \pi/2].$$

Then the inverse function is defined on the interval $[-1, 1]$. *Call it*

$$g(x) = \arcsin x.$$

Then g is differentiable in the open interval $-1 < x < 1$, *and*

$$g'(x) = \frac{1}{\sqrt{1 - x^2}}.$$

VII, §3. EXERCISES

1. Viewing the cosine as defined only on the interval $[0, \pi]$, prove that the inverse function arccos exists. On what interval is it defined? Sketch the graph.

2. What is the derivative of arccosine?

3. Let $g(x) = \arcsin x$. Find the following values:
   (a) $g'(1/2)$ (b) $g'(1/\sqrt{2})$ (c) $g(1/2)$
   (d) $g(1/\sqrt{2})$ (e) $g'(\sqrt{3}/2)$ (f) $g(\sqrt{3}/2)$

4. Let $g(x) = \arccos x$. What is $g'(\tfrac{1}{2})$? What is $g'(1/\sqrt{2})$? What is $g(\tfrac{1}{2})$? What is $g(1/\sqrt{2})$?

5. Let sec $x = 1/\cos x$. Define the inverse function of the secant over a suitable interval and obtain a formula for the derivative of this inverse function.

Find the following numbers.

6. $\arcsin(\sin 3\pi/2)$ 7. $\arcsin(\sin 2\pi)$ 8. $\arccos(\cos 3\pi/2)$

9. $\arccos(\cos -\pi/2)$ 10. $\arcsin(\sin -3\pi/4)$

Find the derivatives of the following functions.

11. $\arcsin(x^2 - 1)$ 12. $\arccos(2x + 5)$

13. $\dfrac{1}{\arcsin x}$ 14. $\dfrac{2}{\arccos 2x}$

15. Determine the intervals over which the function arcsin is bending upward, and bending downward.

VII, §4. THE ARCTANGENT

Let $f(x) = \tan x$ and view this function as defined over the interval

$$-\frac{\pi}{2} < x < \frac{\pi}{2}.$$

As x goes from $-\pi/2$ to $\pi/2$, the tangent goes from very large negative values to very large positive values. As x approaches $\pi/2$, the tangent has in fact arbitrarily large positive values, and similarly when x approaches $-\pi/2$, the tangent has arbitrarily large negative values.

We recall that the **graph of the tangent** looks like that in the following figure.

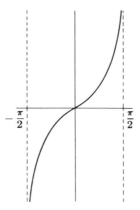

The derivative of $\tan x$ is

$$\frac{d(\tan x)}{dx} = 1 + \tan^2 x.$$

Hence the derivative is always positive, and the tan function is strictly increasing. Furthermore when x ranges over the interval

$$-\pi/2 < x < \pi/2,$$

$\tan x$ ranges from large negative to large positive values. Hence $\tan x$ ranges over all numbers. Therefore the inverse function is defined for all numbers. We call it the **arctangent**.

As with the arcsin and arccos, we may say roughly that $\arctan y$ is the angle whose tangent is y. We put "roughly" in our statement, because as pointed out before, we really mean the number of radians of the angle whose tangent is y, such that this number lies between $-\pi/2$ and $\pi/2$.

Example 1. We have $\arctan(-1/\sqrt{3}) = -\pi/6$, but

$$\arctan(-1/\sqrt{3}) \neq 5\pi/6,$$

even though $\tan 5\pi/6 = -1/\sqrt{3}$.

Example 2. In the same vein, we find that

$$\arctan \tan 5\pi/6 = \arctan(-1/\sqrt{3}) = -\pi/6.$$

The reason why $\arctan \tan x \neq x$ in this case is due to our choice of interval of definition for the tangent, when we wish to have an inverse function, and the fact that $x = 5\pi/6$ does not lie in this interval. Of course, if x lies between $-\pi/2$ and $\pi/2$, then we must have

$$\arctan \tan x = x$$

Thus

$$\arctan \tan(-\pi/6) = -\pi/6.$$

The **graph of the arctan** looks like this:

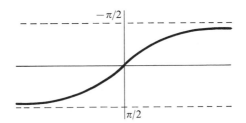

Now for the derivative. Let $x = g(y) = \arctan y$. Then

$$g'(y) = \frac{1}{f'(x)} = \frac{1}{1 + \tan^2 x}$$

so that

$$g'(y) = \frac{1}{1 + y^2}.$$

Here again we are able to get an explicit formula for the derivative of the inverse function.

As with the arcsine, when dealing simultaneously with the function and its inverse function, we have to keep our letters x, y separate. However, we now summarize the properties of the arctan in terms of our usual notation.

Theorem 4.1. *The inverse function of the tangent is defined for all numbers. Call it the arctangent. Then it has a derivative, and that derivative is given by the relation*

$$\frac{d(\arctan x)}{dx} = \frac{1}{1 + x^2}.$$

As x becomes very large positive, arctan *x approaches* $\pi/2$.
As x becomes very large negative, arctan *x approaches* $-\pi/2$.
The arctangent is strictly increasing for all x.

Example 3. Let $h(x) = \arctan 2x$. To find the derivative, we use the chain rule, letting $u = 2x$. Then

$$h'(x) = \frac{1}{1 + (2x)^2} \cdot 2 = \frac{2}{1 + 4x^2}.$$

Example 4. Let g be the arctan function. Then

$$g'(5) = \frac{1}{1 + 5^2} = \frac{1}{26}.$$

Example 5. Find the equation of the tangent line to the curve

$$y = \arctan 2x$$

at the point $x = 1/(2\sqrt{3})$.
Let $h(x) = \arctan 2x$. Then

$$h'\left(\frac{1}{2\sqrt{3}}\right) = \frac{2}{1 + 4\left(\frac{1}{2\sqrt{3}}\right)^2} = \frac{3}{2}.$$

When $x = 1/2\sqrt{3}$ we have

$$y = \arctan\left(2 \cdot \frac{1}{2\sqrt{3}}\right) = \arctan \frac{1}{\sqrt{3}} = \pi/6.$$

Hence we must find the equation of the line with slope $3/2$, passing through the point $(1/2\sqrt{3}, \pi/6)$. We know how to do this; the equation is

$$y - \frac{\pi}{6} = \frac{3}{2}\left(x - \frac{1}{2\sqrt{3}}\right).$$

Example 6. A balloon leaves the ground 100 m from an observer at the rate of 50 m/min. How fast is the angle of elevation of the observer's line of sight increasing when the balloon is at an altitude of 100 m?

The figure is as follows:

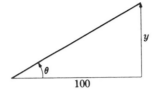

We have to determine $d\theta/dt$. We know that $dy/dt = 50$. We have

$$\frac{y}{100} = \tan\theta, \quad\text{whence}\quad \theta = \arctan\left(\frac{y}{100}\right).$$

Since

$$\frac{d\theta}{dt} = \frac{d\theta}{dy}\frac{dy}{dt}$$

we get:

$$\frac{d\theta}{dt} = \frac{1}{1 + \left(\dfrac{y}{100}\right)^2}\frac{1}{100}\, 50.$$

Hence

$$\left.\frac{d\theta}{dt}\right|_{y=100} = \frac{1}{1 + 1^2}\frac{1}{100}\, 50 = \frac{1}{4}\text{ rad/min.}$$

VII, §4. EXERCISES

1. Let g be the arctan function. What is $g(1)$? What is $g(1/\sqrt{3})$? What is $g(-1)$? What is $g(\sqrt{3})$?

2. Let g be the arctan function. What is $g'(1)$? What is $g'(1/\sqrt{3})$? What is $g'(-1)$? What is $g'(\sqrt{3})$?

3. Suppose you were to define an inverse function for the tangent in the interval $\pi/2 < x < 3\pi/2$. What would be the derivative of this inverse function?

4. What is
   (a) $\arctan(\tan 3\pi/4)$?   (b) $\arctan(\tan 2\pi)$?
   (c) $\arctan(\tan 5\pi/6)$?   (d) $\arctan(\tan(-5\pi/6))$?

Find the derivatives of the following functions.

5. $\arctan 3x$ 6. $\arctan \sqrt{x}$ 7. $\arcsin x + \arccos x$

8. $x \arcsin x$ 9. $\arctan(\sin 2x)$ 10. $x^2 \arctan 2x$

11. $\dfrac{\sin x}{\arcsin x}$ 12. $\arcsin(\cos x - x^2)$ 13. $\arctan \dfrac{1}{x}$

14. $\arctan \dfrac{1}{2x}$ 15. $(1 + \arcsin 3x)^3$ 16. $(\arcsin 2x + \arctan x^2)^{3/2}$

Find the equation of the tangent line at the indicated point for the following curves.

17. $y = \arcsin x$, $x = 1/\sqrt{2}$ 18. $y = \arccos x$, $x = 1/\sqrt{2}$

19. $y = \arctan 2x$, $x = \sqrt{3}/2$ 20. $y = \arctan x$, $x = -1$

21. $y = \arcsin x$, $x = -\frac{1}{2}$

22. A balloon leaves the ground 500 ft from an observer at the rate of 200 ft/min. How fast is the angle of elevation of the observer's line of sight increasing when the balloon is at an altitude of 1000 ft?

23. An airplane at an altitude of 4400 ft is flying horizontally directly away from an observer. When the angle of elevation is $\pi/4$, the angle is decreasing at the rate of 0.05 rad/sec. How fast is the airplane flying at that instant?

24. A man is walking along a sidewalk at the rate of 5 ft/sec. A searchlight on the ground 30 ft from the walk is kept trained on him. At what rate is the searchlight revolving when the man is 20 ft from the point on the sidewalk nearest the light?

25. A tower stands at the end of a street. A man drives toward the tower at the rate of 50 ft/sec. The tower is 500 ft tall. How fast is the angle subtended by the tower at the man's eye increasing when the man is 1000 ft from the tower?

26. A police car approaches an intersection at 80 ft/sec. When it is 200 ft from the intersection, a car crosses the intersection traveling at a right angle from the police car at the rate of 60 ft/sec. If the policeman directs his beam of light on this second car, how fast is the light beam turning 2 sec later, assuming that both cars continue at their original rate?

27. A weight is drawn along a level floor by means of a rope which passes over a hook 6 ft above the floor. If the rope is pulled over the hook at the rate of 1 ft/sec find an expression for the rate of change of the angle θ between the rope and the floor as a function of the angle θ.

28. A man standing at a fixed point on a wharf pulls in a small boat. The wharf is 20 ft above the level of the water. If he is pulling the rope at 2 ft/sec, how fast is the angle that the rope makes with the water increasing when the distance from the man to the boat is 50 ft?

29. A helicopter leaves the ground 1000 ft from an observer and rises vertically at 20 ft/sec. At what rate is the observer's angle of elevation of the helicopter changing when the helicopter is 800 ft above the ground?

30. Determine those intervals where arctan is bending upward and bending downward.

31. A helicopter leaves a base, rising straight up at a speed of 15 ft/sec. At the same time that the helicopter leaves, an observer starts from a point 100 ft away from the base, and moves on a straight line away from the base at a speed of 80 ft/sec. How fast is the angle of elevation from the observer to the helicopter increasing when the observer is (a) 400 ft from the base? (b) 600 ft from the base?

32. A train is moving on a straight line away from the station at a speed of 20 ft/sec. A cameraman starts from a point 50 ft away from the station at the same time that the train leaves, and, directing the camera toward the train, moves away from the station perpendicularly to the line made by the tracks, at a speed of 10 ft/sec. At what rate is the angle of the camera turning after the train has moved (a) 80 ft? (b) 100 ft?

33. A car is moving on a straight line toward the point where a rocket is being launched. The car is traveling at 50 ft/sec. When the car is 300 ft from the launching site, the rocket starts going up, and its height is given as function of time by $y = t^3$ ft. A person in the car is photographing the rocket. How fast is the angle of elevation of the camera turning 5 seconds after the rocket has started?

CHAPTER VIII

Exponents and Logarithms

We remember that we had trouble at the very beginning with the function 2^x (or 3^x, or 10^x). It was intuitively very plausible that there should be such functions, satisfying the fundamental equation

$$2^{x+y} = 2^x 2^y$$

for all numbers x, y, and $2^0 = 1$, but we had difficulties in saying what we meant by $2^{\sqrt{2}}$ (or 2^{π}).

It is the purpose of this chapter to study this function, and others like it.

VIII, §1. THE EXPONENTIAL FUNCTION

If n is a positive integer, we know what 2^n means: It is the product of 2 with itself n times. For instance, 2^8 is the product of 2 with itself eight times.

Furthermore, we also know that $2^{1/n}$ is the n-th root of 2; it is that number whose n-th power is 2. Thus $2^{1/8}$ is that number whose 8-th power is 2.

If $x = m/n$ is a quotient of two positive integers, then

$$2^{m/n} = (2^{1/n})^m = (2^m)^{1/n}$$

can be expressed in terms of roots and powers, so fractional powers of 2 are also easily understood. The problem arises in understanding 2^x when

x is not a quotient of two positive integers. We leave this problem aside for the moment, and assume that there is a function defined for all x, denoted by 2^x, which is differentiable. We shall now see how to find its derivative.

We form the Newton quotient. It is

$$\frac{2^{x+h} - 2^x}{h}.$$

Using the fundamental equation we see that this quotient is equal to

$$\frac{2^x 2^h - 2^x}{h} = 2^x \frac{2^h - 1}{h}.$$

As h approaches 0, 2^x remains fixed, but it is very difficult to see what happens to

$$\frac{2^h - 1}{h}.$$

It is not at all clear that this quotient approaches a limit. Roughly speaking, we meet a difficulty which is analogous to the one we met when we tried to find the derivative of $\sin x$. However, in the present situation, a direct approach would lead to much greater difficulties than those which we met when we discussed

$$\lim_{h \to 0} \frac{\sin h}{h}.$$

It is, in fact, true that

$$\lim_{h \to 0} \frac{2^h - 1}{h}$$

exists. We see that it does not depend on x. It depends only on 2.

If we tried to take the derivative of 10^x, we would end up with the problem of determining the limit

$$\lim_{h \to 0} \frac{10^h - 1}{h},$$

which is also independent of x.

In general, we shall assume the following.

Let a be a number >1. *There exists a function a^x, defined for all numbers x, satisfying the following properties:*

Property 1. The fundamental equation

$$a^{x+y} = a^x a^y$$

holds for all numbers x, y.

Property 2. If x is a rational number, $x = m/n$ with m, n positive integers, then $a^{m/n}$ has the usual meaning:

$$a^{m/n} = (a^{1/n})^m = (a^m)^{1/n}.$$

Property 3. The function a^x is differentiable.

The function a^x is called an **exponential function**.
We can then apply the same procedure to a^x that we applied to 2^x. We form the Newton quotient

$$\frac{a^{x+h} - a^x}{h} = \frac{a^x a^h - a^x}{h} = a^x\left(\frac{a^h - 1}{h}\right).$$

Since we assumed that a^x is differentiable, it follows that

$$\boxed{\frac{da^x}{dx} = \lim_{h\to 0} \frac{a^{x+h} - a^x}{h} = a^x \lim_{h\to 0} \frac{a^h - 1}{h}.}$$

Thus we meet the mysterious limit

$$\lim_{h\to 0} \frac{a^h - 1}{h}.$$

This is a similar situation to the differentiation of $\sin x$, but previously we were able to find the limit of $(\sin h)/h$ as h approaches 0. Here we cannot take a direct approach. The limit will be clarified later when we study the log.

However, we can analyze this limit a little more. Let

$$f(x) = a^x.$$

We claim that

$$a^0 = 1.$$

This is because

$$a = a^{1+0} = a \cdot a^0.$$

If we multiply by a^{-1} on both sides, we get $1 = a^0$.
    Similarly, we find

$$a^{-x} = \frac{1}{a^x},$$

because

$$1 = a^0 = a^{x-x} = a^x a^{-x}.$$

Now if we put $x = 0$ in the formula

$$f'(x) = a^x \lim_{h \to 0} \frac{a^h - 1}{h},$$

then we find that

$$f'(0) = \lim_{h \to 0} \frac{a^h - 1}{h}$$

because $a^0 = 1$. *Consequently the mysterious limit on the right-hand side is the slope of the curve $y = a^x$ at $x = 0$.*

Let us try to get a feeling for curves like 2^x, or 3^x, or 10^x by plotting points. We give a table of values for 2^x.

x	2^x
1	2
2	4
3	8
4	16
5	32
10	1024
20	1048576

x	2^x
-1	1/2
-2	1/4
-3	1/8
-4	1/16
-5	1/32
-10	1/1024
-20	1/1048576

We see that the value $y = 2^x$ increases rapidly when x becomes large, and approaches 0 rapidly when x becomes large negative, as illustrated on the figure.

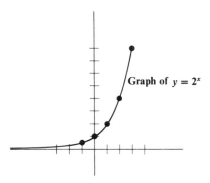

Graph of $y = 2^x$

The behavior when x becomes large negative is due to the relation $2^{-x} = 1/2^x$. For instance, $2^{-10} = 1/2^{10}$ which is small. We can write

$$\lim_{x \to -\infty} 2^x = 0.$$

Observe that $2^x > 0$ for all numbers x. Similarly, if $a > 0$, then

$$\boxed{a^x > 0 \qquad \text{for all } x.}$$

We can even prove this from what we have assumed explicitly. For suppose $a^c = 0$ with some number c. Then for all x we get

$$a^x = a^{x-c+c} = a^{x-c}a^c = 0,$$

which is not true since $a^1 = a \neq 0$.

Exercise. Make a similar table for 3^x, 10^x, and $(3/2)^x$.

Next suppose that $1 < a < b$. It is plausible that the curve b^x has a bigger slope than the curve a^x. We are especially interested in the slope when $x = 0$. If b is very large, then the curve $y = b^x$ will have a very steep slope at $x = 0$. If a is > 1 but close to 1, then the curve $y = a^x$ will have a small slope at $x = 0$. We have drawn these curves on the next figure.

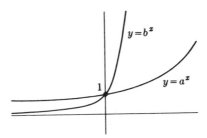

Try for yourself plotting some points on the curves 2^x, 3^x, 10^x to see what happens in these concrete cases. It is plausible that as a increases fron numbers close to 1 (and > 1) to very large numbers, the slope of a^x at $x = 0$ increases continuously from values close to 0 to large values, and therefore for some value of a, which we call e, this slope is precisely equal to 1. Thus in this naive approach, e is the number such that the slope of e^x at $x = 0$ is equal to 1, that is for $f(x) = e^x$, we have

$$\lim_{h \to 0} \frac{f(0 + h) - f(0)}{h} = \lim_{h \to 0} \frac{e^h - e^0}{h} = 1.$$

So in addition to the three properties stated previously, we assume:

Property 4. There is a number $e > 1$ such that

$$\frac{de^x}{dx} = e^x,$$

or equivalently,

$$\lim_{h \to 0} \frac{e^h - 1}{h} = 1.$$

This number e is called the **natural base for exponential functions**.

Warning. Do not confuse the functions 2^x and x^2. The derivative of x^2 is $2x$. The derivative of 2^x is

$$\frac{d(2^x)}{dx} = 2^x \lim_{h \to 0} \frac{2^h - 1}{h}.$$

Similarly, when a is a fixed number, do not confuse the function a^x and x^a, where x is the variable.

At first we have no idea how big or small the number e may be. In Exercises 16 through 20, you will learn a very efficient way of finding a decimal expansion, or approximations for e by rational numbers. It turns out that e lies between 2 and 3, and in particular is approximately equal to $2.7183\ldots$.

Assuming the basic properties of e^x as we have done, we can apply some of our previous techniques in the context of this exponential function. First we show that e^x is the only function equal to its own derivative, up to a constant factor.

Theorem 1.1. *Let $g(x)$ be a function defined for all numbers and such that $g'(x) = g(x)$. Then there is a constant C such that $g(x) = Ce^x$.*

Proof. We have to prove that $g(x)/e^x$ is constant. We know how to do this. It suffices to prove that the derivative is 0. But we find:

$$\frac{d}{dx}\left(\frac{g(x)}{e^x}\right) = \frac{e^x g'(x) - g(x)e^x}{e^{2x}}$$

$$= \frac{e^x g(x) - g(x)e^x}{e^{2x}}$$

$$= 0.$$

Hence there is a constant C such that $g(x)/e^x = C$. Multiplying both sides by e^x we get

$$g(x) = Ce^x,$$

thus proving the theorem.

As a special case of the theorem we have:

Let g be a differentiable function such that $g'(x) = g(x)$ and $g(0) = 1$. Then $g(x) = e^x$.

Proof. Since $g(x) = Ce^x$ we get $g(0) = Ce^0 = C$. Hence $C = 1$ and $g(x) = e^x$.

Thus there is one and only one function g which is equal to its own derivative and such that $g(0) = 1$. This function is called **the exponential function**, and is sometimes denoted by **exp**. We may write

$$\exp'(x) = \exp(x) \qquad \text{and} \qquad \exp(0) = 1.$$

But usually we use the notation e^x as before, instead of $\exp(x)$.

There are several ways of *proving* the existence of a function $g(x)$ such that $g'(x) = g(x)$ and $g(0) = 1$, rather than giving the plausibility arguments as above.

In Chapter XIV we shall give a proof by infinite series. On the other hand, when we study the logarithm in §6, we shall first show that there exists a function $L(x)$ such that $L'(x) = 1/x$ and $L(1) = 0$. Then we can define the inverse function, and it is easy to see that this inverse function g satisfies $g'(y) = g(y)$ and $g(0) = 1$. Anyone interested in such theory can suit their tastes and look up these later sections as they see fit.

We now give examples and applications involving the function e^x.

Example. Find the derivative of e^{3x^2}.
We use the chain rule, with $u = 3x^2$. Then

$$\frac{d(e^u)}{dx} = \frac{de^u}{du}\frac{du}{dx} = e^{3x^2} \cdot 6x.$$

Example. Let $f(x) = e^{\cos 2x}$. We find the derivative of f by the chain rule, namely

$$f'(x) = e^{\cos 2x}(-\sin 2x)2.$$

There is no point simplifying this expression.

Example. Find the equation of the tangent line to the curve $y = e^x$ at $x = 2$.
Let $f(x) = e^x$. Then $f'(x) = e^x$ and $f'(2) = e^2$. When $x = 2$, $y = e^2$. Hence we must find the equation of the line with slope e^2, passing through the point $(2, e^2)$. This equation is

$$y - e^2 = e^2(x - 2).$$

Graph of e^x

Let us sketch the graph of e^x. We justify our statements by using only the four properties listed above. Since

$$\frac{de^x}{dx} = e^x > 0 \qquad \text{for all } x,$$

we conclude that the function $f(x) = e^x$ is strictly increasing. Since

$$f''(x) = f'(x) = f(x) > 0 \qquad \text{for all } x,$$

we conclude that the function is bending up.

Since $f(0) = 1$ and the function is strictly increasing, we conclude that

$$f(1) = e > 1.$$

Hence when n is a positive integer, $n = 1, 2, 3, \ldots,$ the powers e^n become large as n becomes large. Since e^x is strictly increasing for all x, this also shows that e^x becomes large when x is a large real number.

We had also seen that

$$e^{-x} = (e^x)^{-1}.$$

Hence when x is large, the inverse

$$(e^x)^{-1} = 1/e^x$$

is small (positive).

Thus we may write:

$$\boxed{\begin{aligned} &\text{If } \quad x \to \infty \qquad \text{then} \quad e^x \to \infty. \\ &\text{If } \quad x \to -\infty \quad \text{then} \quad e^x \to 0. \end{aligned}}$$

We are now in a position to see that the graph of e^x looks like this:

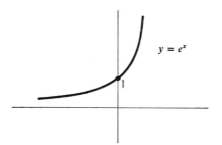

$y = e^x$

VIII, §1. EXERCISES

1. What is the equation of the tangent line to the curve $y = e^{2x}$ at the point whose x-coordinate is (a) 1, (b) -2, (c) 0?

2. What is the equation of the tangent line to the curve $y = e^{x/2}$ at the point whose x-coordinate is (a) -4, (b) 1, (c) 0?

3. What is the equation of the tangent line to the curve $y = xe^x$ at the point whose x-coordinate is 2?

4. Find the derivatives of the following functions:
   (a) $e^{\sin 3x}$ (b) $\sin(e^x + \sin x)$
   (c) $\sin(e^{x+2})$ (d) $\sin(e^{4x-5})$

5. Find the derivatives of the following functions:
   (a) $\arctan e^x$ (b) $e^x \cos(3x + 5)$
   (c) $e^{\sin 2x}$ (d) $e^{\arccos x}$
   (e) $1/e^x$ (f) x/e^x
   (g) e^{e^x} (h) $e^{-\arcsin x}$
   (i) $\tan(e^x)$ (j) $\arctan e^{2x}$
   (k) $1/(\sin e^x)$ (l) $\arcsin(e^x + x)$
   (m) $e^{\tan x}$ (n) $\tan e^x$

6. (a) Show that the n-th derivative of xe^x is $(x + n)e^x$ for $n = 1, 2, 3, 4, 5$.
   (b) Show that the n-th derivative of xe^{-x} is $(-1)^n(x - n)e^{-x}$ for $n = 1, 2, 3, 4, 5$.
   (c) Suppose you have already proved the above formulas for the n-th derivative of xe^x and xe^{-x}. How would you proceed to prove these formulas for the $(n + 1)$-th derivative?

7. Let $f(x)$ be a function such that $f'(x) = f(x)$ and $f(0) = 2$. Determine f completely in terms of e^x.

8. (a) Let $f(x)$ be a differentiable function over some interval satisfying the relation $f'(x) = Kf(x)$ for some constant K. Show that there is a constant C such that $f(x) = Ce^{Kx}$. [*Hint*: Show that the function $f(x)/e^{Kx}$ is constant.]
   (b) Let f be a differentiable function such that $f'(x) = -2xf(x)$. Show that there is a constant C such that $f(x) = Ce^{-x^2}$.
   (c) In general suppose there is a function h such that $f'(x) = h'(x)f(x)$. Show that $f(x) = Ce^{h(x)}$. [*Hint*: Show that the function $f(x)/e^{h(x)}$ is constant.] The technique of this exercise will be used in applications in the last section.

Find the tangent line to the curve at the indicated point.

9. $y = e^{2x}$, $x = 1$ 10. $y = xe^x$, $x = 2$

11. $y = xe^x$, $x = 5$ 12. $y = xe^{-x}$, $x = 0$

13. $y = e^{-x}$, $x = 0$ 14. $y = x^2e^{-x}$, $x = 1$

15. Prove that there is a unique number x such that $e^x + x = 0$. [*Hint*: Show that the function is strictly increasing, and has positive and negative values.]

16. Prove the inequalities for $x > 0$:

   (a) $1 < e^x$ (b) $1 + x < e^x$ (c) $1 + x + \dfrac{x^2}{2} < e^x$

   [*Hint*: Using the method of Chapter V, §2, first prove (a). Then prove (b) using (a). Then prove (c) using (b).]

17. Let $x = 1$ in Exercise 16. Show that $2 < e$. Also show that $2.5 < e$.

18. Prove for $n = 3, 4, 5, 6$ that for $x > 0$ we have

$$1 + x + \frac{x^2}{2!} + \cdots + \frac{x^n}{n!} < e^x.$$

By $n!$ (read n **factorial**) we mean the product of the first n integers. For instance:

$$\begin{aligned} 1! &= 1 & 4! &= 24 \\ 2! &= 2 & 5! &= 120 \\ 3! &= 6 & 6! &= 720 \end{aligned}$$

19. For $x > 0$ prove:

(a) $1 - x < e^{-x}$ (b) $e^{-x} < 1 - x + \frac{x^2}{2}$

(c) $1 - x + \frac{x^2}{2} - \frac{x^3}{3 \cdot 2} < e^{-x}$

(d) $e^{-x} < 1 - x + \frac{x^2}{2} - \frac{x^3}{3 \cdot 2} + \frac{x^4}{4 \cdot 3 \cdot 2}$

20. (a) Let $x = 1/2$ in Exercise 19(a). Show that $e < 4$.
    (b) Let $x = 1$ in Exercise 19(c). Show that $e < 3$.

Hyperbolic functions

21. (a) Define the functions **hyperbolic cosine** and **hyperbolic sine** by the formulas

$$\cosh t = \frac{e^t + e^{-t}}{2} \quad \text{and} \quad \sinh t = \frac{e^t - e^{-t}}{2}.$$

   Show that their derivatives are given by

$$\cosh' = \sinh \quad \text{and} \quad \sinh' = \cosh.$$

(b) Show that for all t we have

$$\cosh^2 t - \sinh^2 t = 1.$$

Note: We see that the functions $\cosh t$ and $\sinh t$ satisfy the equation of a hyperbola, in a way similar to ordinary sine and cosine which satisfy the equation of a circle, namely

$$\sin^2 t + \cos^2 t = 1.$$

That is the reason $\cosh t$ and $\sinh t$ are called hyperbolic cosine and hyperbolic sine respectively.

22. Sketch the graph of the function

$$f(x) = \frac{e^x + e^{-x}}{2}.$$

Plot at least six points on this graph.

23. Sketch the graph of the function

$$f(x) = \frac{e^x - e^{-x}}{2}.$$

Plot at least six points on this graph.

24. Let $f(x) = \frac{1}{2}(e^x + e^{-x}) = \cosh x = y$.
    (a) Show that f is strictly increasing for $x \geq 0$.
        Then the inverse function exists for this interval. Denote this inverse function by $x = \text{arccosh } y = g(y)$.
    (b) For which numbers y is $\text{arccosh } y$ defined?
    (c) Show that

$$g'(y) = \frac{1}{\sqrt{y^2 - 1}}.$$

25. Let $f(x) = \frac{1}{2}(e^x - e^{-x}) = \sinh x = y$.
    (a) Show that f is strictly increasing for all x. Let $x = \text{arcsinh } y$ be the inverse function.
    (b) For which numbers y is $\text{arcsinh } y$ defined?
    (c) Let $g(y) = \text{arcsinh } y$. Show that

$$g'(y) = \frac{1}{\sqrt{1 + y^2}}.$$

VIII, §2. THE LOGARITHM

If $e^x = y$, then we define $x = \log y$. *Thus the* log, *here and thereafter, is what some call the natural* log. We don't deal with any other log. By definition, we therefore have:

$$\boxed{e^{\log x} = x \quad \text{and} \quad \log e^x = x.}$$

Thus log is the inverse function of the exponential function e^x. Since e^x is strictly increasing, the inverse function exists.

Examples. We have

$$\log e^2 = 2, \qquad \log e^{-\sqrt{2}} = -\sqrt{2},$$

$$\log e^{-3} = -3, \qquad \log e^{\pi} = \pi.$$

And the other way:

$$e^{\log 2} = 2, \qquad e^{\log \pi} = \pi.$$

Furthermore, the relation $e^0 = 1$ means that

$$\boxed{\log 1 = 0.}$$

Since all values e^x are positive for all numbers x, it follows that

$$\log y \quad \textit{is defined only for positive numbers } y.$$

The rule $e^{a+b} = e^a e^b$ translates into a rule for the log, as follows.

Theorem 2.1. *If* u, v *are* > 0, *then*

$$\boxed{\log uv = \log u + \log v.}$$

Proof. Let $a = \log u$ and $b = \log v$. Then

$$e^{a+b} = e^a e^b = e^{\log u} e^{\log v} = uv.$$

By definition, the relation

$$e^{a+b} = uv$$

means that

$$\log uv = a + b = \log u + \log v$$

as was to be shown.

Theorem 2.2. *If* $u > 0$, *then*

$$\boxed{\log u^{-1} = -\log u.}$$

Proof. We have $1 = uu^{-1}$. Hence

$$0 = \log 1 = \log(uu^{-1}) = \log u + \log u^{-1}.$$

Adding $-\log u$ to both sides proves the theorem.

Examples. We have

$$\log(1/2) = -\log 2$$

$$\log(2/3) = \log 2 - \log 3.$$

Of course, we can take the log of a product with more than two terms, just as we can take the exponential of a sum of more than two terms. For instance

$$e^{a+b+c} = e^{a+b}e^c = e^a e^b e^c.$$

Similarly, if n is a positive integer, then

$$e^{na} = e^{a+a+\cdots+a} = e^a e^a \cdots e^a = (e^a)^n,$$

where the product on the right is taken n times.

We have the corresponding rule for the log, namely

$$\boxed{\log(u^n) = n \log u.}$$

For instance, by Theorem 2.1, we find:

$$\log(u^2) = \log(u \cdot u) = \log u + \log u = 2 \log u.$$

$$\log(u^3) = \log(u^2 u) = \log u^2 + \log u$$

$$= 2 \log u + \log u$$

$$= 3 \log u.$$

And so forth, to get $\log u^n = n \log u$.

It now follows that if n is a positive integer, then

$$\boxed{\log u^{1/n} = \frac{1}{n} \log u.}$$

Proof. Let $v = u^{1/n}$. Then $v^n = u$, and we have already seen that

$$\log v^n = n \log v.$$

Hence

$$\log v = \frac{1}{n} \log v^n,$$

which is precisely the relation $\log u^{1/n} = 1/n \log u$.

The same type of rule holds for fractional exponents, that is:

If m, n are positive integers, then

$$\log u^{m/n} = \frac{m}{n} \log u.$$

Proof. We write $u^{m/n} = (u^m)^{1/n}$. Then

$$\log u^{m/n} = \log(u^m)^{1/n}$$

$$= \frac{1}{n} \log u^m$$

$$= \frac{m}{n} \log u$$

by using the two cases separately.

Just to give you a feeling for the behavior of the log, we give a few approximate values:

$$\log 10 = 2.3,\ldots, \qquad \log 10,000 = 9.2,\ldots,$$

$$\log 100 = 4.6,\ldots, \qquad \log 100,000 = 11.5,\ldots,$$

$$\log 1000 = 6.9,\ldots, \qquad \log 1,000,000 = 13.8,\ldots.$$

You can see that if x grows like a geometric progression, then $\log x$ grows like an arithmetic progression. The above values illustrate the rule

$$\log 10^n = n \log 10,$$

where $\log 10$ is approximately 2.3.

In Exercises 17 and 19 of §1 you should have proved that $2.5 < e < 3$. Make up a table of the values e^n and $\log e^n = n$. You can then compare the growth of e^n with $\log e^n$ in a similar way. For positive integers you can then see that $\log e^n$ grows very slowly compared to e^n. For instance,

$$\log e^3 = 3,$$

$$\log e^4 = 4,$$

$$\log e^5 = 5,$$

$$\log e^{10} = 10.$$

Using the fact that e lies between 2 and 3, you can see that powers like e^5 or e^{10} are quite large compared to the values of the log, which are 5 and 10, respectively, in these cases. For instance, since $e > 2$ we have

$$e^{10} > 2^{10} > 1{,}000.$$

We have the same phenomenon in the opposite direction for negative powers of e. For instance:

$$\log \frac{1}{e} = -1,$$

$$\log \frac{1}{e^2} = -2,$$

$$\log \frac{1}{e^3} = -3,$$

$$\log \frac{1}{e^{10}} = -10.$$

Put $h = 1/e^y$. As h approaches 0, y becomes large positive, but rather slowly. Make a similar table for $\log(1/10^n)$ with $n = 1, 2, 3, 4, 5, 6$ to get a feeling for numerical examples.

Observe that if x is a small positive number and we write $x = e^y$ then $y = \log x$ is large negative. For instance

if $x = 1/e^{10^6} = e^{-10^6}$ then $\log x = -10^6,$

if $x = 1/e^{10^{100}} = e^{-10^{100}}$ then $\log x = -10^{100}.$

In short:

$$\boxed{\text{If}\quad x \to 0 \quad \text{then}\quad \log x \to -\infty.}$$

The reason comes from the behavior of e^y. If $y \to -\infty$ then $e^y \to 0$. Similarly, if $y \to \infty$ then $e^y \to \infty$. This translates into the corresponding property of the inverse function:

$$\boxed{\text{If}\quad x \to \infty \quad \text{then}\quad \log x \to \infty.}$$

The derivative of log

Next we consider the differentiation properties of the log function. Let

$$y = e^x \quad \text{and} \quad x = \log y.$$

By the rule for differentiating inverse functions, we find:

$$\frac{dx}{dy} = \frac{1}{dy/dx} = \frac{1}{e^x} = \frac{1}{y}.$$

Hence we have the formula:

Theorem 2.3.

$$\boxed{\frac{d \log y}{dy} = \frac{1}{y}.}$$

From the graph of e^x, we see that e^x takes on all values > 0. Hence the inverse function log is defined for all positive real numbers, and by the general way of finding the graph of an inverse function, we see that its graph looks like that in the figure.

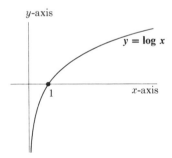

In the figure, the graph crosses the horizontal axis at 1, because

$$e^0 = 1 \quad \text{means} \quad \log 1 = 0.$$

Note that the derivative satisfies

$$\frac{d \log x}{dx} = \frac{1}{x} > 0 \quad \text{for all} \quad x > 0,$$

so the log function is strictly increasing.

Furthermore

$$\frac{d^2 \log x}{dx^2} = -\frac{1}{x^2} < 0.$$

We conclude that the log function is bending down as shown.

Remark. We shall sometimes consider composite functions of the type $\log(f(x))$. Since the log is not defined for numbers < 0, the expression $\log(f(x))$ is defined only for numbers x such that $f(x) > 0$. This is to be understood whenever we write such an expression.

Thus when we write $\log(x - 2)$, this is defined only when $x - 2 > 0$, in other words $x > 2$. When we write $\log(\sin x)$, this is meaningful only when $\sin x > 0$. It is not defined when $\sin x \leq 0$.

Example. Find the tangent line to the curve $y = \log(x - 2)$ at the point $x = 5$.

Let $f(x) = \log(x - 2)$. Then $f'(x) = 1/(x - 2)$, and

$$f'(5) = 1/3.$$

When $x = 5$, $\log(x - 2) = \log 3$. We must find the equation of the line with slope $1/3$, passing through $(5, \log 3)$. This is easy, namely:

$$y - \log 3 = \tfrac{1}{3}(x - 5).$$

Example. Sketch the graph of the function $f(x) = x^2 + \log x$, for $x > 0$.

We begin by taking the derivative, namely

$$f'(x) = 2x + \frac{1}{x}.$$

The function f has a critical point precisely when $2x = -1/x$, that is $2x^2 = -1$. This can never be the case. Hence there is no critical point. When $x > 0$, the derivative is positive. Hence in this interval, the function is strictly increasing.

When x becomes large positive, both x^2 and $\log x$ become large positive. Hence

$$\text{if} \quad x \to \infty \quad \text{then} \quad f(x) \to \infty.$$

As x approaches 0 from the right, x^2 approaches 0, but $\log x$ becomes large negative. Hence

$$\text{if} \quad x \to 0 \quad \text{and} \quad x > 0 \quad \text{then} \quad f(x) \to -\infty.$$

Finally, to determine the regions where f is bending up or down, we take the second derivative, and find

$$f''(x) = 2 - \frac{1}{x^2} = \frac{2x^2 - 1}{x^2}.$$

Then:

$$f''(x) > 0 \quad \Leftrightarrow \quad 2x^2 - 1 > 0 \quad \Leftrightarrow \quad x > 1/\sqrt{2}$$

$$\Leftrightarrow \quad f \text{ is bending up.}$$

Similarly,

$$f''(x) < 0 \quad \Leftrightarrow \quad 2x^2 - 1 < 0 \quad \Leftrightarrow \quad x < 1/\sqrt{2}$$

$$\Leftrightarrow \quad f \text{ is bending down.}$$

Hence $1/\sqrt{2}$ is an inflection point. We claim that

$$f(1/\sqrt{2}) > 0.$$

Indeed,

$$f(1/\sqrt{2}) = \frac{1}{2} - \log(\sqrt{2}) = \frac{1}{2} - \frac{1}{2}\log 2.$$

But the log is strictly increasing, and $2 < e$ so

$$\log 2 < \log e = 1.$$

Therefore $1 - \log 2 > 0$. This proves that $f(1/\sqrt{2}) > 0$. It follows that the graph of f looks like this.

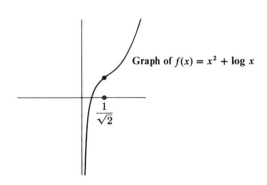

Graph of $f(x) = x^2 + \log x$

$\dfrac{1}{\sqrt{2}}$

VIII, §2. EXERCISES

1. What is the tangent line to the curve $y = \log x$ at the point whose x-coordinate is (a) 2, (b) 5, (c) $\frac{1}{2}$?

2. What is the equation of the tangent line of the curve $y = \log(x^2 + 1)$ at the point whose x-coordinate is (a) -1, (b) 2, (c) -3?

3. Find the derivatives of the following functions:

   (a) $\log(\sin x)$ (b) $\sin(\log(2x + 3))$ (c) $\log(x^2 + 5)$ (d) $\dfrac{\log 2x}{\sin x}$

4. What is the equation of the tangent line of the curve $y = \log(x + 1)$ at the point where x-coordinate is 3?

5. What is the equation of the tangent line of the curve $y = \log(2x - 5)$ at the point whose x-coordinate is 4?

6. (a) Prove that $\log(1 + x) < x$ for all $x > 0$. [*Hint*: Let $f(x) = x - \log(1 + x)$, find $f(0)$, and show that f is strictly increasing for $x \geq 0$.]
   (b) For $x > 0$ show that

$$\frac{x}{1 + x} < \log(1 + x).$$

Find the tangent line to the curve at the indicated point.

7. $y = \log x$,   at $x = e$ 8. $y = x \log x$,   at $x = e$

9. $y = x \log x$,   at $x = 2$ 10. $y = \log(x^3)$,   at $x = e$

11. $y = \dfrac{1}{\log x}$,   at $x = e$ 12. $y = \dfrac{1}{\log x}$,   at $x = 2$

Differentiate the following functions.

13. $\log(2x + 5)$ 14. $\log(x^2 + 3)$ 15. $\dfrac{1}{\log x}$

16. $\dfrac{x}{\log x}$ 17. $x(\log x)^{1/3}$ 18. $\log \sqrt{1 - x^2}$

19. Sketch the curve $y = x + \log x$, $x > 0$.

20. Prove that there is a unique number $x > 0$ such that $\log x + x = 0$. [*Hint*: Show that the function is strictly increasing and takes on positive and negative values. Use the intermediate value theorem.]

VIII, §3. THE GENERAL EXPONENTIAL FUNCTION

Let a be a number > 0. In §1 we listed four properties of the function a^x, with x the variable. We shall now list one more:

Property 5. For all numbers x, y we have

$$(a^x)^y = a^{xy}.$$

For example, if $x = n$ and $y = m$ are positive integers, then $(a^m)^n$ is the product of a^m with itself n times, which is equal to a^{mn}. We shall now deduce some consequences from this property.

First, from the preceding section, we know that

$$a = e^{\log a}.$$

Therefore

$$a^x = (e^{\log a})^x.$$

By Property 5, this yields

$$\boxed{a^x = e^{x \log a}}$$

because $(\log a)x = x \log a$. Thus for instance,

$$2^x = e^{x \log 2}, \qquad \pi^x = e^{x \log \pi}, \qquad 10^x = e^{x \log 10}.$$

The above formula allows us to find the derivative of a^x. Remember that a is viewed as *constant*.

Theorem 3.1. *We have*

$$\boxed{\frac{d(a^x)}{dx} = a^x(\log a).}$$

Proof. We use the chain rule. Let $u = (\log a)x$. Then $du/dx = \log a$ and $a^x = e^u$. Hence

$$\frac{d(a^x)}{dx} = \frac{d(e^u)}{du} \frac{du}{dx} = e^u(\log a) = a^x \log a$$

as desired.

Example.

$$\frac{d(2^x)}{dx} = 2^x \log 2.$$

Warning. The derivative of x^x is **NOT** $x^x \log x$. Work it out in Exercise 7. The difference between x^x and a^x (like 2^x, or 10^x) is that a is constant, whereas in the expression x^x, the variable x appears twice.

The result of Theorem 3.1 clarifies the mysterious limit

$$\lim_{h \to 0} \frac{2^h - 1}{h}$$

which we encountered in §1. We shall now see that this limit is equal to $\log 2$. More generally, let $a > 0$ and let

$$f(x) = a^x.$$

In §1 we gave the argument that

$$f'(x) = a^x \lim_{h \to 0} \frac{a^h - 1}{h}.$$

By Theorem 3.1 we know also that

$$f'(x) = a^x \log a.$$

Therefore

$$\boxed{\lim_{h \to 0} \frac{a^h - 1}{h} = \log a.}$$

Since the log is strictly increasing, there is only one number a such that $\log a = 1$, and that number is $a = e$. Thus the exponential function e^x is the only one among all possible exponential functions a^x whose derivative is equal to itself. By Theorem 3.1, since

$$\frac{da^x}{dx} = a^x \log a,$$

if $a \neq e$ then we get the factor $\log a \neq 1$ coming into the formula for the derivative of a^x.

As an application of our theory of the exponential function, we also can take care of the general power function (which we had left dangling in Chapter III).

Theorem 3.2. *Let c be any number, and let*

$$f(x) = x^c$$

be defined for $x > 0$. Then $f'(x)$ exists and is equal to

$$f'(x) = cx^{c-1}.$$

Proof. Put $u = c \log x$. By definition,

$$f(x) = e^{c \log x} = e^u.$$

Then

$$\frac{du}{dx} = \frac{c}{x}.$$

Using the chain rule, we see that

$$f'(x) = e^u \cdot \frac{du}{dx} = e^{c \log x} \cdot \frac{c}{x} = x^c \cdot \frac{c}{x} = cx^c x^{-1} = cx^{c-1}.$$

This proves our theorem.

Warning. the number c in Theorem 3.2 is constant, and x is the variable. **Do not confuse**

$$\frac{dc^x}{dx} = c^x \log c \qquad \text{and} \qquad \frac{dx^c}{dx} = cx^{c-1}.$$

Example. Find the tangent line to the curve $y = x^x$ at $x = 2$. We can write the function x^x in the form

$$y = f(x) = e^{x \log x}.$$

Then

$$f'(x) = e^{x \log x}\left(x \cdot \frac{1}{x} + \log x\right)$$

$$= x^x(1 + \log x).$$

In particular, we get the derivative (slope) at $x = 2$,

$$f'(2) = 2^2(1 + \log 2) = 4(1 + \log 2).$$

We have $f(2) = 2^2 = 4$. Therefore the equation of the tangent line at $x = 2$ is

$$y - 4 = 4(1 + \log 2)(x - 2).$$

Definition. When x, y are two numbers such that $y = 2^x$, it is customary to say that x is the **log of** y **to the base** 2. Similarly, if a is a number > 0, and $y = a^x = e^{x \log a}$, we say that x is the **log of** y **to the base** a. When $y = e^x$, we simply say that $x = \log y$.

The log to the base a is sometimes written $\log_a$.
We conclude this section by discussing limits, which are now very easy to handle.
First, we have

Limit 1.
$$\lim_{h \to 0} \frac{1}{h} \log(1 + h) = 1.$$

Indeed, the limit on the left is nothing else but the limit of the Newton quotient

$$\lim_{h \to 0} \frac{\log(1 + h) - \log 1}{h} = \log'(1).$$

Since $\log'(x) = 1/x$, it follows that $\log'(1) = 1$, as desired.

Limit 2.
$$\lim_{h \to 0} (1 + h)^{1/h} = e.$$

Proof. We have
$$(1 + h)^{1/h} = e^{(1/h)\log(1 + h)}$$

because by definition $a^x = e^{x \log a}$. We have just seen that

$$\frac{1}{h} \log(1 + h) \to 1 \qquad \text{as} \quad h \to 0.$$

Hence
$$e^{(1/h)\log(1 + h)} \to e^1 = e \qquad \text{as} \quad h \to 0.$$

This proves the desired limit.

Remark. We are using here the *continuity* of the function e^x.

If x approaches a number x_0 then e^x approaches e^{x_0}.

That the function e^x is continuous follows from our assumption that e^x is differentiable.

For instance, if x approaches $\sqrt{2}$ then e^x approaches $e^{\sqrt{2}}$.
If x approaches 1 then e^x approaches $e^1 = e$.
If x approaches 0 then e^x approaches $e^0 = 1$.

Let us go back to Limit 2. We reformulate this limit. Write $h = 1/x$. When h approaches 0 then x becomes large, that is

$$h \to 0 \quad \text{if and only if} \quad x \to \infty.$$

Therefore we find the limit:

Limit 3.
$$\lim_{x \to \infty} \left(1 + \frac{1}{x} \right)^x = e.$$

In the exercises, you will deduce easily from this limit that for $r > 0$, we have

$$\lim_{x \to \infty} \left(1 + \frac{r}{x} \right)^x = e^r.$$

This has an interesting application.

Example. Compound interest. Let an amount of A dollars be invested at yearly compound interest of $100r$ per cent where $r > 0$. Thus r is the ratio of the rate of interest of 100 per cent. Then this original amount increases to the following after the indicated number of years:

After 1 year: $A + rA = (1 + r)A$.
After 2 years: $(1 + r)A + r(1 + r)A = (1 + r)^2 A$.
After 3 years: $(1 + r)^2 A + r(1 + r)^2 A = (1 + r)^3 A$.

Continuing in this way, we conclude that after n years the amount is

$$A_n = (1 + r)^n A.$$

Now suppose that this same interest rate $100r$ per cent is compounded every $1/m$ years, where m is a positive integer. This is equivalent to saying that the rate is $100r/m$ per cent per $1/m$ years, compounded every $1/m$ years. Let us apply the preceding formula to the case where the unit of time is $1/m$ year. Then q years are equal to $qm \cdot 1/m$ years. Therefore, if the interest is compounded every $1/m$ years, after q years the amount is

$$A_{q,m} = \left(1 + \frac{r}{m}\right)^{qm} A.$$

The amount one gets after q years if the interest is compounded continuously is the limit of $A_{q,m}$ as $m \to \infty$. In light of the limit which you should have determined, we see that after q years of continuous compounding, the amount is

$$\lim_{m \to \infty} \left(1 + \frac{r}{m}\right)^{qm} A = e^{rq} A.$$

To give a numerical case, suppose 1,000 dollars return 15 per cent compounded continuously. Then $r = 15/100$. After 10 years, the amount will be

$$e^{\frac{15}{100} 10} \cdot 1,000 = e^{1.5} 1,000.$$

Since e is approximately 2.7 you can get a definite numerical answer.

VIII, §3. EXERCISES

1. What is the derivative of 10^x? 7^x?

2. What is the derivative of 3^x? π^x?

3. Sketch the curves $y = 3^x$ and $y = 3^{-x}$. Plot at least five points.

4. Sketch the curves $y = 2^x$ and $y = 2^{-x}$. Plot at least five points.

5. Find the equation of the tangent line to the curve $y = 10^x$ at $x = 0$.

6. Find the equation of the tangent line to the curve of $y = \pi^x$ at $x = 2$.

7. (a) What is the derivative of the function x^x (defined for $x > 0$)? [*Hint*: $x^x = e^{x \log x}$]

   (b) What is the derivative of the function $x^{(x^x)}$?

8. Find the equation of the tangent line to the curve $y = x^x$
   (a) at the point $x = 1$ (b) at $x = 2$ (c) at $x = 3$.

Find the tangent lines of the following curves:

9. $y = x^{\sqrt{x}}$ (a) at $x = 2$ (b) at $x = 5$

10. $y = x^{\sqrt[3]{x}}$ (a) at $x = 2$ (b) at $x = 5$

11. If a is a number > 1 and $x > 0$, show that

$$x^a - 1 \geqq a(x - 1).$$

12. Let a be a number > 0. Find the critical points of the function $f(x) = x^2/a^x$.

13. Let $0 < r$. Using Limit 3, prove the limit

$$\lim_{x \to \infty} \left(1 + \frac{r}{x}\right)^x = e^r.$$

[*Hint*: Let $x = ry$ and let $y \to \infty$.]

14. Show that

$$\lim_{n \to \infty} n(\sqrt[n]{a} - 1) = \log a.$$

[*Hint*: Let $h = 1/n$.] This exercise shows how approximations of the log can be obtained just by taking ordinary n-th roots. In fact, if we take $n = 2^k$ and use large integers k, we obtain arbitrarily good approximations of the log by extracting a succession of square roots. Do it on a pocket calculator to check it out.

VIII, §4. SOME APPLICATIONS

It is known (from experimental data) that when a piece of radium is left to disintegrate, the rate of disintegration is proportional to the amount of radium left. Two quantities are proportional when one is a constant multiple of the other.

Suppose that at time $t = 0$ we have 10 grams of radium and let $f(t)$ be the amount of radium left at time t. Then

$$\boxed{\frac{df}{dt} = Kf(t)}$$

for some constant K. We take K negative since the physical interpretation is that the amount of substance decreases.

Let us show that there is a constant C such that

$$\boxed{f(t) = Ce^{Kt}.}$$

If we take the derivative of the quotient

$$\frac{f(t)}{e^{Kt}}$$

and use the rule for the derivative of a quotient, we find

$$\frac{d}{dt}\left(\frac{f(t)}{e^{Kt}}\right) = \frac{e^{Kt}f'(t) - Ke^{Kt}}{e^{2Kt}} = 0$$

because $f'(t) = Kf(t)$. Since the derivative is 0, the quotient $f(t)/e^{Kt}$ is constant, or equivalently, there is a constant C such that

$$f(t) = Ce^{Kt}.$$

Let $t = 0$. Then $f(0) = C$. Thus $C = 10$, if we assumed that we started with 10 grams.

In general, if $f(t) = Ce^{Kt}$ is the function giving the amount of substance as a function of time, then

$$\boxed{f(0) = C,}$$

and C is interpreted as the amount of substance when $t = 0$, that is the original amount.

Similarly, consider a chemical reaction. It is frequently the case that the rate of the reaction is proportional to the quantity of reacting substance present. If $f(t)$ denotes the amount of substance left after time t, then

$$\frac{df}{dt} = Kf(t)$$

for some constant K (determined experimentally in each case). We are therefore in a similar situation as before, and

$$f(t) = Ce^{Kt},$$

where C is the amount of substance at $t = 0$.

Example 1. Suppose $f(t) = 10e^{Kt}$ where K is constant. Assume that $f(3) = 5$. Find K.

We have

$$5 = 10e^{K3}$$

and therefore

$$e^{3K} = \tfrac{5}{10} = \tfrac{1}{2},$$

whence

$$3K = \log(1/2) \qquad \text{and} \qquad K = \frac{-\log 2}{3}.$$

Example 2. Sugar in water decomposes at a rate proportional to the amount still unchanged. If 50 lb of sugar reduce to 15 lb in 3 hr, when will 20 per cent of the sugar be decomposed?

Let $S(t)$ be the amount of sugar undecomposed, at time t. Then by hypothesis,

$$S(t) = Ce^{-kt},$$

for suitable constants C and k. Furthermore, since $S(0) = C$, we have $C = 50$. Thus

$$S(t) = 50e^{-kt}.$$

We also have

$$S(3) = 50e^{-3k} = 15$$

so

$$e^{-3k} = \tfrac{15}{50} = \tfrac{3}{10}.$$

Thus we can solve for k, namely we take the log and get

$$-3k = \log(3/10),$$

whence

$$-k = \tfrac{1}{3}\log(3/10).$$

When 20 per cent has decomposed then 80 per cent is left. Note that 80 per cent of 50 is 40. We want to find t such that

$$40 = 50e^{-kt},$$

or in other words,

$$e^{-kt} = \tfrac{40}{50} = \tfrac{4}{5}.$$

We obtain

$$-kt = \log(4/5),$$

whence

$$t = \frac{\log(4/5)}{-k} = 3\,\frac{\log(4/5)}{\log(3/10)}.$$

This is our answer.

Remark. It does not make any difference whether originally we let

$$S(t) = Ce^{-kt} \qquad \text{or} \qquad S(t) = Ce^{Kt}.$$

We could also have worked the problem the other way. For applications, when substances decrease, it is convenient to use a convention such that $k > 0$ so that the expression e^{-kt} decreases when t increases. But mathematically the procedures are equivalent, putting $K = -k$.

Example 3. A radioactive substance disintegrates proportionally to the amount of substance present at a given time, say

$$f(t) = Ce^{-kt}$$

for some positive constant k. At what time will there be exactly 1/4-th of the original amount left?

To do this, we want to know the value of t such that

$$f(t) = C/4.$$

Thus we want to solve

$$Ce^{-kt} = C/4.$$

Note that we can cancel C to get $e^{-kt} = 1/4$. Taking logs yields

$$-kt = -\log 4,$$

whence

$$t = \frac{\log 4}{k}.$$

Observe that the answer is independent of the original amount C. Experiments also allow us to determine the constant k. For instance, if we can analyze a sample, and determine that 1/4-th is left after 1000 years, then we find that

$$k = \frac{\log 4}{1000}.$$

Example 4. Exponential growth also reflects population explosion. If $P(t)$ is the population as a function of time t, then its rate of increase is proportional to the total population, in other words,

$$\frac{dP}{dt} = KP(t)$$

for some positive constant K. It then follows that

$$P(t) = Ce^{Kt}$$

for some constant C which is the population at time $t = 0$.

Suppose we ask at what time the population will double. We must then find t such that

$$Ce^{Kt} = 2C,$$

or equivalently

$$e^{Kt} = 2.$$

Taking the log yields

$$Kt = \log 2,$$

whence

$$t = \frac{\log 2}{K}.$$

Note that this time depends only on the rate of change of the population, not on the original value of C.

VIII, §4. EXERCISES

1. Let $f(t) = 10e^{Kt}$ for some constant K. Suppose you know that $f(1/2) = 2$. Find K.

2. Let $f(t) = Ce^{2t}$. Suppose that you know $f(2) = 5$. Determine the constant C.

3. One gram of radium is left to disintegrate. After one million years, there is 0.1 gram left. What is the formula giving the rate of disintegration?

4. A certain chemical substance reacts in such a way that the rate of reaction is equal to the quantity of substance present. After one hour, there are 20 grams of substance left. How much substance was there at the beginning?

5. A radioactive substance disintegrates proportionally to the amount of substance present at a given time, say

$$f(t) = Ce^{Kt}.$$

At what time will there be exactly half the original amount left?

6. Suppose $K = -4$ in the preceding exercise. At what time will there be one-third of the substance left?

7. If bacteria increase in number at a rate proportional to the number present, how long will it take before 1,000,000 bacteria increase to 10,000,000 if it takes 12 minutes to increase to 2,000,000?

8. A substance decomposes at a rate proportional to the amount present. At the end of 3 minutes, 10 per cent of the original substance has decomposed. When will half the original amount have decomposed?

9. Let f be a function of a variable t and increasing at the rate $df/dt = kf$ where k is a constant. Let $a_n = f(nt_1)$ where t_1 is a fixed value of t, $t_1 > 0$. Show that $a_0, a_1, a_2, \ldots$ is a geometric progression.

10. In 1900 the population of a city was 50,000. In 1950 it was 100,000. If the rate of increase of the population is proportional to the population, what is the population in 1984? In what year is it 200,000?

11. Assume that the rate of change with respect to height of atmospheric pressure at any height is proportional to the pressure there. If the barometer reads 30 at sea level and 24 at 6000 ft above sea level, find the barometric reading 10,000 ft above sea level.

12. Sugar in water decomposes at a rate proportional to the amount still unchanged. If 30 lb of sugar reduces to 10 lb in 4 hr, when will 95 per cent of the sugar be decomposed?

13. A particle moves with speed $s(t)$ satisfying $ds/dt = -ks$, where k is some constant. If the initial speed is 16 units/min and if the speed is halved in 2 min, find the value of t when the speed is 10 units/min.

14. Assume that the difference x between the temperature of a body and that of surrounding air decreases at a rate proportional to this difference. If $x = 100°$ when $t = 0$, and $x = 40°$ when $t = 40$ minutes, find t (a) when $x = 70°$, (b) when $x = 16°$, (c) the value of x when $t = 20$.

15. A moron loses money in gambling at a rate equal to the amount he owns at any given time. At what time t will he have lost half of his initial capital?

16. It is known that radioactive carbon has a half-life of 5568 years, meaning that it takes that long for one-half of the original amount to decompose. Also, the rate of decomposition is proportional to the amount present, so that by what we have seen in the text, we have the formula

$$f(t) = Ce^{Kt}$$

for this amount, where C, are constants.
(a) Find the constant K explicitly.
(b) Some decomposed carbon is found in a cave, and an analysis shows that one-fifth of the original amount has decomposed. How long has the carbon been in the cave?

VIII, §5. ORDER OF MAGNITUDE

In this section we analyze more closely what we mean when we say that e^x grows much faster than x, and $\log x$ grows much slower than x, when x becomes large positive.

We consider the quotient

$$\frac{e^x}{x}$$

as x becomes large positive. Both the numerator and the denominator become large, and the question is, what is the behavior of the quotient?

First let us make a table for simple values $2^n/n$ when n is a *positive integer*, to see that $2^n/n$ becomes large as n becomes large, experimentally. We agree to the convention that n always denotes a positive integer, unless otherwise specified.

n	2^n	$2^n/n$
1	2	2
2	4	2
3	8	8/3
4	16	4
5	32	$32/5 > 6$
10	1,024	$102.4 > 100$
20	1,048,576	$52,428.8 > 5 \times 10^4$

Since $2 < e$, we have $2^n/n < e^n/n$, and we see *experimentally* that e^n/n becomes large. We now wish to *prove* this fact. We first prove some inequalities for e^x. We use techniques from the exercises of §1. We proceed stepwise. We consider $x \geq 0$.

(a) We first show that

$$1 + x < e^x \qquad \text{for} \quad x > 0.$$

Let $f_1(x) = e^x - (1 + x)$. Then $f'_1(x) = e^x - 1$. Since $e^x > 1$ for $x > 0$, we conclude that

$$f'_1(x) > 0 \qquad \text{for} \quad x > 0.$$

Therefore $f_1(x)$ is strictly increasing for $x \geq 0$. Since $f_1(0) = 0$, we conclude that $f_1(x) > 0$ for $x > 0$, which means

$$e^x - (1 + x) > 0, \qquad \text{or in other words,} \qquad e^x > 1 + x,$$

as was to be shown.

(b) Next we show that

$$1 + x + \frac{x^2}{2} < e^x \qquad \text{for} \quad x > 0.$$

Let $f_2(x) = e^x - (1 + x + x^2/2)$. Then $f_2(0) = 0$. Furthermore,

$$f'_2(x) = e^x - (1 + x).$$

By part (a), we know that $f'_2(x) > 0$ for $x > 0$. Hence f_2 is strictly increasing, and it follows that $f_2(x) > 0$ for $x > 0$, or in other words,

$$e^x - \left(1 + x + \frac{x^2}{2}\right) > 0 \qquad \text{for} \quad x > 0.$$

This proves the desired inequality.

Theorem 5.1. *The function e^x/x becomes large as x becomes large.*

Proof. We divide both sides of inequality (b) by x we obtain

$$\frac{1}{x} + 1 + \frac{x}{2} < \frac{e^x}{x}.$$

As x becomes large, so does the left-hand side, and Theorem 5.1 is proved.

Theorem 5.2. *The function e^x/x^2 becomes large as x becomes large. More generally, let m be a positive integer. Then*

$$\frac{e^x}{x^m} \to \infty \qquad \text{as} \quad x \to \infty.$$

Proof. We use the same method. First we prove the inequality

(c)
$$\boxed{\; 1 + x + \frac{x^2}{2!} + \frac{x^3}{3!} < e^x \qquad \text{for} \quad x > 0. \;}$$

Recall that by definition, $2! = 2$ and $3! = 3 \cdot 2 = 6$. This time we let

$$f_3(x) = e^x - \left(1 + x + \frac{x^2}{2!} + \frac{x^3}{3!}\right).$$

Then $f_3(0) = 0$. Furthermore, using inequality (b) we find

$$f'_3(x) = e^x - \left(1 + x + \frac{x^2}{2!}\right) = f_2(x) > 0 \qquad \text{for} \quad x > 0.$$

Hence $f_3(x)$ is strictly increasing, and therefore $f_3(x) > 0$ for $x > 0$. This proves inequality (c).

If we divide both sides of inequality (c) by x^2, then we find

$$\frac{1}{x^2} + \frac{1}{x} + \frac{1}{2} + \frac{x}{6} < \frac{e^x}{x^2}.$$

As x becomes large, the left-hand side becomes large, so e^x/x^2 becomes large. This proves the first statement of Theorem 5.2.

We can continue the same method to prove the general statement about e^x/x^n. First you should prove that

(d)
$$1 + x + \frac{x^2}{2!} + \frac{x^3}{3!} + \frac{x^4}{4!} < e^x \qquad \text{for} \quad x > 0$$

in order to get a good feeling for the stepwise procedure used. We shall now prove the general step using an arbitrary integer n. In general, let

$$P_n(x) = 1 + x + \frac{x^2}{2!} + \frac{x^2}{3!} + \cdots + \frac{x^n}{n!}.$$

Suppose we have already proved that

$$1 + x + \frac{x^2}{2!} + \cdots + \frac{x^n}{n!} < e^x \qquad \text{for} \quad x > 0,$$

or in other words, that

$$P_n(x) < e^x \qquad \text{for} \quad x > 0.$$

We shall then prove

$$P_{n+1}(x) < e^x \qquad \text{for} \quad x > 0.$$

To do this, we let

$$f_{n+1}(x) = e^x - P_{n+1}(x) \qquad \text{and} \qquad f_n(x) = e^x - P_n(x).$$

Then $f_{n+1}(0) = 0$ and $f'_{n+1}(x) = f_n(x) > 0$ for $x > 0$. Hence f_{n+1} is strictly increasing, and therefore $f_{n+1}(x) > 0$ for $x > 0$, as desired.

Therefore, given our integer m, we have an inequality

$$1 + x + \frac{x^2}{2} + \cdots + \frac{x^{m+1}}{(m+1)!} < e^x \qquad \text{for} \quad x > 0.$$

We divide both sides of this inequality by x^m. Then the left-hand side consists of a sum of positive terms, the last of which is

$$\frac{x}{(m+1)!}.$$

Hence we obtain the inequality

$$\frac{x}{(m+1)!} < \frac{e^x}{x^m} \qquad \text{for} \quad x > 0.$$

Since the left-hand side becomes large when x becomes large, so does the right-hand side, and Theorem 5.2 is proved.

Example. We sketch the graph of $f(x) = xe^x$. We have

$$f'(x) = xe^x + e^x = e^x(x + 1).$$

Since $e^x > 0$ for all x, we get:

$$f'(x) = 0 \quad \Leftrightarrow \quad x + 1 = 0 \quad \Leftrightarrow \quad x = -1,$$

$$f'(x) > 0 \quad \Leftrightarrow \quad x + 1 > 0 \quad \Leftrightarrow \quad x > -1,$$

$$f'(x) < 0 \quad \Leftrightarrow \quad x + 1 < 0 \quad \Leftrightarrow \quad x < -1.$$

There is only one critical point at $x = -1$, and the other inequalities give us the regions of increase and decrease for f.

As to the bending up or down:

$$f''(x) = e^x \cdot 1 + e^x(x + 1) = e^x(x + 2).$$

Therefore:

$$f''(x) = 0 \quad \Leftrightarrow \quad x = -2,$$

$$f''(x) > 0 \quad \Leftrightarrow \quad x > -2 \quad \Leftrightarrow \quad f \text{ is bending up,}$$

$$f''(x) < 0 \quad \Leftrightarrow \quad x < -2 \quad \Leftrightarrow \quad f \text{ is bending down.}$$

If $\quad x \to \infty \quad$ then $\quad e^x \to \infty \quad$ so $\quad f(x) \to \infty.$

If $\quad x \to -\infty \quad$ then we put $\quad x = -y \quad$ with $\quad y \to \infty.$

By Theorem 5.1,

$$xe^x = -ye^{-y} \to 0 \qquad \text{if} \quad y \to \infty.$$

Finally, $f(0) = 0$, $f(-1) = -1/e$, $f(-2) = -2/e^2$. Hence the graph looks like this.

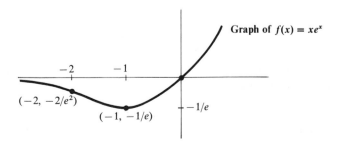

Graph of $f(x) = xe^x$

Example. Let $0 < a < 1$ and a is a fixed number. Find the maximum of the function

$$f(x) = xa^x.$$

First, take the derivative:

$$f'(x) = x \cdot a^x \log a + a^x$$

$$= a^x(x \log a + 1).$$

Since $a^x > 0$ for all x, we see that

$$f'(x) = 0 \quad \Leftrightarrow \quad x = -\frac{1}{\log a}.$$

Thus the function has exactly one critical point. Furthermore,

$$f'(x) > 0 \quad \Leftrightarrow \quad x \log a + 1 > 0 \quad \Leftrightarrow \quad x < -\frac{1}{\log a}$$

$$f'(x) < 0 \quad \Leftrightarrow \quad x \log a + 1 < 0 \quad \Leftrightarrow \quad x > -\frac{1}{\log a}.$$

(Remember that $0 < a < 1$, so that $\log a$ is negative.) Consequently the function is increasing on the interval to the left of the critical point, and decreasing on the interval to the right of the critical point. Therefore the critical point is the desired maximum. The value of f at this critical point is equal to

$$f(-1/\log a) = -\frac{1}{\log a} a^{-1/\log a} = -\frac{1}{\log a} e^{-\log a/\log a}$$

$$= -\frac{1}{e \log a}.$$

Example. Show that the equation $3^x = 5x$ has at least one solution.

Let $f(x) = 3^x - 5x$. Then $f(0) = 1$, and by trial and error we find a value where f is negative, namely

$$f(2) = 9 - 10 < 0.$$

By the intermediate value theorem, there exists some number x between 2 and 0 such that $f(x) = 0$, and this number fulfills our requirement.

From Theorems 5.1 and 5.2, by means of a change of variable, we can analyze what happens when comparing $\log x$ with powers of x.

Theorem 5.3. *As x becomes large, the quotient x/log x also becomes large.*

Proof. Our strategy is to reduce this statement to Theorem 5.1. We make a change of variables. Let $y = \log x$. Then $x = e^y$ and our quotient has the form

$$\frac{x}{\log x} = \frac{e^y}{y}.$$

We know that $y = \log x$ becomes large when x becomes large. So does e^y/y by Theorem 5.1. This proves the theorem.

Corollary 5.4. *As x becomes large, the function x − log x also becomes large.*

Proof. We write

$$x - \log x = x\left(1 - \frac{\log x}{x}\right),$$

that is we factor x in the expression $x - \log x$. By Theorem 5.3, $(\log x)/x$ approaches 0 as x becomes large. Hence the factor

$$1 - \frac{\log x}{x}$$

approaches 1. The factor x becomes large. Hence the product becomes large. This proves the corollary.

Remark. We have just used the same factoring technique that was used in analyzing the behavior of polynomials, as when we wrote

$$x^3 - 2x^2 + 5 = x^3\left(1 - \frac{2}{x} + \frac{5}{x^3}\right)$$

to see that the x^3 term determines the behavior of the polynomial when x becomes large.

Corollary 5.5. *As x becomes large, $x^{1/x}$ approaches 1 as a limit.*

Proof. We write

$$x^{1/x} = e^{(\log x)/x}.$$

By Theorem 5.3 we know that $(\log x)/x$ approaches 0 when x becomes large. Hence

$$e^{(\log x)/x}$$

approaches 1, as desired.

Remark. In Corollary 5.5 we used the fact that the function e^u is continuous, because any differentiable function is continuous. If u approaches u_0 then e^u approaches e^{u_0}. Thus if $u = (\log x)/x$, then u approaches 0 as x becomes large, so e^u approaches $e^0 = 1$.

VIII, §5. EXERCISES

1. Sketch the graph of the curve $y = xe^{2x}$. In this and other exercises, you may treat the convexity properties as optional, but it usually comes out easily.

   Sketch the graphs of the following functions. (In Exercises 6 through 8, $x \neq 0$.)

2. xe^{-x} 3. xe^{-x^2} 4. $x^2 e^{-x^2}$

5. $x^2 e^{-x}$ 6. e^x/x 7. e^x/x^2

8. e^x/x^3 9. $e^x - x$ 10. $e^x + x$

11. $e^{-x} + x$

12. Sketch the graph of $f(x) = x - \log x$.

13. Show that the equation $e^x = ax$ has at least one solution for any number a except when $0 \leq a < e$.

14. (a) Give values of $x \log x$ when $x = \frac{1}{2}, \frac{1}{4}, \frac{1}{8}, \ldots$, in general when $x = 1/2^n$ for some positive integer n.
    (b) Does $x \log x$ approach a limit as $x \to 0$? What about $x^2 \log x$? [*Hint*: Let $x = e^{-y}$ and let y become large.]

15. Let n be a positive integer. Prove that $x(\log x)^n \to 0$ as $x \to 0$.

16. Prove that $(\log x)^n/x \to 0$ as $x \to \infty$.

17. Sketch the following curves for $x > 0$.
    (a) $y = x \log x$ (b) $y = x^2 \log x$
    (c) $y = x(\log x)^2$ (d) $y = x/\log x$

18. Show that the function $f(x) = x^x$ is strictly increasing for $x > 1/e$.

19. Sketch the curve $f(x) = x^x$ for $x > 0$.

20. Sketch the curve $f(x) = x^{-x}$ for $x > 0$.

21. Let $f(x) = 2^x x^x$. Show that f is strictly increasing for $x > 1/2e$.

22. Find the following limits as $n \to \infty$
    (a) $(\log n)^{1/n}$ (b) $[(\log n)/n]^{1/n}$
    (c) $(n/e^n)^{1/n}$ (d) $(n \log n)^{1/n}$

VIII, §6. THE LOGARITHM AS THE AREA UNDER THE CURVE $1/x$

The present section is interesting for its own sake, because it gives us further insight into the logarithm. It also provides a very nice and concrete introduction to integration which is going to be covered in the next part. We shall give an interpretation of the logarithm as the area under a curve.

We define a function $L(x)$ to be the area under the curve $1/x$ between 1 and x if $x \geq 1$, and the negative of the area under the curve $1/x$ between 1 and x if $0 < x < 1$. In particular, $L(1) = 0$.

The shaded portion of the picture that follows represents the area under the curve between 1 and x. On the left we have taken $x > 1$.

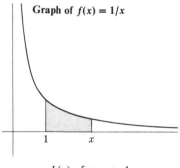
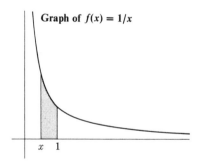

$L(x)$ for $x > 1$ $L(x)$ for $0 < x < 1$

If $0 < x < 1$, we would have the picture shown on the right. If $0 < x < 1$, we have said that $L(x)$ is equal to the negative of the area. Thus $L(x) < 0$ if $0 < x < 1$ and $L(x) > 0$ if $x > 1$.

We shall prove:

1. $L'(x) = 1/x$.
2. $L(x) = \log x$.

The first assertion that $L'(x) = 1/x$ is independent of everything else in this chapter, and we state it as a separate theorem.

Theorem 6.1. *The function $L(x)$ is differentiable, and*

$$\frac{dL(x)}{dx} = \frac{1}{x}.$$

Proof. We form the Newton quotient

$$\frac{L(x + h) - L(x)}{h}$$

and have to prove that it approaches $1/x$ as a limit when h approaches 0.

Let us take $x \geq 1$ and $h > 0$ for the moment. Then $L(x + h) - L(x)$ is the area under the curve between x and $x + h$. Since the curve $1/x$ is decreasing, this area satisfies the following inequalities:

$$h \frac{1}{x + h} < L(x + h) - L(x) < h \frac{1}{x}.$$

Indeed, $1/x$ is the height of the big rectangle as drawn on the next figure, and $1/(x + h)$ is the height of the small rectangle. Since h is the base of the rectangle, and since the area under the curve $1/x$ between x and

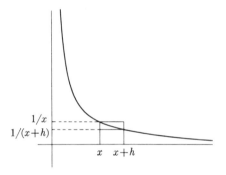

$x + h$ is in between the two rectangles, we see that it satisfies our inequalities. We divide both sides of our inequalities by the positive number h. Then the inequalities are preserved, and we get

$$\frac{1}{x + h} < \frac{L(x + h) - L(x)}{h} < \frac{1}{x}.$$

As h approaches 0, our Newton quotient is squeezed between $1/(x + h)$ and $1/x$ and consequently approaches $1/x$. This proves our theorem in case $h > 0$.

When $h < 0$ we use an entirely similar argument, which we leave as an exercise. (You have to pay attention to the sign of L. Also when you divide an inequality by h and $h < 0$, then the inequality gets reversed. However, you will again see that the Newton quotient is squeezed between $1/x$ and $1/(x + h)$.)

Theorem 6.2. *The function $L(x)$ is equal to $\log x$.*

Proof. Both functions $L(x)$ and $\log x$ have the same derivative, namely $1/x$ for $x > 0$. Hence there is a constant C such that

$$L(x) = \log x + C.$$

This is true for all $x > 0$. In particular, let $x = 1$. We obtain

$$0 = L(1) = \log 1 + C.$$

But $\log 1 = 0$. Hence $C = 0$, and the theorem is proved.

The identification of the log with the area under the curve $1/x$ can be used to give inequalities for the log. This is simple, and is given as an exercise. We can also obtain an estimate for e.

Example. The area under the curve $1/x$ between 1 and 2 is less than the area of a rectangle whose base is the interval $[1, 2]$ and whose height is 1, as shown on the following figure.

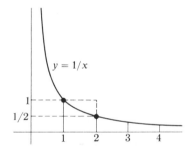

Hence we get the inequality

$$\log 2 < 1.$$

Since $\log e = 1$, it follows that $2 < e$. This gives us a lower estimate for e.
  Similarly we obtain an upper estimate as follows. The area under the curve $1/x$ between 1 and 2 is greater than the area of the rectangle whose base is the interval $[1, 2]$ and whose height is $1/2$, as shown on the above figure. Hence we get the inequality

$$\log 2 > \frac{1}{2}.$$

Then

$$\log 4 = \log(2^2) = 2 \log 2 > 2 \cdot \frac{1}{2} = 1.$$

Since $\log e = 1$ it follows that $e < 4$.

In Exercises 16 through 20 of §1 we had another method to obtain estimates for *e*. The method with the area under the curve can be used in other contexts and is independently useful.

VIII, §6. EXERCISES

1. Let *h* be a positive number. Compare the area under the curve $1/x$ between 1 and $1 + h$ with the area of suitable rectangles to show that

$$\frac{h}{1 + h} < \log(1 + h) < h.$$

2. Prove by using Exercise 1 that

$$\lim_{h \to 0} \frac{1}{h} \log(1 + h) = 1.$$

3. Prove by comparing areas, that for every positive integer *n*, we have

$$\frac{1}{n + 1} < \log\left(1 + \frac{1}{n}\right) < \frac{1}{n}.$$

4. Instead of using $\log 4 = \log(2^2)$ as in the text, use two rectangles under the graph of $1/x$, with bases $[1, 2]$ and $[2, 4]$ to show that $\log 4 > 1$.

VIII, APPENDIX. SYSTEMATIC PROOF OF THE THEORY OF EXPONENTIALS AND LOGARITHMS

Instead of assuming the five basic properties of the exponential function as in §1, and §3, we might have given a development of the log and exponential as follows. This is intended only for those interested in theory.

We *start* by defining $L(x)$ as we did in §6. The proof that $L'(x) = 1/x$ is self-contained, and yields a function L defined for all $x > 0$ and satisfying

$$L'(x) = 1/x \quad \text{and} \quad L(0) = 1.$$

Since $1/x > 0$ for all $x > 0$, it follows that the function L is strictly increasing, and so has an inverse function, which we denote by $x = E(y)$. Then using the rule for derivative of an inverse function, we find:

$$E'(y) = \frac{1}{L'(x)} = \frac{1}{1/x} = x = E(y).$$

Thus we have found a function E such that $E'(y) = E(y)$ for all y. In other words, we have found a function equal to its own derivative.

Since $L(0) = 1$ we find that $E(1) = 0$.

Next we prove:

For all numbers a, $b > 0$ we have

$$\boxed{L(ab) = L(a) + L(b).}$$

Proof. Fix the number a and let $f(x) = L(ax)$. By the chain rule, we obtain

$$f'(x) = \frac{1}{ax} \cdot a = \frac{1}{x}.$$

Since L and f have the same derivative, there is a constant C such that $f(x) = L(x) + C$ for all $x > 0$. In particular, for $x = 1$ we get

$$L(a) = f(1) = L(1) + C = 0 + C = C.$$

So $L(a) = C$ and $L(ax) = L(x) + L(a)$. This proves the first property.

Since $L'(x) = 1/x > 0$ for $x > 0$ it follows that L is strictly increasing. Since $L''(x) = -1/x^2 < 0$, it follows that L is bending down.

Let $a > 0$. We get:

$$L(a^2) = L(a) + L(a) = 2L(a),$$

$$L(a^3) = L(a^2 a) = L(a^2) + L(a) = 2L(a) + L(a) = 3L(a).$$

Continuing in this way, we get for all positive integers n:

$$L(a^n) = nL(a)$$

In particular, take $a > 1$. Since $L(1) = 0$ we conclude that $L(a) > 0$ because L is strictly increasing. Hence $L(a^n) \to \infty$ as $n \to \infty$. Again since L is strictly increasing, it follows that $L(x) \to \infty$ as $x \to \infty$.

From the formula

$$0 = L(1) = L(aa^{-1}) = L(a) + L(a^{-1})$$

we conclude that

$$\boxed{L(a^{-1}) = -L(a).}$$

Next, let x approach 0. Write $x = 1/y$ where $y \to \infty$. Then

$$L(x) = -L(y) \to -\infty \quad \text{as} \quad y \to \infty$$

so $L(x) \to -\infty$ as $x \to 0$.

Now let E be the inverse function of L. We already proved that $E' = E$. The inverse function of L is defined on the set of values of L, which is all numbers. The set of values of E is the domain of definition of L, which is the set of positive numbers. So $E(y) > 0$ for all y. Thus E is strictly increasing, and $E''(y) = E(y)$ for all y shows that the graph of E bends up.

We have $E(0) = 1$ because $L(1) = 0$.

Then we can prove

$$\boxed{E(u + v) = E(u)E(v).}$$

Namely, let $a = E(u)$ and $b = E(v)$. By the meaning of inverse function, $u = L(a)$ and $v = L(b)$. Then:

$$L(ab) = L(a) + L(b) = u + v.$$

Hence

$$E(u)E(v) = ab = E(u + v),$$

as was to be shown.

We now **define** $e = E(1)$. Since E is the inverse function of L we have $L(e) = 1$. From the rule

$$E(u + v) = E(u)E(v)$$

we now get for any positive integer n that

$$E(n) = E(1 + 1 + \cdots + 1) = E(1)^n = e^n.$$

Similarly,

$$E(nu) = E(u)^n.$$

Put $u = 1/n$. Then

$$e = E(1) = E\left(n \cdot \frac{1}{n}\right) = E\left(\frac{1}{n}\right)^n.$$

Hence $E(1/n)$ is the n-th root of e. From now on we write

$$e^u \quad \text{instead of} \quad E(u).$$

Next we deal with the general exponential function.

Let a be a positive number, and x any number. We **define**

$$a^x = e^{x \log a}.$$

Thus

$$a^{\sqrt{2}} = e^{\sqrt{2} \log a}.$$

If we put $u = x \log a$ and use $\log e^u = u$, we find the formula

$$\log a^x = x \log a.$$

For instance,

$$\log 3^{\sqrt{2}} = \sqrt{2} \log 3.$$

Having made the general definition of a^x, we must be sure that in those cases when we have a preconceived idea of what a^x should be, for instance when $x = n$ is a positive integer, then

$$e^{n \log a} \quad \text{is the product of } a \text{ with itself } n \text{ times.}$$

For instance, take $x = 2$. Then

$$e^{2 \log a} = e^{\log a + \log a} = e^{\log a} e^{\log a} = a \cdot a,$$

$$e^{3 \log a} = e^{\log a + \log a + \log a} = e^{\log a} e^{\log a} e^{\log a} = a \cdot a \cdot a$$

and so forth. For any positive integer n we have

$$e^{n \log a} = e^{\log a + \log a + \cdots + \log a}$$

$$= e^{\log a} e^{\log a} \cdots e^{\log a}$$

$$= a \cdot a \cdots a \qquad \text{(product taken } n \text{ times).}$$

Therefore, if n is a positive integer, $e^{n \log a}$ means the product of a with itself n times.

Similarly,

$$(e^{(1/n) \log a})^n = e^{(1/n) \log a} e^{(1/n) \log a} \cdots e^{(1/n) \log a} \qquad \text{(product taken } n \text{ times)}$$

$$= e^{(1/n) \log a + (1/n) \log a + \cdots + (1/n) \log a}$$

$$= e^{\log a}$$

$$= a.$$

Hence the n-th power of $e^{(1/n)\log a}$ is equal to a, so

$$e^{(1/n)\log a} \quad \text{is the } n\text{-th root of } a.$$

This shows that $e^{x\log a}$ is what we expect when x is a positive integer or a fraction.

Next we prove other properties of the function a^x. First:

$$\boxed{a^0 = 1.}$$

Proof. By definition, $a^0 = e^{0\log a} = e^0 = 1$.

For all numbers x, y we have

$$\boxed{a^{x+y} = a^x a^y.}$$

Proof. We start with the right-hand side to get:

$$a^x a^y = e^{x\log a}e^{y\log a} = e^{x\log a + y\log a}$$

$$= e^{(x+y)\log a}$$

$$= a^{x+y}.$$

This proves the formula.

For all numbers x, y,

$$\boxed{(a^x)^y = a^{xy}.}$$

Proof.

$$(a^x)^y = e^{y\log a^x} \quad \text{(because } u^y = e^{y\log u} \quad \text{for } u > 0)$$

$$= e^{yx\log a} \quad \text{(because } \log a^x = x\log a)$$

$$= a^{xy} \quad \text{(because } a^t = e^{t\log a}, \text{ with the special value } t = xy = yx)$$

thus proving the desired property.

At this point we have recovered all five properties of the general exponential function which were used in §1, §2, and §3.

VIII, APPENDIX. EXERCISE

Suppose you did not know anything about the exponential and log functions. You are given a function E such that

$$E'(x) = E(x) \qquad \text{for all numbers } x, \qquad \text{and} \qquad E(0) = 1.$$

Prove:

(a) $E(x) \neq 0$ for all x. [*Hint*: Differentiate the product $E(x)E(-x)$ to show that this product is constant. Using $E(0) = 1$, what is this constant?]

(b) Let f be a function such that $f'(x) = f(x)$ for all x. Show that there exists a constant C such that $f(x) = CE(x)$.

(c) For all numbers u, v the function E satisfies

$$E(u + v) = E(u)E(v).$$

[*Hint*: Fix the number u and let $f(x) = E(u + x)$. Then apply (b).]

Part Three

Integration

CHAPTER IX

Integration

In this chapter, we solve, more or less simultaneously, the following problems:

(1) Given a function $f(x)$, find a function $F(x)$ such that

$$F'(x) = f(x).$$

This is the inverse of differentiation, and is called integration.

(2) Given a function $f(x)$ which is $\geqq 0$, give a definition of the area under the curve $y = f(x)$ which does not appeal to geometric intuition.

Actually, in this chapter, we give the ideas behind the solutions of our two problems. The techniques which allow us to compute effectively when specific data are given will be postponed to the next chapter.

In carrying out (2) we shall follow an idea of Archimedes. It is to approximate the function f by horizontal functions, and the area under f by the sum of little rectangles.

IX, §1. THE INDEFINITE INTEGRAL

Let $f(x)$ be a function defined over some interval.

Definition. An **indefinite integral** for f is a function F such that

$$F'(x) = f(x) \qquad \text{for all } x \text{ in the interval.}$$

If $G(x)$ is another indefinite integral of f, then $G'(x) = f(x)$ also. Hence the derivative of the difference $F - G$ is 0:

$$(F - G)'(x) = F'(x) - G'(x) = f(x) - f(x) = 0.$$

Consequently, by Corollary 3.3 of Chapter V, there is a constant C such that

$$F(x) = G(x) + C$$

for all x in the interval.

Example 1. An indefinite integral for $\cos x$ would be $\sin x$. But $\sin x + 5$ is also an indefinite integral for $\cos x$.

Example 2. $\log x$ is an indefinite integral for $1/x$. So is $\log x + 10$ or $\log x - \pi$.

In the next chapter, we shall develop techniques for finding indefinite integrals. Here, we merely observe that every time we prove a formula for a derivative, it has an analogue for the integral.

It is customary to denote an indefinite integral of a function f by

$$\int f \quad\text{or}\quad \int f(x)\, dx.$$

In this second notation, the dx is meaningless by itself. It is only the full expression $\int f(x)\, dx$ which is meaningful. When we study the method of substitution in the next chapter, we shall get further confirmation for the practicality of our notation.

We shall now make a table of some indefinite integrals, using the information which we have obtained about derivatives.

Let n be an integer, $n \neq -1$. Then we have

$$\int x^n\, dx = \frac{x^{n+1}}{n+1}.$$

If $n = -1$, then

$$\int \frac{1}{x}\, dx = \log\, x.$$

(This is true only in the interval $x > 0$.)

In the interval $x > 0$ we also have

$$\int x^c \, dx = \frac{x^{c+1}}{c+1}$$

for any number $c \neq -1$.

The following indefinite integrals are valid for all x.

$$\int \cos x \, dx = \sin x, \qquad \int \sin x \, dx = -\cos x,$$

$$\int e^x \, dx = e^x, \qquad \int \frac{1}{1+x^2} \, dx = \arctan x.$$

Finally, for $-1 < x < 1$, we have

$$\int \frac{1}{\sqrt{1-x^2}} \, dx = \arcsin x.$$

In practice, one frequently omits mentioning over what interval the various functions we deal with are defined. However, in any specific problem, one has to keep it in mind. For instance, if we write

$$\int x^{-1/3} \, dx = \tfrac{3}{2} \cdot x^{2/3},$$

this is valid for $x > 0$ and is also valid for $x < 0$. But 0 cannot be in any interval of definition of our function. Thus we could have

$$\int x^{-1/3} \, dx = \tfrac{3}{2} \cdot x^{2/3} + 5$$

when $x < 0$ and

$$\int x^{-1/3} \, dx = \tfrac{3}{2} \cdot x^{2/3} - 2$$

when $x > 0$. Any other constants besides 5 and -2 could also be used.

We agree throughout that indefinite integrals are defined only over intervals. Thus in considering the function $1/x$, we have to consider **separately** the cases $x > 0$ and $x < 0$. For $x > 0$, we have already remarked that $\log x$ is an indefinite integral. It turns out that for the interval $x < 0$ we can also find an indefinite integral, and in fact we have

$$\int \frac{1}{x} \, dx = \log(-x) \qquad \text{for} \quad x < 0.$$

Observe that when $x < 0$, $-x$ is positive, and thus $\log(-x)$ is meaningful. The derivative of $\log(-x)$ is equal to $1/x$, by the chain rule, namely, let $u = -x$. Then $du/dx = -1$, and

$$\frac{d \log(-x)}{dx} = \frac{1}{-x} (-1) = \frac{1}{x}.$$

For $x < 0$, any other indefinite integral is given by

$$\log(-x) + C,$$

where C is a constant.

It is sometimes stated that in all cases,

$$\int \frac{1}{x} \, dx = \log|x| + C.$$

With our conventions, we do not attribute any meaning to this, because our functions are not defined over intervals (the missing point 0 prevents this). In any case, the formula would be **false**. Indeed, for $x < 0$ we have

$$\int \frac{1}{x} \, dx = \log|x| + C_1,$$

and for $x > 0$ we have

$$\int \frac{1}{x} \, dx = \log|x| + C_2.$$

However, the two constants need not be equal, and hence we cannot write

$$\int \frac{1}{x} \, dx = \log|x| + C$$

in all cases. This formula is true only over an interval not containing 0.

IX, §1. EXERCISES

Find indefinite integrals for the following functions:

1. $\sin 2x$ 2. $\cos 3x$ 3. $\dfrac{1}{x+1}$ 4. $\dfrac{1}{x+2}$

(In these last two problems, specify the intervals over which you find an indefinite integral.)

IX, §2. CONTINUOUS FUNCTIONS

Definition. Let $f(x)$ be a function. We shall say that f is **continuous** if

$$\lim_{h \to 0} f(x + h) = f(x)$$

for all x at which the function is defined.

It is understood that in taking the limit, only values of h for which $f(x + h)$ is defined are considered. For instance, if f is defined on an interval

$$a \leqq x \leqq b$$

(assuming $a < b$), then we would say that f is continuous at a if

$$\lim_{\substack{h \to 0 \\ h > 0}} f(a + h) = f(a).$$

We cannot take $h < 0$, since the function would not be defined for $a + h$ if $h < 0$.

Geometrically speaking, a function is continuous if there is no break in its graph. All differentiable functions are continuous. We have already remarked this fact, because if a quotient

$$\frac{f(x + h) - f(x)}{h}$$

has a limit, then the numerator $f(x + h) - f(x)$ must approach 0, because

$$\lim_{h \to 0} f(x + h) - f(x) = \lim_{h \to 0} h \, \frac{f(x + h) - f(x)}{h}$$

$$= \lim_{h \to 0} h \lim_{h \to 0} \frac{f(x + h) - f(x)}{h} = 0.$$

The following are graphs of functions which are not continuous.

In Fig. 1, we have the graph of a function like

$$f(x) = -1 \qquad \text{if} \quad x \leqq 0,$$
$$f(x) = 1 \qquad \text{if} \quad x > 0.$$

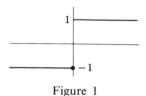

Figure 1

We see that

$$f(a + h) = f(h) = 1$$

for all $h > 0$. Hence

$$\lim_{\substack{h \to 0 \\ h > 0}} f(a + h) = 1,$$

which is unequal to $f(0)$.

A similar phenomenon occurs in Fig. 2 where there is a break. (Cf. Example 6 of Chapter III, §2.)

Figure 2

IX, §3. AREA

Let $a < b$ be two numbers, and let $f(x)$ be a continuous function defined on the interval $a \leqq x \leqq b$. This closed interval is denoted by $[a, b]$.

We wish to find a function $F(x)$ which is differentiable in this interval, and such that

$$F'(x) = f(x).$$

In this section, we appeal to our geometric intuition concerning area. *We assume that $f(x) \geqq 0$ for all x in the interval.* We let:

$F(x)$ = numerical measure of the area under the graph of f between a and x.

The following figure illustrates this.

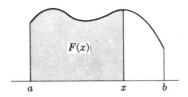

We thus have $F(a) = 0$. The area between a and a is 0.

Theorem 3.1. *The function $F(x)$ is differentiable, and its derivative is $f(x)$.*

Proof. Since we defined F geometrically, we shall have to argue geometrically.

We have to consider the Newton quotient

$$\frac{F(x + h) - F(x)}{h}.$$

Suppose first that x is unequal to the end point b, and also suppose that we consider only values of $h > 0$.

Then $F(x + h) - F(x)$ is the area between x and $x + h$. A picture may look like this.

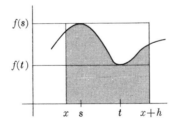

The shaded area represents $F(x + h) - F(x)$.

We let s be a point in the closed interval $[x, x + h]$ which is a maximum for our function f in *that small interval*. We let t be a point in the same closed interval which is a minimum for f in that interval. Thus

$$f(t) \leqq f(u) \leqq f(s)$$

for all u satisfying

$$x \leqq u \leqq x + h.$$

(We are forced to use another letter, u, since x is already being used.)

The area under the curve between x and $x + h$ is bigger than the area of the small rectangle in the figure above, i.e. the rectangle having base h and height $f(t)$.

The area under the curve between x and $x + h$ is smaller than the area of the big rectangle, i.e. the rectangle having base h and height $f(s)$. This gives us

$$h \cdot f(t) \leq F(x + h) - F(x) \leq h \cdot f(s).$$

Dividing by the positive number h yields

$$f(t) \leq \frac{F(x + h) - F(x)}{h} \leq f(s).$$

Since s, t are between x and $x + h$, as h approaches 0 both $f(s)$ and $f(t)$ approach $f(x)$. Hence the Newton quotient for F is squeezed between two numbers which approach $f(x)$. It must therefore approach $f(x)$ itself, and we have proved Theorem 3.1, when $h > 0$.

The proof is essentially the same as the proof which we used to get the derivative of $\log x$. The only difference in the present case is that we pick a maximum and a minimum without being able to give an explicit value for it, the way we could for the function $1/x$. Otherwise, there is no difference in the arguments.

If $x = b$, we look at negative values for h. The argument in that case is entirely similar to the one we have written down in detail, and we find again that the Newton quotient of F is squeezed between $f(s)$ and $f(t)$. We leave it as an exercise.

We now know that if $F(x)$ denotes the area under the graph of f between a and x then

$$F'(x) = f(x).$$

We can compute the area in practice by the following property:

Let G be any function on the interval $a \leq x \leq b$ such that

$$G'(x) = f(x).$$

Suppose $f(x) \geq 0$ for all x. Then the area under the graph of f between $x = a$ and $x = b$ is equal to

$$G(b) - G(a).$$

Proof. Since $F'(x) = G'(x)$ for all x, the two functions F and G have the same derivative on the interval. Hence there is a constant C such that

$$F(x) = G(x) + C \qquad \text{for all } x.$$

Let $x = a$. We get

$$0 = F(a) = G(a) + C.$$

This shows that $C = -G(a)$. Hence letting $x = b$ yields

$$F(b) = G(b) - G(a).$$

Thus the area under the curve between a and b is $G(b) - G(a)$. This is very useful to know in practice, because we can usually guess the function G.

Example 1. Find the area under the curve $y = x^2$ between $x = 1$ and $x = 2$.

Let $f(x) = x^2$. If $G(x) = x^3/3$ then $G'(x) = f(x)$. Hence the area under the curve between 1 and 2 is

$$G(2) - G(1) = \frac{2^3}{3} - \frac{1^3}{3} = \frac{7}{3}.$$

Example 2. Find the area under one arch of the function $\sin x$.

We have to find the area under the curve between 0 and π. Let

$$G(x) = -\cos x.$$

Then $G'(x) = \sin x$. Hence the area is

$$G(\pi) - G(0) = -\cos \pi - (-\cos 0)$$
$$= -(-1) + 1$$
$$= 2.$$

Note how remarkable this is. The arch of the sine curve going from 0 to π seems to be a very irrational curve, and yet the area turns out to be the integer 2!

IX, §3. EXERCISES

Find the area under the given curves between the given bounds.

1. $y = x^3$ between $x = 1$ and $x = 5$.

2. $y = x$ between $x = 0$ and $x = 2$.

3. $y = \cos x$, one arch.

4. $y = 1/x$ between $x = 1$ and $x = 2$.

5. $y = 1/x$ between $x = 1$ and $x = 3$.

6. $y = x^4$ between $x = -1$ and $x = 1$.

7. $y = e^x$ between $x = 0$ and $x = 1$.

IX, §4. UPPER AND LOWER SUMS

To show the existence of the integral, we use the idea of approximating our curves by constant functions.

Example. Consider the function $f(x) = x^2$. Suppose that we want to find the area between its graph and the x-axis, from $x = 0$ to $x = 1$. We cut up the interval $[0, 1]$ into smaller intervals, and approximate the function by constant functions, as shown on the next figure.

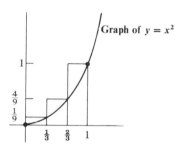

We have used three intervals, of length $1/3$, and on each of these intervals, we take the constant function whose value is the square of the right end point of the interval. These values are, respectively,

$$f(1/3) = 1/9, \qquad f(2/3) = 4/9, \qquad f(1) = 1.$$

Thus we obtain three rectangles, lying above the curve $y = x^2$. Each rectangle has a base of length $1/3$. The sum of the areas of these rectangles is equal to

$$\tfrac{1}{3}(\tfrac{1}{9} + \tfrac{4}{9} + 1) = \tfrac{14}{27}.$$

We could also have taken rectangles lying below the curve, using the values of f at the left end points of the intervals. The picture is as follows:

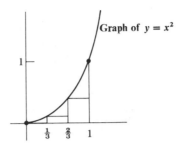

Graph of $y = x^2$

The heights of the three rectangles thus obtained are, respectively,

$$f(0) = 0, \qquad f(1/3) = 1/9, \qquad f(2/3) = 4/9.$$

The sum of their areas is equal to

$$\tfrac{1}{3}(0 + \tfrac{1}{9} + \tfrac{4}{9}) = \tfrac{5}{27}.$$

Thus we know that the area under the curve $y = x^2$ between $x = 0$ and $x = 1$ lies between 5/27 and 14/27. This is not a very good approximation to this area, but we can get a better approximation by using smaller intervals, say of lengths 1/4, or 1/5, or 1/6, or in general $1/n$. Let us write down the approximation with intervals of length $1/n$. The end points of the intervals will then be

$$0, \frac{1}{n}, \frac{2}{n}, \frac{3}{n}, \ldots, \frac{n-1}{n}, \frac{n}{n} = 1.$$

If we approximate the curve from above, then we get rectangles of heights respectively equal to

$$f\!\left(\frac{1}{n}\right) = \frac{1}{n^2}, \quad f\!\left(\frac{2}{n}\right) = \frac{2^2}{n^2}, \quad \ldots, \quad f\!\left(\frac{n}{n}\right) = \frac{n^2}{n^2}.$$

The general term for height of such a rectangle is

$$f\!\left(\frac{k}{n}\right) = \frac{k^2}{n^2} \qquad \text{for} \quad k = 1, 2, \ldots, n.$$

We have drawn these rectangles on the next figure.

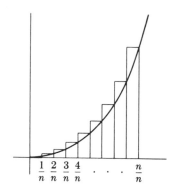

We see pictorially that the approximation to the curve is already much better. The area of each rectangle is equal to

$$\frac{1}{n} \cdot \frac{k^2}{n^2}, \qquad k = 1,\ldots,n,$$

because it is equal to the base times the altitude. The sum of these areas is equal to

$$\frac{1}{n^3} \sum_{k=1}^{n} k^2 = \frac{1}{n^3} (1 + 2^2 + 3^2 + \cdots + n^2).$$

Such a sum is called an **upper sum**, because we took the maximum of the function x^2 over each interval $\left[\frac{k-1}{n}, \frac{k}{n}\right]$. When we take n larger and larger, it is plausible that such sums will approximate the area under the graph of x^2 between 0 and 1. In any case, this upper sum is bigger than the area.

We shall now write down in general the sums which approximate the area under a curve. Note that we can take rectangles lying above the curve or below the curve, thus giving rise to upper sums and lower sums.

Let a, b be two numbers, with $a \leq b$. Let f be a continuous function in the interval $a \leq x \leq b$.

Definition. A **partition of the interval** $[a, b]$ is a sequence of numbers

$$a = x_0 \leq x_1 \leq x_2 \leq \cdots \leq x_n = b \quad \text{also written} \quad \{a = x_0,\ldots,x_n = b\}$$

between a and b, such that $x_i \leq x_{i+1}$ ($i = 0, 1,\ldots,n - 1$). For instance, we could take just two numbers,

$$x_0 = a \qquad \text{and} \qquad x_1 = b.$$

This will be called the **trivial partition**.

A partition divides our interval in a lot of smaller intervals $[x_i, x_{i+1}]$.

$$a = x_0 \quad x_1 \qquad x_2 \quad x_3 \quad \cdots \quad x_{n-1} \quad x_n = b$$

Given any number between a and b, in addition to $x_0, .., x_n$, we can add it to the partition to get a new partition having one more small interval. If we add enough intermediate numbers to the partition, then the intervals can be made arbitrarily small.

Let f be a function defined on the interval

$$a \leq x \leq b$$

and continuous. If c_i is a point between x_i and x_{i+1}, then we form the sum

$$f(c_0)(x_1 - x_0) + f(c_1)(x_2 - x_1) + \cdots + f(c_{n-1})(x_n - x_{n-1}).$$

Such a sum will be called a **Riemann sum**. Each value $f(c_i)$ can be viewed as the height of a rectangle, and each $(x_{i+1} - x_i)$ can be viewed as the length of the base.

Let s_i be a point between x_i and x_{i+1} such that f has a **maximum** in this small interval $[x_i, x_{i+1}]$ at s_i. In other words,

$$f(x) \leq f(s_i) \qquad \text{for} \quad x_i \leq x \leq x_{i+1}.$$

The rectangles then look like those in the next figure. In the figure, s_0 happens to be equal to x_1, $s_1 = s_1$ as shown, $s_2 = x_2$, $s_3 = x_4$, $s_4 = s_4$ as shown.

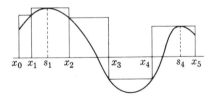

The main idea which we are going to carry out is that, as we make the intervals of our partitions smaller and smaller, the sum of the areas of the rectangles will approach a limit, and this limit can be used to define the area under the curve.

If P is the partition given by the numbers

$$x_0 \leq x_1 \leq x_2 \leq \cdots \leq x_n,$$

then the sum

$$f(s_0)(x_1 - x_0) + f(s_1)(x_2 - x_1) + \cdots + f(s_{n-1})(x_n - x_{n-1})$$

will be called the **upper sum** associated with the function f, and the partition P of the interval $[a, b]$. We shall denote it by the symbols

$$U_a^b(P, f) \qquad \text{or simply} \qquad U(P, f).$$

Observe, however, that when $f(x)$ becomes negative, the value $f(s_i)$ may be negative. Thus the corresponding rectangle gives a negative contribution

$$f(s_i)(x_{i+1} - x_i)$$

to the sum. Also, it is tiresome to write the sum by repeating each term, and so we shall use the abbreviation

$$\sum_{i=0}^{n-1} f(s_i)(x_{i+1} - x_i)$$

to mean the sum when i ranges from 0 to $n - 1$ of the terms $f(s_i)(x_{i+1} - x_i)$. Thus we have:

Definition. The **upper sum of f with respect to the partition** is

$$U_a^b(P, f) = \sum_{i=0}^{n-1} f(s_i)(x_{i+1} - x_i),$$

where $f(s_i)$ is the maximum of f on the interval $[x_i, x_{i+1}]$. Note that the **indices** i range from 0 to $n - 1$. Thus the sum is taken for $i = 0, \ldots, n - 1$.

By definition, we let

$$\max_{[x_i, x_{i+1}]} f = f(s_i) = \text{maximum of } f \text{ on the interval} \quad x_i \leqq x \leqq x_{i+1}.$$

Then the sum could also be written with the notation

$$\boxed{U_a^b(P, f) = \sum_{i=0}^{n-1} \left(\max_{[x_i, x_{i+1}]} f \right)(x_{i+1} - x_i).}$$

Instead of taking a maximum s_i in the interval $[x_i, x_{i+1}]$ we could have taken a minimum. Let t_i be a point in this interval, such that

$$f(t_i) \leqq f(x) \qquad \text{for} \quad x_i \leqq x \leqq x_{i+1}.$$

We call the sum

$$f(t_0)(x_1 - x_0) + f(t_1)(x_2 - x_1) + \cdots + f(t_{n-1})(x_n - x_{n-1})$$

the **lower sum** associated with the function f, and the partition P of the interval $[a, b]$. The lower sum will be denoted by

$$L_a^b(P, f) \qquad \text{or simply} \qquad L(P, f).$$

Thus we have the

Definition. The **lower sum of f with respect to the partition** is the sum

$$L_a^b(P, f) = \sum_{i=0}^{n-1} f(t_i)(x_{i+1} - x_i),$$

where $f(t_i)$ is the minimum of f on the interval $[x_i, x_{i+1}]$.

On the next figure, we have drawn a typical term of the sum.

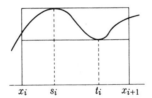

We can rewrite this lower sum using a notation similar to that used in the upper sum. Namely, we let

$$\min_{[x_i, x_{i+1}]} f = \text{minimum of } f \text{ on the interval } [x_i, x_{i+1}]$$

$$= f(t_i).$$

Then

$$L_a^b(P, f) = \sum_{i=0}^{n-1} \left(\min_{[x_i, x_{i+1}]} f \right)(x_{i+1} - x_i).$$

For all numbers x in the interval $[x_i, x_{i+1}]$ we have

$$f(t_i) \leq f(x) \leq f(s_i).$$

Since $x_{i+1} - x_i$ is ≥ 0, it follows that each term of the lower sum is less than or equal to each term of the upper sum. Therefore

$$L_a^b(P, f) \leq U_a^b(P, f).$$

Furthermore, any Riemann sum taken with points c_i (which are not necessarily maxima or minima) is between the lower and upper sum.

Example. Let $f(x) = x^2$ and let the interval be $[0, 1]$. Write out the upper and lower sums for the partition consisting of $\{0, \frac{1}{2}, 1\}$.

The minimum of the function in the interval $[0, \frac{1}{2}]$ is at 0, and $f(0) = 0$. The minimum of the function in the interval $[\frac{1}{2}, 1]$ is at $\frac{1}{2}$ and $f(\frac{1}{2}) = \frac{1}{4}$. Hence the lower sum is

$$L_0^1(P, f) = f(0)(\tfrac{1}{2} - 0) + f(\tfrac{1}{2})(1 - \tfrac{1}{2}) = \tfrac{1}{4} \cdot \tfrac{1}{2} = \tfrac{1}{8}.$$

The maximum of the function in the interval $[0, \frac{1}{2}]$ is at $\frac{1}{2}$ and the maximum of the function in the interval $[\frac{1}{2}, 1]$ is at 1. Thus the upper sum is

$$U_0^1(P, f) = f(\tfrac{1}{2})(\tfrac{1}{2} - 0) + f(1)(1 - \tfrac{1}{2}) = \tfrac{1}{8} + \tfrac{1}{2} = \tfrac{5}{8}.$$

We have given a numerical value for the upper and lower sums. Unless we wanted to compare them explicitly, we could have left the answer in the shape that it has on the left-hand side of these equalities.

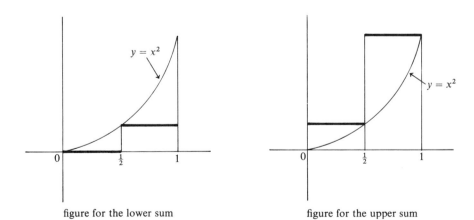

figure for the lower sum   figure for the upper sum

Example. We shall write down the lower sums in another special case when the function is positive and negative in the interval.

Let $f(x) = \sin x$, let the interval be $[0, 2\pi]$, and let the partition be

$$P = \left\{ 0, \frac{\pi}{4}, \frac{2\pi}{4}, \frac{3\pi}{4}, \frac{4\pi}{4}, \frac{5\pi}{4}, \frac{6\pi}{4}, \frac{7\pi}{4}, \frac{8\pi}{4} \right\}.$$

We illustrate this on the next figure.

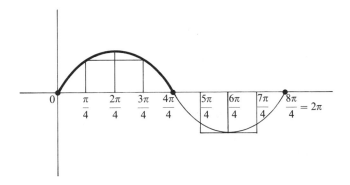

Each small interval of the partition has length $\pi/4$. Let us write down the lower sum:

$$L(P, f) = 0 \, \frac{\pi}{4} + \left(\sin \frac{\pi}{4} \right) \frac{\pi}{4} + \left(\sin \frac{3\pi}{4} \right) \frac{\pi}{4} + 0 \, \frac{\pi}{4}$$

$$+ \left(\sin \frac{5\pi}{4} \right) \frac{\pi}{4} + \left(\sin \frac{6\pi}{4} \right) \frac{\pi}{4} + \left(\sin \frac{6\pi}{4} \right) \frac{\pi}{4} + \left(\sin \frac{7\pi}{4} \right) \frac{\pi}{4}$$

$$= \frac{1}{\sqrt{2}} \frac{\pi}{4} + \frac{1}{\sqrt{2}} \frac{\pi}{4} - \frac{1}{\sqrt{2}} \frac{\pi}{4} - \frac{\pi}{4} - \frac{\pi}{4} - \frac{1}{\sqrt{2}} \frac{\pi}{4}.$$

Observe that the first term of the lower sum is 0 because the minimum of the function $\sin x$ on the interval $[0, \pi/4]$ is equal to 0. Similarly, the fourth term is also 0, and so can be omitted since $0 + A = A$ for all numbers A.

Also observe that:

$$\text{minimum of } \sin x \text{ on the interval } [5\pi/4, 6\pi/4] = \sin 6\pi/4$$

$$= -1.$$

With negative numbers, we have for instance

$$-1 = \sin 6\pi/4 < -\frac{1}{\sqrt{2}} = \sin 5\pi/4.$$

Thus the lower sum $L(P, f)$ contains positive terms and negative terms.

The negative terms represent minus the area of certain rectangles, as shown on the figure.

What happens to our sums when we add a new point to a partition? We shall see that the lower sum increases and the upper sum decreases.

Theorem 4.1. *Let f be a continuous function on the interval* $[a, b]$*. Let*

$$P = (x_0, \ldots, x_n)$$

be a partition of $[a, b]$*. Let $\bar{x}$ be any number in the interval, and let Q be the partition obtained from P by adding $\bar{x}$ to* $(x_0, \ldots, x_n)$*. Then*

$$L_a^b(P, f) \leq L_a^b(Q, f) \leq U_a^b(Q, f) \leq U_a^b(P, f).$$

Proof. Let us look at the lower sums, for example. Suppose that our number $\bar{x}$ is between x_j and x_{j+1}:

$$x_j \leq \bar{x} \leq x_{j+1}.$$

When we form the lower sum for P, it will be the same as the lower sum for Q except that the term

$$f(t_j)(x_{j+1} - x_j)$$

will now be replaced by two terms. If u is a minimum for f in the interval between x_j and $\bar{x}$, and v is a minimum for f in the interval between $\bar{x}$ and x_{j+1}, then these two terms are

$$f(u)(\bar{x} - x_j) + f(v)(x_{j+1} - \bar{x}).$$

We can write $f(t_j)(x_{j+1} - x_j)$ in the form

$$f(t_j)(x_{j+1} - x_j) = f(t_j)(\bar{x} - x_j) + f(t_j)(x_{j+1} - \bar{x}).$$

Since $f(t_j) \leq f(u)$ and $f(t_j) \leq f(v)$ (because t_j was a minimum in the whole interval between x_j and x_{j+1}), it follows that

$$f(t_j)(x_{j+1} - x_j) \leq f(u)(\bar{x} - x_j) + f(v)(x_{j+1} - \bar{x}).$$

Thus when we replace the term in the sum for P by the two terms in the sum for Q, the value of the contribution of these two terms increases. Since all other terms are the same, our assertion is proved.

The assertion concerning the fact that the upper sum decreases is left as an exercise. The proof is very similar.

As a consequence of our theorem, we obtain:

Corollary 4.2. *Every lower sum is less than or equal to every upper sum.*

Proof. Let P and Q be two partitions. If we add to P all the points of Q and add to Q all the points of P, we obtain a partition R such that every point of P is a point of R and every point of Q is a point of R. Thus R is obtained by adding points to P and to Q. Consequently, we have the inequalities

$$L_a^b(P, f) \leqq L_a^b(R, f) \leqq U_a^b(R, f) \leqq U_a^b(Q, f).$$

This proves our assertion.

It is now a very natural question to ask whether **there is a unique number between the lower sums and the upper sums**. The answer is yes.

Theorem 4.3. *Let f be a continuous function on $[a, b]$. There exists a unique number which is greater than or equal to every lower sum and less than or equal to every upper sum.*

Definition. The definite integral of f between a and b

$$\int_a^b f$$

is the unique number which is greater than or equal to every lower sum, and less than or equal to every upper sum.

We shall also use notation like

$$\int_a^b f(x)\, dx \qquad \text{or} \qquad \int_a^b f(t)\, dt.$$

It does not matter which letter we use, but it should be the same letter in both occurrences, i.e. in $f(x)\, dx$ or $f(t)\, dt$, or $f(u)\, du$, etc.

We shall not give the details of the proof of Theorem 4.3, which are tedious. The technique involved will not be used anywhere else in the course.

There is another statement which it is illuminating to know. Let P be a partition,

$$x_0 \leqq x_1 \leqq \cdots \leqq x_n.$$

The maximum length of the intervals $[x_i, x_{i+1}]$ is called the **size** of the partition. For example, if we cut up the interval $[0, 1]$ into n small intervals of the same length $1/n$, then the size of this partition is $1/n$.

Theorem 4.4. *Let f be a continuous function on $[a, b]$. Then the lower sums $L_a^b(P, f)$ and the upper sums $U_a^b(P, f)$ come arbitrarily close to the integral*

$$\int_a^b f$$

if the size of the partition P is sufficiently small.

Again, we shall not prove this theorem. But intuitively, it tells us that the upper and lower sums are good approximations to the integral when the size of the partition is taken sufficiently small.

Example. We give a physical example illustrating the application of upper and lower sums, relating density to mass.

Consider an interval $[a, b]$ with $0 \leq a < b$. We think of this interval as a rod, and let f be a continuous positive function defined on this interval. We interpret f as a density on the rod, so that $f(x)$ is the density at x.

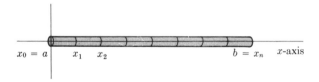

Given

$$a \leq c \leq d \leq b,$$

we denote by $M_c^d(f)$ the mass of the rod between c and d, corresponding to the given density f. We wish to determine a mathematical notion to represent $M_c^d(f)$. If f is a constant density, with constant value $K \geq 0$ on $[c, d]$, then the mass $M_c^d(f)$ should be $K(d - c)$. On the other hand, if g is another density such that

$$f(x) \leq g(x),$$

then certainly we should have $M_c^d(f) \leq M_c^d(g)$. In particular, if k, K are constants ≥ 0 such that

$$k \leq f(x) \leq K$$

for x in the interval $[c, d]$, then the mass should satisfy

$$k(d - c) \leq M_c^d(f) \leq K(d - c).$$

Finally, the mass should be additive, that is the mass of two disjoint pieces should be the sum of the mass of the pieces. In particular,

$$M_a^c(f) + M_c^d(f) = M_a^d(f).$$

We shall now see that the mass of the rod is given by the integral of the density, that is

$$M_a^b(f) = \int_a^b f(x)\, dx.$$

Let P be a partition of the interval $[a, b]$:

$$a = x_0 \leq x_1 \leq x_2 \leq \cdots \leq x_n = b.$$

Let $f(t_i)$ be the minimum for f on the small interval $[x_i, x_{i+1}]$, and let $f(s_i)$ be the maximum for f on this same small interval. Then the mass of each piece of the rod between x_i and x_{i+1} satisfies the inequality

$$f(t_i)(x_{i+1} - x_i) \leq M_{x_i}^{x_{i+1}}(f) \leq f(s_i)(x_{i+1} - x_i).$$

Adding these together, we find

$$\sum_{i=0}^{n-1} f(t_i)(x_{i+1} - x_i) \leq M_a^b(f) \leq \sum_{i=0}^{n-1} f(s_i)(x_{i+1} - x_i).$$

The expressions on the left and right are lower and upper sums for the integral, respectively. Since the integral is the unique number between the lower sums and upper sum, it follows that

$$M_a^b(f) = \int_a^b f(x)\, dx,$$

as we wanted to show.

IX, §4. EXERCISES

Write out the lower and upper sums for the following functions and intervals. Use a partition such that the length of each small interval is (a) $\frac{1}{2}$, (b) $\frac{1}{3}$, (c) $\frac{1}{4}$, (d) $1/n$.

1. $f(x) = x^2$ in the interval $[1, 2]$.   2. $f(x) = 1/x$ in the interval $[1, 3]$.

3. $f(x) = x$ in the interval $[0, 2]$.   4. $f(x) = x^2$ in the interval $[0, 2]$.

5. Let $f(x) = 1/x$ and let the interval be $[1, 2]$. Let n be a positive integer. Write out the upper and lower sum, using the partition such that the length of each small interval is $1/n$.

6. Using the definition of a definite integral, prove that

$$\frac{1}{n+1} + \frac{1}{n+2} + \cdots + \frac{1}{n+n} \leq \log 2 \leq \frac{1}{n} + \frac{1}{n+1} + \cdots + \frac{1}{2n-1}.$$

7. Let $f(x) = \log x$. Let n be a positive integer. Write out the upper and lower sums, using the partition of the interval between 1 and n consisting of the integers from 1 to n, i.e. the partition $(1, 2, \ldots, n)$.

IX, §5. THE FUNDAMENTAL THEOREM

The integral satisfies two basic properties which are very similar to those satisfied by area. We state them explicitly.

Property 1. *If M, m are two numbers such that*

$$m \leq f(x) \leq M$$

for all x in the interval $[b, c]$, *then*

$$m(c - b) \leq \int_b^c f \leq M(c - b).$$

Property 2. *We have*

$$\int_a^b f + \int_b^c f = \int_a^c f.$$

We shall not give the details of the proofs of these properties but we make some comments which we hope make them clear.

For Property 1, suppose we want to verify the inequality on the left-hand side. We may take the trivial partition of the interval $[b, c]$ consisting just of this interval. Then a lower sum is certainly $\geq m(c - b)$. Since the lower sums increase when we take a finer partition, and since the lower sums are at most equal to the integral, we see that the left-hand inequality

$$m(c - b) \leq \int_b^c f$$

is true. The right-hand inequality of Property 1 is proved in the same way.

For Property 2, suppose that $a \leq b \leq c$. Let P be a partition of suffi-ciently small size, such that the lower sum $L_a^c(P, f)$ approximates the integral $\int_a^c f$ very closely. The point b may not be in this partition. We may take a finer partition by inserting this point b, as shown on the figure.

Then P together with b form partitions P' and P'' of the intervals $[a, b]$ and $[b, c]$. If P has sufficiently small size, then P' and P'' have small size, and the lower sums

$$L_a^b(P', f) \qquad \text{and} \qquad L_b^c(P'', f)$$

give good approximations to the integrals

$$\int_a^b f \qquad \text{and} \qquad \int_b^c f,$$

respectively. But we have

$$L_a^c(P', P'', f) = L_a^b(P', f) + L_b^c(P'', f).$$

Since $L_a^c(P', P'', f)$ is an approximation of the integral

$$\int_a^c f,$$

one can see by passing to a limit that

$$\int_a^c f = \int_a^b f + \int_b^c f.$$

Property 2 was formulated when $a < b < c$. We now want to formu-late it when a, b, c are taken in any order. For this, suppose a, b are numbers in an interval where f is continuous, and $b < a$. We **define**

$$\boxed{\int_a^b f = -\int_b^a f.}$$

Then we have **Property 2 in general**:

Let a, b, c be three numbers in an interval where f is continuous. Then

$$\int_a^c f = \int_a^b f + \int_b^c f.$$

Proof. We have to distinguish cases. Suppose for instance that $b < a < c$. Then by the original property, for this ordering we get

$$\int_b^c = \int_b^a + \int_a^c = -\int_a^b + \int_a^c \qquad \text{by definition.}$$

Adding $\int_a^b$ to both sides proves the desired relation. All other cases can be proved similarly.

Theorem 5.1. *Let f be continuous on an interval $[a, b]$. Let*

$$F(x) = \int_a^x f.$$

Then F is differentiable and its derivative is

$$F'(x) = f(x).$$

Proof. We have to form the Newton quotient

$$\frac{F(x + h) - F(x)}{h} = \frac{1}{h} \left(\int_a^{x+h} f - \int_a^x f \right),$$

and see if it approaches a limit as $h \to 0$. (If $x = a$, then it is to be understood that $h > 0$, and if $x = b$, then $h < 0$. If $a < x < b$, then h may be positive or negative. The proof then shows that f is right differentiable at a and left differentiable at b.)

Assume for the moment that $h > 0$. By Property 2, applied to the numbers a, x, $x + h$ we conclude that our Newton quotient is equal to

$$\frac{1}{h} \left(\int_a^x f + \int_x^{x+h} f - \int_a^x f \right) = \frac{1}{h} \int_x^{x+h} f.$$

This reduces our investigation of the Newton quotient to the interval between x and $x + h$.

Let s be a point between x and $x + h$ such that f reaches a maximum in this small interval $[x, x + h]$ and let t be a point in this interval such that f reaches a minimum. We let

$$m = f(t) \quad \text{and} \quad M = f(s)$$

and apply Property 1 to the interval $[x, x + h]$. We obtain

$$f(t)(x + h - x) \leq \int_x^{x+h} f \leq f(s)(x + h - x),$$

which we can rewrite as

$$f(t) \cdot h \leq \int_x^{x+h} f \leq f(s) \cdot h.$$

Dividing by the positive number h preserves the inequalities, and yields

$$f(t) \leq \frac{\int_x^{x+h} f}{h} \leq f(s).$$

Since s, t lie between x and $x + h$, we must have (by continuity)

$$\lim_{h \to 0} f(s) = f(x) \quad \text{and} \quad \lim_{h \to 0} f(t) = f(x).$$

Thus our Newton quotient is squeezed between two numbers which approach $f(x)$. It must therefore approach $f(x)$, and our theorem is proved when $h > 0$.

The argument when $h < 0$ is entirely similar. We omit it.

IX, §5. EXERCISES

1. Using Theorem 5.1 prove that if f is continuous on an open interval containing 0, then

$$\lim_{h \to 0} \frac{1}{h} \int_0^h f = f(0).$$

[*Hint*: Can you interpret the left-hand side as the limit of a Newton quotient?]

2. Let f be continuous on the interval $[a, b]$. Prove that there exists some number c in the interval such that

$$f(c)(b - a) = \int_a^b f(t)\, dt.$$

[*Hint*: Apply the mean value theorem to $\int_a^x f(t)\, dt = F(x)$.]

CHAPTER X

Properties of the Integral

This is a short chapter. It shows how the integral combines with addition and inequalities. There is no good formula for the integral of a product. The closest thing is integration by parts, which is postponed to the next chapter.

Connecting the integral with the derivative is what allows us to compute integrals. The fact that two functions having the same derivative differ by a constant is again exploited to the hilt.

X, §1. FURTHER CONNECTION WITH THE DERIVATIVE

Let f be a continuous function on some interval. Let a, b be two points of the interval such that $a < b$, and let F be a function which is differentiable on the interval and whose derivative is f.

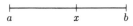

By the fundamental theorem, the functions

$$F(x) \qquad \text{and} \qquad \int_a^x f$$

have the same derivative. Hence there is a constant C such that

$$\int_a^x f = F(x) + C \qquad \text{for } all \text{ } x \text{ in the interval.}$$

What is this constant? If we put $x = a$, we get

$$0 = \int_a^a f = F(a) + C,$$

whence $C = -F(a)$. We also have

$$\int_a^b f = F(b) + C.$$

From this we obtain: *If $dF/dx = f(x)$, then*

$$\boxed{\int_a^b f = F(b) - F(a).}$$

This is extremely useful in practice, because we can usually guess the function F, and once we have guessed it, we can then compute the integral by means of this relation.

Furthermore, it is also practical to use the notation

$$F(x)\Big|_a^b = F(b) - F(a).$$

Remark. The argument which we gave to compute C shows that the value $F(b) - F(a)$ does not depend on the choice of function F such that $F'(x) = f(x)$. But you may want to see this another way. Suppose that $G'(x) = f(x)$ also, for all x in the interval. Then there is a constant C such that

$$G(x) = F(x) + C \qquad \text{for all } x \text{ in the interval.}$$

Then

$$G(b) - G(a) = F(b) + C - [F(a) + C]$$

$$= F(b) - F(a) \qquad \text{because } C \text{ cancels.}$$

Finally, we shall usually call the **indefinite integral** such as

$$\int \sin x \, dx, \qquad \text{or} \qquad \int \frac{1}{1 + x^2} \, dx$$

simply an **integral**, since the context makes clear what is meant. When we deal with a **definite integral**

$$\int_a^b$$

the numbers a and b are sometimes called the **lower limit** and **upper limit** respectively.

Example. We want to find the integral

$$\int_0^\pi \sin x \, dx.$$

Here we have $f(x) = \sin x$, and the indefinite integral is

$$\int \sin x \, dx = F(x) = -\cos x.$$

Hence

$$\int_0^\pi \sin x \, dx = -\cos x \Big|_0^\pi = -\cos \pi - (-\cos 0) = 2.$$

Example. Suppose we want to find

$$\int_1^3 x^2 \, dx.$$

Let $F(x) = x^3/3$. Then $F'(x) = x^2$. Hence

$$\int_1^3 x^2 \, dx = \frac{x^3}{3} \Big|_1^3 = \frac{27}{3} - \frac{1}{3} = \frac{26}{3}.$$

Example. Let us find

$$\int_0^1 \frac{1}{1 + x^2} \, dx.$$

Since $d \arctan x/dx = 1/1 + x^2$, we have the indefinite integral

$$\int \frac{1}{1 + x^2} \, dx = \arctan x.$$

Hence

$$\int_0^1 \frac{1}{1+x^2}\, dx = \arctan x \Big|_0^1$$

$$= \arctan 1 - \arctan 0$$

$$= \pi/4.$$

Example. Prove the inequality

$$\frac{1}{2} + \frac{1}{3} + \cdots + \frac{1}{n} \leq \log n.$$

To do this, we try to identify the left-hand side with a lower sum, and the right-hand side with a corresponding integral. We have the indefinite integral

$$\log x = \int \frac{1}{x}\, dx.$$

We let $f(x) = 1/x$.

We let the interval $[a, b]$ be $[1, n]$, that is all x with $1 \leq x \leq n$.

We let the partition $P = \{1, 2, \ldots, n\}$ consist of the positive integers from 1 to n.

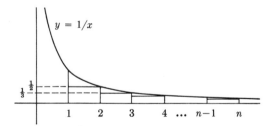

Then

$$L(P, f) = \frac{1}{2} + \frac{1}{3} + \cdots + \frac{1}{n}$$

because the length of the base of each rectangle is equal to 1. The value of the integral is

$$\int_1^n \frac{1}{x}\, dx = \log x \Big|_1^n = \log n - \log 1 = \log n.$$

Thus we obtain the desired inequality, because a lower sum is $\leq$ the integral.

In working out and proving similar inequalities, you should give:

The function $f(x)$;
The interval $[a, b]$ and the value of the definite integral

$$\int_a^b f(x)\,dx;$$

The partition P of the interval $[a, b]$.

You should then identify the sum with a lower sum (or upper sum, as the case may be) with respect to the above data, thus obtaining a comparison with the integral of the desired type.

Example. By a similar method, one can give an inequality having considerable practical interest. Let $n!$ denote the product of the first n integers. Thus

$$n! = 1 \cdot 2 \cdot 3 \cdots n.$$

We have the first few:

$$1! = 1,$$

$$2! = 1 \cdot 2 = 2,$$

$$3! = 1 \cdot 2 \cdot 3 = 6,$$

$$4! = 1 \cdot 2 \cdot 3 \cdot 4 = 6 \cdot 4 = 24,$$

$$5! = 24 \cdot 5 = 120.$$

Exercise 10 will show you how to prove the inequality

$$\boxed{(n - 1)! \leqq n^n e^{-n} e \leqq n!}$$

It is fun to work it out, so we don't do it here in the text.

The principle of these examples applies to comparing sums of functions with integrals, and the functions may be decreasing, as, for instance, the functions

$$\frac{1}{x}, \quad \frac{1}{x^{1/2}}, \quad \frac{1}{x^4}, \quad \text{etc.,}$$

or they may be increasing, as for instance the functions

$$x, \quad x^2, \quad x^4, \quad x^{1/3}.$$

The graphs may look like this, say on the interval $[1, n]$ where n is a positive integer, and the partition

$$P = \{1,\ldots,n\}$$

consists of the positive integers from 1 to n.

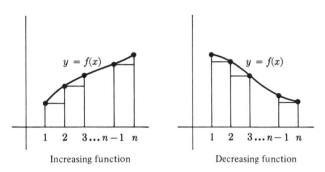

Increasing function   Decreasing function

In such a case, the base of each rectangle has length 1. Hence we obtain inequalities, for f increasing:

$$f(1) + f(2) + \cdots + f(n-1) \leqq \int_1^n f(x)\,dx \leqq f(2) + \cdots + f(n),$$

and for f decreasing:

$$f(2) + f(3) + \cdots + f(n) \leqq \int_1^n f(x)\,dx \leqq f(1) + \cdots + f(n-1).$$

X, §1. EXERCISES

Find the following integrals:

1. $\displaystyle\int_1^2 x^5\,dx$

2. $\displaystyle\int_{-1}^1 x^{1/3}\,dx$

3. $\displaystyle\int_{-\pi}^{\pi} \sin x\,dx$

4. $\displaystyle\int_0^{\pi} \cos x\,dx$

5. Prove the following inequalities:

(a) $\dfrac{1}{2} + \dfrac{1}{3} + \cdots + \dfrac{1}{n} \leqq \log n \leqq \dfrac{1}{1} + \dfrac{1}{2} + \dfrac{1}{3} + \cdots + \dfrac{1}{n-1}$

(b) $\dfrac{1}{2^{1/2}} + \dfrac{1}{3^{1/2}} + \cdots + \dfrac{1}{n^{1/2}} \leqq 2(\sqrt{n} - 1)$

(c) $2(\sqrt{n} - 1) \leqq 1 + \dfrac{1}{2^{1/2}} + \dfrac{1}{3^{1/2}} + \cdots + \dfrac{1}{(n-1)^{1/2}}$

6. Prove the following inequalities:

(a) $1^2 + 2^2 + \cdots + (n-1)^2 \leqq \dfrac{n^3}{3} \leqq 1^2 + 2^2 + \cdots + n^2$

(b) $1^3 + 2^3 + \cdots + (n-1)^3 \leqq \dfrac{n^4}{4} \leqq 1^3 + 2^3 + \cdots + n^3$

(c) $1^{1/4} + 2^{1/4} + \cdots + (n-1)^{1/4} \leqq \frac{4}{5}n^{5/4} \leqq 1^{1/4} + 2^{1/4} + \cdots + n^{1/4}$

7. Give similar inequalities as in Exercise 6, for the sums:

(a) $\displaystyle\sum_{k=1}^{n-1} k^4$ (b) $\displaystyle\sum_{k=1}^{n-1} k^{1/3}$ (c) $\displaystyle\sum_{k=1}^{n} k^5$ (d) $\displaystyle\sum_{k=1}^{n} \dfrac{1}{k^4}$

8. Prove the inequalities

$$\frac{1}{n} \sum_{k=1}^{n} \frac{n^2}{n^2 + k^2} \leqq \frac{\pi}{4} \leqq \frac{1}{n} \sum_{k=0}^{n-1} \frac{n^2}{n^2 + k^2}.$$

$\left[\textit{Hint: Write } \dfrac{n^2}{n^2 + k^2} = \dfrac{1}{1 + k^2/n^2}. \right]$ What is the interval? What is the partition?

9. Prove the inequality

$$\frac{1}{n}\left[\left(\frac{1}{n}\right)^2 + \left(\frac{2}{n}\right)^2 + \cdots + \left(\frac{n-1}{n}\right)^2 \right] \leqq \frac{1}{3}.$$

10. For this exercise, verify first that if we let

$$F(x) = x \log x - x$$

then $F'(x) = \log x$.

(a) Evaluate the integral

$$\int_1^n \log x \, dx.$$

(b) Compare this integral with the upper and lower sum associated with the partition $P = \{1, 2, \ldots, n\}$ of the interval $[1, n]$.

(c) In part (b), you will have found certain inequalities of the form

$$A \leqq B \leqq C.$$

Using the fact that

$$e^A \leqq e^B \leqq e^C,$$

prove the following inequality:

$$(n-1)! \leqq n^n e^{-n} e \leqq n!.$$

Here we denote by $n!$ the product of the first n integers.

X, §2. SUMS

Let $f(x)$ and $g(x)$ be two functions defined over some interval, and let $F(x)$ and $G(x)$ be (indefinite) integrals for f and g, respectively. This means $F'(x) = f(x)$ and $G'(x) = g(x)$. Since the derivative of a sum is the sum of the derivatives, we see that $F + G$ is an integral for $f + g$; in other words,

$$(F + G)'(x) = F'(x) + G'(x),$$

and therefore

$$\int [f(x) + g(x)] \, dx = \int f(x) \, dx + \int g(x) \, dx.$$

Similarly, let c be a number. The derivative of $cF(x)$ is $cf(x)$. Hence

$$\int cf(x) \, dx = c \int f(x) \, dx.$$

A constant can be taken in and out of an integral.

Example. Find the integral of $\sin x + 3x^4$.

We have

$$\int (\sin x + 3x^4) \, dx = \int \sin x \, dx + \int 3x^4 \, dx$$

$$= -\cos x + 3x^5/5.$$

Any formula involving the indefinite integral yields a formula for the definite integral. Using the same notation as above, suppose we have to find

$$\int_a^b [f(x) + g(x)] \, dx.$$

We know that it is

$$[F(x) + G(x)] \Big|_a^b$$

which is equal to

$$F(b) + G(b) - F(a) - G(a).$$

Thus we get the formula

$$\int_a^b [f(x) + g(x)]\, dx = \int_a^b f(x)\, dx + \int_a^b g(x)\, dx.$$

Similarly, for any constant c,

$$\int_a^b cf(x)\, dx = c \int_a^b f(x)\, dx.$$

Example. Find the integral

$$\int_0^\pi [\sin x + 3x^4]\, dx.$$

This (definite) integral is equal to

$$-\cos x + 3x^5/5 \Big|_0^\pi = -\cos \pi + 3\pi^5/5 - (-\cos 0 + 0)$$

$$= 1 + 3\pi^5/5 + 1$$

$$= 2 + 3\pi^5/5.$$

In some applications, one meets a slightly wider class of functions than continuous ones. Let f be a function defined on an interval $[a, b]$. We shall say that f is **piecewise continuous** on $[a, b]$ if there exist numbers

$$a = a_0 < a_1 < \cdots < a_n = b$$

and on each interval $[a_{i-1}, a_i]$ there is a continuous function f_i such that $f(x) = f_i(x)$ for $a_{i-1} < x < a_i$. If this is the case, then we define the integral of f from a to b to be the sum

$$\int_a^b f = \int_{a_0}^{a_1} f_1 + \int_{a_1}^{a_2} f_2 + \cdots + \int_{a_{n-1}}^{a_n} f_n.$$

A piecewise continuous function may look like this:

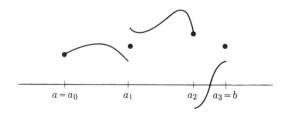

Example. Let f be the function defined on the interval $[0, 2]$ by the conditions:

$$f(x) = x \qquad \text{if} \quad 0 \leqq x \leqq 1,$$
$$f(x) = 2 \qquad \text{if} \quad 1 < x \leqq 2.$$

The graph of f looks like this:

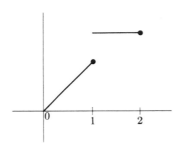

To find the integral of f between 0 and 2, we have

$$\int_0^2 f = \int_0^1 x \, dx + \int_1^2 2 \, dx = \frac{x^2}{2}\Big|_0^1 + 2x \Big|_1^2$$
$$= \tfrac{1}{2} + (4 - 2) = \tfrac{5}{2}.$$

We can also find

$$\int_0^x f(t) \, dt \qquad \text{for} \quad 0 \leqq x \leqq 2.$$

If $0 \leqq x \leqq 1$:

$$\int_0^x f(t) \, dt = \int_0^x t \, dt = \frac{x^2}{2}.$$

If $1 \leqq x \leqq 2$:

$$\int_0^x f(t) \, dt = \int_0^1 f(t) \, dt + \int_1^x f(t) \, dt$$
$$= \tfrac{1}{2} + \int_1^x 2 \, dt = \tfrac{1}{2} + 2x - 2.$$

Example. Let $f(x)$ be defined for $0 \leqq x \leqq \pi$ by the formulas:

$$f(x) = \sin x \qquad \text{if} \quad 0 \leqq x < \pi/2,$$
$$f(x) = \cos x \qquad \text{if} \quad \pi/2 \leqq x \leqq \pi.$$

Then the integral of f from 0 to π is given by:

$$\int_0^\pi f = \int_0^{\pi/2} \sin x \, dx + \int_{\pi/2}^\pi \cos x \, dx$$

$$= -\cos x \Big|_0^{\pi/2} + \sin x \Big|_{\pi/2}^\pi = 0.$$

The graph of f looks like this:

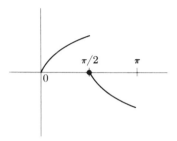

> *The integral of a function represents the area between the graph of the function and the x-axis only when the function is positive. If the function is negative, then this area is represented by minus the integral.*

Example. The function $\sin x$ is **negative** on the interval $[\pi, 2\pi]$. The **area** between the curve $y = \sin x$ and the x-axis over this interval is given by **minus the integral**:

$$-\int_\pi^{2\pi} \sin x \, dx = -(-\cos x) \Big|_\pi^{2\pi} = 2.$$

The **area** between the graph of $\sin x$ and the x-axis between 0 and 2π is equal to twice the area of one of the loops, and is therefore equal to 4. On the other hand,

$$\int_0^{2\pi} \sin x \, dx = -\cos x \Big|_0^{2\pi} = 0.$$

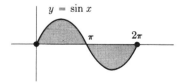

Example. Find the area between the curve

$$y = f(x) = x(x-1)(x-2)$$

and the x-axis,

The curve looks like this:

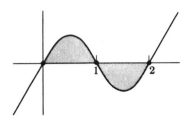

There are two portions between the curve and the x-axis, corresponding to the intervals $[0, 1]$ and $[1, 2]$. However, the function is negative between $x = 1$ and $x = 2$, so that to find the sum of the areas of the two regions, we have to take the absolute value of the integral over the second one. We therefore compute these areas separately.

First we expand the product giving $f(x)$, and get

$$f(x) = x^3 - 3x^2 + 2x.$$

The first integral is equal to

$$\int_0^1 f(x)\, dx = \frac{x^4}{4} - 3\frac{x^3}{3} + 2\frac{x^2}{2}\Big|_0^1$$

$$= \tfrac{1}{4} - 1 + 1 = \tfrac{1}{4}.$$

The second integral is equal to

$$\int_1^2 f(x)\, dx = \frac{x^4}{4} - x^3 + x^2\Big|_1^2$$

$$= \tfrac{16}{4} - 8 + 4 - (\tfrac{1}{4} - 1 + 1) = -\tfrac{1}{4}.$$

Hence the **area** of the two regions is equal to

$$\tfrac{1}{4} + \tfrac{1}{4} = \tfrac{1}{2}.$$

Example. Find the integral

$$\int_0^{3\pi} |\sin x|\, dx.$$

Note that because of the absolute value sign, the graph of the function $|\sin x|$ looks like this:

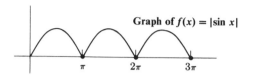

Graph of $f(x) = |\sin x|$

If $\sin x \geq 0$ on an interval, then $|\sin x| = \sin x$.

If $\sin x \leq 0$ on an interval, then $|\sin x| = -\sin x$.

Hence

$$\int_0^{3\pi} |\sin x|\, dx = \int_0^{\pi} \sin x\, dx - \int_{\pi}^{2\pi} \sin x\, dx + \int_{2\pi}^{3\pi} \sin x\, dx$$

$$= 2 + 2 + 2 = 6.$$

Warning. Of course, this comes out three times the area under one arch of the graph, because of symmetries. But if you try to use symmetries in such integrals, be sure to *prove* that they are valid.

Example. Find the area between the curves $y = x$ and $y = \sin x$ for $0 \leq x \leq \pi/4$.

The graphs are as follows.

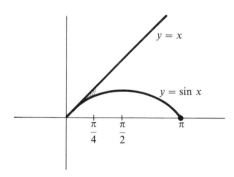

You should know from inequalities proved in Chapter V, §2 that $\sin x \leq x$ for $0 \leq x$. The area between the two curves between $x = 0$ and $x = \pi/4$ is the difference of the areas under the bigger curve and the smaller one, that is:

$$\int_0^{\pi/4} (x - \sin x)\, dx = \int_0^{\pi/4} x\, dx - \int_0^{\pi/4} \sin x\, dx$$

$$= \frac{x^2}{2}\Big|_0^{\pi/4} + \cos x\Big|_0^{\pi/4}$$

$$= \frac{\pi^2}{32} + \frac{1}{\sqrt{2}} - 1.$$

In general, if $f(x)$ and $g(x)$ are two continuous functions such that $f(x) \geq g(x)$ on an interval $[a, b]$, then the area between the two curves, from a to b, is

$$\int_a^b (f(x) - g(x))\, dx.$$

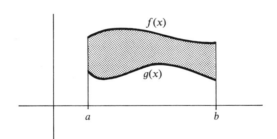

X, §2. EXERCISES

Find the following integrals:

1. $\displaystyle\int 4x^3\, dx$

2. $\displaystyle\int (3x^4 - x^5)\, dx$

3. $\displaystyle\int (2 \sin x + 3 \cos x)\, dx$

4. $\displaystyle\int (3x^{2/3} + 5 \cos x)\, dx$

5. $\displaystyle\int \left(5e^x + \frac{1}{x}\right) dx$

6. $\displaystyle\int_{-\pi}^{\pi} (\sin x + \cos x)\, dx$

7. $\displaystyle\int_{-1}^{1} 2x^5\, dx$

8. $\displaystyle\int_{-1}^{2} e^x\, dx$

9. $\displaystyle\int_{-1}^{3} 4x^2\, dx$

10. Find the area between the curves $y = x$ and $y = x^2$ from 0 to their first point of intersection for $x > 0$.

11. Find the area between the curves $y = x$ and $y = x^3$.

12. Find the area between the curves $y = x^2$ and $y = x^3$.

13. Find the area between the curve $y = (x - 1)(x - 2)(x - 3)$ and the x-axis. (Sketch the curve.)

14. Find the area between the curve $y = (x + 1)(x - 1)(x + 2)$ and the x-axis.

15. Find the area between the curves $y = \sin x$, $y = \cos x$, the y-axis, and the first point where these curves intersect for $x > 0$.

In each of the following problems 16 through 25:
(a) Sketch the graph of the function $f(x)$.
(b) Find the integral of the function over the given interval.

16. On $[-1, 1]$, $f(x) = x$ if $-1 \leq x < 0$ and $f(x) = 5$ if $0 \leq x \leq 1$.

17. On $[-1, 1]$, $f(x) = x^2$ if $-1 \leq x \leq 0$ and $f(x) = -x$ if $0 < x \leq 1$.

18. On $[-1, 1]$, $f(x) = x - 1$ if $-1 \leq x < 0$ and $f(x) = x + 1$ if $0 \leq x \leq 1$.

19. On $[-\pi, \pi]$, $f(x) = \sin x$ if $-\pi \leq x \leq 0$, and $f(x) = x$ if $0 < x \leq \pi$.

20. On $[-\pi, \pi]$, $f(x) = |\sin x|$.   21. On $[-\pi, \pi]$, $f(x) = |\cos x|$.

22. On $[-1, 1]$, $f(x) = |x|$.   23. On $[-\pi, \pi]$, $f(x) = \sin x + |\cos x|$.

24. On $[-\pi, \pi]$, $f(x) = x - |x|$.   25. On $[-\pi, \pi]$, $f(x) = \sin x + |\sin x|$.

26. Find the value of the integrals (a) $\int_0^{7\pi} |\sin x|\, dx$, (b) $\int_0^{7\pi} |\cos x|\, dx$. (c) For any positive integer n, $\int_0^{n\pi} |\sin x|\, dx$.

X, §3. INEQUALITIES

You may wait to read this section until it is used to estimate remainder terms in Taylor's formula.

Theorem 3.1. *Let a, b be two numbers, with $a \leq b$. Let f, g be two continuous functions on the interval $[a, b]$ and assume that $f(x) \leq g(x)$ for all x in the interval. Then*

$$\int_a^b f(x)\, dx \leq \int_a^b g(x)\, dx.$$

Proof. Since $g(x) - f(x) \geq 0$, we can use the basic Property 1 of Chapter IX, §5 (with $m = 0$) to conclude that

$$\int_a^b (g - f) \geq 0.$$

But

$$\int_a^b (g - f) = \int_a^b g - \int_a^b f.$$

Transposing the second integral on the right in our inequality, we obtain

$$\int_a^b g \geq \int_a^b f,$$

as desired.

Theorem 3.1 will be used mostly when $g(x) = |f(x)|$. Since a negative number is always $\leq$ a positive number, we know that

$$f(x) \leq |f(x)|$$

and

$$-f(x) \leq |f(x)|.$$

Theorem 3.2. *Let a, b be two numbers, with $a \leq b$. Let f be a continuous function on the interval $[a, b]$. Then*

$$\left| \int_a^b f(x) \, dx \right| \leq \int_a^b |f(x)| \, dx.$$

Proof. We simply let $g(x) = |f(x)|$ in the preceding theorem. Then

$$f(x) \leq |f(x)|$$

and also $-f(x) \leq |f(x)|$. The absolute value of the integral on the left is equal to

$$\int_a^b f(x) \, dx \qquad \text{or} \qquad -\int_a^b f(x) \, dx.$$

We can apply Theorem 3.1 either to $f(x)$ or $-f(x)$ to get Theorem 3.2.

We make one other application of Theorem 3.2.

Theorem 3.3. *Let a, b be two numbers and f a continuous function on the closed interval between a and b. (We do not necessarily assume that $a < b$.) Let M be a number such that $|f(x)| \leq M$ for all x in the interval. Then*

$$\left| \int_a^b f(x) \, dx \right| \leq M|b - a|.$$

Proof. If $a \leqq b$, we can use Theorem 3.2 to get

$$\left| \int_a^b f(x)\, dx \right| \leqq \int_a^b M\, dx = M \int_a^b dx = M(b - a).$$

If $b < a$, then

$$\int_a^b f = - \int_b^a f.$$

Taking the absolute value gives us the estimate $M(a - b)$. Since

$$a - b = |b - a|$$

in case $b < a$, we have proved our theorem.

Theorem 3.4. *Let f be a continuous function on the interval $[a, b]$ with $a < b$. Assume that $f(x) \geqq 0$ for every x in this interval, and $f(x) > 0$ for some x in this interval. Then*

$$\int_a^b f(x)\, dx > 0.$$

Proof. Let c be a number of the interval such that $f(c) > 0$, and suppose for simplicity that $c \neq b$.

The geometric idea behind the proof is quite simple, in terms of area. Since the function f is assumed $\geqq 0$ everywhere, and > 0 at the point c, then it is greater than some fixed positive number [taken to be $f(c)/2$, say] in some interval near c. This means that we can insert a small rectangle of height > 0 between the curve $y = f(x)$ and the x-axis. Then the area under the curve is at least equal to the area of this rectangle, which is > 0.

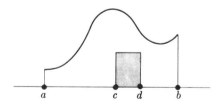

This "proof" can be phrased in terms of the formal properties of the integral as follows. Since f is continuous, there exists some number d close to c in the interval, with $c < d \leqq b$ such that $f(x)$ is close to $f(c)$ for all x satisfying

$$c \leqq x \leqq d.$$

In particular, we have

$$f(x) \geqq \frac{f(c)}{2}, \qquad c \leqq x \leqq d.$$

Then

$$\int_a^b f(x)\,dx = \int_a^c f + \int_c^d f + \int_d^b f$$

$$\geqq \int_c^d f(x)\,dx \geqq \int_c^d \frac{f(c)}{2}\,dx$$

$$\geqq \frac{f(c)}{2}(d - c) > 0.$$

This proves our theorem if $c \neq b$. If $c = b$, we take $d < c$ and argue similarly.

Theorem 3.4 will not be used in the rest of this book except in a couple of exercises, but it is important in subsequent applications.

X, §4. IMPROPER INTEGRALS

Example 1. We start with an example. Let $0 < c < 1$. We look at the integral

$$\int_c^1 \frac{1}{x}\,dx = \log x \Big|_c^1 = \log 1 - \log c = -\log c.$$

The figure illustrates this integral.

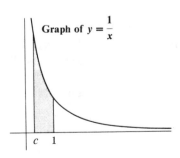

Graph of $y = \dfrac{1}{x}$

We have shaded the portion of the area under the graph lying between c and 1. As c approaches 0, we see that the area becomes arbitrarily large, because

$$-\log c \to \infty \qquad \text{as} \quad c \to 0.$$

Example 2. However, it is remarkable that an entirely different situation will occur when we consider the area under the curve $1/\sqrt{x} = x^{-1/2}$. We take $x > 0$, of course. Let $0 < c < 1$. We compute the integral:

$$\int_c^1 \frac{1}{x^{1/2}} \, dx = \int_c^1 x^{-1/2} \, dx = 2x^{1/2} \Big|_c^1$$

$$= 2 - 2c^{1/2}.$$

Then

$$\text{as} \quad c \to 0, \qquad 2c^{1/2} \to 0$$

and therefore

$$\int_c^1 x^{-1/2} \, dx \to 2 \qquad \text{as} \quad c \to 0.$$

We can illustrate the graph of $1/\sqrt{x}$ on the following figure. Note that at first sight, it does not differ so much from that of the preceding example, but the computation of the area shows the existence of a fundamental difference.

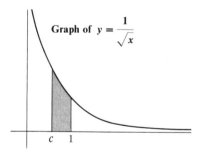

Graph of $y = \dfrac{1}{\sqrt{x}}$

Both in Example 1 and Example 2 we are looking at an infinite chimney, as $c \to 0$. But in Example 1 the area becomes arbitrarily large, whereas in Example 2, the area approaches the limit 2.

Definition. In Example 2, we say that the integral

$$\int_0^1 x^{-1/2} \, dx$$

exists, or **converges** even though the function $x^{-1/2}$ is not defined at 0 and is not continuous in the **closed** interval $[0, 1]$.

In general, suppose we have two numbers a, b with, say, $a < b$. Let f be a continuous function in the interval $a < x \leqq b$. This means that for every number c with $a < c < b$, the function f is continuous on the interval

$$c \leq x \leq b.$$

Let F be any function such that $F'(x) = f(x)$.
We can then evaluate the integral as usual:

$$\int_c^b f(x)\, dx = F(b) - F(c).$$

Definition. If the limit

$$\lim_{c \to a} F(c)$$

exists, then we say that the **improper integral**

$$\int_a^b f(x)\, dx$$

exists, and then we define

$$\int_a^b f(x)\, dx = \lim_{c \to a} \int_c^b f(x)\, dx = F(b) - \lim_{c \to a} F(c).$$

We make similar definitions when we deal with an interval $a \leq x < b$ and a function f which is continuous on this interval. If the limit

$$\lim_{c \to b} \int_a^c f(x)\, dx$$

exists, then we say that the **improper integral exists**, and it is equal to this limit.

Example 3. Show that the improper integral

$$\int_0^1 \frac{1}{x^2}\, dx$$

does not exist.
Let $0 < c < 1$. We first evaluate the integral:

$$\int_c^1 x^{-2}\, dx = \frac{x^{-1}}{-1}\Big|_c^1 = -1 - \left(-\frac{1}{c}\right) = -1 + \frac{1}{c}.$$

But $1/c \to \infty$ as $c \to 0$, and hence the improper integral does not exist.

Example 4. Determine whether the integral

$$\int_1^3 \frac{1}{x-1}\, dx$$

exists, and if it does, find its value.

Let $1 < c < 3$. Then the function $1/(x-1)$ is not continuous on the interval $[1, 3]$ but is continuous on the interval $[c, 3]$. Furthermore

$$\int_c^3 \frac{1}{x-1}\, dx = \log(x-1)\Big|_c^3 = \log 2 - \log(c-1).$$

But

$$-\log(c-1) \to \infty \qquad \text{as} \quad c \to 1 \quad \text{and} \quad c > 1.$$

Hence the integral does not exist.

There is another type of improper integral, dealing with large values.

Let a be a number and f a continuous function defined for $x \geq a$. Consider the integral

$$\int_a^B f(x)\, dx$$

for some number $B > a$. If $F(x)$ is any indefinite integral of f, then our integral is equal to $F(B) - F(a)$. If it approaches a limit as B becomes very large, then we **define**

$$\int_a^\infty f(x)\, dx \quad \text{or} \quad \int_a^\infty f = \lim_{B \to \infty} \int_a^B f(x)\, dx,$$

and say that the **improper integral converges**, or **exists**.

Thus

$$\int_a^\infty f \quad \text{exists if} \quad \lim_{B \to \infty} \int_a^B f \quad \text{exists,}$$

and is equal to the limit. Otherwise, we say that the improper integral **does not converge**, or **does not exist**.

Example 5. Determine whether the improper integral $\int_1^\infty 1/x\, dx$ exists, and if it does, finds its value.

Let B be a number > 1. Then

$$\int_1^B \frac{1}{x}\,dx = \log B - \log 1 = \log B.$$

As B becomes large, so does $\log B$, and hence the improper integral does not exist.

Let us look at the function $1/x^2$. Its graph looks like that in the next figure. At first sight, there seems to be no difference between this function and $1/x$, except that $1/x^2 < 1/x$ when $x > 1$. However, intuitively speaking, we shall find that $1/x^2$ approaches 0 sufficiently faster than $1/x$ to guarantee that the area under the curve between 1 and B approaches a limit as B becomes large.

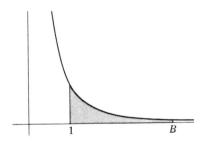

Example 6. Determine whether the improper integral

$$\int_1^\infty \frac{1}{x^2}\,dx$$

exists, and if it does, find its value.

Let B be a number > 1. Then

$$\int_1^B \frac{1}{x^2}\,dx = \frac{-1}{x}\Big|_1^B = -\frac{1}{B} + 1.$$

As B becomes large, $1/B$ approaches 0. Hence the limit as B becomes large exists and is equal to 1, which is the value of our integral. We thus have by definition

$$\int_1^\infty \frac{1}{x^2}\,dx = \lim_{B \to \infty}\left(-\frac{1}{B} + 1\right) = 1.$$

X, §4. EXERCISES

Determine whether the following improper integrals exist or not.

1. $\displaystyle\int_2^\infty \frac{1}{x^{3/2}}\,dx$ 2. $\displaystyle\int_1^\infty \frac{1}{x^{2/3}}\,dx$ 3. $\displaystyle\int_0^\infty \frac{1}{1+x^2}\,dx$

4. $\displaystyle\int_0^5 \frac{1}{5-x}\,dx$ 5. $\displaystyle\int_2^3 \frac{1}{x-2}\,dx$ 6. $\displaystyle\int_1^4 \frac{1}{x-1}\,dx$

7. $\displaystyle\int_0^2 \frac{1}{x-2}\,dx$ 8. $\displaystyle\int_2^3 \frac{1}{(x-2)^2}\,dx$

9. $\displaystyle\int_2^3 \frac{1}{(x-2)^{3/2}}\,dx$ 10. $\displaystyle\int_1^4 \frac{1}{(x-1)^{2/3}}\,dx$

In the preceding exercises, you evaluated improper integrals of the form

$$\int_a^b \frac{1}{x^s}\,dx$$

in special cases. **The next two exercises are important because they tell you in general when such integrals exist or not.**

11. (a) Let s be a number < 1. Show that the improper integral

$$\int_0^1 \frac{1}{x^s}\,dx$$

exists.
(b) If $s > 1$, show that the integral does not exist.
(c) Does the integral exist when $s = 1$?

12. (a) If $s > 1$ show that the following integral exists.

$$\int_1^\infty \frac{1}{x^s}\,dx$$

(b) *If* $s < 1$ show that the integral does not exist.

Determine whether the following integrals exist, and if so find their values.

13. $\displaystyle\int_1^\infty e^{-x}\,dx$ 14. $\displaystyle\int_1^\infty e^x\,dx$

15. Let B be a number > 2. Find the area under the curve $y = e^{-2x}$ between 2 and B. Does this area approach a limit when B becomes very large? If so, what limit?

CHAPTER XI

Techniques of Integration

The purpose of this chapter is to teach you certain basic tricks to find indefinite integrals. It is of course easier to look up integral tables, but you should have a minimum of training in standard techniques.

XI, §1. SUBSTITUTION

We shall formulate the analogue of the chain rule for integration.

Suppose that we have a function $u(x)$ and another function f such that $f(u(x))$ is defined. (All these functions are supposed to be defined over suitable intervals.) We wish to evaluate an integral having the form

$$\int f(u) \frac{du}{dx} \, dx,$$

where u is a function of x. We shall first work out examples to learn the mechanics for finding the answer.

Example 1. Find $\int (x^2 + 1)^3 (2x) \, dx$.

Put $u = x^2 + 1$. Then $du/dx = 2x$ and our integral is in the form

$$\int f(u) \frac{du}{dx} \, dx,$$

the function f being $f(u) = u^3$. We abbreviate $(du/dx)\,dx$ by du, **as if we could cancel** dx. Then we can write the integral as

$$\int f(u)\,du = \int u^3\,du = \frac{u^4}{4} = \frac{(x^2 + 1)^4}{4}.$$

We can check this by differentiating the expression on the right, using the chain rule. We get

$$\frac{d}{dx}\frac{(x^2 + 1)^4}{4} = \frac{4}{4}(x^2 + 1)^3 2x = (x^2 + 1)^3(2x),$$

as desired.

Example 2. Find $\int \sin(2x)(2)\,dx$.

Put $u = 2x$. Then $du/dx = 2$. Hence our integral is in the form

$$\int \sin u\,\frac{du}{dx}\,dx = \int \sin u\,du = -\cos u = -\cos(2x).$$

Observe that

$$\int \sin(2x)\,dx \neq -\cos(2x).$$

If we differentiate $-\cos(2x)$, we get $\sin(2x) \cdot 2$.

The integral in Example 2 could also be written

$$\int 2\sin(2x)\,dx.$$

It does not matter, of course, where we place the 2.

Example 3. Find $\int \cos(3x)\,dx$.

Let $u = 3x$. Then $du/dx = 3$. There is no extra 3 in our integral. However, we can take a constant in and out of an integral. Our integral is equal to

$$\frac{1}{3}\int 3\cos(3x)\,dx = \frac{1}{3}\int \cos u\,du = \frac{1}{3}\sin u = \frac{1}{3}\sin(3x).$$

It is convenient to use a purely formal notation which allows us to make a substitution $u = g(x)$, as in the previous examples. Thus instead of writing

$$\frac{du}{dx} = 2x$$

in Example 1, we would write $du = 2x\,dx$. Similarly, in Example 2, we would write $du = 2\,dx$, and in Example 3 we would write $du = 3\,dx$. We do not attribute any meaning to this. It is merely a device of a type used in programming a computing machine. A machine does not think. One simply adjusts certain electric circuits so that the machine performs a certain operation and comes out with the right answer. The fact that writing

$$du = \frac{du}{dx}\,dx$$

makes us come out with the right answer will be proved in a moment.

Example 4. Find

$$\int (x^3 + x)^9 (3x^2 + 1)\,dx.$$

Let

$$u = x^3 + x.$$

Then

$$du = (3x^2 + 1)\,dx.$$

Hence our integral is of type $\int f(u)\,du$ and is equal to

$$\int u^9\,du = \frac{u^{10}}{10} = \frac{(x^3 + x)^{10}}{10}.$$

We now show how the above procedure, which can be checked in each case by differentiation, actually must give the right answer in all cases.

We suppose therefore that we wish to evaluate an integral of the form

$$\int f(u)\frac{du}{dx}\,dx,$$

where u is a function of x. Let F be a function such that

$$F'(u) = f(u).$$

Thus F is an indefinite integral

$$F(u) = \int f(u)\, du.$$

If we use the chain rule, we get

$$\frac{dF(u(x))}{dx} = \frac{dF}{du}\frac{du}{dx} = f(u)\frac{du}{dx}.$$

Thus we have proved that

$$\int f(u)\frac{du}{dx}\, dx = F(u(x))$$

as desired.

We should also observe that the formula for integration by substitution applies to the definite integral. We can state this formally as follows.

Let g be a differentiable function on the interval $[a, b]$, whose derivative is continuous. Let f be a continuous function on an interval containing the values of g. Then

$$\int_a^b f(g(x))\frac{dg}{dx}\, dx = \int_{g(a)}^{g(b)} f(u)\, du.$$

The proof is immediate. If F is an indefinite integral for f, then $F(g(x))$ is an indefinite integral for

$$f(g(x))\frac{dg}{dx}$$

by the chain rule. Hence the left-hand side of our formula is equal to

$$F(g(b)) - F(g(a)),$$

which is also the value of the right-hand side.

Example 5. Suppose that we consider the integral

$$\int_0^1 (x^3 + x)^9 (3x^2 + 1)\, dx,$$

with $u = x^3 + x$. When $x = 0$, $u = 0$, and when $x = 1$, $u = 2$. Thus our definite integral is equal to

$$\int_0^2 u^9 \, du = \frac{u^{10}}{10}\Big|_0^2 = \frac{2^{10}}{10}.$$

Example 6. Evaluate

$$\int_0^{\sqrt{\pi}} x \sin x^2 \, dx.$$

We let $u = x^2$, $du = 2x \, dx$. When $x = 0$, $u = 0$. When $x = \sqrt{\pi}$, $u = \pi$. Thus our integral is equal to

$$\tfrac{1}{2} \int_0^\pi \sin u \, du = \tfrac{1}{2}(-\cos u)\Big|_0^\pi = \tfrac{1}{2}(-\cos \pi + \cos 0) = 1.$$

Example 7. Evaluate

$$I = \int x^5 \sqrt{1 - x^2} \, dx.$$

We let $u = 1 - x^2$ so $du = -2x \, dx$. Then $x^2 = 1 - u$, $x^4 = (1 - u)^2$, and

$$I = -\tfrac{1}{2} \int (1 - u)^2 u^{1/2} \, du = -\tfrac{1}{2} \int (1 - 2u + u^2) u^{1/2} \, du$$

$$= -\tfrac{1}{2} \int (u^{1/2} - 2u^{3/2} + u^{5/2}) \, du$$

$$= -\tfrac{1}{2}\left(\frac{u^{3/2}}{3/2} - 2\frac{u^{5/2}}{5/2} + \frac{u^{7/2}}{7/2}\right).$$

Substituting $u = 1 - x^2$ gives the answer in terms of x.

XI, §1. EXERCISES

Find the following integrals.

1. $\displaystyle\int x e^{x^2} \, dx$

2. $\displaystyle\int x^3 e^{-x^4} \, dx$

3. $\displaystyle\int x^2(1 + x^3) \, dx$

4. $\displaystyle\int \frac{\log x}{x} \, dx$

5. $\int \dfrac{1}{x(\log x)^n} \, dx$ $(n = \text{integer})$ 6. $\int \dfrac{2x + 1}{x^2 + x + 1} \, dx$

7. $\int \dfrac{x}{x + 1} \, dx$ 8. $\int \sin x \cos x \, dx$

9. $\int \sin^2 x \cos x \, dx$ 10. $\int_0^\pi \sin^5 x \cos x \, dx$

11. $\int_0^\pi \cos^4 x \, dx$ 12. $\int \dfrac{\sin x}{1 + \cos^2 x} \, dx$

13. $\int \dfrac{\arctan x}{1 + x^2} \, dx$ 14. $\int_0^1 x^3 \sqrt{1 - x^2} \, dx$

15. $\int_0^{\pi/2} x \sin(2x^2) \, dx$

16. (a) $\int \sin 2x \, dx$ (b) $\int \cos 2x \, dx$

     (c) $\int \sin 3x \, dx$ (d) $\int \cos 3x \, dx$

     (e) $\int e^{4x} \, dx$ (f) $\int e^{5x} \, dx$ (g) $\int e^{-5x} \, dx$

In the next problems, you will use limits from the theory of the exponential function in Chapter VIII, §5, namely

$$\lim_{B \to \infty} \frac{B}{e^B} = 0.$$

17. Find the area under the curve $y = xe^{-x^2}$ between 0 and a number $B > 0$. Does this area approach a limit as B becomes very large? If so, what limit?

18. Find the area under the curve $y = x^2 e^{-x^3}$ between 0 and a number $B > 0$. Does this area approach a limit as B becomes very large? If so, what limit?

XI, §1. SUPPLEMENTARY EXERCISES

Find the following integrals.

1. $\int \dfrac{x^3}{x^4 + 2} \, dx$ 2. $\int \sqrt{3x + 1} \, dx$

3. $\int \sin^4 x \cos x \, dx$ 4. $\int \dfrac{e^x - e^{-x}}{e^x + e^{-x}} \, dx$

5. $\int \dfrac{x}{\sqrt{x^2 - 1}} \, dx$ 6. $\int x^3 \sqrt{x^4 + 1} \, dx$

7. $\displaystyle\int \frac{x}{(3x^2 + 5)^2}\, dx$

8. $\displaystyle\int (x^2 + 3)^4 x^3\, dx$

9. $\displaystyle\int \frac{\cos x}{\sin^3 x}\, dx$

10. $\displaystyle\int e^x \sqrt{e^x + 1}\, dx$

11. $\displaystyle\int (x^3 + 1)^{7/5} x^5\, dx$

12. $\displaystyle\int \frac{x}{(x^2 - 4)^{3/2}}\, dx$

13. $\displaystyle\int \sin 3x\, dx$

14. $\displaystyle\int \cos 4x\, dx$

15. $\displaystyle\int e^x \sin e^x\, dx$

16. $\displaystyle\int x^2 \sqrt{x^3 + 1}\, dx$

17. $\displaystyle\int \frac{1}{x \log x}\, dx$

18. $\displaystyle\int \frac{\sin x}{1 + \cos^2 x}\, dx$

19. $\displaystyle\int \frac{e^x}{e^x + 1}\, dx$

20. $\displaystyle\int \frac{(\log x)^4}{x}\, dx$

21. $\displaystyle\int_0^{\pi/2} \sin^3 x \cos x\, dx$

22. $\displaystyle\int_{-3}^{-1} \frac{1}{(x - 1)^2}\, dx$

23. $\displaystyle\int_0^1 \sqrt{2 - x}\, dx$

24. $\displaystyle\int_0^\pi \sin^2 x \cos x\, dx$

25. $\displaystyle\int_0^{\pi/2} \frac{\cos x}{1 + \sin^2 x}\, dx$

26. $\displaystyle\int_0^{2\pi} \frac{\sin x}{1 + \cos^2 x}\, dx$

27. $\displaystyle\int_0^{1/2} \frac{\arcsin x}{\sqrt{1 - x^2}}\, dx$

28. $\displaystyle\int_0^1 \frac{\arctan x}{1 + x^2}\, dx$

29. $\displaystyle\int_0^1 \frac{1 + e^{2x}}{e^x}\, dx$

30. $\displaystyle\int_0^{\pi/2} x \sin x^2\, dx$

XI, §2. INTEGRATION BY PARTS

If f, g are two differentiable functions of x, then

$$\frac{d(fg)}{dx} = f(x)\frac{dg}{dx} + g(x)\frac{df}{dx}.$$

Hence

$$f(x)\frac{dg}{dx} = \frac{d(fg)}{dx} - g(x)\frac{df}{dx}.$$

Using the formula for the integral of a sum, which is the sum of the integrals, and the fact that

$$\int \frac{d(fg)}{dx}\, dx = f(x)g(x),$$

we obtain

$$\int f(x)\frac{dg}{dx}\, dx = f(x)g(x) - \int g(x)\frac{df}{dx}\, dx,$$

which is called the formula for **integrating by parts**.

If we let $u = f(x)$ and $v = g(x)$, then the formula can be abbreviated in our shorthand notation as follows:

$$\int u\, dv = uv - \int v\, du.$$

Example 1. Find the integral $\int \log x\, dx$.

Let $u = \log x$ and $dv = dx$. Then $du = (1/x)\, dx$ and $v = x$. Hence our integral is in the form $\int u\, dv$ and is equal to

$$uv - \int v\, du = x \log x - \int 1\, dx = x \log x - x.$$

Example 2. Find the integral $\int xe^x\, dx$.

Let $u = x$ and $dv = e^x\, dx$. Then $du = dx$ and $v = e^x$. So

$$\int xe^x\, dx = \int u\, dv = xe^x - \int e^x\, dx = xe^x - e^x.$$

Observe how du/dx is a simpler function than u itself, whereas v and dv/dx are the same, in the present case. A similar procedure works for integrals of the form $\int x^n e^x\, dx$ when n is a positive integer, putting $u = x^n$, $du = nx^{n-1}\, dx$. See Exercises 7 and 8.

The two preceding examples illustrate a general fact: Passing from the function u to du/dx in the process of integrating by parts will work provided that going up from dv to v does not make this side of the procedure too much worse. In the first example, with $u = \log x$, then $du/dx = 1/x$ is a simpler function (a power of x), while going up from $dv = dx$ to $v = x$ still only contributes powers of x to the procedure.

Next we give an example where we have to integrate by parts twice before getting an answer.

Example 3. Find $\int e^x \sin x \, dx$.

Let $u = e^x$ and $dv = \sin x \, dx$. Then

$$du = e^x \, dx \qquad \text{and} \qquad v = -\cos x.$$

If we call our integral I, then

$$I = -e^x \cos x - \int -e^x \cos x \, dx$$

$$= -e^x \cos x + \int e^x \cos x \, dx.$$

This looks as if we were going around in circles. Don't lose heart. Rather, repeat the same procedure on $e^x \cos x$. Let

$$t = e^x \qquad \text{and} \qquad dz = \cos x \, dx.$$

Then,

$$dt = e^x \, dx \qquad \text{and} \qquad z = \sin x.$$

The second integral becomes

$$\int t \, dz = tz - \int z \, dt = e^x \sin x - \int e^x \sin x \, dx.$$

We have come back to our original integral

$$\int e^x \sin x \, dx$$

but with a minus sign! Thus

$$I = -e^x \cos x + e^x \sin x - I.$$

Hence

$$2I = e^x \sin x - e^x \cos x,$$

and dividing by 2 gives us the value

$$I = \frac{e^x \sin x - e^x \cos x}{2}.$$

We give an example where we first make a substitution *before* integrating by parts.

Example 4. Find the integral $\int e^{-\sqrt{x}}\, dx$.

We let $x = u^2$ so that $dx = 2u\, du$. Then

$$\int e^{-\sqrt{x}}\, dx = \int e^{-u} 2u\, du = 2 \int u e^{-u}\, du.$$

This can now be integrated by parts with $u = u$ and $dv = e^{-u}\, du$. Then $v = -e^{-u}$, and you can do the rest.

Remark. It can be shown, although not easily, that no procedure will allow you to express the integral

$$\int e^{-x^2}\, dx$$

in terms of the standard functions: powers of x, trigonometric functions, exponential and log function, sums, products, or composites of these.

XI, §2. EXERCISES

Find the following integrals.

1. $\int \arcsin x\, dx$

2. $\int \arctan x\, dx$

3. $\int e^{2x} \sin 3x\, dx$

4. $\int e^{-4x} \cos 2x\, dx$

5. $\int (\log x)^2\, dx$

6. $\int (\log x)^3\, dx$

7. $\int x^2 e^x\, dx$

8. $\int x^2 e^{-x}\, dx$

9. $\int x \sin x\, dx$

10. $\int x \cos x\, dx$

11. $\int x^2 \sin x\, dx$

12. $\int x^2 \cos x\, dx$

13. $\int x^3 \cos x^2\, dx$

14. $\int x^5 \sqrt{1 - x^2}\, dx$

[*Hint*: In Exercise 13, make first the substitution $u = x^2$, $du = 2x\, dx$. In Exercise 14, let $u = 1 - x^2$, $x^2 = 1 - u$.]

15. $\int x^2 \log x\, dx$

16. $\int x^3 \log x\, dx$

17. $\displaystyle\int x^2(\log x)^2\,dx$ 18. $\displaystyle\int x^3 e^{-x^2}\,dx$

19. $\displaystyle\int \frac{x^7}{(1-x^4)^2}\,dx$ 20. $\displaystyle\int_{-\pi}^{\pi} x^2\cos x\,dx$

In the next improper integrals, we shall use limits from the theory of exponentials and logarithms, Chapter VIII, §5. These limits are as follows. If n is a positive integer, then

$$\lim_{B\to\infty}\frac{B^n}{e^B}=0.$$

Also

$$\lim_{a\to 0} a\log a = 0.$$

The first limit states that the exponential function becomes large faster than any polynomial. The second limit can be deduced from the first, namely write $a=e^{-B}$. Then $a\to 0$ if and only if $B\to\infty$. But $\log a = -B$. So

$$a\log a = \frac{-B}{e^B},$$

whence we see that the first limit implies the second by taking $n=1$.

21. Let B be a number >0. Find the area under the curve $y=xe^{-x}$ between 0 and B. Does this area approach a limit as B becomes very large? If yes, what limit?

22. Does the improper integral $\int_1^\infty x^2 e^{-x}\,dx$ exist? If yes, what is its value?

23. Does the improper integral $\int_1^\infty x^3 e^{-x}\,dx$ exist? If yes, what is its value?

24. Let B be a number >2. Find the area under the curve

$$y=\frac{1}{x(\log x)^2}$$

between 2 and B. Does this area approach a limit as B becomes very large? If so, what limit?

25. Does the improper integral

$$\int_3^\infty \frac{1}{x(\log x)^4}\,dx$$

exist? If yes, what is its value?

26. Does the improper integral

$$\int_0^2 \log x \, dx$$

exist? If yes, what is its value?

XI, §2. SUPPLEMENTARY EXERCISES

Find the following integrals, using substitutions and integration by parts, as needed.

1. $\int x \arctan x \, dx$

2. (a) $\int \sqrt{1 - x^2} \, dx$ [*Hint*: Use $u = \sqrt{1 - x^2}$ and $dv = dx$.]

   (b) $\int x \arcsin x \, dx$

3. $\int x \arccos x \, dx$ 4. $\int x^3 e^{2x} \, dx$

5. $\int_{-1}^0 \arcsin x \, dx$ 6. $\int_1^2 x^3 \log x \, dx$

7. $\int_0^{1/2} x \arcsin 2x \, dx$ 8. $\int_1^2 \sqrt{x} \log x \, dx$

9. $\int_{-1}^1 x e^x \, dx$ 10. $\int_0^1 x^3 \sqrt{1 - x^2} \, dx$

11. $\int x e^{-\sqrt{x}} \, dx$ 12. $\int \sqrt{x} e^{-\sqrt{x}} \, dx$

Prove the formulas, where m, n are positive integers.

13. $\int (\log x)^n \, dx = x(\log x)^n - n \int (\log x)^{n-1} \, dx$

14. $\int x^n e^x \, dx = x^n e^x - n \int x^{n-1} e^x \, dx$

15. $\int x^m (\log x)^n \, dx = \dfrac{x^{m+1}(\log x)^n}{m + 1} - \dfrac{n}{m + 1} \int x^m (\log x)^{n-1} \, dx$

*16. Let $n! = n(n - 1) \cdots 1$ be the product of the first n integers. Show that

$$\int_0^\infty x^n e^{-x} \, dx = n!.$$

[*Hint*: First find the indefinite integral $\int x^n e^{-x} \, dx$ in terms of $\int x^{n-1} e^{-x} \, dx$ as in Exercise 14. Then evaluate between 0 and B, and let B become large. If I_n denotes the desired integral, you should find $I_n = n I_{n-1}$. You can then continue stepwise, until you are reduced to evaluating $I_0 = \int_0^\infty e^{-x} \, dx$.]

XI, §3. TRIGONOMETRIC INTEGRALS

We shall investigate integrals involving sine and cosine. It will be useful to have the following formulas:

$$\sin^2 x = \frac{1 - \cos 2x}{2}, \qquad \cos^2 x = \frac{1 + \cos 2x}{2}.$$

These are easily proved, using

$$\cos 2x = \cos^2 x - \sin^2 x \qquad \text{and} \qquad \sin^2 x + \cos^2 x = 1.$$

Example. We use the first boxed formula and get:

$$\int \sin^2 x \, dx = \int \frac{1}{2} \, dx - \frac{1}{2} \int \cos 2x \, dx = \frac{x}{2} - \frac{1}{4} \sin 2x.$$

The next two examples deal with *odd* powers of sine or cosine. A method can be used in this case, but cannot be used when there are only *even* powers.

Example. We wish to find

$$\int \sin^3 x \, dx.$$

We replace $\sin^2 x$ by $1 - \cos^2 x$, so that

$$\int \sin^3 x \, dx = \int (\sin x)(1 - \cos^2 x) \, dx$$

$$= \int \sin x \, dx - \int (\cos^2 x)(\sin x) \, dx.$$

The second of these last integrals can be evaluated by the substitution

$$u = \cos x, \qquad du = -\sin x \, dx.$$

We therefore find that

$$\int \sin^3 x \, dx = -\cos x + \frac{\cos^3 x}{3}.$$

For low powers of the sine and cosine, the above means are the easiest. Especially if we have to integrate an *odd* power of the sine or cosine, we can use a similar method.

Example. Find $I = \displaystyle\int \sin^5 x \cos^2 x \, dx$.

We replace

$$\sin^4 x = (\sin^2 x)^2 = (1 - \cos^2 x)^2,$$

so that

$$\int \sin^5 x \cos^2 x \, dx = \int (1 - \cos^2 x)^2 \cos^2 x \sin x \, dx.$$

Now we put

$$u = \cos x \qquad \text{and} \qquad du = -\sin x \, dx.$$

Then

$$I = -\int (1 - u^2)^2 u^2 \, du = -\int (1 - 2u^2 + u^4)u^2 \, du$$

$$= -\int (u^2 - 2u^4 + u^6) \, du$$

$$= -\left(\frac{u^3}{3} - 2\frac{u^5}{5} + \frac{u^7}{7}\right).$$

If you want the answer in terms of x, you substitute back $u = \cos x$ in this last expression.

In the two preceding examples, we had an *odd* power of the sine. Then by using the identity $\sin^2 x = 1 - \cos^2 x$ and the substitution

$$u = \cos x, \qquad du = -\sin x \, dx,$$

we transform the integral into a sum of integrals of the form

$$\int u^n \, du,$$

where n is a positive integer.

This method works only when there is an odd power of sine or cosine. In general, one has to use another method to integrate arbitrary powers.

There is a general way in which one can integrate $\sin^n x$ for any positive integer n: integrating by parts. Let us take first an example.

Example. Find the integral $\int \sin^3 x \, dx$.

We write the integral in the form

$$I = \int \sin^2 x \sin x \, dx.$$

Let $u = \sin^2 x$ and $dv = \sin x \, dx$. Then

$$du = 2 \sin x \cos x \, dx \qquad \text{and} \qquad v = -\cos x.$$

Thus

$$I = -(\sin^2 x)(\cos x) - \int - \cos x (2 \sin x \cos x) \, dx$$

$$= -\sin^2 x \cos x + 2 \int \cos^2 x \sin x \, dx.$$

This last integral could then be determined by substitution, for instance

$$t = \cos x \qquad \text{and} \qquad dt = -\sin x \, dx.$$

The last integral becomes $-2 \int t^2 \, dt$, and hence

$$I = \int \sin^3 x \, dx = -\sin^2 x \cos x - \tfrac{2}{3} \cos^3 x.$$

To deal with an arbitrary positive integer n, we shall show how to reduce the integral $\int \sin^n x \, dx$ to the integral $\int \sin^{n-2} x \, dx$. Proceeding stepwise downwards will give a method for getting the full answer.

Theorem 3.1. *For any integer $n \geq 2$, we have*

$$\boxed{\int \sin^n x \, dx = -\frac{1}{n} \sin^{n-1} x \cos x + \frac{n-1}{n} \int \sin^{n-2} x \, dx.}$$

Proof. We write the integral as

$$I_n = \int \sin^n x \, dx = \int \sin^{n-1} x \sin x \, dx.$$

Let $u = \sin^{n-1} x$ and $dv = \sin x \, dx$. Then

$$du = (n-1) \sin^{n-2} x \cos x \, dx \qquad \text{and} \qquad v = -\cos x.$$

Thus

$$I_n = -\sin^{n-1} x \cos x - \int -(n-1)\cos x \sin^{n-2} x \cos x \, dx$$

$$= -\sin^{n-1} x \cos x + (n-1) \int \sin^{n-2} x \cos^2 x \, dx.$$

We replace $\cos^2 x$ by $1 - \sin^2 x$ and get

$$I_n = -\sin^{n-1} x \cos x + (n-1) \int \sin^{n-2} x \, dx - (n-1) \int \sin^n x \, dx.$$

Therefore

$$I_n = -\sin^{n-1} x \cos x + (n-1)I_{n-2} - (n-1)I_n,$$

whence

$$nI_n = -\sin^{n-1} x \cos x + (n-1)I_{n-2}.$$

Dividing by n gives us our formula.

We leave the proof of the analogous formula for cosine as an exercise.

$$\int \cos^n x \, dx = \frac{1}{n}\cos^{n-1} x \sin x + \frac{n-1}{n}\int \cos^{n-2} x \, dx.$$

Integrals involving tangents can be done by a similar technique, because

$$\frac{d \tan x}{dx} = 1 + \tan^2 x.$$

These functions are less used than sine and cosine, and hence we don't write out the formulas, to lighten this printed page which would otherwise become oppressive.

Mixed powers of sine and cosine

One can integrate mixed powers of sine and cosine by replacing $\sin^2 x$ by $1 - \cos^2 x$, for instance.

Example. Find $\int \sin^2 x \cos^2 x \, dx$.

Replacing $\cos^2 x$ by $1 - \sin^2 x$, we see that our integral is equal to

$$\int \sin^2 x \, dx - \int \sin^4 x \, dx,$$

which we know how to integrate. We could also use a special trick for this case, making the substitutions at the beginning of the section. Thus

$$(\sin^2 x)(\cos^2 x) = \frac{1 - \cos^2 2x}{4},$$

which reduces the powers inside the integral. Another application of this same type, namely

$$\cos^2 2x = \frac{1 + \cos 4x}{2}$$

reduces our problem still further. Take your pick and work out the integral completely as an exercise.

Warning. Because of trigonometric identities like

$$\sin^2 x = -\tfrac{1}{2} \cos 2x + \tfrac{1}{2}$$

different forms for the answers are possible. They will differ by a constant of integration.

When we meet an integral involving a square root, we can frequently get rid of the square root by making a trigonometric substitution.

Example. Find the area of a circle of radius 3.

The equation of the circle is

$$x^2 + y^2 = 9,$$

and the portion of the circle in the first quadrant is described by the function

$$y = \sqrt{3^2 - x^2}.$$

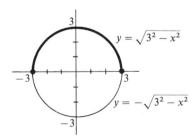

One-fourth of the area is therefore given by the integral

$$\int_0^3 \sqrt{3^2 - x^2}\, dx.$$

For such integrals, we want to get rid of the horrible square root sign, so we try to make the expression under the integral into a perfect square. We use the substitution

$$x = 3 \sin t \quad \text{and} \quad dx = 3 \cos t \, dt, \quad \text{with} \quad 0 \leqq t \leqq \pi/2.$$

When $t = 0$ then $x = 0$ and when $t = \pi/2$, then $x = 3$. Hence our integral becomes

$$\int_0^{\pi/2} \sqrt{3^2 - 3^2 \sin^2 t} \, 3 \cos t \, dt = \int_0^{\pi/2} 9 \cos t \cos t \, dt$$

$$= 9 \int_0^{\pi/2} \cos^2 t \, dt$$

$$= \frac{9\pi}{4}.$$

Note that on the stated interval $0 \leqq t \leqq \pi/2$, the cosine $\cos t$ is positive, and so

$$\sqrt{1 - \sin^2 t} = \sqrt{\cos^2 t} = \cos t.$$

If we picked an interval where the cosine is negative, then when taking the square root, we would need to use an extra minus sign. That is, if $\cos t < 0$, then

$$\sqrt{\cos^2 t} = -\cos t.$$

Since the integral above represented 1/4-th of the area of the circle, it follows that the total area of the circle is 9π.

In general, an integral involving expressions like

$$\sqrt{1 - x^2}$$

can sometimes be evaluated by using the substitution

$$x = \sin \theta, \quad dx = \cos \theta \, d\theta,$$

because the expression under the square root then becomes a perfect square, namely $1 - \sin^2 \theta = \cos^2 \theta$.

In making this substitution, we usually let

$$-\pi/2 \leqq \theta \leqq \pi/2 \quad \text{and} \quad -1 \leqq x \leqq 1.$$

This is the range where

$$x = \sin \theta \quad \text{has the inverse function } \theta = \arcsin x.$$

Negative powers of sine and cosine

It is usually a pain to integrate negative powers of sine and cosine, although it can be done. You should be aware that the following formula exists.

$$\int \frac{1}{\cos \theta} d\theta = \int \sec \theta \, d\theta = \log(\sec \theta + \tan \theta).$$

This is done by substitution. We have

$$\frac{1}{\cos \theta} = \sec \theta = \frac{(\sec \theta)(\sec \theta + \tan \theta)}{\sec \theta + \tan \theta}.$$

Let $u = \sec \theta + \tan \theta$. Then the integral is in the form

$$\int \frac{1}{u} du.$$

(This is a good opportunity to emphasize that the formula we just obtained is valid on any interval such that $\cos \theta \neq 0$ and

$$\sec \theta + \tan \theta > 0.$$

Otherwise the symbols are meaningless. Determine such an interval as an exercise.) The expression in the above formula is sufficiently complicated that **you should not memorize it**. Plug into it when needed. There is a similar formula for the integral of $1/\sin \theta$, which is obtained by using the prefix co- on the right-hand side. The formula is:

$$I = \int \frac{1}{\sin \theta} d\theta = \int \csc \theta \, d\theta = -\log(\csc \theta + \cot \theta),$$

which is similar to the answer given previously for $\int (1/\cos \theta) \, d\theta$. Of course, the answer is over an interval where the expression inside the logarithm is positive. Otherwise, one has to take the absolute value of this expression. The proof is similar, and you can also check it by differentiating the right-hand side to get $1/\sin \theta$.

Warning. On the left-hand side we have sine instead of cosine, so the prefix co- is deleted. On the right-hand side, we have cosecant and cotangent, so the prefix co- is added. *You should remember that there is such a symmetry,* but always check exactly what the correct relation is before using it, because in using this symmetry, *certain minus signs appear* just as a minus sign now appeared on the right-hand side. Do not attempt to memorize when such minus signs occur. Check each time that you need a similar formula, or look it up in integral tables.

Example. Let us evaluate the integral

$$I = \int \frac{1}{x\sqrt{1 - x^2}}\, dx.$$

Let $x = \sin \theta$, $dx = \cos \theta\, d\theta$. Then

$$I = \int \frac{1}{\sin \theta \sqrt{\cos^2 \theta}} \cos \theta\, d\theta.$$

Over an interval where $\cos \theta$ is positive, we have

$$\sqrt{\cos^2 \theta} = \cos \theta,$$

and hence

$$I = \int \frac{1}{\sin \theta}\, d\theta.$$

This integral was evaluated in the above box.

XI, §3. EXERCISES

Find the following integrals.

1. $\int \sin^4 x\, dx$ 2. $\int \cos^3 x\, dx$ 3. $\int \sin^2 x \cos^3 x\, dx$

Find the area of the region enclosed by the following curves.

4. $x^2 + \dfrac{y^2}{9} = 1$ 5. $\dfrac{x^2}{4} + \dfrac{y^2}{16} = 1$ 6. $\dfrac{x^2}{a^2} + \dfrac{y^2}{b^2} = 1$

7. Find the area of a circle of radius $r > 0$.

8. Find the integrals.

(a) $\int \sqrt{1 - \cos \theta} \, d\theta$ (b) $\int \sqrt{1 + \cos \theta} \, d\theta$

[*Hint*: Write $\theta = 2u$. This should help you make the expression under the square root into a perfect square.]

9. For any integers m, n prove the formulas:

$$\sin mx \sin nx = \tfrac{1}{2}[\cos(m - n)x - \cos(m + n)x],$$
$$\sin mx \cos nx = \tfrac{1}{2}[\sin(m + n)x + \sin(m - n)x],$$
$$\cos mx \cos nx = \tfrac{1}{2}[\cos(m + n)x + \cos(m - n)x].$$

[*Hint*: Expand the right-hand side and cancel as much as you can. Use the addition formulas of Chapter IV, §3.]

10. Use the preceding exercise to do this one.
    (a) Show that

$$\int_{-\pi}^{\pi} \sin 3x \cos 2x \, dx = 0.$$

(b) Show that

$$\int_{-\pi}^{\pi} \cos 5x \cos 2x \, dx = 0.$$

11. Show in general that for any positive integers m, n we have

$$\int_{-\pi}^{\pi} \sin mx \cos nx \, dx = 0.$$

12. Show in general that for positive integers m, n,

$$\int_{-\pi}^{\pi} \sin mx \sin nx \, dx = \begin{cases} 0 & \text{if } m \neq n, \\ \pi & \text{if } m = n. \end{cases}$$

[*Hint*: If $m \neq n$, use Exercise 9. If $m = n$, use $\sin^2 nx = \tfrac{1}{2}(1 - \cos 2nx)$.]

13. Find $\int \tan x \, dx$.

Find the following integrals.

14. $\int \dfrac{1}{\sqrt{9 - x^2}} \, dx$ 15. $\int \dfrac{1}{\sqrt{3 - x^2}} \, dx$

16. $\int \dfrac{1}{\sqrt{2 - 4x^2}} \, dx$ 17. $\int \dfrac{1}{\sqrt{a^2 - b^2 x^2}} \, dx$

18. Let f be a continuous function on the interval $[-\pi, \pi]$. We define numbers c_0, a_n, b_n for positive integers n by the formulas

$$c_0 = \frac{1}{2\pi} \int_{-\pi}^{\pi} f(x)\, dx,$$

$$a_n = \frac{1}{\pi} \int_{-\pi}^{\pi} f(x) \cos nx\, dx, \qquad b_n = \frac{1}{\pi} \int_{-\pi}^{\pi} f(x) \sin nx\, dx.$$

These numbers c_0, a_n, b_n are called the **Fourier coefficients** of f.

Example. Let $f(x) = x$. Then the 0-th Fourier coefficient of f is the integral:

$$c_0 = \frac{1}{2\pi} \int_{-\pi}^{\pi} x\, dx = \frac{1}{2\pi} \frac{x^2}{2} \Big|_{-\pi}^{\pi} = 0.$$

The coefficients a_n and b_n are given by the integrals

$$a_n = \frac{1}{\pi} \int_{-\pi}^{\pi} x \cos nx\, dx \qquad \text{and} \qquad b_n = \frac{1}{2\pi} \int_{-\pi}^{\pi} x \sin nx\, dx.$$

You should be able to evaluate these integrals using integration by parts. Compute the Fourier coefficients of the following functions. (If you do 19 first, you might have less work.)

(a) $f(x) = x$
(b) $f(x) = x^2$
(c) $f(x) = |x|$
(d) $f(x) = \cos x$
(e) $f(x) = \sin x$
(f) $f(x) = \sin^2 x$
(g) $f(x) = \cos^2 x$
(h) $f(x) = |\sin x|$
(i) $f(x) = |\cos x|$
(j) $f(x) = 1$

19. (a) Let f be an even function [that is $f(x) = f(-x)$]. Show that its Fourier coefficients b_n are all equal to 0.
   (b) Let f be an odd function [that is $f(x) = -f(-x)$]. What can you say about its Fourier coefficients?

XI, §3. SUPPLEMENTARY EXERCISES

1. $\displaystyle\int \frac{\cos^3 x}{\sin x}\, dx$

2. $\displaystyle\int \tan^2 x\, dx$

3. $\displaystyle\int e^x \sin e^x\, dx$

4. $\displaystyle\int \frac{1}{1 - \cos x}\, dx$

5. $\displaystyle\int_0^{\pi/2} \cos^2 x\, dx$

6. $\displaystyle\int_0^{\pi/3} \sin^6 x\, dx$

7. $\displaystyle\int_{-\pi}^{\pi} \sin^2 x \cos^2 x\, dx$

8. $\displaystyle\int_0^{2\pi} \sin^3 2x\, dx$

9. $\displaystyle\int_0^{\pi/2} \sin^2 2x \cos^2 2x \, dx$

10. $\displaystyle\int_0^{\pi/4} \cos^4 x \, dx$

11. $\displaystyle\int x^2 \sqrt{1 - x^2} \, dx$

12. $\displaystyle\int \frac{1}{(x^2 + 1)^2} \, dx$

13. $\displaystyle\int_0^1 \frac{1}{\sqrt{1 - x^2}} \, dx$

14. $\displaystyle\int \frac{\sqrt{1 - x^2}}{x^2} \, dx$

15. $\displaystyle\int \frac{x^3}{\sqrt{16 - x^2}} \, dx$

16. $\displaystyle\int \frac{x^3}{\sqrt{1 + x^2}} \, dx$

In the next exercises, we let a be a positive number.

17. $\displaystyle\int \frac{1}{x\sqrt{a^2 - x^2}} \, dx$

18. $\displaystyle\int \frac{x^2}{\sqrt{a^2 - x^2}} \, dx$

19. $\displaystyle\int \frac{1}{x^3\sqrt{a^2 - x^2}} \, dx$

20. $\displaystyle\int \frac{1}{x^2\sqrt{a^2 - x^2}} \, dx$

21. $\displaystyle\int \frac{\sqrt{1 - x^2}}{x} \, dx$

22. $\displaystyle\int \frac{\sqrt{a^2 - x^2}}{x^2} \, dx$

23. $\displaystyle\int \frac{x^2}{(a^2 - x^2)^{3/2}} \, dx$

24. $\displaystyle\int_0^a x^4 \sqrt{a^2 - x^2} \, dx$

XI, §4. PARTIAL FRACTIONS

We want to study the integrals of quotients of polynomials.

Let $f(x)$ and $g(x)$ be two polynomials. We want to investigate the integral

$$\int \frac{f(x)}{g(x)} \, dx.$$

Using long division, one can reduce the problem to the case when the degree of f is less than the degree of g. The following example illustrates this reduction.

Example. Consider the two polynomials $f(x) = x^3 - x + 1$ and

$$g(x) = x^2 + 1.$$

Dividing f by g (you should know how from high school) we obtain a quotient of x with remainder $-2x + 1$. Thus

$$x^3 - x + 1 = (x^2 + 1)x + (-2x + 1).$$

Hence

$$\frac{f(x)}{g(x)} = x + \frac{-2x + 1}{x^2 + 1}.$$

To find the integral of $f(x)/g(x)$ we integrate x, and the quotient on the right, which has the property that the degree of the numerator is less than the degree of the denominator.

From now on, *we assume throughout that when we consider a quotient $f(x)/g(x)$, the degree of f is less than the degree of g.* We assume this because the method we shall describe works only in this case. Factoring out a constant if necessary, we also assume that $g(x)$ can be written

$$g(x) = x^d + \text{lower terms}.$$

We shall begin by discussing special cases, and then describe afterwards how the general case can be reduced to these.

First part. Linear factors in the denominator

Case 1. If a is a number, and n an integer ≥ 1, then

$$\int \frac{1}{(x-a)^n} \, dx = \frac{1}{-n+1} \frac{1}{(x-a)^{n-1}} \qquad \text{if} \quad n \neq 1,$$

$$= \log(x - a) \qquad\qquad \text{if} \quad n = 1.$$

This is an old story. We know how to do it. In fact, we have

$$\int \frac{1}{(x-a)^n} \, dx = \int (x-a)^{-n} \, dx = \int u^{-n} \, du.$$

Suppose $n \neq 1$. Then by substitution $u = x - a$, $du = dx$, we get

$$\int (x-a)^{-n} \, dx = \frac{(x-a)^{-n+1}}{-n+1} = \frac{1}{-n+1} \frac{1}{(x-a)^{n-1}}$$

because

$$u^{-n+1} = u^{-(n-1)} = \frac{1}{u^{n-1}}.$$

Suppose $n = 1$. Then the integral has the form

$$\int \frac{1}{u} \, du = \log u,$$

and hence

$$\int \frac{1}{x - a}\, dx = \log(x - a).$$

Case 2. Next we consider integrals of expressions like

$$\int \frac{1}{(x - 2)(x - 3)}\, dx \quad \text{or} \quad \int \frac{x + 1}{(x - 1)^2(x - 2)}\, dx,$$

where the denominator consists of a product of terms of the form

$$(x - a_1) \cdots (x - a_n)$$

for some numbers, $a_1, \ldots, a_n$ which need not be distinct. The procedure amounts to writing the expression under the integral as a sum of terms, as in Case 1.

Example. We wish to find the integral

$$\int \frac{1}{(x - 2)(x - 3)}\, dx.$$

To do this, we want to write

$$\frac{1}{(x - 2)(x - 3)} = \frac{c_1}{x - 2} + \frac{c_2}{x - 3}$$

with some numbers c_1 and c_2, for which we have to solve. Put the expression on the right over a common denominator. We find

$$\frac{c_1}{x - 2} + \frac{c_2}{x - 3} = \frac{c_1(x - 3) + c_2(x - 2)}{(x - 2)(x - 3)}.$$

Thus $(x - 2)(x - 3)$ is the common denominator, and

$$\text{numerator} = c_1(x - 3) + c_2(x - 2) = (c_1 + c_2)x - 3c_1 - 2c_2.$$

We want the fraction to be equal to $1/(x - 2)(x - 3)$. Thus the numerator must be equal to 1, that is we must have

$$(c_1 + c_2)x - 3c_1 - 2c_2 = 1.$$

Therefore it suffices to solve the simultaneous equations

$$c_1 + c_2 = 0,$$

$$-3c_1 - 2c_2 = 1.$$

Solving for c_1 and c_2 gives $c_2 = 1$ and $c_1 = -1$. Hence

$$\int \frac{1}{(x-2)(x-3)} \, dx = \int \frac{-1}{(x-2)} \, dx + \int \frac{1}{(x-3)} \, dx$$

$$= -\log(x-2) + \log(x-3).$$

Example. Find the integral

$$\int \frac{x+1}{(x-1)^2(x-2)} \, dx.$$

We want to find numbers c_1, c_2, c_3 such that

$$\frac{x+1}{(x-1)^2(x-2)} = \frac{c_1}{x-1} + \frac{c_2}{(x-1)^2} + \frac{c_3}{x-2}.$$

Note that $(x-1)^2$ appears in the denominator of the original quotient. To take this into account, it is necessary to include two terms with $(x-1)$ and $(x-1)^2$ in their denominators, appearing above as

$$\frac{c_1}{x-1} + \frac{c_2}{(x-1)^2}.$$

On the other hand, $(x-2)$ appears only in the first power in the original quotient, so it gives rise only to one term

$$\frac{c_3}{x-2}$$

in the partial fraction decomposition. (The general rule is stated at the end of the section.)

We now describe how to find the constants c_1, c_2, c_3, satisfying the relation

$$\frac{x+1}{(x-1)^2(x-2)} = \frac{c_1}{x-1} + \frac{c_2}{(x-1)^2} + \frac{c_3}{x-2}$$

$$= \frac{c_1(x-1)(x-2) + c_2(x-2) + c_3(x-1)^2}{(x-1)^2(x-2)}.$$

Here we put the fraction on the right over the common denominator

$$(x - 1)^2(x - 2).$$

We have

$$x + 1 = \text{numerator} = c_1(x - 1)(x - 2) + c_2(x - 2) + c_3(x - 1)^2$$

$$= (c_1 + c_3)x^2 + (-3c_1 + c_2 - 2c_3)x + 2c_1 - 2c_2 + c_3.$$

Thus to find the constants c_1, c_2, c_3 satisfying the desired relation, we have to solve the simultaneous equations

$$c_1 \quad\quad + c_3 = 0,$$

$$-3c_1 + c_2 - 2c_3 = 1,$$

$$2c_1 - 2c_2 + c_3 = 1.$$

This is a system of three linear equations in three unknowns, which you can solve to determine c_1, c_2, and c_3. One finds $c_1 = -3$, $c_2 = -2$, $c_3 = 3$. Hence

$$\int \frac{x + 1}{(x - 1)^2(x - 2)} \, dx = \int \frac{-3}{x - 1} \, dx + \int \frac{-2}{(x - 1)^2} \, dx + \int \frac{3}{(x - 2)} \, dx$$

$$= -3 \log(x - 1) + \frac{2}{x - 1} + 3 \log(x - 2).$$

It is a theorem in algebra that if you follow the above procedure to write a fraction in terms of simpler fractions according to the method illustrated in the examples, you will always be able to solve for the coefficients $c_1, c_2, c_3, \ldots$. The *proof* cannot be given at the level of this course, but in practice, unless you or I make a mistake, we just solve numerically in each case. If we have higher powers of some factor in the denominator, then we have to use higher powers also in the simpler fractions on the right-hand side.

Example. We can decompose

$$\frac{x + 1}{(x - 1)^3(x - 2)} = \frac{c_1}{x - 1} + \frac{c_2}{(x - 1)^2} + \frac{c_3}{(x - 1)^3} + \frac{c_4}{x - 2}.$$

Putting the right-hand side over a common denominator, and equating the numerator with $x + 1$, we can solve for the coefficients

$$c_1, c_2, c_3, c_4.$$

Second part. Quadratic factors in the denominator

Case 3. We want to find the integral

$$I_n = \int \frac{1}{(x^2 + 1)^n} \, dx$$

when n is a positive integer. If $n = 1$, there is nothing new, the integral is arctan x. If $n > 1$ we shall use integration by parts. There will be a slight twist on the usual procedure, because if we integrate I_n by parts in the natural way, we find that the exponent n increases by one unit. Let us do the case $n = 1$ as an example, so we start with

$$I_1 = \int \frac{1}{x^2 + 1} \, dx.$$

Let

$$u = \frac{1}{x^2 + 1} = (x^2 + 1)^{-1}, \qquad dv = dx,$$

$$du = \frac{-2x}{(x^2 + 1)^2} \, dx, \qquad\qquad v = x.$$

Then

$$I_1 = \frac{x}{x^2 + 1} - \int \frac{-2x^2}{(x^2 + 1)^2} \, dx$$

(*) $$= \frac{x}{x^2 + 1} + 2 \int \frac{x^2}{(x^2 + 1)^2} \, dx.$$

In the last integral on the right, write $x^2 = x^2 + 1 - 1$. Then

$$\int \frac{x^2}{(x^2 + 1)^2} \, dx = \int \frac{x^2 + 1 - 1}{(x^2 + 1)^2} \, dx = \int \frac{1}{x^2 + 1} \, dx - \int \frac{1}{(x^2 + 1)^2} \, dx$$

$$= \text{arctan } x - I_2.$$

If we now substitute this in expression (*) we obtain

$$I_1 = \frac{x}{x^2 + 1} + 2 \text{ arctan } x - 2I_2.$$

Therefore we can solve for I_2 in terms of I_1, and we find

$$2I_2 = \frac{x}{x^2 + 1} + 2 \arctan x - I_1$$

$$= \frac{x}{x^2 + 1} + 2 \arctan x - \arctan x$$

$$= \frac{x}{x^2 + 1} + \arctan x.$$

Dividing by 2 yields the value for I_2:

$$\boxed{\int \frac{1}{(x^2 + 1)^2}\, dx = \frac{1}{2}\frac{x}{x^2 + 1} + \frac{1}{2} \arctan x.}$$

The same method works in general. We want to reduce I_n to finding I_{n-1}, where

$$I_{n-1} = \int \frac{1}{(x^2 + 1)^{n-1}}\, dx.$$

Let

$$u = \frac{1}{(x^2 + 1)^{n-1}} \qquad \text{and} \qquad dv = dx.$$

Then

$$du = -(n - 1)\frac{2x}{(x^2 + 1)^n}\, dx \qquad \text{and} \qquad v = x.$$

Thus

$$I_{n-1} = \frac{x}{(x^2 + 1)^{n-1}} + 2(n - 1) \int \frac{x^2}{(x^2 + 1)^n}\, dx.$$

We write $x^2 = x^2 + 1 - 1$. We obtain

$$I_{n-1} = \frac{x}{(x^2 + 1)^{n-1}} + 2(n - 1) \int \frac{1}{(x^2 + 1)^{n-1}}\, dx - 2(n - 1) \int \frac{1}{(x^2 + 1)^n}\, dx$$

or in other words:

$$I_{n-1} = \frac{x}{(x^2 + 1)^{n-1}} + 2(n - 1)I_{n-1} - 2(n - 1)I_n.$$

Therefore

$$2(n-1)I_n = \frac{x}{(x^2+1)^{n-1}} + (2n-3)I_{n-1},$$

whence

$$\int \frac{1}{(x^2+1)^n}\, dx = \frac{1}{2(n-1)} \frac{x}{(x^2+1)^{n-1}}$$
$$+ \frac{(2n-3)}{2(n-1)} \int \frac{1}{(x^2+1)^{n-1}}\, dx,$$

or using the abbreviation I_n, we find:

$$I_n = \frac{1}{2n-2} \frac{x}{(x^2+1)^{n-1}} + \frac{2n-3}{2n-2} I_{n-1}.$$

This gives us a recursion formula which lowers the exponent n in the denominator until we reach $n=1$. In that case, we know that

$$\int \frac{1}{x^2+1}\, dx = \arctan x.$$

If you want to find I_3, use the formula to reduce it to I_2, then use the formula again to reduce it to I_1, which is arctan x. This gives a complete formula for I_3. To get a complete formula for I_n takes n steps. Of course you should not memorize the above formula; you should only remember the method by which it is obtained to apply it to special cases, say to finding I_3, I_4.

Eliminating extra constants by substitution

Sometimes we meet an integral which is a slight variation of the one just considered, with an extra constant. For instance, if b is a number, find

$$\int \frac{1}{(x^2+b^2)^n}\, dx.$$

Using the substitution $x = bz$, $dx = b\,dz$ reduces the integral to

$$\int \frac{1}{(b^2z^2 + b^2)^n}\, b\, dz = \int \frac{b}{b^{2n}(z^2 + 1)^n}\, dz$$

$$= \frac{1}{b^{2n-1}} \int \frac{1}{(z^2 + 1)^n}\, dz.$$

We have

$$\int \frac{1}{(z^2 + 1)^n}\, dz = \int \frac{1}{(x^2 + 1)^n}\, dx$$

because the two integrals differ only by a change of letters. This shows how to use a substitution to reduce the computation of the integral with b to the integral when $b = 1$ treated above.

Case 4. Find the integral

$$\int \frac{x}{(x^2 + b^2)^n}\, dx.$$

This is an old story. We make the substitution

$$u = x^2 + b^2 \qquad \text{and} \qquad du = 2x\, dx.$$

Then

$$\int \frac{x}{(x^2 + b^2)^n}\, dx = \frac{1}{2} \int \frac{1}{u^n}\, du,$$

which we know how to evaluate, and thus we find

$$\int \frac{x}{(x^2 + b^2)^n}\, dx = \begin{cases} \frac{1}{2} \log(x^2 + b^2) & \text{if } n = 1, \\ \dfrac{1}{2(-n + 1)} \dfrac{1}{(x^2 + b^2)^{n-1}} & \text{if } n \neq 1. \end{cases}$$

Example. Find

$$\int \frac{5x - 3}{(x^2 + 5)^2}\, dx.$$

We write

$$\int \frac{5x - 3}{(x^2 + 5)^2}\, dx = 5 \int \frac{x}{(x^2 + 5)^2}\, dx - 3 \int \frac{1}{(x^2 + 5)^2}\, dx.$$

Then:

$$5 \int \frac{x}{(x^2 + 5)^2} \, dx = \frac{5}{2} \int \frac{2x}{(x^2 + 5)^2} \, dx = \frac{5}{2} \int \frac{1}{u^2} \, du = \frac{5}{2} \frac{u^{-1}}{-1} = -\frac{5}{2} \frac{1}{x^2 + 5}.$$

For the second integral on the right, we may put

$$x = \sqrt{5} t \quad\text{and}\quad dx = \sqrt{5} \, dt.$$

Then:

$$\int \frac{1}{(x^2 + 5)^2} \, dx = \int \frac{1}{(5t^2 + 5)^2} \sqrt{5} \, dt = \frac{\sqrt{5}}{25} \int \frac{1}{(t^2 + 1)^2} \, dt$$

and we have previously computed

$$\int \frac{1}{(t^2 + 1)^2} \, dt = \frac{1}{2} \left(\frac{t}{t^2 + 1} + \arctan t \right).$$

Putting everything together, using $t = x/\sqrt{5}$, we find:

$$\int \frac{5x - 3}{(x^2 + 5)^2} \, dx = -\frac{5}{2} \frac{1}{x^2 + 5} - 3 \frac{\sqrt{5}}{25} \frac{1}{2} \left(\frac{x/\sqrt{5}}{(x/\sqrt{5})^2 + 1} + \arctan \frac{x}{\sqrt{5}} \right).$$

Third part. The general quotient $f(x)/g(x)$

If you are given a polynomial of type $x^2 + bx + c$, then you **factor or complete the square**. The polynomial can thus be written in the form

$$(x - \alpha)(x - \beta) \quad\text{or}\quad (x - \alpha)^2 + \beta^2$$

with suitable numbers α, β. Two cases arise. For example:

Case 1. $x^2 - x - 6 = (x + 2)(x - 3).$

Case 2. $x^2 - 2x + 5 = (x - 1)^2 + 2^2.$

In Case 1, we have factored the polynomial into two factors, and each factor has degree 1.

In Case 2, we have not factored the polynomial. By a change of variables, we can turn it into an expression $t^2 + 1$. Namely, let

$$x - 1 = 2t \quad\text{so}\quad x = 2t + 1.$$

Then

$$(x - 1)^2 + 2^2 = 2^2 t^2 + 2^2 = 2^2(t^2 + 1).$$

We made the change of variables so that 2^2 would come out as a factor. We note that in Case 2, we cannot factor the polynomial any further.

The following general result can be proved, but the proof is long, and cannot be given in this course.

Let $g(x)$ be a polynomial with real numbers as coefficients. Then $g(x)$ can always be written as a product of terms of type

$$(x - \alpha)^n \qquad and \qquad [(x - \beta)^2 + \gamma^2]^m,$$

n, m being integers $\geqq 0$, and some constant factor.

This can be quite difficult to do explicitly, but in the exercises, the situation is fixed up so that it is easy.

Example. By completing the square, we write

$$x^2 + 2x + 3 = (x + 1)^2 + 2 = (x + 1)^2 + (\sqrt{2})^2.$$

We can then evaluate the integral:

$$\int \frac{1}{x^2 + 2x + 3} \, dx$$

Let $x + 1 = \sqrt{2}\,t$ and $dx = \sqrt{2} \, dt$. Then

$$\int \frac{1}{x^2 + 2x + 3} \, dx = \int \frac{1}{(x + 1)^2 + (\sqrt{2})^2} \, dx = \frac{1}{2} \int \frac{1}{t^2 + 1} \sqrt{2} \, dt$$

$$= \frac{\sqrt{2}}{2} \arctan t$$

$$= \frac{\sqrt{2}}{2} \arctan \frac{x + 1}{\sqrt{2}}.$$

Example. Let us find

$$\int \frac{x}{x^2 + 2x + 3} \, dx.$$

We write

$$\int \frac{x}{x^2 + 2x + 3} \, dx = \frac{1}{2} \int \frac{2x + 2 - 2}{x^2 + 2x + 3} \, dx$$

$$= \frac{1}{2} \int \frac{2x + 2}{x^2 + 2x + 3} \, dx - \int \frac{1}{x^2 + 2x + 3} \, dx.$$

Then:

$$\frac{1}{2}\int \frac{2x+2}{x^2+2x+3}\,dx = \frac{1}{2}\int \frac{1}{u}\,du = \frac{1}{2}\log(x^2+2x+3).$$

Putting this together with the previous example, we find:

$$\int \frac{x}{x^2+2x+3}\,dx = \tfrac{1}{2}\log(x^2+2x+3) - \frac{\sqrt{2}}{2}\arctan\!\left(\frac{x+1}{\sqrt{2}}\right).$$

Example. Find the integral

$$\int \frac{2x+5}{(x^2+1)^2(x-3)}\,dx.$$

We can find numbers c_1, c_2,...such that the quotient is equal to

$$\frac{2x+5}{(x^2+1)^2(x-3)} = \frac{c_1+c_2x}{x^2+1} + \frac{c_3+c_4x}{(x^2+1)^2} + \frac{c_5}{x-3}$$

$$= \frac{c_1}{x^2+1} + c_2\frac{x}{x^2+1} + c_3\frac{1}{(x^2+1)^2}$$

$$+ c_4\frac{x}{(x^2+1)^2} + c_5\frac{1}{x-3}.$$

It is a theorem of algebra that you can always solve for the constants c_1, c_2, c_3, c_4, c_5 to get such a decomposition of the original fraction into the sum on the right, which is called the **partial fraction decomposition.** Observe that corresponding to the term with x^2+1 you need several terms on the right-hand side, especially those with an x in the numerator. If you do not include these, then you would get an incomplete formula, which would not work out. You could not compute the constants.

We now compute the constants. We put the right-hand side of the decomposition over the common denominator

$$(x^2+1)^2(x-3).$$

The numerator is equal to

$$2x + 5 = c_1(x^2 + 1)(x - 3) + c_2 x(x^2 + 1)(x - 3) + c_3(x - 3)$$

$$+ c_4 x(x - 3) + c_5(x^2 + 1)^2.$$

We equate the coefficients of x^4, x^3, x^2, x and the respective constants, and get a system of five linear equations in five unknowns, which can be solved. It is tedious to do it here and we leave it as an exercise, but we write down the equations:

$$c_2 \qquad\qquad\qquad + c_5 = 0 \qquad \text{(coefficient of } x^4\text{)},$$

$$c_1 - 3c_2 \qquad\qquad\qquad = 0 \qquad \text{(coefficient of } x^3\text{)},$$

$$-3c_1 + c_2 \qquad + c_4 + 2c_5 = 0 \qquad \text{(coefficient of } x^2\text{)},$$

$$c_1 - 3c_2 + c_3 - 3c_4 \qquad = 2 \qquad \text{(coefficient of } x\text{)},$$

$$-3c_1 \qquad - 3c_3 \qquad + c_5 = 5 \qquad \text{(coefficient of 1)}.$$

For the integral, we then obtain:

$$\int \frac{2x + 5}{(x^2 + 1)^2(x - 3)} \, dx = c_1 \arctan x + \tfrac{1}{2}c_2 \log(x^2 + 1)$$

$$+ c_3 \int \frac{1}{(x^2 + 1)^2} \, dx - \tfrac{1}{2}c_4 \frac{1}{x^2 + 1} + c_5 \log(x - 3).$$

The integral which we left standing is just that of Case 3, so we have shown how to find the desired integral.

Example. There is a partial fraction decomposition.

$$\frac{x^4 + 2x - 1}{(x^2 + 2)^3(x - 5)^2} = \frac{c_1 + c_2 x}{(x^2 + 2)} + \frac{c_3 + c_4 x}{(x^2 + 2)^2} + \frac{c_5 + c_6 x}{(x^2 + 2)^3}$$

$$+ \frac{c_7}{x - 5} + \frac{c_8}{(x - 5)^2}.$$

It would be tedious to compute the constants, and we don't do it.

The general rule is as follows: Suppose we have a quotient $f(x)/g(x)$ with degree of $f <$ degree of g. We factor g as far as possible into terms like

$$(x - \alpha)^n \qquad \text{and} \qquad [(x - \beta)^2 + \gamma^2]^m,$$

n, m being integers ≥ 0. Then

$$\frac{f(x)}{g(x)} = \text{sum of terms of the following type:}$$

$$\frac{c_1}{x - \alpha} + \frac{c_2}{(x - \alpha)^2} + \cdots + \frac{c_n}{(x - \alpha)^n}$$

$$+ \frac{d_1 + e_1 x}{(x - \beta)^2 + \gamma^2} + \cdots + \frac{d_m + e_m x}{[(x - \beta)^2 + \gamma^2]^m}$$

with suitable constants $c_1, c_2, \ldots, d_1, d_2, \ldots, e_1, e_2, \ldots$.

Once the quotient $f(x)/g(x)$ is written as above, then Cases 1, 2, and 3 allow us to integrate each term. We then find that the integral involves functions of the following type:

A rational function
Log terms
Arctangent terms.

XI, §4. EXERCISES

Find the following integrals.

1. $\displaystyle\int \frac{2x - 3}{(x - 1)(x + 7)} \, dx$

2. $\displaystyle\int \frac{x}{(x^2 - 3)^2} \, dx$

3. (a) $\displaystyle\int \frac{1}{(x - 3)(x + 2)} \, dx$ (b) $\displaystyle\int \frac{1}{(x + 2)(x + 1)} \, dx$ (c) $\displaystyle\int \frac{1}{x^2 - 1} \, dx$

4. $\displaystyle\int \frac{x}{(x + 1)(x + 2)(x + 3)} \, dx$

5. $\displaystyle\int \frac{x + 2}{x^2 + x} \, dx$

6. $\displaystyle\int \frac{x}{(x + 1)^2} \, dx$

7. $\displaystyle\int \frac{x}{(x + 1)(x + 2)^2} \, dx$

8. $\displaystyle\int \frac{2x - 3}{(x - 1)(x - 2)} \, dx$

9. Write out in full the integral

$$\int \frac{1}{(x^2 + 1)^2} \, dx.$$

10. Either by doing the integration by parts repeatedly or by plugging into the general formula in the text, write out in full the following integrals:

(a) $\int \dfrac{1}{(x^2 + 1)^3}\, dx$ (b) $\int \dfrac{1}{(x^2 + 1)^4}\, dx.$

Find the following integrals.

11. $\int \dfrac{2x - 3}{(x^2 + 1)^2}\, dx$

12. $\int \dfrac{x + 1}{(x^2 + 9)^2}\, dx$

13. $\int \dfrac{4}{(x^2 + 16)^2}\, dx$

14. $\int \dfrac{1}{(x + 1)(x^2 + 1)}\, dx$

15. Find the constants in the expression from the example in the text:

$$\frac{2x + 5}{(x^2 + 1)^2(x - 3)} = \frac{c_1 + c_2 x}{x^2 + 1} + \frac{c_3 + c_4 x}{(x^2 + 1)^2} + \frac{c_5}{x - 3}.$$

16. Using substitution, prove the two formulas:

(a)
$$\int \frac{1}{x^2 + b^2}\, dx = \frac{1}{b} \arctan \frac{x}{b}.$$

(b)
$$\int \frac{1}{(x + a)^2 + b^2}\, dx = \frac{1}{b} \arctan \frac{x + a}{b}.$$

For the next problems, factor $x^3 - 1$ and $x^4 - 1$ into irreducible factors.

17. (a) $\int \dfrac{1}{x^4 - 1}\, dx$ (b) $\int \dfrac{x}{x^4 - 1}\, dx$

18. (a) $\int \dfrac{1}{x^3 - 1}\, dx$ (b) $\int \dfrac{1}{x(x^2 + x + 1)}\, dx$

19. $\int \dfrac{x^2 - 2x - 2}{x^3 - 1}\, dx$

XI, §5. EXPONENTIAL SUBSTITUTIONS

This section has several purposes.

First, we expand our techniques of integration, by using the exponential function.

Second, this gives practice in the exponential function and the logarithm in a new context, which will make you learn these functions better for having used them.

Third, we shall introduce two new functions

$$\frac{e^x + e^{-x}}{2} \quad \text{and} \quad \frac{e^x - e^{-x}}{2}.$$

In the next chapter you will see these functions applied to some physical situations, to describe the equation of a hanging cable, or a soap film between two rings. Such functions will also be used to find the integrals which give the length of various curves. Here we just use them systematically to find integrals.

We start by showing how to make a simple substitution.

Example. Let us find

$$I = \int \sqrt{1 - e^x} \, dx.$$

We put $u = e^x$, $du = e^x \, dx$ so that $dx = du/u$. Then

$$I = \int \sqrt{1 - u} \, \frac{1}{u} \, du.$$

Now put $1 - u = v^2$ and $-du = 2v \, dv$ to get rid of the square root sign. Then $u = 1 - v^2$ and we obtain

$$I = \int \frac{v}{1 - v^2} (-2v) \, dv = 2 \int \frac{v^2}{v^2 - 1} \, dv$$

$$= 2 \int \frac{v^2 - 1 + 1}{v^2 - 1} \, dv$$

$$= 2 \left[\int dv + \int \frac{1}{v^2 - 1} \, dv \right]$$

$$= 2v + 2 \int \frac{1}{(v + 1)(v - 1)} \, dv.$$

This last integral can be integrated by partial fractions, to give the final answer.

We have learned how to integrate expressions involving

$$\sqrt{1 - x^2}.$$

We substitute $x = \sin \theta$ to make the expression under the square root sign into a perfect square. But what if we have to deal with an integral like

$$\int \sqrt{1 + x^2} \, dx?$$

We need to make a substitution which makes the expression under the square root sign into a perfect square. There are two possible types of functions which we can use. First, let us try to substitute $x = \tan \theta$ to get rid of the square root. We find

$$\sqrt{1 + \tan^2 \theta} = \sec \theta = \frac{1}{\cos \theta}$$

over an interval where $\cos \theta$ is positive. Already a negative power of cosine is not so nice. Even worse,

$$dx = \sec^2 \theta \, d\theta,$$

so

$$\int \sqrt{1 + x^2} \, dx = \int \sec^3 \theta \, d\theta = \int \frac{1}{\cos^3 \theta} \, d\theta,$$

which can be done, but not pleasantly, so we don't do it.

Here we give a better way of getting rid of the horrible square root sign. We need a better pair of functions $f_1(t)$ and $f_2(t)$ such that

$$1 + f_1(t)^2 = f_2(t)^2.$$

Such functions are easily found by using the exponential function e^t. Namely, we let

$$f_1(t) = \frac{e^t - e^{-t}}{2} \quad \text{and} \quad f_2(t) = \frac{e^t + e^{-t}}{2}.$$

If you multiply out, you will find immediately that these functions satisfy the desired relation. These functions have a name: they are called the **hyperbolic sine** and **hyperbolic cosine**, and are denoted by **sinh** and **cosh.** (sinh is pronounced cinch, while cosh is pronounced cosh.) Thus we **define**

$$\boxed{\sinh t = \frac{e^t - e^{-t}}{2} \quad \text{and} \quad \cosh t = \frac{e^t + e^{-t}}{2}.}$$

Carry out the manipulation which shows that

$$\cosh^2 t - \sinh^2 t = 1.$$

Furthermore, the standard rules for differentiation show that

$$\frac{d \cosh t}{dt} = \sinh t \qquad \text{and} \qquad \frac{d \sinh t}{dt} = \cosh t.$$

These formulas are very similar to those for the ordinary sine and cosine, except for some reversals of sign. They allow us to treat some cases of integrals which could not be done before, and in particular get rid of square root signs as follows.

Example. Find

$$I = \int \sqrt{1 + x^2}\, dx.$$

We make the substitution

$$x = \sinh t \qquad \text{and} \qquad dx = \cosh t\, dt.$$

Then $1 + \sinh^2 t = \cosh^2 t$, so that $\sqrt{1 + x^2} = \sqrt{\cosh^2 t} = \cosh t$. Hence

$$I = \int \cosh t \cosh t\, dt$$

$$= \int \frac{e^t + e^{-t}}{2} \frac{e^t + e^{-t}}{2}\, dt$$

$$= \frac{1}{4} \int (e^{2t} + 2 + e^{-2t})\, dt$$

$$= \frac{1}{4} \left(\frac{e^{2t}}{2} + 2t - \frac{e^{-2t}}{2} \right).$$

The answer is, of course, given in terms of t. If we want it in terms of x, then we need to study the *inverse function*, which we may call **arcsinh (hyperbolic arcsine)**, and we may write

$$t = \operatorname{arcsinh} x.$$

At first it seems that we are in a situation similar to that of sine and cosine, when we could not give explicitly a formula for the inverse function. We just called those inverse functions arcsine and arccosine. It is remarkable that here, we can give a formula as follows.

$$\text{If } \quad x = \sinh t \quad then \quad t = \log(x + \sqrt{x^2 + 1}).$$

Proof. We have

$$x = \tfrac{1}{2}(e^t - e^{-t}).$$

Let $u = e^t$. Then

$$x = \frac{1}{2}\left(u - \frac{1}{u}\right).$$

We multiply this equation by $2u$ and get the equation

$$u^2 - 2ux - 1 = 0.$$

We can then solve for u in terms of x by the quadratic formula, and get

$$u = \frac{2x \pm \sqrt{4x^2 + 4}}{2}$$

so

$$u = x \pm \sqrt{x^2 + 1}.$$

But $u = e^t > 0$ for all t. Since $\sqrt{x^2 + 1} > x$, it follows that we cannot have the minus sign in this relation. Hence finally

$$e^t = u = x + \sqrt{x^2 + 1}.$$

Now we take the log to find

$$t = \log(x + \sqrt{x^2 + 1}).$$

This proves the desired formula.

Thus unlike the case of sine and cosine, we get here an explicit formula for the inverse function of the hyperbolic sine.

If we now substitute $t = \log(x + \sqrt{x^2 + 1})$ in the indefinite integral found above, we get the explicit answer:

$$\int \sqrt{1 + x^2}\, dx = \frac{1}{4}\left[\frac{1}{2}(x + \sqrt{x^2 + 1})^2 + 2 \log(x + \sqrt{x^2 + 1})\right.$$
$$\left. - \frac{1}{2}(x + \sqrt{x^2 + 1})^{-2}\right].$$

We may also want to find a definite integral.

Example. Let $B > 0$. Find

$$\int_0^B \sqrt{1 + x^2}\, dx.$$

We substitute B in the indefinite integral, we substitute 0, and subtract, to find:

$$\int_0^B \sqrt{1 + x^2}\, dx = \frac{1}{4}\left[\frac{e^{2t}}{2} + 2t - \frac{e^{-2t}}{2}\right]_0^{\log(B + \sqrt{B^2 + 1})}$$

$$= \tfrac{1}{4}[\tfrac{1}{2}(B + \sqrt{B^2 + 1})^2 + 2 \log(B + \sqrt{B^2 + 1})$$

$$- \tfrac{1}{2}(B + \sqrt{B^2 + 1})^{-2}]$$

because when we substitute 0 for t in the expression in brackets we find 0.

In cases when you have to use an inverse function for cosh, you can rely on the following assertion.

For $x \geq 0$, the function $x = \cosh t$ has an inverse function, which is given by

$$t = \log(x + \sqrt{x^2 - 1}).$$

This is proved just like the similar statement for sinh. Do Exercises 5 and 6, which are actually worked out in the answer section. But do them before looking up the answer section, you will learn the subject better for doing so.

Remark. Integrals like

$$\int \sqrt{1 + x^3}\, dx \qquad \text{and} \qquad \int \sqrt{1 + x^4}\, dx$$

are much more complicated, and cannot be found by means of the elementary functions of this course.

XI, §5. EXERCISES

Find the integrals.

1. $\displaystyle\int \sqrt{1 + e^x}\, dx$

2. $\displaystyle\int \frac{1}{1 + e^x}\, dx$

3. $\displaystyle\int \frac{1}{e^x + e^{-x}}\, dx$

4. $\displaystyle\int \frac{1}{\sqrt{e^x + 1}}\, dx$

5. Let $f(x) = \frac{1}{2}(e^x - e^{-x}) = \sinh x = y$.
   (a) Show that f is strictly increasing for all x.
   (b) Sketch the graph of f.
   Let $x = \operatorname{arcsinh} y$ be the inverse function.
   (c) For which numbers y is $\operatorname{arcsinh} y$ defined?
   (d) Let $g(y) = \operatorname{arcsinh} y$. Show that

$$g'(y) = \frac{1}{\sqrt{1 + y^2}}.$$

It was shown in the text that $x = g(y) = \log(y + \sqrt{y^2 + 1})$.

6. Let $f(x) = \frac{1}{2}(e^x + e^{-x}) = \cosh x = y$.
   (a) Show that f is strictly increasing for $x \geq 0$.
       Then the inverse function exists for this interval. Denote this inverse function by $x = \operatorname{arccosh} y$.
   (b) Sketch the graph of f.
   (c) For which numbers y is $\operatorname{arccosh} y$ defined?
   (d) Let $g(y) = \operatorname{arccosh} y$. Show that

$$g'(y) = \frac{1}{\sqrt{y^2 - 1}}.$$

   (e) Show that $x = g(y) = \log(y + \sqrt{y^2 - 1})$. Thus you can actually give an explicit expression for this inverse function in terms of the logarithm. This is another way in which the hyperbolic functions behave more simply than sine and cosine, because we could not give an explicit formula for the arcsine and arccosine.

Find the following integrals.

7. $\int \dfrac{x^2}{\sqrt{x^2 + 4}}\, dx$ 8. $\int \dfrac{1}{\sqrt{x^2 + 1}}\, dx$

9. $\int \dfrac{x^2 + 1}{x - \sqrt{x^2 + 1}}\, dx$ 10. $\int \sqrt{x^2 - 1}\, dx$

11. Find the area between the x-axis and the hyperbola

$$x^2 - y^2 = 1$$

in the first quadrant between $x = 1$ and $x = B$, with $B > 1$.

For the graph of the hyperbola, see Chapter II, §9.

12. Find the area between the x-axis and the hyperbola

$$y^2 - x^2 = 1$$

in the first quadrant, between $x = 0$ and $x = B$.

13. Let a be a positive number, and let $y = a \cosh(x/a)$. Show that

$$\frac{d^2 y}{dx^2} = \frac{1}{a}\sqrt{1 + \left(\frac{dy}{dx}\right)^2}.$$

[This is the differential equation of the hanging cable. See the appendix after §3 of the next chapter.]

14. Verify that for any number $a > 0$ we have

$$\int \sqrt{a^2 + x^2}\, dx = \tfrac{1}{2}[x\sqrt{a^2 + x^2} + a^2 \log(x + \sqrt{a^2 + x^2})].$$

CHAPTER XII

Applications of Integration

Mathematics consists in discovering and describing certain objects and structures. It is essentially impossible to give an all-encompassing description of these. Hence, instead of such a definition, we simply state that the objects of study of mathematics as we know it are those which you will find described in the mathematical journals of the past two centuries, and leave it at that. There are many reasons for studying these objects, among which are aesthetic reasons (some people like them), and practical reasons (some mathematics can be applied).

Physics, on the other hand, consists in describing the empirical world by means of mathematical structures. The empirical world is the world with which we come into contact through our senses, through experiments, measurements, etc. What makes a good physicist is the ability to choose, among many mathematical structures and objects, the ones which can be used to describe the empirical world. I should of course immediately qualify the above assertion in two ways: First, the description of physical situations by mathematical structures can only be done within the degree of accuracy provided by the experimental apparatus. Second, the description should satisfy certain aesthetic criteria (simplicity, elegance). After all, a complete listing of all results of all experiments performed is a description of the physical world, but is quite a distinct thing from giving at one single stroke a general principle which will account simultaneously for the results of all these experiments.

For psychological reasons, it is impossible (for most people) to learn certain mathematical theories without seeing first a geometric or physical interpretation. Hence in this book, before introducing a mathematical notion, we frequently introduce one of its geometric or physical interpretations. These two, however, should not be confused. Thus we might make two columns, as shown on the following page.

As far as the logical development of our course is concerned, we could omit the second column entirely. The second column is used, however, for many purposes: To motivate the first column (because our brain is made up in such a way that to understand something in the first column, it needs the second). To provide applications for the first column, other than pure aesthetic satisfaction (granting that you like the subject).

Mathematics	Physics and geometry
Numbers	Points on a line
Derivative	Slope of a curve Rate of change
$\dfrac{df}{dx} = Kf(x)$	Exponential decay
Integral	Length Area Volume Work

Nevertheless, it is important to keep in mind that the derivative, as the limit

$$\lim_{h \to 0} \frac{f(x + h) - f(x)}{h},$$

and the integral, as a unique number between upper and lower sums, are not to be confused with a slope or an area, respectively. It is simply our mind which interprets the mathematical notion in physical or geometric terms. Besides, we frequently assign several such interpretations to the same mathematical notion (viz. the integral being interpreted as an area, or as the work done by a force).

And by the way, the above remarks which are about physics and mathematics belong neither to physics nor to mathematics. They belong to philosophy.

Experience shows that for a course which deals with integration and Taylor's formula in one term, time is lacking to cover *all* the applications of integration given in the book, as well as to cover the computations associated with Taylor's formula and an estimate of its remainder. The basic applications like length of curve, volume of revolution, area in polar coordinates cannot be omitted. One then has to make a choice about the others, which deal with geometric concepts (area of revolution) or physical concepts (work). As stated already in the foreword, my feeling is that except for doing the section on work, if time is lacking, it is best to omit other applications in order to have plenty of time to handle the computations resulting from Taylor's formula.

XII, §1. VOLUMES OF REVOLUTION

We start our applications with volumes of revolutions. The main reason is that the integrals to be evaluated come out easier than in other applications. But ultimately, we derive systematically the lengths, areas, and volumes of all the standard geometric figures.

Let $y = f(x)$ be a continuous function of x on some interval $a \leq x \leq b$. Assume that $f(x) \geq 0$ in this interval. If we revolve the curve $y = f(x)$ around the x-axis, we obtain a solid, whose volume we wish to compute.

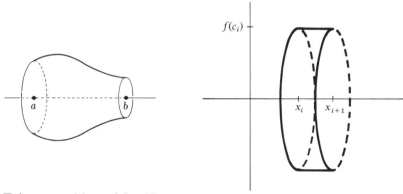

Take a partition of $[a, b]$, say

$$a = x_0 \leq x_1 \leq \cdots \leq x_n = b.$$

Let c_i be a minimum of f in the interval $[x_i, x_{i+1}]$ and let a_i be a maximum of f in that interval. Then the solid of revolution in that small interval lies between a small cylinder and a big cylinder. The width of these cylinders is $x_{i+1} - x_i$ and the radius is $f(c_i)$ for the small cylinder and $f(d_i)$ for the big one. Hence the volume of revolution, denoted by V, satisfies the inequalities

$$\sum_{i=0}^{n-1} \pi f(c_i)^2 (x_{i+1} - x_i) \leq V \leq \sum_{i=0}^{n-1} \pi f(d_i)^2 (x_{i+1} - x_i).$$

It is therefore reasonable to define this volume to be

$$V = \int_a^b \pi f(x)^2 \, dx.$$

Example. Compute the volume of the sphere of radius 1.

We take the function $y = \sqrt{1 - x^2}$ between 0 and 1. If we rotate this curve around the x-axis, we shall get half the sphere. Its volume is therefore

$$\int_0^1 \pi(1 - x^2)\, dx = \tfrac{2}{3}\pi.$$

The volume of the full sphere is therefore $\tfrac{4}{3}\pi$.

Example. Find the volume obtained by rotating the region between $y = x^3$ and $y = x$ in the first quadrant around the x-axis.

The graph of the region is illustrated on the figure. We take only that part in the first quadrant, so $0 \leq x \leq 1$. The required volume V is equal to the difference of the volumes obtained by rotating $y = x$ and $y = x^3$. Let $f(x) = x$ and $g(x) = x^3$. Then

$$V = \pi \int_0^1 f(x)^2\, dx - \pi \int_0^1 g(x)^2\, dx$$

$$= \pi \int_0^1 x^2\, dx - \pi \int_0^1 x^6\, dx$$

$$= \frac{\pi}{3} - \frac{\pi}{7}.$$

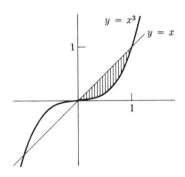

Example. We can make infinite solid chimneys and see if they have finite volume. Consider the function

$$f(x) = 1/\sqrt{x}.$$

Let

$$0 < a < 1.$$

The volume of revolution of the curve

$$y = 1/\sqrt{x}$$

between $x = a$ and $x = 1$ is given by the integral

$$\int_a^1 \pi \frac{1}{x}\, dx = \pi \log x \Big|_a^1$$

$$= -\pi \log a.$$

As a approaches 0, $\log a$ becomes very large negative, so that $-\log a$ becomes very large positive, and the volume becomes arbitrarily large. We illustrate the chimney in the following figure.

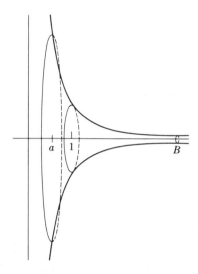

However, if you compute the volume of the curve

$$y = \frac{1}{x^{1/4}}$$

between a and 1, you will find that it approaches a limit, as $a \to 0$. Do Exercise 12.

In the above computation, we determined the volume of a chimney near the y-axis. We can also find the volume of the chimney going off to the right, say between 1 and a number $B > 1$. Suppose the chimney is defined by $y = 1/\sqrt{x}$. The volume of revolution between 1 and B is given by the integral

$$\int_1^B \pi \left(\frac{1}{\sqrt{x}}\right)^2 dx = \int_1^B \pi \frac{1}{x} dx = \pi \log B.$$

As $B \to \infty$, we see that this volume becomes arbitrarily large. However, using another function, as for instance in Exercise 13, you will find a finite volume for the infinite chimney!

XII, §1. EXERCISES

1. Find the volume of a sphere of radius r.

Find the volumes of revolution of the following:

2. $y = 1/\cos x$ between $x = 0$ and $x = \pi/4$

3. $y = \sin x$ between $x = 0$ and $x = \pi/4$

4. $y = \cos x$ between $x = 0$ and $x = \pi/4$

5. The region between $y = x^2$ and $y = 5x$

6. $y = xe^{x/2}$ between $x = 0$ and $x = 1$

7. $y = x^{1/2}e^{x/2}$ between $x = 1$ and $x = 2$

8. $y = \log x$ between $x = 1$ and $x = 2$

9. $y = \sqrt{1 + x}$ between $x = 1$ and $x = 5$

10. (a) Let B be a number > 1. What is the volume of revolution of the curve $y = e^{-x}$ between 1 and B? Does this volume approach a limit as B becomes large? If so, what limit?
    (b) Same question for the curve $y = e^{-2x}$.
    (c) Same question for the curve $y = \sqrt{x}e^{-x^2}$.

11. Find the volume of a cone whose base has radius r, and of height h, by rotating a straight line passing through the origin around the x-axis. What is the equation of the straight line?

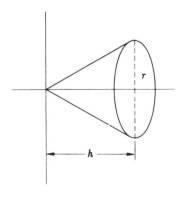

12. Compute the volume of revolution of the curve

$$y = \frac{1}{x^{1/4}}$$

between a and 1. Determine the limit as $a \to 0$.

13. Compute the volume of revolution of the curve

$$y = 1/x^2$$

between $x = 2$ and $x = B$, for any number $B > 2$. Does this volume approach a limit as $B \to \infty$? If yes, what limit?

14. For which numbers $c > 0$ will the volume of revolution of the curve

$$y = 1/x^c$$

between 1 and B approach a limit as $B \to \infty$? Find this limit in terms of c.

15. For which numbers $c > 0$ will the volume of revolution of the curve

$$y = 1/x^c$$

between a and 1 approach a limit as $a \to 0$? Find this limit in terms of c.

XII, §1. SUPPLEMENTARY EXERCISES

1. Find the volume of a doughnut as shown on the figure. The doughnut is obtained by rotating a circle of radius a about a straight line, say the x-axis.

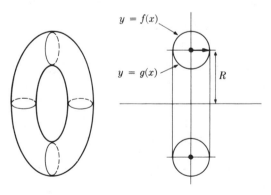

(a) The doughnut   (b) Cross section of
                                the doughnut

Let R be the distance from the line to the center of the disc. We assume $R > a$. You can reduce this problem to the case discussed in the section as follows. Let $y = f(x)$ be the function whose graph is the upper half of the circle, and let $y = g(x)$ be the function whose graph is the lower half of the circle. Write down f and g explicitly. You then have to subtract the volume obtained by rotating the lower semicircle from the volume obtained by rotating the upper semicircle.

Find the volumes of the solid obtained by rotating each region as indicated, around the x-axis.

2. $y = x^2$, between $y = 0$ and $x = 2$

3. $y = \dfrac{4}{x + 1}$, $x = -5, x = -2, y = 0$

4. $y = \sqrt{x}$, the x-axis and $x = 2$

5. $y = 1/x$, $x = 1$, $x = 3$ and the x-axis

6. $y = \sqrt{x}$, $y = x^3$

7. The region bounded by the line $x + y = 1$ and the coordinate axes

8. The ellipse $a^2x^2 + b^2y^2 = a^2b^2$

9. $y = e^{-x}$, between $x = 1$ and $x = 5$

10. $y = \log x$, between $x = 1$ and $x = 2$

11. $y = \tan x$, $x = \pi/3$ and the x-axis

In the next problems, you are asked to find a volume of revolution of a region between certain bounds, and determine whether this volume approaches a limit when the bound B becomes very large. If it does, give this limit.

12. The region bounded by $1/x$, the x-axis, between $x = 1$ and $x = B$ for $B > 1$.

13. The region bounded by $1/x^2$ and the x-axis, between $x = 1$ and $x = B$ for $B > 1$.

14. The region bounded by $y = 1/\sqrt{x}$, the x-axis, between $x = 1$ and $x = B$ for $B > 1$.

In the next problems, find the volume of revolution, determined by bounds involving a number $a > 0$, and find whether this volume approaches a limit as a approaches 0. If it does, state what limit.

15. The region bounded by $y = 1/\sqrt{x}$, the x-axis, between $x = a$ and $x = 1$ for $0 < a < 1$.

16. The region bounded by $y = 1/x$, the x-axis, between $x = a$ and $x = 1$, with $0 < a < 1$.

17. The region bounded by $(\cos x)/\sqrt{\sin x}$, the x-axis, between $x = a$ and $x = \pi/4$, with $0 < a < \pi/4$.

XII, §2. AREA IN POLAR COORDINATES

Suppose we are given a continuous function

$$r = f(\theta)$$

which is defined in some interval $a \leq \theta \leq b$. We assume that $f(\theta) \geq 0$ and $b \leq a + 2\pi$.

We wish to find an integral expression for the area encompassed by the curve $r = f(\theta)$ between the two bounds a and b.

Let us take a partition of $[a, b]$, say

$$a = \theta_0 \leq \theta_1 \leq \cdots \leq \theta_n = b.$$

The picture between θ_i and θ_{i+1} might look like this:

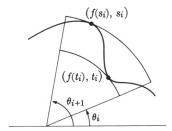

We let s_i be a number between θ_i and θ_{i+1} such that $f(s_i)$ is a maximum in that interval, and we let t_i be a number such that $f(t_i)$ is a minimum in that interval. In the figure, we have drawn the circles (or rather the sectors) of radius $f(s_i)$ and $f(t_i)$, respectively. Let

$$A_i = \text{area between} \quad \theta = \theta_i, \quad \theta = \theta_{i+1}, \quad \text{and bounded by the curve}$$

$$= \text{area of the set of points } (r, \theta) \text{ in polar coordinates such that}$$

$$\theta_i \leq \theta \leq \theta_{i+1} \quad \text{and} \quad 0 \leq r \leq f(\theta).$$

The area of a sector having angle $\theta_{i+1} - \theta_i$ and radius R is equal to the fraction

$$\frac{\theta_{i+1} - \theta_i}{2\pi}$$

of the total area of the circle of radius R, namely πR^2. Hence we get the inequality

$$\frac{\theta_{i+1} - \theta_i}{2\pi} \pi f(t_i)^2 \leq A_i \leq \frac{\theta_{i+1} - \theta_i}{2\pi} \pi f(s_i)^2.$$

Let $G(\theta) = \frac{1}{2}f(\theta)^2$. We see that the sum of the small pieces of area A_i satisfies the inequalities

$$\sum_{i=0}^{n-1} G(t_i)(\theta_{i+1} - \theta_i) \leq \sum_{i=0}^{n-1} A_i \leq \sum_{i=0}^{n-1} G(s_i)(\theta_{i+1} - \theta_i).$$

Thus the desired area lies between the upper sum and lower sum associated with the partition. Thus it is reasonable that the area in polar coordinates is given by

$$A = \int_a^b \frac{1}{2}f(\theta)^2 \, d\theta.$$

Example. Find the area bounded by one loop of the curve

$$r^2 = 2a^2 \cos 2\theta \qquad (a > 0).$$

If $-\frac{\pi}{4} \leq \theta \leq \frac{\pi}{4}$, then $\cos 2\theta \geq 0$. Thus we can write

$$r = \sqrt{2a}\sqrt{\cos 2\theta}.$$

The area is therefore

$$\int_{-\pi/4}^{\pi/4} \frac{1}{2}2a^2 \cos 2\theta \, d\theta = a^2.$$

Example. Find the area bounded by the curve

$$r = 2 + \cos \theta,$$

in the first quadrant.

First we sketch the area in the first quadrant, i.e. for θ between 0 and $\pi/2$. It looks like this:

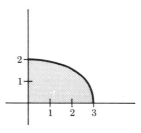

The area is given by the integral

$$\frac{1}{2}\int_0^{\pi/2} (2 + \cos \theta)^2 \, d\theta = \frac{1}{2}\int_0^{\pi/2} (4 + 4\cos \theta + \cos^2 \theta) \, d\theta.$$

Each term is easily integrated. The final answer is

$$\frac{1}{2}\left(2\pi + 4 + \frac{\pi}{4}\right).$$

Example. Let us find the area enclosed by the curve given in polar coordinates by

$$r = 1 + 2\sin\theta.$$

Note that for $-\pi/6 \leqq \theta \leqq 7\pi/6$ and only for those θ is

$$1 + 2\sin\theta \geqq 0.$$

The curve looks like the figure.

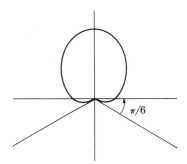

The area is therefore equal to

$$A = \tfrac{1}{2}\int_{-\pi/6}^{7\pi/6} (1 + 2\sin\theta)^2 \, d\theta$$

$$= 2\cdot\tfrac{1}{2}\int_{-\pi/6}^{\pi/2} (1 + 4\sin\theta + 4\sin^2\theta) \, d\theta.$$

We use the identity

$$\sin^2\theta = \frac{1 - \cos 2\theta}{2}.$$

The integral is then easily evaluated, and we leave this to the reader.

XII, §2. EXERCISES

Find the area enclosed by the following curves:

1. $r = 2(1 + \cos \theta)$

2. $r^2 = a^2 \sin 2\theta \quad (a > 0)$

3. $r = 2a \cos \theta$

4. $r = \cos 3\theta, \quad -\pi/6 \leq \theta \leq \pi/6$

5. $r = 1 + \sin \theta$

6. $r = 1 + \sin 2\theta$

7. $r = 2 + \cos \theta$

8. $r = 2 \cos 3\theta, \quad -\pi/6 \leq \theta \leq \pi/6$

XII, §2. SUPPLEMENTARY EXERCISES

Find the areas of the following regions, bounded by the curve given in polar coordinates.

1. $r = 10 \cos \theta$

2. $r = 1 - \cos \theta$

3. $r = \sqrt{1 - \cos \theta}$

4. $r = 2 + \sin 2\theta$

5. $r = \sin^2 \theta$

6. $r = 1 - \sin \theta$

7. $r = 1 + 2 \sin \theta$

8. $r = 1 + \sin 2\theta$

9. $r = \cos 3\theta$

10. $r = 2 + \cos \theta$

Find the area between the following curves, given in rectangular coordinates.

11. $y = 4 - x^2$, $y = 0$, between $x = -2$ and $x = 2$

12. $y = 4 - x^2$, $y = 8 - 2x^2$, between $x = -2$ and $x = 2$

13. $y = x^3 + x^2$, $y = x^3 + 1$, between $x = -1$ and $x = 1$

14. $y = x - x^2$, $y = -x$, between $x = 0$ and $x = 2$

15. $y = x^2$, $y = x + 1$, between the two points where the two curves intersect.

16. $y = x^3$ and $y = x + 6$ between $x = 0$ and the value of $x > 0$ where the two curves intersect.

XII, §3. LENGTH OF CURVES

Let $y = f(x)$ be a differentiable function over some interval $[a, b]$ (with $a < b$) and assume that its derivative f' is continuous. We wish to determine the length of the curve described by the graph. The main idea is to approximate the curve by small line segments and add these up.

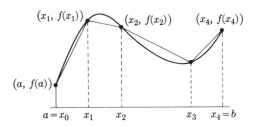

Consequently, we consider a partition of our interval:

$$a = x_0 \leq x_1 \leq \cdots \leq x_n = b.$$

For each x_i we have the point $(x_i, f(x_i))$ on the curve $y = f(x)$. We draw the line segments between two successive points. The length of such a segment is the length of the line between

$$(x_i, f(x_i)) \qquad \text{and} \qquad (x_{i+1}, f(x_{i+1})),$$

and is equal to

$$\sqrt{(x_{i+1} - x_i)^2 + (f(x_{i+1}) - f(x_i))^2}.$$

By the mean value theorem, we conclude that

$$f(x_{i+1}) - f(x_i) = (x_{i+1} - x_i)f'(c_i)$$

for some number c_i between x_i and x_{i+1}. Using this, we see that the length of our line segment is

$$\sqrt{(x_{i+1} - x_i)^2 + (x_{i+1} - x_i)^2 f'(c_i)^2}.$$

We can factor out $(x_{i+1} - x_i)^2$ and we see that the sum of the length of these line segments is

$$\sum_{i=0}^{n-1} \sqrt{1 + f'(c_i)^2}(x_{i+1} - x_i).$$

Let $G(x) = \sqrt{1 + f'(x)^2}$. Then $G(x)$ is continuous, and we see that the sum we have just written down is

$$\sum_{i=0}^{n-1} G(c_i)(x_{i+1} - x_i).$$

The value $G(c_i)$ satisfies the inequalities

$$\min_{[x_i, x_{i+1}]} G \leq G(c_i) \leq \max_{[x_i, x_{i+1}]} G$$

that is $G(c_i)$ lies between the minimum and the maximum of G on the interval $[x_i, x_{i+1}]$. Thus the sum we have written down lies between a lower sum and an upper sum for the function G. We called such sums Riemann sums. This is true for every partition of the interval. We know from the basic theory of integration that there is exactly one number lying between every upper sum and every lower sum, and that number is the definite integral. Therefore it is very reasonable to **define**:

---

length of our curve between a and b

$$= \int_a^b \sqrt{1 + \left(\frac{dy}{dx}\right)^2} \, dx = \int_a^b \sqrt{1 + f'(x)^2} \, dx.$$

Example. We wish to set up the integral for the length of the curve $y = x^2$ between $x = 0$ and $x = 1$. From the definition above, we see that the integral is

$$\int_0^1 \sqrt{1 + (2x)^2} \, dx = \int_0^1 \sqrt{1 + 4x^2} \, dx.$$

This integral is of the same type as that considered in Chapter XI, §5. First let

$$u = 2x, \qquad du = 2 \, dx.$$

When $x = 0$, then $u = 0$, and when $x = 1$, then $u = 2$. Hence

$$\int_0^1 \sqrt{1 + 4x^2} \, dx = \int_0^2 \sqrt{1 + u^2} \, \tfrac{1}{2} \, du = \tfrac{1}{2} \int_0^2 \sqrt{1 + u^2} \, du.$$

The answer then comes from Chapter XI, §5.

Example. We want the length of the curve $y = e^x$ between $x = 1$ and $x = 2$. We have $dy/dx = e^x$ and $(dy/dx)^2 = e^{2x}$ so by the general formula, the length is given by the integral

$$\int_1^2 \sqrt{1 + e^{2x}} \, dx.$$

This can be evaluated more rapidly, and we carry out the computation. Make the substitution

$$1 + e^{2x} = u^2.$$

Then

$$2e^{2x} \, dx = 2u \, du.$$

Since $e^{2x} = u^2 - 1$, we obtain

$$\int \sqrt{1 + e^{2x}} \, dx = \int u \, \frac{u \, du}{u^2 - 1} = \int \frac{u^2}{u^2 - 1} \, du$$

$$= \int \frac{u^2 - 1 + 1}{u^2 - 1} \, du$$

$$= \int 1 \, du + \int \frac{1}{u^2 - 1} \, du.$$

But

$$\frac{1}{u^2 - 1} = \frac{1}{2} \left(\frac{1}{u - 1} - \frac{1}{u + 1} \right).$$

Hence

$$\int \sqrt{1 + e^{2x}} \, dx = u + \frac{1}{2} \left[\log \frac{u - 1}{u + 1} \right].$$

When $x = 1$, $u = \sqrt{1 + e^2}$. When $x = 2$, $u = \sqrt{1 + e^4}$. Hence the length of the curve over the given interval is equal to

$$\int_1^2 \sqrt{1 + e^{2x}} \, dx = u + \frac{1}{2} \left[\log \frac{u - 1}{u + 1} \right] \Bigg|_{\sqrt{1 + e^2}}^{\sqrt{1 + e^4}}$$

$$= \sqrt{1 + e^4} + \frac{1}{2} \log \frac{\sqrt{1 + e^4} - 1}{\sqrt{1 + e^4} + 1}$$

$$- \sqrt{1 + e^2} - \frac{1}{2} \log \frac{\sqrt{1 + e^2} - 1}{\sqrt{1 + e^2} + 1}.$$

A little complicated, but it is an explicit answer.

APPENDIX. THE HANGING CABLE

We want to show here how one can determine the equation of a hanging
cable as shown on the figure.

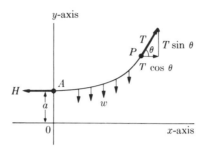

We suppose that the cable is fixed on the left to a wall, and is sub-
jected to a tension T in a certain direction at the other end. We want to
find explicitly the height of the cable

$$y = f(x).$$

The answer is as follows.

---

*If a is the height on the left, and the cable has horizontal slope
where it touches the wall, then*

$$y = a \cosh(x/a).$$

We shall show this by first deriving the differential equation satisfied
by the cable. Thus we first show that

$$\frac{d^2y}{dx^2} = \frac{1}{a} \sqrt{1 + \left(\frac{dy}{dx}\right)^2}.$$

The cable is fixed to the wall at point A, submitted to a horizontal
tension which is constant, and denoted by H. Let the weight be w per
unit length, and let the length be s. Then the weight W of cable of
length s is ws.

The tension at P has to balance the horizontal tension H, and the
weight W which pulls down. This tension has a horizontal component
and a vertical component, which are given by $T \cos \theta$ and $T \sin \theta$, respectively. Thus we must have

$$T \cos \theta = H, \qquad T \sin \theta = W = ws.$$

Dividing, we get

$$\frac{T \sin \theta}{T \cos \theta} = \tan \theta = \frac{W}{H}.$$

But $\tan \theta$ is the slope of the cable at the point $P = (x, y)$. Therefore

$$\frac{dy}{dx} = \frac{w}{H} s.$$

On the other hand, we know that

$$\frac{ds}{dx} = \sqrt{1 + \left(\frac{dy}{dx}\right)^2}.$$

Therefore

$$\frac{d^2 y}{dx^2} = \frac{w}{H} \frac{ds}{dx} = \frac{w}{H} \sqrt{1 + \left(\frac{dy}{dx}\right)^2}.$$

This is the differential equation which we wanted.

You may have done an exercise in Chapter XI, §5 showing that the function

$$y = a \cosh(x/a)$$

satisfies the equation

(∗)
$$\frac{d^2 y}{dx^2} = \frac{1}{a} \sqrt{1 + \left(\frac{dy}{dx}\right)^2}.$$

We now prove the converse.

Theorem 3.1. *If $y = f(x)$ satisfies* (∗), *and also*

$$f(0) = a, \qquad f'(0) = 0,$$

then $y = a \cosh(x/a)$.

The condition $f(0) = a$ means that the cable is hooked on the left at height a above the x-axis. The condition $f'(0) = 0$ means that at this point, the cable is horizontal.

Let

$$\frac{dy}{dx} = u.$$

Our differential equation can then be written

$$\frac{du}{dx} = \frac{1}{a}\sqrt{1 + u^2}.$$

Thus

$$\frac{1}{\sqrt{1 + u^2}} du = \frac{1}{a} dx$$

and we integrate by a substitution as in Chapter XI, §5. We put

$$u = \sinh t, \qquad du = \cosh t \, dt.$$

We have

$$\sqrt{1 + u^2} = \sqrt{1 + \sinh^2 t} = \sqrt{\cosh^2 t} = \cosh t,$$

because $\cosh t > 0$ for all t. Hence

$$\int \frac{1}{\sqrt{1 + u^2}} du = \int \frac{\cosh t}{\cosh t} dt = t.$$

Therefore

$$t = \frac{1}{a}x + C$$

for some constant C. Hence

$$u = \sinh t = \sinh\!\left(\frac{x}{a} + C\right).$$

But $u = dy/dx = f'(x)$, and $f'(0) = 0$ by assumption. Hence

$$\sinh C = 0,$$

which means that $C = 0$. Therefore

$$\frac{dy}{dx} = f'(x) = \sinh\!\left(\frac{x}{a}\right).$$

Integrating once more, we get

$$y = f(x) = a \cosh\!\left(\frac{x}{a}\right) + K$$

for some constant K. But $f(0) = a$, and $\cosh 0 = 1$, so that

$$a = f(0) = a \cdot 1 + K.$$

Hence $K = 0$, and thus finally

$$y = f(x) = a \cosh(x/a)$$

as desired.

XII, §3. EXERCISES

Find the lengths of the following curves:

1. $y = x^{3/2}, \quad 0 \leq x \leq 4$
   2. $y = \log x, \quad \frac{1}{2} \leq x \leq 2$

3. $y = \log x, \quad 1 \leq x \leq e^2$
   4. $y = 4 - x^2, \quad -2 \leq x \leq 2$

5. $y = e^x$ between $x = 0$ and $x = 1$.

6. $y = x^{3/2}$ between $x = 1$ and $x = 3$.

7. $y = \frac{1}{2}(e^x + e^{-x})$ between $x = -1$ and $x = 1$.

8. $y = \log(1 - x^2), \quad 0 \leq x \leq \frac{3}{4}$

9. $y = \frac{1}{2}(e^x + e^{-x}), \quad -1 \leq x \leq 0$

10. $y = \log \cos x, \quad 0 \leq x \leq \pi/3$

XII, §4. PARAMETRIC CURVES

There is one other way in which we can describe a curve. Suppose that we look at a point which moves in the plane. Its coordinates can be given as a function of time t. Thus, when we give two functions of t, say

$$x = f(t), \qquad y = g(t),$$

we may view these as describing a point moving along a curve. The functions f and g give the coordinates of the point as functions of t.

Example 1. Let $x = \cos \theta$ and $y = \sin \theta$. Then

$$(x, y) = (\cos \theta, \sin \theta)$$

is a point on the circle.

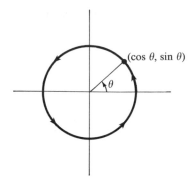

As θ increases, we view the point as moving along the circle, in counter-clockwise direction. The choice of letter θ really does not matter, and we could use t instead. In practice, the angle θ is itself expressed as a function of time. For example, if a bug moves around the circle with uniform (constant) angular speed, then we can write

$$\theta = \omega t,$$

where ω is constant. Then

$$x = \cos(\omega t) \quad \text{and} \quad y = \sin(\omega t).$$

This describes the motion of a bug around the circle with angular speed ω.

When (x, y) is described by two functions of t as above, we say that we have a **parametrization of the curve** in terms of the **parameter** t.

Example 2. Sketch the curve $x = t^2$, $y = t^3$.
We can make a table of values as usual.

t	x	y
0	0	0
1	1	1
2	4	8
−1	1	−1
−2	4	−8

Thus for each number t, we can plot the corresponding point (x, y). We also investigate when x and y are increasing or decreasing functions of t. For instance, taking the derivative, we get

$$\frac{dx}{dt} = 2t, \quad \text{and} \quad \frac{dy}{dt} = 3t^2.$$

Thus x increases when $t > 0$ and decreases when $t < 0$. The y-coordinate is increasing since $t^2 > 0$ (unless $t = 0$). Furthermore, the x-coordinate is always positive (unless $t = 0$). Thus the graph looks like this:

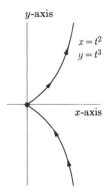

The parametric expression for the x- and y-coordinates is often useful to describe a motion of a bug (or a particle), whose coordinates are given as a function of time t. The arrows drawn in the figure suggest such a motion.

We can sometimes transform a curve given parametrically into a curve defined by an equation, possibly with some additional inequalities.

Example 3. The points (t^2, t^3) satisfy an "ordinary" equation

$$y^2 = x^3, \qquad \text{or} \qquad y = x^{3/2}.$$

However, we might also have written down the equation

$$y^4 = x^6,$$

which is satisfied by all points on our curve. In this case, however, there are solutions of this equation which are not given by our parametrization, corresponding to negative values of x, for instance,

$$x = -2, \qquad y = \pm 2\sqrt{2}.$$

Thus if we want to describe the set of all points on the parametrized curve by this latter relation, we must add an inequality $x \geq 0$. It is then correct to say that the set of all points on the parametrized curve is the set of all solutions of the equation

$$y^4 = x^6$$

satisfying the inequality $x \geq 0$.

Similarly, it is also correct to say that the set of all points on the parametrized curve is the set of all solutions of the equation

$$y^8 = x^{12}$$

satisfying the inequality $x \geq 0$. And so on.

The graph of the equation $y^4 = x^6$ is as shown on the figure.

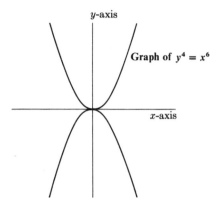

It is symmetric about both the x-axis and the y-axis. However, in the parametrized curve,

$$x = t^2, \qquad y = t^3,$$

only the right-hand portion of this graph occurs.

Example 4. Let

$$x(t) = \tfrac{1}{2}(e^t + e^{-t}) \qquad \text{and} \qquad y(t) = \tfrac{1}{2}(e^t - e^{-t}).$$

Either you have already verified that

$$x(t)^2 - y(t)^2 = 1,$$

or you should do so now by a straightforward multiplication and subtraction. Then you see that the point

$$(x(t), y(t)) = (\tfrac{1}{2}(e^t + e^{-t}), \tfrac{1}{2}(e^t - e^{-t}))$$

lies on the hyperbola defined by the equation

$$x^2 - y^2 = 1.$$

But note that $x(t) > 0$, in other words the x-coordinate given by the above function of t is always positive. Thus our functions

$$(x(t), y(t)) = (\tfrac{1}{2}(e^t + e^{-t}), \tfrac{1}{2}(e^t - e^{-t}))$$

describe a point on the right-hand branch of the hyperbola.

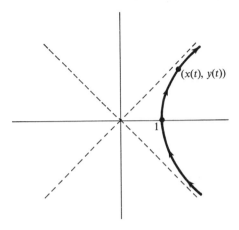

When t is large negative, then $x(t)$ is large positive, and $y(t)$ is large negative. When t is large positive, then $x(t)$ is large positive, and $y(t)$ is large positive.

As t increases the y-coordinate $y(t)$ increases from large negative to large positive. Thus a bug moving along the hyperbola according to the above parametrization is moving up, on the right-hand part of the hyperbola.

Length of parametrized curves

We shall now determine the length of a curve given by a parametrization.

Suppose that our curve is given by

$$x = f(t), \qquad y = g(t),$$

with $a \leq t \leq b$, and assume that both f, g have continuous derivatives. We consider a partition of the t-interval $[a, b]$:

$$a = t_0 \leq t_1 \leq \cdots \leq t_n = b.$$

We then obtain points

$$(x_i, y_i) = (f(t_i), g(t_i))$$

on the curve. The distance between two successive points is

$$\sqrt{(y_{i+1} - y_i)^2 + (x_{i+1} - x_i)^2} = \sqrt{(f(t_{i+1}) - f(t_i))^2 + (g(t_{i+1}) - g(t_i))^2}.$$

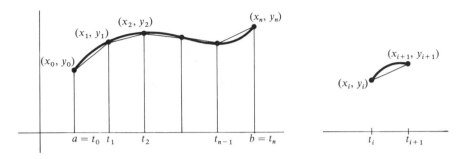

The sum of the lengths of the line segments gives an approximation of the length of the curve when the partition is sufficiently fine, that is when the numbers t_i, t_{i+1} are close together. Thus the sum

$$\sum_{i=0}^{n-1} \sqrt{(f(t_{i+1}) - f(t_i))^2 + (g(t_{i+1}) - g(t_i))^2}$$

gives an approximation to the length of the curve. We use the mean value theorem for f and g. There are numbers c_i and d_i between t_i and t_{i+1} such that

$$f(t_{i+1}) - f(t_i) = f'(c_i)(t_{i+1} - t_i),$$

$$g(t_{i+1}) - g(t_i) = g'(d_i)(t_{i+1} - t_i).$$

Substituting these values and factoring out $(t_{i+1} - t_i)$, we see that the sum of the lengths of our line segments is equal to

$$\sum_{i=0}^{n-1} \sqrt{f'(c_i)^2 + g'(d_i)^2}(t_{i+1} - t_i).$$

Let

$$G(t) = \sqrt{f'(t)^2 + g'(t)^2}.$$

Then our sum is almost equal to

$$\sum_{i=0}^{n-1} G(c_i)(t_{i+1} - t_i),$$

which would be a Riemann sum for G. It is not, because it is not necessarily true that $c_i = d_i$. Nevertheless, what we have done makes it very reasonable to **define the length of our curve (in parametric form) to be**

$$\ell_a^b = \int_a^b \sqrt{f'(t)^2 + g'(t)^2} \, dt.$$

A complete justification that this integral is a limit, in a suitable sense, of our sums would require some additional theory, which is irrelevant anyway since we just want to make it reasonable that the above integral should represent what we mean physically by length.

Observe that when a curve is given in usual form $y = f(x)$ we can let

$$t = x = g(t) \qquad \text{and} \qquad y = f(t).$$

This shows how to view the usual form as a special case of the parametric form. In that case, $g'(t) = 1$ and the formula for the length in parametric form is seen to be the same as the formula we obtained before for a curve $y = f(x)$.

It is also convenient to put the formula in the other standard notation for the derivative. We have

$$\frac{dx}{dt} = f'(t) \qquad \text{and} \qquad \frac{dy}{dt} = g'(t).$$

Hence the length of the curve can be written in the form

$$\ell_a^b = \int_a^b \sqrt{\left(\frac{dx}{dt}\right)^2 + \left(\frac{dy}{dt}\right)^2} \, dt.$$

It is customary to let

$$s(t) = \text{length of the curve as function of } t.$$

Thus we may write

$$s(t) = \int_a^t \sqrt{\left(\frac{dx}{du}\right)^2 + \left(\frac{dy}{du}\right)^2} \, du.$$

This yields

$$\frac{ds}{dt} = \sqrt{\left(\frac{dx}{dt}\right)^2 + \left(\frac{dy}{dt}\right)^2} = \sqrt{f'(t)^2 + g'(t)^2}.$$

Sometimes one writes symbolically

$$(ds)^2 = (dx)^2 + (dy)^2,$$

to suggest the Phythagoras theorem.

Example. Find the length of the curve

$$x = \cos t, \qquad y = \sin t$$

between $t = 0$ and $t = \pi$.

The length is the integral

$$\int_0^\pi \sqrt{(-\sin t)^2 + (\cos t)^2} \, dt.$$

In view of the relation $(-\sin t)^2 = (\sin t)^2$ and a basic formula relating sine and cosine, we get

$$\int_0^\pi dt = \pi.$$

If we integrated between 0 and 2π we would get 2π. This is the length of the circle of radius 1.

Example. Find the length of the curve

$$x = \cos^3 \theta, \qquad y = \sin^3 \theta$$

for $0 \leq \theta \leq \pi/2$.

We have

$$\frac{dx}{d\theta} = 3 \cos^2 \theta(-\sin \theta) \qquad \text{and} \qquad \frac{dy}{d\theta} = 3 \sin^2 \theta \cos \theta.$$

Hence

$$\ell_0^{\pi/2} = \int_0^{\pi/2} \sqrt{9 \cos^4 \theta \sin^2 \theta + 9 \sin^4 \theta \cos^2 \theta} \, d\theta$$

$$= 3 \int_0^{\pi/2} \sqrt{\cos^2 \theta \sin^2 \theta} \, d\theta \qquad \text{(because } \cos^2 \theta + \sin^2 \theta = 1)$$

$$= 3 \int_0^{\pi/2} \sin \theta \cos \theta \, d\theta \qquad \text{(because } \sin \theta \cos \theta \geq 0 \text{ for } 0 \leq \theta \leq \pi/2).$$

We integrate this by letting $u = \sin \theta$, $du = \cos \theta \, d\theta$, so that the integral is of the form

$$\int u \, du = u^2/2.$$

Then

$$\ell_0^{\pi/2} = 3 \left.\frac{\sin^2 \theta}{2}\right|_0^{\pi/2} = 3/2.$$

Example. Find the length of the same curve as in the preceding example, but for $0 \leq \theta \leq 2\pi$.

The same argument as before leads to the length formula

$$\ell_0^{\pi} = 3 \int_0^{2\pi} \sqrt{\cos^2 \theta \sin^2 \theta} \, d\theta.$$

However, if A is a number, **the formula**

$$\sqrt{A^2} = A$$

is true only if A is positive. If A is negative, then

$$\sqrt{A^2} = -A = |A|.$$

Thus in taking the square root, we must be careful of the intervals where $\cos \theta \sin \theta$ is positive or negative. We have to split the integral into a sum:

$$\ell_0^{2\pi} = 3 \int_0^{\pi/2} \cos \theta \sin \theta \, d\theta - 3 \int_{\pi/2}^{\pi} \cos \theta \sin \theta \, d\theta$$

$$+ 3 \int_{\pi}^{3\pi/2} \cos \theta \sin \theta \, d\theta - 3 \int_{3\pi/2}^{2\pi} \cos \theta \sin \theta \, d\theta.$$

These can now be easily evaluated as before to give the final answer 6. On the other hand, observe that

$$\int_0^{2\pi} \cos \theta \sin \theta \, d\theta = \left.\tfrac{1}{2} \sin^2 \theta\right|_0^{2\pi} = 0.$$

Here you get the value 0 because the function is sometimes positive and sometimes negative over the larger interval $0 \leq \theta \leq 2\pi$, and there are cancellations.

Polar coordinates

Let us now find a formula for the length of curves given in polar coordinates. Say the curve is

$$r = f(\theta),$$

with $\theta_1 \leqq \theta \leqq \theta_2$. We know that

$$
\boxed{
\begin{aligned}
x &= r \cos \theta = f(\theta) \cos \theta, \\
y &= r \sin \theta = f(\theta) \sin \theta.
\end{aligned}
}
$$

This puts the curve in parametric form, just as in the preceding considerations. Consequently we can apply the definition as before, and we see that the length is

$$\int_{\theta_1}^{\theta_2} \sqrt{\left(\frac{dx}{d\theta}\right)^2 + \left(\frac{dy}{d\theta}\right)^2}\, d\theta.$$

You can compute $dx/d\theta$ and $dy/d\theta$ using the rule for the derivative of a product. If you do this, you will find that many terms cancel, and you obtain:

The length of a curve expressed in polar coordinates by $r = f(\theta)$ is given by the formula

$$
\boxed{
\int_{\theta_1}^{\theta_2} \sqrt{f(\theta)^2 + f'(\theta)^2}\, d\theta.
}
$$

You should work it out for yourself, but for the record, we also do it in full here. Try not to look at it before you do it on your own.

We have:

$$\frac{dx}{d\theta} = -f(\theta) \sin \theta + f'(\theta) \cos \theta,$$

$$\frac{dy}{d\theta} = f(\theta) \cos \theta + f'(\theta) \sin \theta.$$

Hence

$$
\begin{aligned}
\left(\frac{dx}{d\theta}\right)^2 + \left(\frac{dy}{d\theta}\right)^2 &= f(\theta)^2 \sin^2 \theta - 2f(\theta)f'(\theta) \sin \theta \cos \theta + f'(\theta)^2 \cos^2 \theta \\
&\quad + f(\theta)^2 \cos^2 \theta + 2f(\theta)f'(\theta) \cos \theta \sin \theta + f'(\theta)^2 \sin^2 \theta \\
&= f(\theta)^2 + f'(\theta)^2
\end{aligned}
$$

because $\sin^2 \theta + \cos^2 \theta = 1$ and the middle terms cancel. The formula then follows by plugging in.

Example. Find the length of the curve given in polar coordinates by $r = \sin \theta$, between $\theta = 0$ and $\theta = \pi/2$.

We use the formula just derived, and see that this length is given by the integral

$$\int_0^{\pi/2} \sqrt{\sin^2 \theta + \cos^2 \theta} \, d\theta = \int_0^{\pi/2} d\theta = \pi/2.$$

Example. Find the length of the curve given in polar coordinates by $r = 1 - \cos \theta$ for $0 \leq \theta \leq \pi/4$.

We let $f(\theta) = 1 - \cos \theta$. The formula gives the length as

$$\ell_0^{\pi/4} = \int_0^{\pi/4} \sqrt{f(\theta)^2 + f'(\theta)^2} \, d\theta$$

$$= \int_0^{\pi/4} \sqrt{1 - 2\cos \theta + \cos^2 \theta + \sin^2 \theta} \, d\theta$$

$$= \int_0^{\pi/4} \sqrt{2(1 - \cos \theta)} \, d\theta.$$

Put $\theta = 2u$. Recall the formula $1 - \cos 2u = 2 \sin^2 u$. Then

$$1 - \cos \theta = 2 \sin^2(\theta/2).$$

Hence the integral is

$$\ell_0^{\pi/4} = \int_0^{\pi/4} \sqrt{4 \sin^2(\theta/2)} \, d\theta.$$

$$= 2 \int_0^{\pi/4} \sin\left(\frac{\theta}{2}\right) d\theta \qquad (\text{because} \quad \sin(\theta/2) \geq 0 \quad \text{if} \quad 0 \leq \theta \leq \pi/4)$$

$$= -4 \cos\left(\frac{\theta}{2}\right)\Big|_0^{\pi/4} = 4\left[1 - \cos\left(\frac{\pi}{8}\right)\right].$$

XII, §4. EXERCISES

1. Carry out the computation, giving the length in polar coordinates.

2. Find the length of a circle of radius r.

3. Find the length of the curve $x = e^t \cos t$, $y = e^t \sin t$ between $t = 1$ and $t = 2$.

4. Find the length of the curve $x = \cos^3 t$, $y = \sin^3 t$ (a) between $t = 0$ and $t = \pi/4$. and (b) between $t = 0$ and $t = \pi$.

Find the length of the following curves in the indicated interval.

5. $x = 2t + 1$, $y = t^2$, $0 \leq t \leq 2$

6. $x = 4 + 2t$, $y = \frac{1}{2}t^2 + 3$, $-2 \leq t \leq 2$

7. $x = 9t^2$, $y = 9t^3 - 3t$, $0 \leq t \leq 1/\sqrt{3}$

8. $x = 3t$, $y = 4t - 1$, $0 \leq t \leq 1$

9. $x = 1 - \cos t$, $y = t - \sin t$, $0 \leq t \leq 2\pi$

10. $x = a(1 - \cos t)$, $y = a(t - \sin t)$, with $a > 0$, and $0 \leq t \leq \pi$.

11. Sketch the curve $r = e^{\theta}$ (in polar coordinates), and also the curve $r = e^{-\theta}$.

12. Find the length of the curve $r = e^{\theta}$ between $\theta = 1$ and $\theta = 2$.

13. In general, give the length of the curve $r = e^{\theta}$ between two values θ_1 and θ_2.

Find the length of the following curves given in polar coordinates.

14. $r = 3\theta^2$ from $\theta = 1$ to $\theta = 2$

15. $r = e^{-4\theta}$, from $\theta = 1$ to $\theta = 2$

16. $r = 3 \cos \theta$, from $\theta = 0$ to $\theta = \pi/4$

17. $r = 2/\theta$ from $\theta = \frac{1}{2}$ to $\theta = 4$ [*Hint:* Use $\theta = \sinh t$.]

18. $r = 1 + \cos \theta$ from $\theta = 0$ to $\theta = \pi/4$

19. $r = 1 - \cos \theta$ from $\theta = 0$ to $\theta = \pi$

20. $r = \sin^2 \dfrac{\theta}{2}$ from $\theta = 0$ to $\theta = \pi$

21. Find the length of one loop of the curve $r = 1 + \cos \theta$

22. Same, with $r = \cos \theta$, between $-\pi/2$ and $\pi/2$.

23. Find the length of the curve $r = 2/\cos \theta$ between $\theta = 0$ and $\theta = \pi/3$.

XII, §5. SURFACE OF REVOLUTION

Let $y = f(x)$ be a positive continuously differentiable function on an interval $[a, b]$. We wish to find a formula for the area of the surface of revolution of the graph of f around the x-axis, as illustrated on the figure.

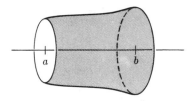

We shall see that the surface area is given by the integral

$$S = \int_a^b 2\pi y \sqrt{1 + \left(\frac{dy}{dx}\right)^2}\, dx.$$

The idea again is to approximate the curve by line segments, as illustrated. We use a partition

$$a = x_0 \leq x_1 \leq x_2 \leq \cdots \leq x_n = b.$$

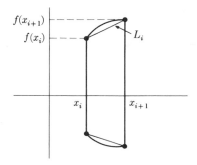

On the small interval $[x_i, x_{i+1}]$ the curve is approximated by the line segment joining the points $(x_i, f(x_i))$ and $(x_{i+1}, f(x_{i+1}))$. Let L_i be the length of the segment. Then

$$L_i = \sqrt{(x_{i+1} - x_i)^2 + (f(x_{i+1}) - f(x_i))^2}.$$

The length of a circle of radius y is $2\pi y$. If we rotate the line segment about the x-axis, then the area of the surface of rotation will be between

$$2\pi f(t_i) L_i \qquad \text{and} \qquad 2\pi f(s_i) L_i,$$

where $f(t_i)$ and $f(s_i)$ are the minimum and maximum of f, respectively, on the interval $[x_i, x_{i+1}]$. This is illustrated on Fig. 1.

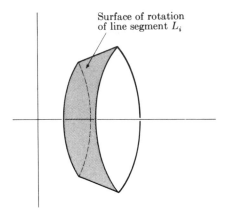

Figure 1

On the other hand, by the mean value theorem, we can write

$$f(x_{i+1}) - f(x_i) = f'(c_i)(x_{i+1} - x_i)$$

for some number c_i between x_i and x_{i+1}. Hence

$$L_i = \sqrt{(x_{i+1} - x_i)^2 + f'(c_i)^2(x_{i+1} - x_i)^2}$$

$$= \sqrt{1 + f'(c_i)^2}(x_{i+1} - x_i).$$

Therefore the expression

$$2\pi f(c_i)\sqrt{1 + f'(c_i)^2}(x_{i+1} - x_i)$$

is an approximation of the surface of revolution of the curve over the small interval $[x_i, x_{i+1}]$. Now take the sum:

$$\sum_{i=0}^{n=1} 2\pi f(c_i)\sqrt{1 + f'(c_i)^2}(x_{i+1} - x_i).$$

This is a Riemann sum, between the upper and lower sums for the integral

$$S = \int_a^b 2\pi f(x)\sqrt{1 + f'(x)^2} \, dx.$$

Thus it is reasonable that the surface area should be defined by this integral, as was to be shown.

Physical example. It occurs frequently in practice that one wants to determine a minimal surface of revolution, obtained by rotating a curve between two given points $P_1 = (x_1, y_1)$ and $P_2 = (x_2, y_2)$ in the plane. This is sometimes called the soap film problem. Indeed, given two rings perpendicular to the x-axis, the problem is to find a soap-film stretching across these two rings.

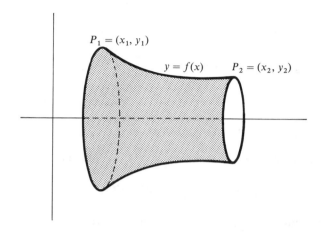

The soap film will realize the minimal surface of revolution. What is the equation for the curve $y = f(x)$? It turns out to be similar to that of the hanging cable, namely

$$y = b \cosh \frac{x - a}{b},$$

where a, b are constants depending on the two points (x_1, y_1) and (x_2, y_2). Here again we see a use of the cosh function.

Area of revolution for parametric curves

As with length, we can also deal with curves given in parametric form. Suppose that

$$x = f(t), \qquad y = g(t), \qquad a \leq t \leq b.$$

We take a partition

$$a = t_0 \leq t_1 \leq t_2 \leq \cdots \leq t_n = b.$$

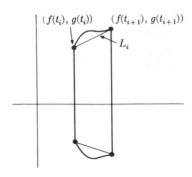

Then the length L_i between $(f(t_i), g(t_i))$ and $(f(t_{i+1}), g(t_{i+1}))$ is given by

$$L_i = \sqrt{(f(t_{i+1}) - f(t_i))^2 + (g(t_{i+1}) - g(t_i))^2}$$

$$= \sqrt{f'(c_i)^2 + g'(d_i)^2}(t_{i+1} - t_i),$$

where c_i, d_i are numbers between t_i and t_{i+1}. Hence

$$2\pi g(c_i)\sqrt{f'(c_i)^2 + g'(d_i)^2}(t_{i+1} - t_i)$$

is an approximation for the surface of revolution of the curve in the small integral $[t_i, t_{i+1}]$. Consequently, it is reasonable that the surface of revolution is given by the integral

$$S = \int_a^b 2\pi y \sqrt{\left(\frac{dx}{dt}\right)^2 + \left(\frac{dy}{dt}\right)^2}\, dt.$$

When $t = x$, this coincides with the formula found previously. It is also useful to write this formula symbolically

$$S = \int 2\pi y\, ds,$$

where symbolically, we had used

$$ds = \sqrt{(dx)^2 + (dy)^2}.$$

When using this symbolic notation, we do not put limits of integration. Only when we use the explicit parameter t over an interval $a \leq t \leq b$ do we put the values a, b for t below and above the integral sign. In this case, the surface area is written

$$S = \int_a^b 2\pi y \frac{ds}{dt}\, dt.$$

Example. We wish to find the area of a sphere of radius $a > 0$. It is best to view the sphere as the area of revolution of a circle of radius a, and to express the circle in parametric form,

$$x = a \cos \theta, \qquad y = a \sin \theta, \qquad 0 \leq \theta \leq \pi.$$

Then the formula yields:

$$S = \int_0^\pi 2\pi a \sin \theta \sqrt{a^2 \sin^2 \theta + a^2 \cos^2 \theta} \, d\theta$$

$$= \int_0^\pi 2\pi a^2 \sin \theta \, d\theta$$

$$= 2\pi a^2 (-\cos \theta) \Big|_0^\pi$$

$$= 4\pi a^2.$$

Let us now look at surfaces of revolution in terms of limits. Let $y = f(x)$ be a positive function as above, defined for all positive numbers x. Let:

V_B = volume of revolution of the graph of f between $x = 1$ and $x = B$;

S_B = area of revolution of the graph of f between $x = 1$ and $x = B$.

It is a fact which is usually very surprising that there may be cases when V_B approaches a finite limit when $B \to \infty$ whereas S_B become arbitrarily large when $B \to \infty$!!

Example. Let $f(x) = 1/x$.

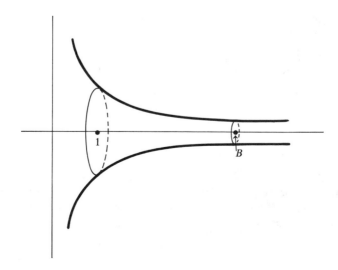

Then using the formulas for volumes and surface of revolution, we find:

$$V_B = \int_1^B \pi \frac{1}{x^2} \, dx = \pi \left(1 - \frac{1}{B}\right) \to \pi \qquad \text{as } B \to \infty.$$

$$S_B = \int_1^B 2\pi \frac{1}{x} \sqrt{1 + f'(x)^2} \, dx$$

Now $f'(x)^2$ is a positive number, and so the expression under the square root sign is ≥ 1. Therefore

$$S_B \geq 2\pi \int_1^B \frac{1}{x} \, dx = 2\pi \log B \to \infty \qquad \text{as } B \to \infty.$$

Here we see how the volume approaches the finite limit π, whereas the surface of revolution becomes arbitrarily large.

In terms of a naive interpretation, suppose you have a bucket of paint containing π cubic units of paint. Then you can fill out the funnel inside the surface of revolution with this paint. But seemingly paradoxically, there isn't enough paint to paint the surface of revolution, as $B \to \infty$. This shows how treacherous naive intuition can be.

XII, §5. EXERCISES

1. Find the area of the surface obtained by rotating the curve

$$x = a \cos^3 \theta, \qquad y = a \sin^3 \theta$$

   around the x-axis. [Sketch the curve. There is some symmetry. Determine the appropriate interval of θ.]

2. Find the area of the surface obtained by rotating the curve $y = x^3$ around the x-axis, between $x = 0$ and $x = 1$.

3. Find the area of the surface obtained by rotating the curve

$$x = \tfrac{1}{2}t^2 + t, \qquad y = t + 1$$

   around the x-axis, from $t = 0$ to $t = 4$.

4. The circle $x^2 + y^2 = a^2$ is rotated around a line tangent to the circle. Find the area of the surface of rotation. [*Hint*: Set up coordinate axes and a parametrization of the circle in a convenient way. Remember what the curve $r = 2a \sin \theta$ in polar coordinates looks like? What if you rotate this curve around the x-axis?]

5. A circle as shown on the figure is rotated around the x-axis to form a torus (fancy name for a doughnut). What is the area of the torus?

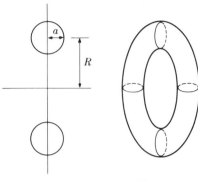

Cross section of torus   The torus

6. Find the area of the surface obtained by rotating an arc of the curve $y = x^{1/2}$ between $(0, 0)$ and $(4, 2)$ around the x-axis.

XII, §6. WORK

Suppose a particle moves on a curve, and that the length of the curve is described by a variable u.

Let $f(u)$ be a function. We interpret f as a force acting on the particle, in the direction of the curve. We want to find an integral expression for the work done by the force between two points on the curve.

Whatever our expression will turn out to be, it is reasonable that the work done should satisfy the following properties:

If a, b, c are three numbers, with $a \leq b \leq c$, then the work done between a and c is equal to the work done between a and b, plus the work done between b and c. If we denote the work done between a and b by $W_a^b(f)$, then we should have

$$W_a^c(f) = W_a^b(f) + W_b^c(f).$$

Furthermore, if we have a constant force M acting on the particle, it is reasonable to expect that the work done between a and b is

$$M(b - a).$$

Finally, if g is a stronger force than f, say $f(u) \leq g(u)$, on the interval $[a, b]$, then we shall do more work with g than with f, meaning

$$W_b^a(f) \leq W_a^b(g).$$

In particular, if there are two constant forces m and M such that

$$m \leq f(u) \leq M$$

throughout the interval $[a, b]$, then

$$m(b - a) \leq W_a^b(f) \leq M(b - a).$$

We shall see below that the work done by the force f between a distance a and a distance b is given by the integral

$$W_a^b(f) = \int_a^b f(u) \, du.$$

If the particle or object happens to move along a straight line, say along the x-axis, then f is given as a function of x, and our integral is simply

$$\int_a^b f(x) \, dx.$$

Furthermore, if the length of the curve u is given as a function of time t (as it is in practice, cf. §3) we see that the force becomes a function of t by the chain rule, namely $f(u(t))$. Thus between time t_1 and t_2 the work done is equal to

$$\boxed{\int_{t_1}^{t_2} f(u(t)) \frac{du}{dt} \, dt.}$$

This is the most practical expression for the work, since curves and forces are most frequently expressed as functions of time.

Let us now see why the work done is given by the integral. Let P be a partition of the interval $[a, b]$:

$$a = u_0 \leq u_1 \leq u_2 \leq \cdots \leq u_n = b.$$

Let $f(t_i)$ be a minimum for f on the small interval $[u_i, u_{i+1}]$, and let $f(s_i)$ be a maximum for f on this same small interval. Then the work done by moving the particle from length u_i to u_{i+1} satisfies the inequalities

$$f(t_i)(u_{i+1} - u_i) \leq W_{u_i}^{u_{i+1}}(f) \leq f(s_i)(u_{i+1} - u_i).$$

Adding these together we find

$$\sum_{i=0}^{n-1} f(t_i)(u_{i+1} - u_i) \leq W_a^b(f) \leq \sum_{i=0}^{n-1} f(s_i)(u_{i+1} - u_i).$$

The expressions on the left and right are lower and upper sums for the integral, respectively. Since the integral is the unique number between the lower sums and upper sums, it follows that

$$W_a^b(f) = \int_a^b f(u)\, du.$$

Example. Find the work done in stretching a spring from its unstretched position to a length of 10 cm longer. You may assume that the force needed to stretch the spring is proportional to the increase in length.

We visualize the spring as being horizontal, on the x-axis. Thus there is a constant K such that the force is given by

$$f(x) = Kx.$$

The work done is therefore

$$\int_0^{10} Kx\, dx = \tfrac{1}{2}K \cdot 100$$

$$= 50\,K.$$

Example. Assume that gravity is a force inversely proportional to the square of the distance from the center of the earth. What work is done lifting a weight of 2 tons from the surface of the earth to a height of 100 mi above the earth? Assume that the radius of the earth is 4000 mi.

By assumption, there is a constant C such that the force of gravity is given by $f(x) = C/(x + 4000)^2$, where x denotes the height above the earth. When $x = 0$, our assumption is that

$$f(0) = 2 \text{ tons} = \frac{C}{(4000)^2}.$$

Hence $C = 32 \times 10^6$ tons. The work done is equal to the integral

$$\int_0^{100} f(x)\, dx = 32 \times 10^6 \left(-\frac{1}{(x + 4000)} \right)\Big|_0^{100}$$

$$= 32 \times 10^6 \left[\frac{1}{4000} - \frac{1}{4100} \right] \text{ ton miles.}$$

XII, §6. EXERCISES

1. A spring is 18 in. long, and a force of 10 lb is needed to hold the spring to a length of 16 in. If the force is given as $f(x) = kx$, where k is a constant, and x is the decrease in length, what is the constant k? How much work is done in compressing the spring from 16 in. to 12 in.?

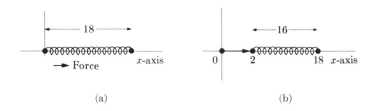

(a)   (b)

2. Assuming that the force is given as $k \sin(\pi x/18)$, in the spring compression problem, answer the two questions of the preceding problem for this force.

3. A particle at the origin attracts another particle with a force inversely proportional to the square of the distance between them. Let C be the proportionality constant. What work is done in moving the second particle along a straight line away from the origin, from a distance r_1 to a distance $r > r_1$ from the origin?

4. In the preceding exercise, determine whether the work approaches a limit as r becomes very large, and find this limit if it exists.

5. Two particles repel each other with a force inversely proportional to the cube of their distance. If one particle is fixed at the origin, what work is done in moving the other along the x-axis from a distance of 10 cm to a distance of 1 cm toward the origin?

6. Assuming that gravity, as usual, is a force inversely proportional to the square of the distance from the center of the earth, what work is done in lifting a weight of 1000 lb from the surface of the earth to a height of 4000 mi above the surface? (Assume the radius of the earth is 4000 mi.)

7. A metal bar has length L and cross section S. If it is stretched x units, then the force $f(x)$ required is given by

$$f(x) = \frac{ES}{L} x$$

where E is a constant. If a bar 12 in. long of uniform cross section 4 in^2 is stretched 1 in. find the work done (in terms of E).

8. A particle of mass M grams at the origin attracts a particle of mass m grams at a point x cm away on the x-axis with a force of CmM/x^2 dynes, where C is a constant. Find the work done by the force
(a) when m moves from $x = 1/100$ to $x = 1/10$;
(b) when m moves from $x = 1$ to $x = 1/10$.

9. A unit positive charge of electricity at 0 repels a positive charge of amount c with a force c/r^2, where r is the distance between the particles. Find the work done by this force when the charge c moves along a straight line through 0 from a distance r_1 to a distance r_2 from 0.

10. Air is confined in a cylindrical chamber fitted with a piston. If the volume of air, at a pressure of 20 pounds/in^2 is 75 in^3, find the work done on the piston when the air expands to twice its original volume. Use the law

$$\text{Pressure} \cdot \text{Volume} = \text{Constant}.$$

XII, §7. MOMENTS AND CENTER OF GRAVITY

Suppose we have masses $m_1, \ldots, m_n$ at points $x_1, \ldots, x_n$ on the x-axis. The total **moment** of these masses is defined to be

$$m_1 x_1 + \cdots + m_n x_n = \sum_{i=1}^{n} m_i x_i.$$

We may think of these masses as being distributed on some rod of uniform density, as shown on the figure.

The total mass is

$$m = m_1 + \cdots + m_n = \sum_{i=1}^{n} m_i.$$

We wish to find the point of the rod such that if we balance the rod at that point, then the rod will not move either up or down. Call this point $\bar{x}$.

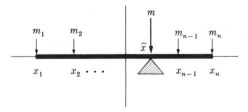

Then $\bar{x}$ is a point such that if the total mass m is placed at $\bar{x}$ it will have the same balancing effect as the other masses m_i at x_i. The equation for this condition is that

$$m\bar{x} = m_1 x_1 + \cdots + m_n x_n = \sum_{i=1}^{n} m_i x_i.$$

Thus we can solve for $\bar{x}$ and get

$$\bar{x} = \frac{\sum m_i x_i}{\sum m_i} = \frac{1}{m} \sum_{i=1}^{n} m_i x_i.$$

This point $\bar{x}$ is called the **center of gravity**, or **center of mass** of the masses $m_1, \ldots, m_n$.

Example. Let $m_1 = 4$ be at the point $x_1 = -3$ and let $m_2 = 7$ be at the point $x_2 = 2$. Then the total mass is

$$m = 4 + 7 = 11$$

and the moment is

$$4 \cdot (-3) + 7 \cdot 2 = 2.$$

Hence the center of gravity is at the point

$$\bar{x} = 2/11.$$

Now suppose that we have a thin rod, placed along the x-axis on an interval $[a, b]$, as on the figure. Suppose that the rod has constant (uniform) density K.

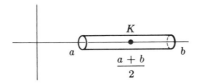

The length of the rod is $(b - a)$. The total mass of the rod is then the density times the length, namely

$$\text{mass} = K(b - a).$$

It is reasonable to define the moment of the rod to be the same as that of the total mass placed at the center of the rod. This center has coordinates at the midpoint of the interval, namely

$$\frac{a + b}{2}.$$

Hence the **moment** of the rod is

$$M_a^b = K \frac{(a + b)}{2} (b - a).$$

Next suppose the density of the rod is not constant, but varies continuously, and so can be represented by a function $f(x)$. We try to find an approximation of what we mean by the moment of the rod. Thus we take a partition of the interval $[a, b]$,

$$a = x_0 \leq x_1 \leq \cdots \leq x_n = b.$$

On each small interval $[x_i, x_{i+1}]$ the density will not vary much, and an approximation for the moment of the piece of rod along this interval is then given by

$$f(c_i)c_i(x_{i+1} - x_i)$$

where

$$c_i = \frac{x_{i+1} + x_i}{2}$$

is the midpoint of the small interval. Let

$$G(x) = xf(x).$$

Taking the sum of the above approximations yields

$$\sum_{i=0}^{n-1} f(c_i)c_i(x_{i+1} - x_i)$$

which is a Riemann sum for the integral

$$\int_a^b G(x)\, dx = \int_a^b xf(x)\, dx.$$

Consequently it is natural to define the **moment** of the rod with variable density as the integral

$$M_a^b(f) = \int_a^b xf(x)\, dx.$$

Let $\bar{x}$ be the coordinate of the center of gravity of the rod. This means that if the mass of the rod is placed at $\bar{x}$ then it has the same moment as the rod itself, and amounts to the equation

$$\bar{x} \cdot \text{total mass of rod} = \int_a^b xf(x)\, dx.$$

Hence we get an expression for the center of gravity, namely

$$\bar{x} = \frac{\int_a^b xf(x)\,dx}{\int_a^b f(x)\,dx}.$$

Example. Suppose a rod of length 5 cm has density proportional to the distance from one end. Find the center of gravity of the rod.

We suppose the rod laid out so that one end is at the origin, as shown on the figure.

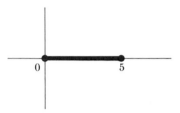

The assumption on the density means that there is a constant C such that the density is given by the function

$$f(x) = Cx.$$

(i)   The total mass is

$$\int_0^5 f(x)\,dx = \int_0^5 Cx\,dx = \frac{25C}{2}.$$

(ii)   The moment is

$$\int_0^5 xf(x)\,dx = \int_0^5 Cx^2\,dx = \frac{125C}{3}.$$

Therefore

$$\bar{x} = \frac{125C/3}{25C/2} = \frac{10}{3}.$$

This center of gravity is such that if we balance the rod on a sharp edge at the point $\bar{x}$, then the rod will not lean either way.

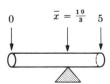

A similar discussion can be carried out in higher dimensional space, for plane areas and solid volumes. It is best for this to wait until we discuss double and triple integrals in two or three variables in the next course.

XII, §7. EXERCISES

1. Suppose the density of a rod is proportional to the square of the distance from the origin, and the rod is 10 cm long, lying along the x-axis between 5 and 15 cm from the origin. Find its center of gravity.

2. Same as in Exercise 1, but assume the rod has constant density C.

3. Same as in Exercise 1, but assume the density of the rod is inversely proportional to the distance from the origin.

Part Four

Taylor's Formula
and Series

In this part we study the approximation of functions by certain sums, called series. The chapter on Taylor's formula shows how to approximate functions by polynomials, and we estimate the error term to see how good an approximation we can get.

Note that the derivation of Taylor's formula is an application of integration by parts.

CHAPTER XIII

Taylor's Formula

We finally come to the point where we develop a method which allows us to compute the values of the elementary functions like sine, exp, and log. The method is to approximate these functions by polynomials, with an error term which is easily estimated. This error term will be given by an integral, and our first task is to estimate integrals. We then go through the elementary functions systematically, and derive the approximating polynomials.

You should review the estimates of Chapter X, §3, which will be used to estimate our error terms.

XIII, §1. TAYLOR'S FORMULA

Let f be a function which is differentiable on some interval. We can then take its derivative f' on that interval. Suppose that this derivative is also differentiable. We need a notation for its derivative. We shall denote it by $f^{(2)}$. Similarly, if the derivative of the function $f^{(2)}$ exists, we denote it by $f^{(3)}$, and so forth. In this system, the first derivative is denoted by $f^{(1)}$. (Of course, we can also write $f^{(2)} = f''$.)

In the d/dx notation, we also write:

$$f^{(2)}(x) = \frac{d^2 f}{dx^2},$$

$$f^{(3)}(x) = \frac{d^3 f}{dx^3},$$

and so forth.

Taylor's formula gives us a polynomial which approximates the function, in terms of the derivatives of the function. Since these derivatives are usually easy to compute, there is no difficulty in computing these polynomials.

For instance, if $f(x) = \sin x$, then $f^{(1)}(x) = \cos x$, $f^{(2)}(x) = -\sin x$, $f^{(3)}(x) = -\cos x$, and $f^{(4)}(x) = \sin x$. From there on, we start all over again.

In the case of e^x, it is even easier, namely $f^{(n)}(x) = e^x$ for all positive integers n.

It is also customary to denote the function f itself by $f^{(0)}$. Thus $f(x) = f^{(0)}(x)$.

We need one more piece of notation before stating Taylor's formula. When we take successive derivatives of functions, the following numbers occur frequently:

$$1, \quad 2 \cdot 1, \quad 3 \cdot 2 \cdot 1, \quad 4 \cdot 3 \cdot 2 \cdot 1, \quad 5 \cdot 4 \cdot 3 \cdot 2 \cdot 1, \quad \text{etc.}$$

These numbers are denoted by

$$1! \quad 2! \quad 3! \quad 4! \quad 5! \quad \text{etc.}$$

Thus

$$1! = 1, \quad 4! = 24,$$
$$2! = 2, \quad 5! = 120,$$
$$3! = 6, \quad 6! = 720.$$

When n is a positive integer, the symbol $n!$ is read n **factorial**. Thus in general,

$$n! = n(n-1)(n-2) \cdots 2 \cdot 1$$

is the product of the first n integers from 1 to n.

It is also convenient to agree that $0! = 1$. This is the convention which makes certain formulas easiest to write.

Let us now look at the case of a polynomial

$$P(x) = c_0 + c_1 x + c_2 x^2 + \cdots + c_n x^n.$$

The numbers $c_0, \ldots, c_n$ are called the **coefficients** of the polynomial. We shall now see that these coefficients can be expressed in terms of the derivatives of $P(x)$ at $x = 0$. You should remember what you did in Chapter III, §7 when you computed higher derivatives. Let k be an integer ≥ 0. Then the k-th derivative of $P(x)$ is given by

$$P^{(k)}(x) = c_k k! + \text{an expression containing } x \text{ as a factor.}$$

The reason is: if we differentiate k times the terms

$$c_0, \ c_1 x, \ldots, c_{k-1} x^{k-1}$$

then we get 0. And if we differentiate k times a power x^j with $j > k$ then some positive power of x will be left. Then if we evaluate the k-th derivative at 0, we get

$$P^{(k)}(0) = c_k k!$$

because when we substitute 0 for x all the other terms give 0. Therefore we find the desired expression of c_k in terms of the k-th derivative:

$$c_k = \frac{P^{(k)}(0)}{k!}.$$

Next, let f be a function which has derivatives up to order n on an interval. We are seeking a polynomial

$$P(x) = c_0 + c_1 x + c_2 x^2 + \cdots + c_n x^n$$

whose derivatives at 0 (up to order n) are the same as the derivatives of f at 0, in other words

$$P^{(k)}(0) = f^{(k)}(0).$$

What must the coefficients $c_0, \ c_1, \ldots, c_n$ be to achieve this? The answer is immediate from our computation of the coefficients of a polynomial, namely we must have

$$k! \ c_k = f^{(k)}(0)$$

for each integer $k = 0, 1, \ldots, n$. Hence we have the desired expression for c_k, namely

$$c_k = \frac{f^{(k)}(0)}{k!}.$$

Definition. The **Taylor polynomial** of degree $\leq n$ for the function f is the polynomial

$$P_n(x) = f(0) + f^{(1)}(0)x + \frac{f^{(2)}(0)}{2!} x^2 + \cdots + \frac{f^{(n)}(0)}{n!} x^n.$$

Example. Let $f(x) = \sin x$. It is easy to work out the derivatives (see §3), and you will find that the Taylor polynomials have the form

$$P_{2m+1}(x) = x - \frac{x^3}{3!} + \frac{x^5}{5!} - \frac{x^7}{7!} + \cdots + (-1)^m \frac{x^{2m+1}}{(2m+1)!}.$$

Only odd values of n occur, so we write

$$n = 2m + 1 \qquad \text{with} \quad m \geq 0.$$

Example. Let $f(x) = e^x$. Then $f^{(k)}(x) = e^x$ for all positive integers k. Hence $f^{(k)}(0) = 1$ for all k, and so the Taylor polynomial has the form

$$P_n(x) = 1 + x + \frac{x^2}{2!} + \frac{x^3}{3!} + \cdots + \frac{x^n}{n!}.$$

We now want to know how good an approximation the polynomial $P_n(x)$ gives to $f(x)$. So we write

$$f(x) = P_n(x) + R_{n+1}(x),$$

where R_{n+1} is called the **remainder**.

We shall have to estimate the remainder term $R_{n+1}(x)$. We shall eventually prove that there is a number c between 0 and x such that

$$\boxed{R_{n+1}(x) = \frac{f^{(n+1)}(c)}{(n+1)!} x^{n+1}.}$$

Thus the remainder term will look very much like the main terms, except that the coefficient

$$\frac{f^{(n+1)}(c)}{(n+1)!}$$

is taken at some intermediate point c instead of being taken at 0.

Since it is easy to estimate the derivatives of the functions $\sin x$, $\cos x$, e^x we shall be able to see that the Taylor polynomials give good approximations to the function. If you are willing to take for granted the expression

$$\boxed{f(x) = f(0) + f^{(1)}(0)x + \frac{f^{(2)}(0)}{2!} x^2 + \cdots + \frac{f^{(n-1)}(0)}{(n-1)!} x^{n-1} + R_n(x),}$$

where

$$R_n(x) = \frac{f^{(n)}(c)}{n!} x^n$$

for some number c between 0 and x, then you can read immediately the later sections, §3, and so on, to get immediately into the applications to the elementary functions.

Observe that it does not make any difference whether we write

$$f(x) = P_n(x) + R_{n+1}(x) \qquad \text{or} \qquad f(x) = P_{n-1}(x) + R_n(x).$$

This amounts merely to a change of indices. We use whichever is more convenient.

Of course, the above assertion does not state anything precise about the number c other than c lies between 0 and x. But the point of the formula, and its remainder term, is that we do not need to know anything more precise in order to *estimate* the remainder term. The polynomial preceding the remainder term gives a value at x. We want to know only how close this value is to $f(x)$. For that purpose, we need only give a bound

$$\frac{|f^{(n)}(c)|}{n!} |x|^n \leq \text{something or other,}$$

and so we need only give a bound for the n-th derivative $|f^{(n)}(c)|$. To give such a bound can be done without knowing an exact value for the n-th derivative at the number c. You can see how this is done in §3 and the following sections, when we deal systematically with all the elementary functions.

We shall now develop the Taylor formula theoretically, and prove that the remainder term has the stated form. It will also be convenient instead of working with the numbers 0 and x to work with arbitrary numbers a and b. We also state Taylor's formula with a somewhat different form for the remainder term, but a form which will arise naturally in the proof. After that, we shall prove that the integral form is equal to the expression stated above.

Theorem 1.1. *Let f be a function defined on a closed interval between two numbers a and b. Assume that the function has n derivatives on this interval, and that all of them are continuous functions. Then*

$$f(b) = f(a) + \frac{f^{(1)}(a)}{1!}(b-a) + \frac{f^{(2)}(a)}{2!}(b-a)^2 + \cdots$$
$$+ \frac{f^{(n-1)}(a)}{(n-1)!}(b-a)^{n-1} + R_n,$$

where R_n (which is called the remainder term) is the integral

$$R_n = \int_a^b \frac{(b-t)^{n-1}}{(n-1)!} \, f^{(n)}(t) \, dt.$$

The remainder term looks slightly complicated. In Theorem 2.1 we shall prove that R_n can be expressed in a form very similar to the other terms, namely

$$R_n = \frac{f^{(n)}(c)}{n!} (b-a)^n$$

for some number c between a and b. Taylor's formula with this form of the remainder is then very easy to memorize,

The most important case of Theorem 1.1 occurs when $a = 0$. In that case, the formula reads

$$f(b) = f(0) + \frac{f'(0)}{1!} b + \cdots + \frac{f^{(n-1)}(0)}{(n-1)!} b^{n-1} + R_n.$$

Furthermore, if x is any number between a and b, the same formula remains valid for this number x instead of b, simply by considering the interval between a and x instead of the interval between a and b. Thus if $a = 0$, then the formula reads

$$f(x) = f(0) + \frac{f'(0)}{1!} x + \cdots + \frac{f^{(n-1)}(0)}{(n-1)!} x^{n-1} + R_n(x),$$

where

$$R_n(x) = f^{(n)}(c) \frac{x^n}{n!}$$

and c is a number between 0 and x. Each derivative $f(0)$, $f'(0), \ldots, f^{(n-1)}(0)$ is a number, and we see that the terms preceding R_n make up a polynomial in x. This is the approximating polynomial.

We shall now prove the theorem. The proof is an application of integration by parts. First to get the idea of the proof, we carry out two special cases.

Special cases. We proceed stepwise. We know that a function is the integral of its derivative. Thus when $n = 1$ we have

$$f(b) - f(a) = \int_a^b f'(t)\, dt.$$

Let $u = f'(t)$ and $dv = dt$. Then $du = f''(t)\, dt$. We are tempted to put $v = t$. This is one case where we choose another indefinite integral, namely $v = -(b - t)$, which differs from t by a constant. We still have $dv = dt$ (the minus signs cancel!). Integrating by parts, we get

$$\int_a^b u\, dv = uv \Big|_a^b - \int_a^b v\, du$$

$$= -f'(t)(b - t) \Big|_a^b - \int_a^b -(b - t)f^{(2)}(t)\, dt$$

$$= f'(a)(b - a) + \int_a^b (b - t)f^{(2)}(t)\, dt.$$

This is precisely the Taylor formula when $n = 2$.

We push it one step further, from 2 to 3. We rewrite the integral just obtained as

$$\int_a^b f^{(2)}(t)(b - t)\, dt.$$

Let $u = f^{(2)}(t)$ and $dv = (b - t)\, dt$. Then

$$du = f^{(3)}(t)\, dt \qquad \text{and} \qquad v = \frac{-(b - t)^2}{2} = \int (b - t)\, dt.$$

Thus, integrating by parts, we find that our integral, which is of the form $\int_a^b u\, dv$, is equal to

$$uv \Big|_a^b - \int_a^b v\, du = -f^{(2)}(t) \frac{(b - t)^2}{2} \Big|_a^b - \int_a^b -\frac{(b - t)^2}{2} f^{(3)}(t)\, dt$$

$$= f^{(2)}(a) \frac{(b - a)^2}{2} + R_3.$$

Here, R_3 is the desired remainder, and the term preceding it is just the proper term in the Taylor formula.

If you need it, do the next step yourself, from 3 to 4. We shall now see how the general step goes, from n to $n + 1$.

General case. Suppose that we have already obtained the first $n - 1$ terms of the Taylor formula, with a remainder term

$$R_n = \int_a^b \frac{(b - t)^{n-1}}{(n-1)!} f^{(n)}(t) \, dt,$$

which we rewrite

$$R_n = \int_a^b f^{(n)}(t) \frac{(b - t)^{n-1}}{(n-1)!} \, dt.$$

Let

$$u = f^{(n)}(t) \qquad \text{and} \qquad dv = \frac{(b - t)^{n-1}}{(n-1)!} \, dt.$$

Then

$$du = f^{(n+1)}(t) \, dt \qquad \text{and} \qquad v = \frac{-(b - t)^n}{n!}.$$

Here we use the fact that b is constant, and

$$\int (b - t)^{n-1} \, dt = -\frac{(b - t)^n}{n}.$$

Note the appearance of a minus sign, from the chain rule. Also we use

$$n(n - 1)! = n!$$

to give the stated value of v. Thus the denominator climbs from $(n - 1)!$ to $n!$.

Integrating by parts, we find:

$$R_n = uv \Big|_a^b - \int_a^b v \, du = -f^{(n)}(t) \frac{(b - t)^n}{n!} \Big|_a^b - \int_a^b -\frac{(b - t)^n}{n!} f^{(n+1)}(t) \, dt$$

$$= f^{(n)}(a) \frac{(b - a)^n}{n!} + \int_a^b \frac{(b - t)^n}{n!} f^{(n+1)}(t) \, dt.$$

Thus we have split off one more term of the Taylor formula, and the new remainder is the desired R_{n+1}. This concludes the proof.

XIII, §1. EXERCISES

1. Let $f(x) = \log(1 + x)$.
   (a) Find a formula for the derivatives of $f(x)$. Start with $f^{(1)}(x) = (x + 1)^{-1}$, $f^{(2)}(x) = -(x + 1)^{-2}$. Work out $f^{(k)}(x)$ for $k = 3, 4, 5$ and then write down the formula for arbitrary k.

(b) Find $f^{(k)}(0)$ for $k = 1, 2, 3, 4, 5$. Then show in general that

$$f^{(k)}(0) = (-1)^{k+1}(k-1)!.$$

(c) Conclude that the Taylor polynomial $P_n(x)$ for $\log(1 + x)$ is

$$P_n(x) = x - \frac{x^2}{2} + \frac{x^3}{3} - \frac{x^4}{4} + \cdots + (-1)^{n+1} \frac{x^n}{n}.$$

2. Find the polynomials $P_n(x)$ for the function $f(x) = \cos x$ and the values $n = 1, 2, 3, 4, 5, 6, 7, 8$.

XIII, §2. ESTIMATE FOR THE REMAINDER

Theorem 2.1. *In Taylor's formula of Theorem 1.1, there exists a number* c *between* a *and* b *such that the remainder* R_n *is given by*

$$R_n = \frac{f^{(n)}(c)(b-a)^n}{n!}.$$

If M_n *is a number such that* $|f^{(n)}(x)| \leq M_n$ *for all* x *in the interval, i.e.* M_n *is an upper bound for* $|f^{(n)}(x)|$, *then*

$$|R_n| \leq \frac{M_n|b-a|^n}{n!}.$$

Proof. The second assertion follows at once from the first, making the estimate

$$|R_n| = \frac{|f^{(n)}(c)||b-a|^n}{n!} \leq M_n \frac{|b-a|^n}{n!}.$$

Let us prove the first assertion. Since $f^{(n)}$ is continuous on the interval, there exists a point u in the interval such that $f^{(n)}(u)$ is a maximum, and a point v such that $f^{(n)}(v)$ is a minimum for all values of $f^{(n)}$ in our interval.

Let us assume that $a < b$. Then for any t in the interval, $b - t$ is ≥ 0, and hence

$$\frac{(b-t)^{n-1}}{(n-1)!} f^{(n)}(v) \leq \frac{(b-t)^{n-1}}{(n-1)!} f^{(n)}(t) \leq \frac{(b-t)^{n-1}}{(n-1)!} f^{(n)}(u).$$

Using Theorem 3.1 of Chapter X, §3, we conclude that similar inequalities hold when we take the integral. However, $f^{(n)}(v)$ and $f^{(n)}(u)$ are fixed numbers which can be taken out of the integral sign. Consequently, we obtain

$$f^{(n)}(v) \int_a^b \frac{(b-t)^{n-1}}{(n-1)!}\, dt \leqq R_n \leqq f^{(n)}(u) \int_a^b \frac{(b-t)^{n-1}}{(n-1)!}\, dt.$$

We now perform the integration, which is very easy, and get

$$\int_a^b (b-t)^{n-1}\, dt = -\frac{(b-t)^n}{n}\bigg|_a^b = \frac{(b-a)^n}{n}.$$

[*Remark*: this is the same integral that already came up in the integration by parts, in the proof of Theorem 1.1.] Therefore

$$f^{(n)}(v)\frac{(b-a)^n}{n!} \leqq R_n \leqq f^{(n)}(u)\frac{(b-a)^n}{n!}.$$

By the intermediate value theorem, the n-th derivative $f^{(n)}(t)$ takes on all values between its minimum and maximum in the interval. Hence

$$f^{(n)}(t)\frac{(b-a)^n}{n!}$$

takes on all values between *its* minimum and maximum in the interval. Hence there is some point c in the interval such that

$$R_n = f^{(n)}(c)\frac{(b-a)^n}{n!},$$

which is what we wanted.

The proof in case $b < a$ is similar, except that certain inequalities get reversed. We omit it.

The estimate of the remainder is particularly useful when b is close to a. In that case, let us rewrite Taylor's formula by setting $b - a = h$. We obtain:

Theorem 2.2. *Assumptions being as in Theorem 1.1, we have*

$$f(a+h) = f(a) + f'(a)h + \cdots + f^{(n-1)}(a)\frac{h^{n-1}}{(n-1)!} + R_n$$

with the estimate

$$|R_n| \leqq M_n \frac{|h|^n}{n!},$$

where M_n is a bound for the absolute value of the n-th derivative of f between a and a + h.

In the following sections, we give several examples. We shall often take $a = 0$, so that we have

$$f(x) = f(0) + f'(0)x + \cdots + \frac{f^{(n-1)}(0)}{(n-1)!} x^{n-1} + R_n(x)$$

with the estimate

$$|R_n(x)| \leq M_n \frac{|x|^n}{n!}$$

if M_n is a bound for the n-th derivative of f between 0 and x.

This means that we have expressed $f(x)$ in terms of a polynomial, and a remainder term. As we already said, the polynomial

$$P_n(x) = c_0 + c_1 x + \cdots + c_n x^n,$$

where

$$c_k = \frac{f^{(k)}(0)}{k!},$$

is called the **Taylor polynomial of degree** $\leq n$ of $f(x)$. We call c_k the k-th **Taylor coefficient** of f. These polynomials will be computed explicitly for all the elementary functions in the next sections.

A polynomial is essentially the easiest function to deal with. Thus it is useful if we can prove that the Taylor polynomials give approximations to the given function. For this to be the case, we have to estimate the remainder, and it is true in the case of the elementary functions that the remainder R_n approaches 0 as $n \to \infty$. This means that the Taylor polynomial $P_n(x)$ approaches $f(x)$ as $n \to \infty$, and we therefore get the desired polynomial approximation.

XIII, §3. TRIGONOMETRIC FUNCTIONS

Let $f(x) = \sin x$ and take $a = 0$ in Taylor's formula. We have already mentioned what the derivatives of $\sin x$ and $\cos x$ are. Thus

$$f(0) = 0, \qquad f^{(2)}(0) = 0,$$
$$f'(0) = 1, \qquad f^{(3)}(0) = -1.$$

The Taylor formula for $\sin x$ is therefore as follows:

$$\sin x = x - \frac{x^3}{3!} + \frac{x^5}{5!} - \cdots + (-1)^{m-1} \frac{x^{2m-1}}{(2m-1)!} + R_{2m+1}(x).$$

We see that all the even terms are 0 because $\sin 0 = 0$.
   We can estimate $\sin x$ and $\cos x$ very simply, because

$$|\sin x| \leqq 1 \qquad \text{and} \qquad |\cos x| \leqq 1$$

for all x. In Theorem 2.2 we take the bound $M_n = 1$, that is

$$|f^{(n)}(c)| \leqq 1$$

for all n, and

$$|R_n(x)| \leqq \frac{|x|^n}{n!}.$$

Thus if we look at all values of x such that $|x| \leqq 1$, we see that $R_n(x)$ approaches 0 when n becomes very large.

Example 1. Compute $\sin(0.1)$ to 3 decimals.

Here we have $x = 0.1$. We want to find n such that

$$\frac{|x|^n}{n!} \leqq 10^{-3}.$$

By inspection, we see that $n = 3$ will work. Indeed, we have

$$|R_3(0.1)| \leqq \frac{(0.1)^3}{3!} = \frac{10^{-3}}{6}.$$

Such an error term would put us within the required range of accuracy. Hence we can just use Taylor's formula,

$$\sin x = x + R_3(x).$$

We find

$$\sin(0.1) = 0.100 + E,$$

with the error term $E = R_3(0.1)$, such that $|E| \leq \frac{1}{6} 10^{-3}$. We see how efficient this is for computing the sine for small values of x.

Definition. We shall say that some expression has the **value** A **with an accuracy of** 10^{-n}, or to n **decimals** if the expression is equal to $A + E$, with an error term E such that

$$|E| \leq 10^{-n}.$$

In the preceding example, we may say that $\sin(0.1)$ has the value 0.1 with an accuracy of 10^{-3}, or to 3 decimals.

Warning. Do not write $\sin(0.1) = 0.1$. **This is false.** Always write the error, so write

$$\sin(0.1) = 0.1 + E,$$

and then give an estimate for $|E|$.

Example 2. Let us compute sine of $10°$ with an accuracy of 10^{-3}. First we must convert degrees to radians, and we have

$$10° = 10 \frac{\pi}{180} = \frac{\pi}{18} \text{ radians.}$$

Thus we have to compute $\sin(\pi/18)$. We assume that π is approximately 3.14159... and in particular, $\pi < 3.2$. This will be shown later. Then

$$\frac{\pi}{18} < \frac{1}{5}.$$

We have to express $\sin(\pi/18)$ with the Taylor polynomial of some degree and a remainder which has to be estimated. This requires trial and error. You should experiment with various possibilities. Here we give right away one that works. We have

$$\sin\left(\frac{\pi}{18}\right) = \frac{\pi}{18} - \frac{1}{6} \left(\frac{\pi}{18}\right)^3 + R_5\left(\frac{\pi}{18}\right).$$

If we know π accurately enough, then we can compute the first two terms to within any desired accuracy, by means of simple arithmetic: addition, subtraction, multiplication and division. Then we have to estimate $R_5(\pi/18)$ to know that it is within the desired accuracy. We have

$$\left| R_5\left(\frac{\pi}{18}\right) \right| \leq \frac{1}{5!} \left(\frac{\pi}{18}\right)^5 < \frac{1}{120} \left(\frac{1}{5}\right)^5 < \frac{1}{3} \times 10^{-5}$$

by easy arithmetic. Hence the first two terms

$$\frac{\pi}{18} - \frac{1}{6}\left(\frac{\pi}{18}\right)^3$$

give an **approximation of** sine 10° **to an accuracy** of 10^{-5}, which is better than what we wanted originally. Now try to see how good an approximation you would get by using just one term:

$$\sin\left(\frac{\pi}{18}\right) = \frac{\pi}{18} + R_3\left(\frac{\pi}{18}\right).$$

Example 3. Compute $\sin\left(\frac{\pi}{6} + 0.2\right)$ to an accuracy of 10^{-4}.

In this case, we use the Taylor formula for $f(a + h)$. We take

$$a = \frac{\pi}{6} \qquad \text{and} \qquad h = 0.2.$$

By trial and error, and guessing, we try for the remainder R_4. Thus

$$\sin(a + h) = \sin a + \cos(a)\frac{h}{1} - \sin(a)\frac{h^2}{2!} - \cos(a)\frac{h^3}{3!} + R_4$$

$$= \frac{1}{2} + \frac{\sqrt{3}}{2}(0.2) - \frac{1}{2}\frac{(0.2)^2}{2} - \frac{\sqrt{3}}{2}\frac{(0.2)^3}{6} + R_4.$$

For R_4 we have the estimate

$$|R_4| \leq \frac{(0.2)^4}{4!} = \frac{16 \cdot 10^{-4}}{24} \leq 10^{-4},$$

which is within the required bounds of accuracy.

Convention. In Examples 2 and 3 we left the answer as a sum of a few terms, plus an error which we estimated. *You do not need to carry out the actual decimal expansion of the first four terms.* If you have a small pocket computer, however, then do it on the computer to see that you can actually get a decimal answer. For such a purpose, the machine becomes better than the brain, but the brain was better to estimate the remainder.

It is still true that the remainder term of the Taylor formula for $\sin x$ approaches 0 when n becomes large, even when x is > 1. For this we need to investigate $x^n/n!$ when x is > 1. The difficulty is that when $x > 1$, then x^n becomes large when $n \to \infty$, and also $n! \to \infty$ when $n \to \infty$. Thus the numerator and denominator fight each other, and we must determine which one wins. Let us deal with an example, just to get a feel for what goes on. Take $x = 2$. What happens to the fraction $2^n/n!$ when $n \to \infty$? Make a table:

n	1	2	3.	4	5
$\dfrac{2^n}{n!}$	2	2	$\dfrac{8}{6} = \dfrac{4}{3}$	$\dfrac{16}{24} = \dfrac{2}{3}$	$\dfrac{32}{120} = \dfrac{4}{15}$

It should now become experimentally clear that

$$\frac{2^n}{n!} \to 0 \qquad \text{as} \quad n \to \infty.$$

Thus we *guess* the answer by experimenting numerically. Next, our task is to *prove* the general result.

Theorem 3.1. *Let c be any number. Then $c^n/n!$ approaches 0 as n becomes very large.*

Proof. We may assume $c > 0$. Let n_0 be an integer such that $n_0 > 2c$. Thus $c < n_0/2$, and $c/n_0 < \frac{1}{2}$. We write

$$\frac{c^n}{n!} = \frac{c \cdot c \cdots c}{1 \cdot 2 \cdots n_0} \frac{c}{(n_0 + 1)} \frac{c}{(n_0 + 2)} \cdots \frac{c}{n}$$

$$\leqq \frac{c^{n_0}}{n_0!} \left(\frac{1}{2}\right) \cdots \left(\frac{1}{2}\right)$$

$$= \frac{c^{n_0}}{n_0!} \left(\frac{1}{2}\right)^{n - n_0}.$$

As n becomes large, $(1/2)^{n - n_0}$ becomes small and our fraction approaches 0. Take for instance $c = 10$. We write

$$\frac{10^n}{n!} = \frac{10 \cdots 10}{1 \cdot 2 \cdots 20} \frac{(10) \cdots (10)}{(21) \cdots (n)} < \frac{10^{20}}{20!} \left(\frac{1}{2}\right)^{n - 20}$$

and $(1/2)^{n - 20}$ approaches 0 as n becomes large.

From the theorem we see that the remainder

$$|R_n(x)| \leq \frac{|x|^n}{n!}$$

approaches 0 as n becomes large.

Sometimes a definite integral cannot be evaluated from an indefinite one, but we can find simple approximations to it by using Taylor's expansion.

In the next example, and in exercises, we shall now use frequently the estimate for an integral given in Theorems 3.2 and 3.3 of Chapter X, that is: Let $a < b$ and let f be continuous on $[a, b]$. Let M be a number such that $|f(x)| \leq M$ for all x in the interval. Then

$$\left| \int_a^b f(x) \, dx \right| \leq \int_a^b |f(x)| \, dx \leq M(b - a).$$

Ron Infante tells me the numerical computations of integrals like the one in the next example occur frequently in the study of communication networks, in connection with square waves.

Example 4. Compute to two decimals the integral

$$\int_0^1 \frac{\sin x}{x} \, dx.$$

We have

$$\sin x = x - \frac{x^3}{3!} + R_5(x) \qquad \text{and} \qquad |R_5(x)| \leq \frac{|x|^5}{5!}.$$

Hence

$$\frac{\sin x}{x} = 1 - \frac{x^2}{3!} + \frac{R_5(x)}{x} \qquad \text{and} \qquad \left| \frac{R_5(x)}{x} \right| \leq \frac{|x|^4}{5!}.$$

Hence

$$\int_0^1 \frac{\sin x}{x} \, dx = x - \frac{x^3}{3 \cdot 3!} \bigg|_0^1 + E, \qquad \text{where} \qquad E = \int_0^1 \frac{R_5(x)}{x} \, dx.$$

The error term E satisfies

$$|E| \leq \int_0^1 \left| \frac{R_5(x)}{x} \right| \, dx \leq \int_0^1 \frac{x^4}{5!} \, dx = \frac{x^5}{5 \cdot 5!} \bigg|_0^1 = \frac{1}{600}.$$

Furthermore

$$x - \frac{x^3}{3 \cdot 3!} \bigg|_0^1 = 1 - \frac{1}{18} = \frac{17}{18}.$$

Hence

$$\int_0^1 \frac{\sin x}{x} \, dx = \frac{17}{18} + E \qquad \text{where} \qquad |E| \leq \frac{1}{600}.$$

Example 5. Let us compute

$$I = \int_0^1 \sin x^2 \, dx.$$

We let $u = x^2$. Then

$$\sin u = u - \frac{u^3}{3!} + R_5(u).$$

Hence

$$I = \int_0^1 \left(x^2 - \frac{x^6}{3!} \right) dx + \int_0^1 R_5(x^2) \, dx$$

$$= \left[\frac{x^3}{3} - \frac{x^7}{7 \cdot 6} \right]_0^1 + E$$

$$= \tfrac{1}{3} - \tfrac{1}{42} + E,$$

where

$$E = \int_0^1 R_5(x^2) \, dx.$$

We know that

$$|R_5(u)| \leq \frac{|u|^5}{5!}.$$

Since $u = x^2$, we find

$$|E| \leq \int_0^1 \frac{x^{10}}{5!} \, dx = \frac{1}{11 \cdot 120} < 10^{-3}.$$

Hence

$$I = \tfrac{1}{3} - \tfrac{1}{42} + E, \qquad \text{with} \quad |E| < 10^{-3}.$$

Remark. Although the *notation* in the preceding example is *similar* to integration by substitution, it should be emphasized that the procedure followed is *not* what we called previously integration by substitution.

We have discussed the sine. The cosine can be discussed in the same way. We have the Taylor formula

$$\cos x = 1 - \frac{x^2}{2!} + \frac{x^4}{4!} - \cdots + (-1)^m \frac{x^{2m}}{(2m)!} + R_{2m+2}(x)$$

and

$$|R_{2m+2}(x)| \leq \frac{|x|^{2m+2}}{(2m+2)!}.$$

Observe that only the even terms appear with non-zero coefficient. In the sine formula, only the odd terms appeared. This is because the odd-order derivatives of cosine are equal to 0 at 0, and the even-order derivatives of the sine are equal to 0 at 0.

Example 6. Suppose we want to find the value of

$$\int_0^1 \frac{\cos x - 1}{x} \, dx,$$

to 2 decimals. We write

$$\cos x = 1 - \frac{x^2}{2} + R_4(x).$$

Then

$$\frac{\cos x - 1}{x} = -\frac{x}{2} + \frac{R_4(x)}{x},$$

and for $0 \leq x \leq 1$,

$$\left| \frac{R_4(x)}{x} \right| \leq \frac{x^4}{4!x} = \frac{x^3}{4!}.$$

We obtain

$$\int_0^1 \frac{\cos x - 1}{x} \, dx = -\int_0^1 \frac{x}{2} \, dx + E \qquad \text{where} \qquad E = \int_0^1 \frac{R_4(x)}{x} \, dx.$$

$$= -\frac{1}{2} \frac{x^2}{2} \Big|_0^1 + E$$

$$= -\tfrac{1}{4} + E.$$

We estimate E:

$$|E| \leq \int_0^1 \frac{x^3}{4!} \, dx = \frac{1}{24} \frac{x^4}{4} \Big|_0^1 = \frac{1}{96}.$$

We just miss the desired estimate by a few percentage points. This means that to get the desired accuracy, you have to use one more term from the Taylor polynomial of $\cos x$. Thus you write

$$\cos x = 1 - \frac{x^2}{2!} + \frac{x^4}{4!} + R_6(x)$$

and

$$\frac{\cos x - 1}{x} = -\frac{x}{2} + \frac{x^3}{4!} + \frac{R_6(x)}{x}.$$

Then

$$\int_0^1 \frac{\cos x - 1}{x} \, dx = \int_0^1 \left[-\frac{x}{2} + \frac{x^3}{4!} \right] dx + E,$$

where

$$E = \int_0^1 \frac{R_6(x)}{x} \, dx.$$

Now we use the estimate

$$\left| \frac{R_6(x)}{x} \right| \leq \frac{x^6}{6! \, x} = \frac{x^5}{6!},$$

and

$$|E| \leq \int_0^1 \frac{x^5}{6!} \, dx = \frac{1}{720} \frac{1}{6}.$$

Thus the error satisfies $|E| < 10^{-3}$, and

$$\int_0^1 \left[-\frac{x}{2} + \frac{x^3}{4!} \right] dx = -\frac{1}{4} + \frac{1}{96}.$$

This gives the desired value, with an accuracy of three decimals.

Warning. The definite integral of the above example **cannot be written as a sum**

$$\int_0^1 \frac{\cos x - 1}{x} \, dx = \int_0^1 \frac{\cos x}{x} \, dx - \int_0^1 \frac{1}{x} \, dx.$$

Although it is true that the integral of a sum is the sum of the integrals, this is true only when the integrals make sense. The integral

$$\int_0^1 \frac{1}{x} \, dx$$

does not make sense. First the function $1/x$ is not continuous on the interval $0 \leq x \leq 1$. It blows up when x approaches 0. Even if we try to interpret this as a limiting value,

$$\int_0^1 \frac{1}{x}\, dx = \lim_{h \to 0} \int_h^1 \frac{1}{x}\, dx = \lim_{h \to 0} (\log 1 - \log h),$$

the limit does not exist because $\log h$ becomes large negative when h approaches 0. So we cannot split the desired integral into a sum.

Also, it can be shown that **there is no simple expression giving an indefinite integral**

$$\int \frac{\cos x - 1}{x}\, dx = F(x),$$

with a function $F(x)$ expressible in terms of elementary functions. On the other hand, as we have seen, we can perfectly well evaluate the definite integral to any desired accuracy.

XIII, §3. EXERCISES

Unless otherwise specified, take the Taylor formula with $a = 0$, $b = x$.

In all the computations, *include* an estimate of the remainder (error) term, which shows that your answer is within the required accuracy.

1. Write down the Taylor polynomial of degree 4 for $\cos x$. Prove the Taylor formula stated in the text for $\cos x$.

2. Give the details for the estimate $|R_{2m+2}(x)| \leq |x|^{2m+2}/(2m + 2)!$ for the function $f(x) = \cos x$.

3. Compute $\cos (0.1)$ to 3 decimals.

4. Estimate the remainder R_3 in the Taylor formula for $\cos x$, for the value $x = 0.1$.

5. Estimate the remainder R_4 in the Taylor formula for $\sin x$, for the value $x = 0.2$.

6. Write down the Taylor polynomial of degree 4 for $\tan x$.

7. Estimate the remainder R_5 in the Taylor formula for $\tan x$, for $0 \leq x \leq 0.2$.

In Exercises 8, 9, 10, and 11 the Taylor formula is used with $a \neq 0$.

8. Compute the following values to 3 places.
   (a) $\sin 31°$ (b) $\cos 31°$ (c) $\sin 47°$
   (d) $\cos 47°$ (e) $\sin 32°$ (f) $\cos 32°$

9. Compute cosine 31 degrees to 3 places.

10. Compute sine 61 degrees to 3 places.

11. Compute cosine 61 degrees to 3 places.

12. Compute the following integrals to three decimals.

(a) $\displaystyle\int_0^1 \frac{\sin x}{x}\, dx$

(b) $\displaystyle\int_0^{0.1} \frac{\cos x - 1}{x}\, dx$

(c) $\displaystyle\int_0^1 \sin x^2\, dx$

(d) $\displaystyle\int_0^1 \frac{\sin x^2}{x}\, dx$

(e) $\displaystyle\int_0^1 \cos x^2\, dx$

(f) $\displaystyle\int_0^1 \frac{\sin x^2}{x^2}\, dx$

13. Compute

$$\int_0^{1/2} \frac{\cos x - 1}{x}\, dx$$

to 5 decimals.

XIII, §4. EXPONENTIAL FUNCTION

All derivatives of e^x are equal to e^x and $e^0 = 1$. Hence the Taylor formula for e^x is

$$e^x = 1 + x + \frac{x^2}{2!} + \cdots + \frac{x^{n-1}}{(n-1)!} + R_n(x).$$

The remainder term satisfies

$$R_n(x) = e^c \frac{x^n}{n!},$$

where c is a number between 0 and x. Hence

$$|R_n(x)| \leqq e^c \frac{|x|^n}{n!}.$$

Note that e^c is always positive, and

$$\text{if } x < 0, \quad \text{then} \quad c < 0 \quad \text{and} \quad 0 < e^c < 1.$$

As with the sine function, Theorem 3.1 shows that the remainder term approaches 0 as n becomes large.

Example. 1. Compute e to 3 decimals.

We have $e = e^1$. From Chapter VIII, §6 we know that $e < 4$. We estimate R_7:

$$|R_7| \leqq e \frac{1}{7!} \leqq 4 \frac{1}{5,040} < 10^{-3}.$$

Thus

$$e = 1 + 1 + \frac{1}{2} + \cdots + \frac{1}{6!} + R_7$$

$$= 2.718\ldots.$$

Of course, the smaller x is, the fewer terms of the Taylor series do we need to approximate e^x.

Remark. We had obtained the naive estimate $e < 4$ in Chapter VIII. Now we have the much finer evaluation, which shows in particular that $e < 3$. Thus using a rough estimate and the Taylor formula allows us to get a precise determination of e. Even if we had started with knowing $e < 3$ (which could also be obtained by an estimate similar to that of Chapter VIII), this would not have helped much more to get the more precise value.

Example 2. How many terms of the Taylor formula do you need to compute $e^{1/10}$ to an accuracy of 10^{-3}?

We certainly have $e^{1/10} < 2$. Thus

$$|R_3(1/10)| \leqq 2 \frac{(1/10)^3}{3!} < \frac{1}{2} \, 10^{-3}.$$

Hence we need just 3 terms (including the 0-th term).

XIII, §4. EXERCISES

In the exercises, when asked to compute a quantity with a certain degree of accuracy, always show how you estimate the error term to prove that you get the desired accuracy.

1. Write down the Taylor polynomial of degree 5 for e^{-x}.

2. Estimate the remainder R_3 in the Taylor formula for e^x for $x = 1/2$.

3. Estimate the remainder R_4 for $x = 10^{-2}$.

4. Estimate the remainder R_3 for $x = 10^{-2}$.

5. Compute e to four decimals, then five decimals, then six decimals. In each
   case write down e as a sum of fractions, plus a remainder term, and estimate
   the remainder term. [This exercise is meant to give you a practical feeling
   for the size of the remainder terms, and how many you will need to get a
   desired accuracy. You can compute the sum of the fractions on a pocket
   calculator.]

6. Compute $1/e$ to 3 decimals, and show which remainder would give you an
   accuracy of 10^{-3}.

7. Estimate the remainder R_4 in the Taylor formula for e^x when
   (a) $x = 2$.　　　　　　　　　　　(b) $x = 3$.

8. Estimate the remainder R_5 in the Taylor formula for e^x when
   (a) $x = 2$.　　　　　　　　　　　(b) $x = 3$.

9. How many terms of the Taylor formula for e^x would you need to compute
   e^2 to
   (a) 4 decimals?　　　　　　　(b) 6 decimals?

10. Compute e to 10 decimals. First give it as a sum of rational numbers. Then
    use some calculating machine to get the decimals. Show your estimate of the
    error term.

11. Compute $1/e^2$ to 4 decimals.

12. Compute the following integrals to 3 or 4 decimals, depending on how much
    you want to exert yourself.

   (a) $\displaystyle\int_0^1 \frac{e^x - 1}{x}\, dx$　　　　(b) $\displaystyle\int_0^1 e^{-x^2}\, dx$　　　　(c) $\displaystyle\int_0^1 e^{x^2}\, dx$

   (d) $\displaystyle\int_0^{0.1} e^{x^2}\, dx$　　　　(e) $\displaystyle\int_0^{0.1} e^{-x^2}\, dx$

XIII, §5. LOGARITHM

We want to get a Taylor formula for the log. We cannot handle the log
just by writing

$$\log x = \log 0 + \log'(0)x + \cdots$$

because $\log 0$ is not defined. Thus for the log, it is better to derive a
formula with $a = 1$, so that

$$\log b = \log 1 + \log'(1)(b - 1) + \log''(1)\frac{(b - 1)^2}{2!} + \cdots.$$

Experience shows that it is then more convenient to let

$$b = 1 + x$$

so $b - 1 = x$. In that way we get a Taylor formula

$$\log(1 + x) = x - \frac{x^2}{2} + \cdots.$$

We leave it to you (see the exercise in §1) to derive this formula in the usual way by computing the derivatives $f^{(k)}(0)$, where $f(x) = \log(1 + x)$.

Here we shall obtain the result by another method, which will also be applicable in the next section and is in some ways more efficient, and makes the series for the log easy to remember.

You should know from high school the geometric series

$$\frac{1}{1 - u} = 1 + u + u^2 + u^3 + \cdots.$$

We shall derive it again below, but for the moment let us work formally and not worry about what the infinite sum means. Replacing u by $-x$ we get

$$\frac{1}{1 + x} = 1 - x + x^2 - x^3 + \cdots.$$

Integrate the left-hand side, and the right-hand side term by term, again without worrying about what the infinite sum means. Then we get

$$\log(1 + x) = x - \frac{x^2}{2} + \frac{x^3}{3} - \frac{x^4}{4} + \cdots.$$

On the right-hand side we see that the signs alternate, and we have just n in the denominator of x^n/n, rather than the $n!$ which occurred for $\sin x$, $\cos x$, and e^x.

Now we must start all over again, to derive the formula with a remainder term which will allow us to estimate values for the log. Also observe that since $\log 0$ is not defined, we shall have to take x in some interval of numbers > -1. It turns out that the Taylor formula will give values for the function only in the interval

$$-1 < x \leqq 1.$$

This contrasts with $\sin x$, $\cos x$, e^x when we obtained values for all x.

Let u be any number $\neq 1$. We wish to justify the series

$$\frac{1}{1 - u} = 1 + u + u^2 + u^3 + \cdots.$$

Don't worry at first what the infinite sum on the right means. Use it formally. If you cross multiply, you obtain

$$(1 - u)(1 + u + u^2 + u^3 + \cdots) = 1 + u + u^2 + u^3 + \cdots$$
$$- u - u^2 - u^3 - \cdots$$
$$= 1.$$

So we have justified the geometric series formally.

Next let us worry about the infinite sum. We don't know how to add infinitely many numbers together, so we must state some relation analogue to the above, but with a finite number of terms. This is based on the formula

$$\frac{1 - u^n}{1 - u} = 1 + u + \cdots + u^{n-1}$$

for any integer $n > 1$. The proof is again obtained by cross-multiplying:

$$(1 - u)(1 + u + u^2 + \cdots + u^{n-1}) = 1 + u + u^2 + \cdots + u^{n-1}$$
$$- u - u^2 - \cdots - u^{n-1} - u^n$$
$$= 1 - u^n.$$

Since

$$\frac{1 - u^n}{1 - u} = \frac{1}{1 - u} - \frac{u^n}{1 - u},$$

we find finally

$$\frac{1}{1 - u} = 1 + u + u^2 + \cdots + u^{n-1} + \frac{u^n}{1 - u}.$$

We want to apply this formula to get an expression for $1/(1 + t)$, because we want ultimately to get an expression for

$$\log(1 + x) = \int_0^x \frac{1}{1 + t} \, dt.$$

Thus we substitute $u = -t$. Then we find

$$\frac{1}{1 + t} = 1 - t + t^2 - t^3 + \cdots + (-1)^{n-1} t^{n-1} + (-1)^n \frac{t^n}{1 + t}.$$

Consider the interval $-1 < x \leq 1$, and take the integral from 0 to x (in this interval). The integrals of the powers of t are well known to you. The integral

$$\int_0^x \frac{1}{1+t}\, dt = \log(1+x)$$

is computed by the substitution $u = 1 + t$, $du = dt$. Thus we get:

Theorem 5.1. *For* $-1 < x \leq 1$, *we have*

$$\log(1 + x) = x - \frac{x^2}{2} + \frac{x^3}{3} - \cdots + (-1)^{n-1} \frac{x^n}{n} + R_{n+1}(x)$$

where the remainder $R_{n+1}(x)$ *is the integral*

$$R_{n+1}(x) = (-1)^n \int_0^x \frac{t^n}{1+t}\, dt.$$

Observe that it was essential that $x > -1$ because the expression $1/(1 + t)$ has no meaning when $t = -1$. The above formula also holds for $x > 1$. However, we shall see that the remainder term approaches 0 only when x lies in the stated interval.

Case 1. $0 < x \leq 1$.

In that case, $1 + t \geq 1$. Thus

$$\frac{t^n}{1+t} \leq t^n,$$

and our integral is bounded by $\int_0^x t^n\, dt$. Thus in that case,

$$|R_{n+1}(x)| \leq \int_0^x t^n\, dt \leq \frac{x^{n+1}}{n+1}.$$

In particular, the remainder approaches 0 as n becomes large.

Remark. In the applications we shall use a device which allows us to deal only with Case 1. *Thus you may omit Case 2 if you wish.*

Case 2. $-1 < x < 0$.

In this case, t lies between 0 and x and is negative, but we still have

$$x \leq t \leq 0,$$

and $0 < 1 + x < 1 + t$. Hence

$$\left| \frac{t^n}{1+t} \right| = \frac{|t|^n}{1+t} \leq \frac{(-t)^n}{1+x}.$$

To estimate the absolute value of the integral, we can invert the limits (we do this because $x \leq 0$) and thus

$$|R_{n+1}(x)| \leq \int_x^0 \frac{(-t)^n}{1+x} \, dt$$

so

$$|R_{n+1}(x)| \leq \frac{(-x)^{n+1}}{(n+1)(1+x)} = \frac{|x|^{n+1}}{(n+1)(1+x)}.$$

Therefore the remainder also approaches 0 in that case. However, when x is negative and $-1 < x < 0$ then $1 + x < 1$ and we cannot estimate $1/(1 + x)$ in the same way as in Case 1, because **we do not have** $1/(1 + x) \leq 1$. Thus Case 2 is disagreeable. That is the reason we avoid it.

Example. Let us compute $\log 2$ to 3 decimals. We know that $\log(1 + x)$ can be computed efficiently only when x is close to 0. But $2 = 1 + 1$. We have to use some auxiliary device, which avoids plugging $x = 1$ in the formula. For this, we write

$$2 = \tfrac{4}{3} \cdot \tfrac{3}{2}.$$

Then

$$\tfrac{4}{3} = 1 + \tfrac{1}{3} \qquad \text{and} \qquad \tfrac{3}{2} = 1 + \tfrac{1}{2}.$$

By using $x = \tfrac{1}{3}$ and $x = \tfrac{1}{2}$ we shall achieve what we want. Indeed,

$$\log 2 = \log(\tfrac{4}{3} \cdot \tfrac{3}{2}) = \log(1 + \tfrac{1}{3}) + \log(1 + \tfrac{1}{2}).$$

To find $\log(1 + \tfrac{1}{3})$ we use $x = \tfrac{1}{3}$ and

$$\log(1 + x) = x - \frac{x^2}{2} + \frac{x^3}{3} - \frac{x^4}{4} + \frac{x^5}{5} + R_6(x).$$

By the estimate of Case 1, we get

$$|R_6(\tfrac{1}{3})| \leqq \tfrac{1}{6} (\tfrac{1}{3})^6 \leqq \tfrac{1}{4} \times 10^{-3}.$$

Therefore

$$\log\frac{4}{3} = \log\left(1 + \frac{1}{3}\right) = \frac{1}{3} - \frac{(1/3)^2}{2} + \frac{(1/3)^3}{3} - \frac{(1/3)^4}{4} + \frac{(1/3)^5}{5} + E_1$$

$$= A_1 + E_1$$

with

$$E_1 = R_6(\tfrac{1}{3}) \quad \text{and} \quad |E_1| \leqq \tfrac{1}{4} \times 10^{-3}.$$

Similarly we get $\log\frac{3}{2}$ by

$$\log\frac{3}{2} = \log\left(1 + \frac{1}{2}\right) = \frac{1}{2} - \frac{(1/2)^2}{2} + \cdots + \frac{(1/2)^7}{7} + E_2$$

$$= A_2 + E_2$$

with the error term $E_2 = R_8(1/2)$. We have the estimate

$$|E_2| = |R_8(\tfrac{1}{2})| \leqq \tfrac{1}{2} \times 10^{-3}.$$

Therefore

$$\log 2 = \log(1 + \tfrac{1}{2}) + \log(1 + \tfrac{1}{3})$$

$$= A_1 + A_2 + E_1 + E_2,$$

where A_1, A_2 are simple expressions which can be computed from fractions using addition, multiplication, and subtraction, and the error term is $E = E_1 + E_2$. We can now estimate E, namely

$$|E| \leqq |E_1| + |E_2| \leqq \tfrac{1}{4} \times 10^{-3} + \tfrac{1}{2} \times 10^{-3} < 10^{-3}.$$

This lies within the desired accuracy. If you have a small pocket computer, you can evaluate a numerical decimal for log 2 easily.

In the above computation, we still needed six or eight terms in the polynomial expression approximating the logarithm to get three decimals accuracy. In Exercise 2, following the same general idea, you will see how to get accurate answers using fewer terms.

Example. Suppose we want to compute $\log\frac{3}{4}$ to 3 decimals. Note that $\frac{3}{4} < 1$, and if we tried to evaluate

$$\log \tfrac{3}{4} = \log(1 - \tfrac{1}{4}),$$

then we would have to use Case 2. This can be avoided by using the general rule

$$\log \frac{1}{a} = - \log a$$

for any positive number a. In particular,

$$\log \tfrac{3}{4} = -\log \tfrac{4}{3}$$
$$= -\frac{1}{3} + \frac{(1/3)^2}{2} - \frac{(1/3)^3}{3} + \frac{(1/3)^4}{4} - \frac{(1/3)^5}{5} + E,$$

and

$$|E| \leqq \tfrac{1}{4} \times 10^{-3}.$$

Example. Compute log 1.1 to three decimals.

To do this, i.e. compute $\log(1 + 0.1)$, we take $n = 2$, and $x = 0.1$ in Case 1. We find that

$$|R_3(x)| \leqq \tfrac{1}{3} \times 10^{-3}.$$

Hence
$$\log(1.1) = 0.1 - 0.005 + E$$

with an error E such that $|E| \leqq \tfrac{1}{3} \times 10^{-3}$.

XIII, §5. EXERCISES

1. Compute the following values up to an accuracy of 10^{-3}, estimating the remainder each time.
   (a) log 1.2   (b) log 0.9   (c) log 1.05
   (d) log $\tfrac{9}{10}$ (e) log $\tfrac{24}{25}$ (f) log $\tfrac{26}{25}$

2. (a) Verify the following formulas:

$$\log 2 = 7 \log \tfrac{10}{9} - 2 \log \tfrac{25}{24} + 3 \log \tfrac{81}{80},$$
$$\log 3 = 11 \log \tfrac{10}{9} - 3 \log \tfrac{25}{24} + 5 \log \tfrac{81}{80}.$$

(b) Compute log 2 and log 3 to five decimals, using these formulas.

You may ask how one finds such formulas. The answer is that someone clever, probably more than 200 years ago, found them by experimenting with numbers, and after that, everybody copies them.

XIII, §6. THE ARCTANGENT

We proceed as with the logarithm, except that we put $u = -t^2$ in the geometric series, and obtain

$$\frac{1}{1 + t^2} = 1 - t^2 + t^4 - \cdots + (-1)^{m-1}t^{2m-2} + (-1)^m \frac{t^{2m}}{1 + t^2}.$$

After integration from 0 to any number x, we obtain:

Theorem 6.1. *The arctan has an expansion*

$$\arctan x = x - \frac{x^3}{3} + \frac{x^5}{5} - \cdots + (-1)^{m-1} \frac{x^{2m-1}}{2m - 1} + R_{2m+1}(x),$$

where

$$R_{2m+1}(x) = (-1)^m \int_0^x \frac{t^{2m}}{1 + t^2} \, dt,$$

and

$$|R_{2m+1}(x)| \leqq \int_0^{|x|} t^{2m} \, dt \leqq \frac{|x|^{2m+1}}{2m + 1}.$$

When $-1 \leqq x \leqq 1$, *the remainder approaches* 0 *as* n *becomes large.*

Observe how only odd powers of x occur in the Taylor formula for arctan x. This is the reason for writing $2m + 1$, or R_{2m+1}. If we put $n = 2m + 1$. then we can write the estimate for the remainder in the form

$$|R_n(x)| \leqq \frac{|x|^n}{n}.$$

This estimate is the same as for the log, except that for arctan, we take only odd integers n.

Remark. If x does not lie in the prescribed interval, i.e. if $|x| > 1$, then the remainder term does not tend to 0 as n becomes large. For instance, if $x = 2$, then the remainder term is bounded by

$$\frac{2^{2m+1}}{2m + 1}.$$

By computing a few values with $m = 1, 2, 3, \ldots$ you will see that this expression grows large quite fast. You should know this anyhow from your study of the exponential function.

From our theorem, we get a cute expression for $\pi/4$:

$$\frac{\pi}{4} = 1 - \frac{1}{3} + \frac{1}{5} - \cdots$$

from the Taylor formula for arctan 1. However, it takes many terms to get a good approximation to $\pi/4$ by this expression. Roughly, it takes 1000 terms to get accuracy to 10^{-3}, which is very inefficient. By using a more clever approach, however, we can find π much faster. This is done as follows. First we have:

Addition formula for the tangent:

$$\tan(x + y) = \frac{\tan x + \tan y}{1 - \tan x \tan y}.$$

Proof. Using the addition formulas for sine and cosine proved in Chapter 4, Theorem 3.1, we have

$$\tan(x + y) = \frac{\sin(x + y)}{\cos(x + y)} = \frac{\sin x \cos y + \cos x \sin y}{\cos x \cos y - \sin x \sin y}.$$

Now divide numerator and denominator on the right by $\cos x \cos y$. The desired formula drops out.

In the addition formula for the tangent, put $u = \tan x$ and $v = \tan y$, so that $x = \arctan u$ and $y = \arctan v$. Since

$$\arctan(\tan(x + y)) = x + y = \arctan u + \arctan v,$$

we obtain the **addition formula for the arctangent:**

$$\boxed{\arctan u + \arctan v = \arctan \frac{u + v}{1 - uv}.}$$

Example. Consider the special values $u = 1/2$ and $v = 1/3$. Simple arithmetic shows that for these values, we get

$$\frac{u + v}{1 - uv} = \frac{1/2 + 1/3}{1 - 1/6} = 1.$$

Since arctan $1 = \pi/4$, we obtain the formula:

$$\frac{\pi}{4} = \arctan 1 = \arctan \frac{1}{2} + \arctan \frac{1}{3}.$$

Next we use the Taylor formula for arctan,

$$\arctan x = x - \frac{x^3}{3} + R_5(x)$$

with

$$|R_5(x)| \leqq \frac{|x|^5}{5}.$$

Then we obtain

(1) $$\arctan \frac{1}{2} = \frac{1}{2} - \frac{1}{24} + R_5\left(\frac{1}{2}\right),$$

and

$$|R_5(\tfrac{1}{2})| \leqq \tfrac{1}{160}.$$

This is not exceptionally good, but it shows you that with just two terms of the polynomial approximating arctan x, we already get about 2 decimals accuracy. Using a couple of more terms, you should be able to get 4 decimals.

Similarly, we get

(2) $$\arctan \frac{1}{3} = \frac{1}{3} - \frac{(1/3)^3}{3} + R_5\left(\frac{1}{3}\right)$$

and

$$\left|R_5\left(\frac{1}{3}\right)\right| \leqq \frac{(1/3)^5}{5} \leqq \frac{1}{2025} < \frac{1}{2} \times 10^{-3}.$$

Hence

$$\frac{\pi}{4} = \frac{1}{2} - \frac{1}{24} + \frac{1}{3} - \frac{1}{3^4} + E = A + E,$$

where $E = R_5(\tfrac{1}{2}) + R_5(\tfrac{1}{3})$, and

$$|E| \leqq |R_5(\tfrac{1}{2})| + |R_5(\tfrac{1}{3})| < 10^{-2}.$$

The expression A is a sum of fractions which you can easily calculate on a pocket computer. Then

$$\pi = 4(\arctan \tfrac{1}{2} + \arctan \tfrac{1}{3}) = 4A + 4E,$$

where
$$|4E| < 4 \times 10^{-2}.$$

Notice that this final factor of 4 makes the estimate of the error term somewhat worse. Hence in determining which remainders to take, you have to make sure that in the final step, when you multiply with 4, the accuracy lies within the desired bound. Experiment with R_7 and R_9 to acquire a good feeling for the size of these remainders.

XIII, §6. EXERCISES

1. Prove the formulas:

$$2 \arctan u = \arctan \frac{2u}{1 - u^2} \quad \text{and} \quad 3 \arctan u = \arctan \frac{3u - u^3}{1 - 3u^2}.$$

2. Prove:
   (a) $\arctan \frac{1}{2} = \arctan \frac{1}{3} + \arctan \frac{1}{7}$
   (b) $\arctan \frac{1}{3} = \arctan \frac{1}{5} + \arctan \frac{1}{8}$
   (c) $\pi/4 = 2 \arctan \frac{1}{5} + \arctan \frac{1}{7} + 2 \arctan \frac{1}{8}$

3. Estimating $R_3(x)$, $R_5(x)$, $R_7(x)$, $R_9(x)$ in the Taylor formula for the arctangent, and using the expression

$$\frac{\pi}{4} = \arctan \frac{1}{2} + \arctan \frac{1}{3},$$

   as well as the expression (c) in Exercise 2, find various decimal approximations for π. Ultimately, do verify that

$$\pi = 3.14159\ldots$$

   to an accuracy of 5 decimals.

4. Prove the formula $\pi/4 = 4 \arctan \frac{1}{5} - \arctan \frac{1}{239}$. How few terms of the Taylor formula do you now need in order to get the above accuracy for a decimal approximation to π?

XIII, §7. THE BINOMIAL EXPANSION

In high school, you should have learned the expansion of $(a + b)^n$ or $(1 + x)^n$. For instance

$$(1 + x)^2 = 1 + 2x + x^2,$$
$$(1 + x)^3 = 1 + 3x + 3x^2 + x^3,$$
$$(1 + x)^4 = 1 + 4x + 6x^2 + 4x^3 + x^4.$$

Just using algebra, one can determine the coefficients for the expansion of $(1 + x)^n$ when n is a positive integer. However, here we shall be interested also in powers $(1 + x)^s$ when s is not a positive integer. For this we shall use the general method of Taylor's formula, which states:

$$f(x) = f(0) + f^{(1)}(0)x + \frac{f^{(2)}(0)}{2!} x^2 + \cdots + \frac{f^{(n)}(0)}{n!} x^n + R_{n+1}(x)$$

$$= \sum_{k=0}^{n} \frac{f^{(k)}(0)}{k!} x^k + R_{n+1}(x).$$

We apply this formula to the function

$$f(x) = (1 + x)^s.$$

Theorem 7.1. *Let n be a positive integer and $x \neq -1$. Then*

$$(1 + x)^s = 1 + sx + \frac{s(s-1)}{2!} x^2 + \frac{s(s-1)(s-2)}{3!} x^3 + \cdots$$

$$+ \frac{s(s-1)(s-2)\cdots(s-n+1)}{n!} x^n + R_{n+1}(x).$$

Proof. Let $f(x) = (1 + x)^s$. Then we compute the derivatives:

$$f^{(1)}(x) = s(1 + x)^{s-1}, \qquad\qquad f^{(1)}(0) = s,$$

$$f^{(2)}(x) = s(s-1)(1 + x)^{s-2}, \qquad f^{(2)}(0) = s(s-1),$$

$$f^{(3)}(x) = s(s-1)(s-2)(1 + x)^{s-3}, \qquad f^{(k)}(0) = s(s-1)(s-2),$$

$$\vdots \qquad\qquad\qquad\qquad\qquad \vdots$$

$$f^{(k)}(x) = s(s-1)\cdots(s-k+1)(1 + x)^{s-k}, \qquad f^{(k)}(0) = s(s-1)\cdots$$
$$(s-k+1).$$

Hence

$$\frac{f^{(k)}(0)}{k!} = \frac{s(s-1)\cdots(s-k+1)}{k!}.$$

By the general Taylor formula, this proves the desired expansion for $(1 + x)^s$.

The general formula for the remainder term is

$$R_n(x) = \frac{f^{(n)}(c)x^n}{n!}.$$

with some number c between 0 and x, and therefore in the present case we find:

$$R_n(x) = \frac{s(s-1)\cdots(s-n+1)}{n!}(1+c)^{s-n}x^n.$$

It can be shown that if $-1 < x < 1$ then $R_n(x) \to 0$ as $n \to \infty$. We shall estimate the remainder when $n = 2$ and $n = 3$. We do not give the proof in general that $R_n(x)$ approaches 0 when $n \to \infty$.

In these estimates, we use repeatedly the fact that

$$|ab| = |a||b|.$$

For example, products like $s(s-1)(s-2)$ occur all the time in these estimates. Then

$$|s(s-1)(s-2)| = |s||s-1||s-2|.$$

If $s = 1/3$, we find

$$\left|\frac{1}{3}\left(\frac{1}{3}-1\right)\left(\frac{1}{3}-2\right)\right| = \frac{1}{3}\left|\frac{1}{3}-1\right|\left|\frac{1}{3}-2\right| = \frac{1}{3}\frac{2}{3}\frac{5}{3} = \frac{10}{27}.$$

Examples involving R_2

Now let us look at R_2. Let

$$f(x) = (1+x)^s,$$

where s is not an integer. We have

$$f^{(2)}(x) = s(s-1)(1+x)^{s-2}.$$

The Taylor formula gives

$$(1+x)^s = 1 + sx + R_2(x),$$

where

$$R_2(x) = f^{(2)}(c)\frac{x^2}{2!}$$

$$= s(s-1)(1+c)^{s-2}\frac{x^2}{2},$$

for some number c between 0 and x.

For small x, this means that $1 + sx$ should be a good approximation to the s power of $1 + x$, if $R_2(x)$ can be proved to be small. This we can do easily. We see that

$$|R_2(x)| = \frac{|s(s-1)|}{2} (1 + c)^{s-2}|x|^2,$$

where c is between 0 and x. By an easy estimate one sees, for instance, that $(1 + x)^{1/2}$ is approximately equal to $1 + \frac{1}{2}x$, and $(1 + x)^{1/3}$ is approximately equal to $1 + \frac{1}{3}x$, for small x.

Example 1. Find $\sqrt{1.2}$ to 2 decimals.
We let $x = 0.2 = 2 \times 10^{-1}$ and $s = 1/2$. Then

$$\sqrt{1.2} = (1 + 0.2)^{1/2} = 1 + \frac{1}{2}0.2 + R_2(0.2)$$

$$= 1 + 0.1 + R_2(\tfrac{1}{5}).$$

We must estimate $R_2(1/5)$. Since $0 \leq c \leq 1/5$ and $s - 2 = -3/2$, we find

$$(1 + c)^{s-2} = \frac{1}{(1 + c)^{3/2}} \leq 1,$$

because the smaller the denominator, the larger the fraction. The only information we have on c is that $0 \leq c \leq 1/5$, and the smallest possible value of the denominator is when $c = 0$. Consequently

$$\left| R_2\left(\frac{1}{5}\right) \right| \leq \frac{1}{2}\left| \frac{1}{2} - 1 \right| \frac{1}{2} (0.2)^2$$

$$\leq \frac{1}{8} 4 \times 10^{-2} = \frac{1}{2} \times 10^{-2}.$$

Therefore the estimate for the error term is within the desired accuracy.

Example 2. Let us compute $\sqrt{0.8}$ to 2 decimals.
We use $s = 1/2$ and $x = -0.2$. Then

$$\sqrt{0.8} = (1 - 0.2)^{1/2} = 1 - \frac{1}{2}0.2 + R_2(-0.2) = 0.9 + R_2(-0.2).$$

Here we have $-0.2 \leq c \leq 0$. Hence $1/(1 + c)^{3/2} \leq 1/(0.8)^{3/2}$, and

$$|R_2(-0.2)| \leq \frac{1}{2}\left| \frac{1}{2} - 1 \right| \frac{1}{2!} \frac{1}{(0.8)^{3/2}} (0.2)^2$$

$$\leq \frac{1}{8} \frac{1}{(0.8)^{3/2}} 4 \times 10^{-2}.$$

The presence of the term $(0.8)^{3/2}$, which is < 1, in the denominator makes it slightly more cumbersome to estimate than in the preceding example, but even then it is not so difficult. Without exerting ourselves, to make the estimate simple we replace $3/2$ by 2. Then

$$\frac{1}{(0.8)^{3/2}} < \frac{1}{(0.8)^2} = \frac{1}{0.64} < \frac{10}{6} = \frac{5}{3}.$$

Hence

$$|R_2(-0.2)| < \frac{1}{8}\frac{5}{3}\, 4 \times 10^{-2} < 10^{-2}.$$

Remark. In the two cases of Example 1 and Example 2, we have met the case when $x > 0$ and $x < 0$. In the estimate for R_2, this gives rise to two different cases:

$$\frac{1}{(1 + c)^{3/2}} \leqq 1 \qquad\qquad \text{whenever} \quad x > 0 \quad\text{and}\quad 0 \leqq c \leqq x;$$

$$\frac{1}{(1 + c)^{3/2}} \leqq \frac{1}{(1 + x)^{3/2}} \qquad \text{whenever} \quad x < 0 \quad\text{and}\quad x \leqq c \leqq 0.$$

The second case is more annoying to treat.

In the preceding examples, we computed roots of $1 + x$ when x is small. To find roots of an arbitrary number we can often use a trick as in the next example, in order to reduce the problem to a root $(1 + x)^s$ with small x.

Example 3. Find the value $\sqrt{26}$ to two decimals.
For this we write

$$26 = 25 + 1 = 25(1 + \tfrac{1}{25}).$$

Then

$$\sqrt{26} = 5(1 + \tfrac{1}{25})^{1/2},$$

and we can apply the binomial Taylor formula to find

$$\left(1 + \frac{1}{25}\right)^{1/2} = 1 + \frac{1}{50} + R_2(x),$$

with $x = \tfrac{1}{25}$ and $s = \tfrac{1}{2}$. We are in the case $c \geqq 0$, so we get

$$\left|R_2\left(\frac{1}{25}\right)\right| \leqq \frac{1}{2}\frac{1}{2}\frac{1}{2}\left(\frac{1}{25}\right)^2 \leqq \frac{1}{8}\frac{1}{625} = \frac{1}{5000}.$$

Hence

$$\sqrt{26} = 5(1 + \tfrac{1}{50}) + 5R_2(\tfrac{1}{25}) = 5.1 + E$$

where

$$|E| = 5|R_2(\tfrac{1}{25})| \leq 10^{-3}.$$

Observe the factor 5 which appeared in the last step, and which multi- plies the estimate for $R_2(1/25)$. To get the final accuracy up to 10^{-3}, you needed an accuracy of $(1/5) \times 10^{-3}$ for $R_2(1/25)$ because of this factor 5.

An example involving R_3

Example 4. Find the value $\sqrt{26}$ to four decimals.
For this we write $26 = 25(1 + 1/25)$ as before. By the binomial Taylor formula, we find

$$\left(1 + \frac{1}{25}\right)^{1/2} = 1 + \frac{1}{2}\frac{1}{25} + \frac{1}{2}\left(\frac{1}{2} - 1\right)\left(\frac{1}{25}\right)^2 + R_3\left(\frac{1}{25}\right).$$

In estimating the remainder, we are in the case with $c \geq 0$, so

$$\left|R_3\left(\frac{1}{25}\right)\right| \leq \frac{1}{2}\left|\frac{1}{2} - 1\right|\left|\frac{1}{2} - 2\right|\frac{1}{3!}\left(\frac{1}{25}\right)^3$$

$$\leq \frac{1}{2}\frac{1}{2}\frac{3}{2}\frac{1}{3!}\left(\frac{1}{25}\right)^3$$

$$< \frac{1}{16}\frac{1}{1.5} \times 10^{-4}$$

$$\leq \frac{1}{24} \times 10^{-4}.$$

Then

$$26^{1/2} = 5(1 + 1/25)^{1/2}$$

$$= 5\left(1 + \frac{1}{50} - \frac{1}{4}\frac{1}{625}\right) + E,$$

where $E = 5R_3(1/25)$ and therefore

$$|E| \leq \frac{5}{24} \times 10^{-4} < 10^{-4}.$$

This is within the desired accuracy. Again note the factor of 5 in the last step.

The method in the above example was to find a perfect square near 26, and then use Taylor's formula for $(1 + x)^{1/2}$ with a small x. In general, one can use a similar method to find the square root of a number. Find a perfect square as close to the number as possible, and then use Taylor's formula. A similar technique works for cube roots or other roots.

The binomial expansion $(1 + x)^n$ and $(a + b)^n$

We conclude this section by showing how the binomial expansion for $(1 + x)^s$ using Taylor's formula becomes simpler when s is a positive integer n. So let n be a positive integer. Let

$$f(x) = (1 + x)^n.$$

We have no difficulty computing the derivatives:

$$f^{(1)}(x) = n(1 + x)^{n-1}, \qquad\qquad f^{(1)}(0) = n,$$

$$f^{(2)}(x) = n(n - 1)(1 + x)^{n-2}, \qquad f^{(2)}(0) = n(n - 1),$$

$$f^{(3)}(x) = n(n - 1)(n - 2)x^{n-3}, \qquad f^{(3)}(0) = n(n - 1)(n - 2),$$

$$\vdots \qquad\qquad\qquad\qquad \vdots$$

$$f^{(n)}(x) = n!, \qquad\qquad\qquad f^{(n)}(0) = n!$$

$$f^{(n+1)}(x) = 0. \qquad\qquad\qquad f^{(n+1)}(0) = 0.$$

The new feature here is that $f^{(n+1)}(x) = 0$. Therefore $f^{(k)}(x) = 0$ for all $k \geq n + 1$, and the remainder after the n-th term is equal to 0. Hence we get the exact expression:

Theorem 7.2. *Let n be a positive integer. For any number x, we have*

$$(1 + x)^n = 1 + nx + \frac{n(n - 1)}{2!} x^2 + \frac{n(n - 1)(n - 2)}{3!} x^3 + \cdots + x^n.$$

The coefficient of x^k on the right-hand side is usually denoted by the symbol

$$\binom{n}{k}$$

and is called a **binomial coefficient**. Thus from the derivatives which we found above, we get

$$\boxed{\binom{n}{k} = \frac{n(n - 1)(n - 2) \cdots (n - k + 1)}{k!}.}$$

The numerator consists of the product of integers in descending order, from n to $n - k + 1$. It differs from $n!$ in that the remaining product from $(n - k)$ down to 1 is missing. In order to have a more symmetric expression for the binomial coefficient, we multiply the numerator and denominator by $(n - k)!$. Observe that

$$n(n - 1)(n - 2) \cdots (n - k + 1)(n - k)! = n!$$

and consequently we can write the binomial coefficient in the form

$$\binom{n}{k} = \frac{n!}{k! \, (n - k)!}.$$

In this formula, we let $0 \leq k \leq n$, and **by convention we let**

$$0! = 1.$$

For example:

$$\binom{3}{0} = \frac{3!}{0! \, 3!} = 1, \qquad \binom{3}{1} = \frac{3!}{1! \, 2!} = 3,$$

$$\binom{3}{2} = \frac{3!}{2! \, 1!} = 3, \qquad \binom{3}{3} = \frac{3!}{3! \, 0!} = 1.$$

The integers 1, 3, 3, 1 are exactly the coefficients of the expansion for

$$(1 + x)^3 = 1 + 3x + 3x^2 + x^3.$$

In exercises, you can work out the coefficients for higher powers.

If we want the expansion of $(a + b)^n$ with arbitrary numbers a and b, and $a \neq 0$, then we let $x = b/a$, and so:

$$(a + b)^n = a^n \left(1 + \frac{b}{a}\right)^n = a^n \sum_{k=0}^{n} \binom{n}{k}\left(\frac{b}{a}\right)^k$$

$$= \sum_{k=0}^{n} \binom{n}{k} a^{n-k} b^k.$$

Thus we can write the binomial expansion in the form

$$\boxed{(a + b)^n = \sum_{k=0}^{n} \binom{n}{k} a^{n-k} b^k.}$$

We found this expansion in a fancy way, by using Taylor's formula in a case when the remainder is 0. In high school, the expansion should have been derived in a much more elementary way. The point is that we needed this more general technique here in order to compute values $(1 + x)^s$ with a more general exponent s which may not be an integer. For instance, we needed to compute

$$(1 + x)^{1/2} \quad \text{or} \quad (1 + x)^{1/3}.$$

Then we have to use the method of Taylor's formula.

Definition. Let s be any real number. We define the **binomial coefficient**

$$\binom{s}{k} = \frac{s(s - 1)(s - 2) \cdots (s - k + 1)}{k!}.$$

Example. Suppose that $s = 1/3$. Then

$$\binom{1/3}{2} = \frac{1}{3}\left(\frac{1}{3} - 1\right)\frac{1}{2!},$$

$$\binom{1/3}{3} = \frac{1}{3}\left(\frac{1}{3} - 1\right)\left(\frac{1}{3} - 2\right)\frac{1}{3!},$$

$$\binom{1/3}{4} = \frac{1}{3}\left(\frac{1}{3} - 1\right)\left(\frac{1}{3} - 2\right)\left(\frac{1}{3} - 3\right)\frac{1}{4!},$$

and so on.

Observe that when s is not an integer, then we cannot multiply numerator and denominator by $(s - k)!$ which does not make sense. We have to leave the binomial coefficient as in the definition.

Using the summation sign, we can also write

$$(1 + x)^s = \sum_{k=0}^{n} \binom{s}{k}x^k + R_{n+1}(x).$$

XIII, §7. EXERCISES

In each of the following cases, when asked to compute a number, include the estimate for the error term to show that it lies within the desired accuracy

1. Estimate the remainder R_2 in the Taylor series for $(1 + x)^{1/4}$:
   (a) when $x = 0.01$, (b) when $x = 0.2$, (c) when $x = 0.1$.

2. Estimate the remainder R_3 in the Taylor series for $(1 + x)^{1/2}$:
(a) when $x = 0.2$, (b) when $x = -0.2$, (c) when $x = 0.1$.

3. Estimate R_2 in the remainder of $(1 + x)^{1/3}$ for x lying in the interval

$$-0.1 \leqq x \leqq 0.1.$$

4. Estimate the remainder R_2 in the Taylor series of $(1 + x)^{1/2}$
(a) when $x = -0.2$, (b) when $x = 0.1$.

5. Compute the cube roots to 4 decimals:
(a) $\sqrt[3]{126}$,   (b) $\sqrt[3]{130}$,   (c) $\sqrt[3]{131}$,   (d) $\sqrt[3]{220}$.

6. Compute the square roots to 4 decimals:
(a) $\sqrt{97}$,   (b) $\sqrt{102}$,   (c) $\sqrt{105}$,   (d) $\sqrt{28}$.

XIII, §8. SOME LIMITS

Limits of quotients of functions can be reduced to limits of quotients of polynomials by using a few terms from Taylor's expansion.
Let us first look at polynomials.

Example 1. Find the limit

$$\lim_{x \to 0} \frac{3x - 2x^2 + 5x^4}{7x}.$$

We divide the numerator and denominator by the *lowest* power of x occurring in each, so that we find:

$$\frac{3x - 2x^2 + 5x^4}{7x} = \frac{x(3 - 2x + 5x^3)}{x \cdot 7}$$

$$= \frac{3 - 2x + 5x^3}{7}.$$

It is now easy to find the limit as x approaches 0; namely the limit is $\frac{3}{7}$.

Example 2. Find the limit

$$\lim_{x \to 0} \frac{\cos x - 1}{x^2}.$$

We replace $\cos x$ by $1 - x^2/2 + R_4(x)$, so that

$$\frac{\cos x - 1}{x^2} = \frac{-\frac{1}{2}x^2 + R_4(x)}{x^2} = -\frac{1}{2} + \frac{R_4(x)}{x^2}.$$

Since $|R_4(x)| \leq |x|^4$, it follows that the limit as $x \to 0$ is equal to $-\frac{1}{2}$.

Example 3. Find the limit

$$\lim_{x \to 0} \frac{\sin x - x + x^3/3!}{x^4}.$$

We have

$$\sin x = x - \frac{x^3}{3!} + R_5(x).$$

Hence

$$\sin x - x + \frac{x^3}{3!} = R_5(x) \qquad \text{and} \qquad |R_5(x)| \leq \frac{|x|^5}{5!}.$$

Hence

$$\left| \frac{\sin x - x + x^3/3!}{x^4} \right| \leq \frac{|R_5(x)|}{|x^4|} \leq \frac{|x|}{5!}.$$

The desired limit is therefore equal to 0.

Example 4. Find the limit

$$\lim_{x \to 0} \frac{\sin x^2}{x \tan x}.$$

To do this, we use the fact that

$$\lim_{u \to 0} \frac{\sin u}{u} = 1,$$

and put $u = x^2$. Also

$$\lim_{x \to 0} \frac{x}{\tan x} = \lim_{x \to 0} \frac{x}{\sin x} \cos x = 1.$$

Hence

$$\lim_{x \to 0} \frac{\sin x^2}{x \tan x} = \lim_{x \to 0} \frac{\sin x^2}{x^2} \frac{x}{\tan x} = 1.$$

Example 5. We want to find the limit

$$\lim_{x \to 0} \frac{\log(1 + x)}{\sin x}.$$

By the Taylor formula, we have

$$\log(1 + x) = x + R_2(x),$$
$$\sin x = x + S_3(x).$$

(We write S_3 instead of R_3 because it is a different remainder than for the log.) In each case, we have the estimate

$$|R_2(x)| \leq C|x|^2 \qquad \text{and} \qquad |S_3(x)| \leq C'|x|^3$$

for some constants C, C', and x sufficiently close to 0. Hence

$$\frac{\log(1 + x)}{\sin x} = \frac{x + R_2(x)}{x + S_3(x)}.$$

Dividing numerator and denominator by x shows that this is

$$= \frac{1 + R_2(x)/x}{1 + S_3(x)/x}.$$

As x approaches 0, each quotient $R_2(x)/x$ and $S_3(x)/x$ approach 0. Hence the limit is 1, as we wanted.

Example 6. Find the limit

$$\lim_{x \to 0} \frac{\sin x - e^x + 1}{x}.$$

Again we write the Taylor formula with a few terms:

$$\sin x = x + R_3(x),$$
$$e^x = 1 + x + S_2(x).$$

Then

$$\frac{\sin x - e^x + 1}{x} = \frac{x - 1 - x + 1 + R_3(x) - S_2(x)}{x}$$

$$= \frac{R_3(x) - S_2(x)}{x}.$$

The right-hand side approaches 0, and so the desired limit is 0.

XIII, §8. EXERCISES

Find the following limits as x approaches 0.

1. $\dfrac{\cos x - 1 + x^2/2!}{x^3}$

2. $\dfrac{\cos x - 1 + x^2/2!}{x^4}$

3. $\dfrac{\sin x + e^x - 1}{x}$

4. $\dfrac{\sin x - e^x + 1}{x}$

5. $\dfrac{e^x - 1}{x}$

6. $\dfrac{\sin (x^2)}{(\sin x)^2}$

7. $\dfrac{\tan x}{\sin x}$

8. $\dfrac{\arctan x}{x}$

9. $\dfrac{\log(1 + x)}{x}$

10. $\dfrac{\log(1 + 2x)}{x}$

11. $\dfrac{e^x - (1 + x)}{x^2}$

12. $\dfrac{\sin x - x}{x^2}$

13. $\dfrac{\cos x - 1}{x^2}$

14. $\dfrac{\log(1 + x^2)}{\sin(x^2)}$

15. $\dfrac{\tan(x^2)}{(\sin x)^2}$

16. $\dfrac{\log(1 + x^2)}{(\sin x)^2}$

17. $\dfrac{\sin x - e^x + 1}{x^2}$

18. $\dfrac{\cos x - e^x}{x}$

19. $\dfrac{e^x - e^{-x}}{x}$

20. $\dfrac{\sin x}{e^x - e^{-x}}$

21. $\dfrac{\sin^2 x}{\sin x^2}$

22. $\dfrac{\tan x^2}{\sin^2 x}$

23. $\dfrac{\log(1 - x)}{\sin x}$

24. $\dfrac{e^x + e^{-x} - 2}{x^2}$

25. $\dfrac{e^x + e^{-x} - 2}{x \sin x}$

26. $\dfrac{\sin x - x}{x^2}$

27. $\dfrac{\sin x - x}{x^3}$

28. $\dfrac{e^x - 1 - x}{x}$

29. $\dfrac{e^x - 1 - x}{x^2}$

30. $\dfrac{\log(1 + x^2)}{x^2}$

31. $\dfrac{(1 + x)^{1/2} - 1 - \frac{1}{2}x}{x^2}$

32. $\dfrac{(1 + x)^{1/3} - 1 - \frac{1}{3}x}{x^2}$

33. Let $f(x)$ be a function which has $n + 1$ continuous derivatives in an open interval containing the origin, and assume that the $(n + 1)$-th derivative is

bounded by a constant M on this interval. Let $P_n(x)$ be the Taylor polynomial of degree n for $f(x)$. What are the following limits?

(a) $\displaystyle\lim_{x \to 0} \frac{f(x) - P_n(x)}{x^2}$ (assuming $n \geq 3$)

(b) $\displaystyle\lim_{x \to 0} \frac{f(x) - P_n(x)}{x^n}$ (c) $\displaystyle\lim_{x \to 0} \frac{f(x) - P_n(x)}{x^{n-1}}$ (assuming $n \geq 2$)

Determine the following limits as x approaches 0.

34. $\dfrac{\sin x + \cos x - 1}{x}$

35. $\dfrac{\sin x + \cos x - 1 - x}{x^2}$

36. $\dfrac{\sin x - x + x^3/3!}{x^4}$

37. $\dfrac{\sin x - x + x^3/3!}{x^5}$

38. $\dfrac{\cos x - 1 - x^2/2!}{x}$

39. $\dfrac{\cos x - 1 - x^2/2!}{x^2}$

CHAPTER XIV

Series

Series are a natural continuation of our study of functions. In the preceding chapter we found how to approximate our elementary functions by polynomials, with a certain error term. Conversely, one can define arbitrary functions by giving a series for them. We shall see how in the sections below.

In practice, very few tests are used to determine convergence of series. Essentially, the comparison test is the most frequent. Furthermore, the most important series are those which converge absolutely. Thus we shall put greater emphasis on these.

XIV, §1. CONVERGENT SERIES

Suppose that we are given a sequence of numbers

$$a_1, a_2, a_3,..,$$

i.e. we are given a number a_n for each integer $n \geq 1$. (We picked the starting place to be 1, but we could have picked any integer.) We form the sums

$$s_n = a_1 + a_2 + \cdots + a_n.$$

It appears to be meaningless to form an infinite sum

$$a_1 + a_2 + a_3 + \cdots$$

because we do not know how to add infinitely many numbers. However, if our sums s_n approach a limit, as n becomes large, then we say that the sum of our sequence **converges**, and we now define its **sum** to be that limit.

The symbols

$$\sum_{n=1}^{\infty} a_n$$

will be called a **series**. We shall say that the **series converges** if the sums s_n approach a limit as n becomes large. Otherwise, we say that it does not converge, or **diverges**. If the series converges, we say that the value of the series is

$$\sum_{n=1}^{\infty} a_n = \lim_{n \to \infty} s_n = \lim_{n \to \infty} (a_1 + \cdots + a_n).$$

The symbols $\lim_{n \to \infty}$ are to be read: "The limit as n becomes large."

Example. Consider the sequence

$$1, \frac{1}{2}, \frac{1}{4}, \frac{1}{8}, \frac{1}{16}, \ldots,$$

and let us form the sums

$$S_n = 1 + \frac{1}{2} + \frac{1}{4} + \cdots + \frac{1}{2^n}.$$

You probably know already that these sums approach a limit and that this limit is 2. To prove it, let $r = \frac{1}{2}$. Then

$$(1 + r + r^2 + \cdots + r^n) = \frac{1 - r^{n+1}}{1 - r} = \frac{1}{1 - r} - \frac{r^{n+1}}{1 - r}.$$

As n becomes large, r^{n+1} approaches 0, whence our sums approach

$$\frac{1}{1 - \frac{1}{2}} = 2.$$

Actually, the same argument works if we take for r any number such that

$$-1 < r < 1.$$

In that case, r^{n+1} approaches 0 as n becomes large, and consequently we can write

$$\sum_{n=0}^{\infty} r^n = \frac{1}{1-r}.$$

Of course, if $|r| > 1$, then the series $\sum r^n$ does not converge. For instance, the partial sums of the series with $r = -3$ are

$$1 - 3 + 3^2 - 3^3 + \cdots + (-1)^n 3^n.$$

Observe that the n-th term $(-1)^n 3^n$ does not even approach 0 as n becomes large.

In view of the fact that the limit of a sum is the sum of the limits, and other standard properties of limits, we get the following theorem.

Theorem 1.1. *Let $\{a_n\}$ and $\{b_n\}$ $(n = 1, 2, \ldots)$ be two sequences and assume that the series*

$$\sum_{n=1}^{\infty} a_n \qquad and \qquad \sum_{n=1}^{\infty} b_n$$

converge. Then $\sum_{n=1}^{\infty} (a_n + b_n)$ also converges, and is equal to the sum of the two series. If c is a number, then

$$\sum_{n=1}^{\infty} c a_n = c \sum_{n=1}^{\infty} a_n.$$

Finally, if

$$s_n = a_1 + \cdots + a_n$$

and

$$t_n = b_1 + \cdots + b_n,$$

then

$$\sum_{n=1}^{\infty} a_n \sum_{n=1}^{\infty} b_n = \lim_{n \to \infty} s_n t_n.$$

In particular, series can be added term by term. Of course, they cannot be multiplied term by term!

We also observe that a similar theorem holds for the difference of two series.

If a series $\sum a_n$ converges, then the numbers a_n must approach 0 as n becomes large. However, there are examples of sequences $\{a_n\}$ for which the series does not converge, and yet

$$\lim_{n \to \infty} a_n = 0.$$

Consider, for instance,

$$1 + \frac{1}{2} + \frac{1}{3} + \cdots + \frac{1}{n} + \cdots.$$

We contend that the partial sums s_n become very large when n becomes large. To see this, we look at partial sums as follows:

$$1 + \frac{1}{2} + \underbrace{\frac{1}{3} + \frac{1}{4}} + \underbrace{\frac{1}{5} + \cdots + \frac{1}{8}} + \underbrace{\frac{1}{9} + \cdots + \frac{1}{16}} + \cdots.$$

In each bunch of terms as indicated, we replace each term by that farthest to the right. This makes our sums smaller. Thus our expression is

$$\geqq 1 + \frac{1}{2} + \underbrace{\frac{1}{4} + \frac{1}{4}} + \underbrace{\frac{1}{8} + \cdots + \frac{1}{8}} + \underbrace{\frac{1}{16} + \cdots + \frac{1}{16}} + \cdots$$

$$\geqq 1 + \frac{1}{2} + \frac{1}{2} + \frac{1}{2} + \frac{1}{2} + \cdots$$

and therefore becomes arbitrarily large when n becomes large.

XIV, §2. SERIES WITH POSITIVE TERMS

Throughout this section, we shall assume that our numbers a_n are $\geqq 0$. Then the partial sums

$$s_n = a_1 + \cdots + a_n$$

are increasing, i.e.

$$s_1 \leqq s_2 \leqq s_3 \leqq \cdots \leqq s_n \leqq s_{n+1} \leqq \cdots.$$

If they are to approach a limit at all, they cannot become arbitrarily large. Thus in that case there is a number B such that

$$s_n \leqq B$$

for all n. Such a number B is called an **upper bound**. By a **least upper bound** we mean a number S which is an upper bound, and such that every upper bound B is $\geq S$. We take for granted that a least upper bound exists. The collection of numbers $\{s_n\}$ has therefore a least upper bound, i.e. there is a smallest number S such that

$$s_n \leq S$$

for all n. In that case, the partial sums s_n approach S as a limit. In other words, given any positive number $\varepsilon > 0$, we have

$$S - \varepsilon \leq s_n \leq S$$

for all n sufficiently large.

This simply expresses the fact that S is the least of all upper bounds for our collection of numbers s_n. We express this as a theorem.

Theorem 2.1. *Let $\{a_n\}$ $(n = 1, 2, \dots)$ be a sequence of numbers ≥ 0 and let*

$$s_n = a_1 + \cdots + a_n.$$

If the sequence of numbers $\{s_n\}$ is bounded, then it approaches a limit S, which is its least upper bound.

Example 1. Prove that the series $\displaystyle\sum_{n=1}^{\infty} 1/n^2$ converges. Let us look at the series:

$$\frac{1}{1^2} + \frac{1}{2^2} + \frac{1}{3^2} + \frac{1}{4^2} + \cdots + \frac{1}{8^2} + \cdots + \frac{1}{16^2} + \cdots + \cdots.$$

We look at the groups of terms as indicated. In each group of terms, if we decrease the denominator in each term, then we increase the fraction. We replace 3 by 2, then 4, 5, 6, 7 by 4, then we replace the numbers from 8 to 15 by 8, and so forth. Our partial sums are therefore less than or equal to

$$1 + \frac{1}{2^2} + \frac{1}{2^2} + \frac{1}{4^2} + \cdots + \frac{1}{4^2} + \frac{1}{8^2} + \cdots + \frac{1}{8^2} + \cdots,$$

and we note that 2 occurs twice, 4 occurs four times, 8 occurs eight times, and so forth. Hence the partial sums are less than or equal to

$$1 + \frac{2}{2^2} + \frac{4}{4^2} + \frac{8}{8^2} + \cdots = 1 + \frac{1}{2} + \frac{1}{4} + \frac{1}{8} + \cdots.$$

Thus our partial sums are less than or equal to those of the geometric series and are bounded. Hence our series converges.

Theorem 2.1 gives us a very useful criterion to determine when a series with positive terms converges:

Theorem 2.2. *Let*

$$\sum_{n=1}^{\infty} a_n \quad and \quad \sum_{n=1}^{\infty} b_n$$

be two series, with $a_n \geqq 0$ for all n and $b_n \geqq 0$ for all n. Assume that there is a number $C > 0$ such that

$$a_n \leqq C b_n$$

for all n, and that $\sum_{n=1}^{\infty} b_n$ converges. Then $\sum_{n=1}^{\infty} a_n$ converges, and

$$\sum_{n=1}^{\infty} a_n \leqq C \sum_{n=1}^{\infty} b_n.$$

Proof. We have

$$a_1 + \cdots + a_n \leqq C b_1 + \cdots + C b_n = C(b_1 + \cdots + b_n) \leqq C \sum_{n=1}^{\infty} b_n.$$

This means that $C \sum_{n=1}^{\infty} b_n$ is a bound for the partial sums

$$a_1 + \cdots + a_n.$$

The least upper bound of these sums is therefore $\leqq C \sum_{n=1}^{\infty} b_n$, thereby proving our theorem.

Theorem 2.2 has an analogue to show that a series does not converge.

Theorem 2.2'. *Let*

$$\sum_{n=1}^{\infty} a_n \quad and \quad \sum_{n=1}^{\infty} b_n$$

be two series, with a_n and b_n and $b_n \geq 0$ for all n. Assume that there is a number $C > 0$ such that

$$a_n \geq Cb_n$$

for all n sufficiently large, and that $\sum\limits_{n=1}^{\infty} b_n$ does not converge. Then $\sum\limits_{n=1}^{\infty} a_n$ diverges.

Proof. Assume $a_n \geq Cb_n$ for $n \geq n_0$. Since $\sum b_n$ diverges, we can make the partial sums

$$\sum_{n=n_0}^{N} b_n = b_{n_0} + \cdots + b_N$$

arbitrarily large as N becomes arbitrarily large. But

$$\sum_{n=n_0}^{N} a_n \geq \sum_{n=n_0}^{N} Cb_n = C \sum_{n=n_0}^{N} b_n.$$

Hence the partial sums

$$\sum_{n=1}^{N} a_n = a_1 + \cdots + a_N$$

are arbitrarily large as N becomes arbitrarily large, and hence $\sum\limits_{n=1}^{\infty} a_n$ diverges, as was to be shown.

Example 2. Determine whether the series

$$\sum_{n=1}^{\infty} \frac{n^2}{n^3 + 1}$$

converges.
We write

$$\frac{n^2}{n^3 + 1} = \frac{1}{n + 1/n^2} = \frac{1}{n}\left(\frac{1}{1 + 1/n^3}\right).$$

Then we see that

$$\frac{n^2}{n^3 + 1} \geq \frac{1}{2n}.$$

Since $\sum 1/n$ does not converge, it follows that the series of Example 2 does not converge either.

Example 3. The series·

$$\sum_{n=1}^{\infty} \frac{n^2 + 7}{2n^4 - n + 3}$$

converges. Indeed, we can write

$$\frac{n^2 + 7}{2n^4 - n + 3} = \frac{n^2(1 + 7/n^2)}{n^4(2 - (1/n)^3 + 3/n^4)} = \frac{1}{n^2} \frac{1 + 7/n^2}{2 - (1/n)^3 + 3/n^4}.$$

For n large, the factor

$$\frac{1 + 7/n^2}{2 - (1/n)^3 + 3/n^4}$$

is certainly bounded, and in fact is near $\frac{1}{2}$. Hence we can compare our series with $1/n^2$ to see that it converges, because $\sum 1/n^2$ converges, and the factor is bounded.

XIV, §2. EXERCISES

1. Show that the series $\sum_{n=1}^{\infty} 1/n^3$ converges.

2. (a) Show that the series $\sum (\log n)/n^3$ converges. [*Hint*: Estimate $(\log n)/n.$]
   (b) Show that the series $\sum (\log n)^2/n^3$ converges.

Test the following series for convergence:

3. $\displaystyle\sum_{n=1}^{\infty} \frac{1}{n^{1/2}}$

4. $\displaystyle\sum_{n=1}^{\infty} \frac{n^2}{n^4 + n}$

5. $\displaystyle\sum_{n=1}^{\infty} \frac{n}{n + 1}$

6. $\displaystyle\sum_{n=1}^{\infty} \frac{n}{n + 5}$

7. $\displaystyle\sum_{n=1}^{\infty} \frac{n^2}{n^3 + n + 2}$

8. $\displaystyle\sum_{n=1}^{\infty} \frac{|\sin n|}{n^2 + 1}$

9. $\displaystyle\sum_{n=1}^{\infty} \frac{|\cos n|}{n^2 + n}$

XIV, §3. THE RATIO TEST

We continue to consider only series with terms ≥ 0. To compare such a series with a geometric series, the simplest test is given by the ratio test.

Ratio test. *Let* $\sum\limits_{n=1}^{\infty} a_n$ *be a series with* $a_n > 0$ *for all* n. *Assume that there is a number* c *with* $0 < c < 1$ *such that*

$$\frac{a_{n+1}}{a_n} \leqq c$$

for all n *sufficiently large. Then the series converges.*

Proof. Suppose that there exists some integer N such that

$$\frac{a_{n+1}}{a_n} \leqq c$$

if $n \geqq N$. Then

$$a_{N+1} \leqq c a_N,$$

$$a_{N+2} \leqq c a_{N+1} \leqq c^2 a_N$$

and in general, by induction,

$$a_{N+k} \leqq c^k a_N.$$

Thus

$$\sum_{n=N}^{N+k} a_n \leqq a_N + c a_N + c^2 a_N + \cdots + c^k a_N$$

$$\leqq a_N(1 + c + \cdots + c^k) \leqq a_N \frac{1}{1-c}.$$

Thus in effect, we have compared our series with a geometric series, and we know that the partial sums are bounded. This implies that our series converges.

The ratio test is usually used in the case of a series with positive terms a_n such that

$$\lim_{n \to \infty} \frac{a_{n+1}}{a_n} = c < 1.$$

Example. Show that the series

$$\sum_{n=1}^{\infty} \frac{n}{3^n}$$

converges.

We let $a_n = n/3^n$. Then

$$\frac{a_{n+1}}{a_n} = \frac{n+1}{3^{n+1}} \frac{3^n}{n} = \frac{n+1}{n} \frac{1}{3}.$$

This ratio approaches $\frac{1}{3}$ as $n \to \infty$, and hence the ratio test is applicable: the series converges.

XIV, §3. EXERCISES

Determine whether the following series converge:

1. $\sum n2^{-n}$

2. $\sum n^2 2^{-n}$

3. $\sum \dfrac{1}{\log n}$

4. $\sum \dfrac{\log n}{2^n}$

5. $\sum \dfrac{\log n}{n}$

6. $\sum \dfrac{n^{10}}{3^n}$

7. $\sum \dfrac{1}{\sqrt{n(n+1)}}$

8. $\sum \dfrac{\sqrt{n^3 + 1}}{e^n}$

9. $\sum \dfrac{n+1}{\sqrt{n^4 + n + 1}}$

10. $\sum \dfrac{n+1}{2^n}$

11. $\sum \dfrac{n}{(4n-1)(n+15)}$

12. $\sum \dfrac{1 + \cos(\pi n/2)}{e^n}$

13. $\sum \dfrac{1}{(\log n)^{10}}$

14. $\sum n^2 e^{-n^2}$

15. $\sum n^2 e^{-n^3}$

16. $\sum n^5 e^{-n^2}$

17. $\sum n^4 e^{-n}$

18. $\sum \dfrac{n^n}{n! 3^n}$

19. Let $\{a_n\}$ be a sequence of positive numbers, and assume that

$$\frac{a_{n+1}}{a_n} \geq 1 - \frac{1}{n}$$

for all n. Show that the series $\sum a_n$ diverges.

20. A ratio test can be applied in the opposite direction to determine when a series diverges. Prove the following statement: Let a_n be a sequence of positive numbers, and let $c \geq 1$. If $a_{n+1}/a_n \geq c$ for all n sufficiently large, then the series $\sum a_n$ diverges.

XIV, §4. THE INTEGRAL TEST

You must already have felt that there is an analogy between the convergence of an improper integral and the convergence of a series. We shall now make this precise.

Theorem 4.1. *Let f be a function which is defined and positive for all* $x \geq 1$, *and decreasing. The series*

$$\sum_{n=1}^{\infty} f(n)$$

converges if and only if the improper integral

$$\int_{1}^{\infty} f(x)\, dx$$

converges.

We visualize the situation in the following diagram.

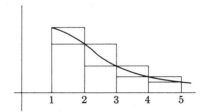

Consider the partial sums

$$f(2) + \cdots + f(n)$$

and assume that our improper integral converges. The area under the curve between 1 and 2 is greater than or equal to the area of the rectangle whose height is $f(2)$ and whose base is the interval between 1 and 2. This base has length 1. Thus

$$f(2) \leq \int_{1}^{2} f(x)\, dx.$$

Again, since the function is decreasing, we have a similar estimate between 2 and 3:

$$f(3) \leq \int_{2}^{3} f(x)\, dx.$$

We can continue up to n, and get

$$f(2) + f(3) + \cdots + f(n) \leq \int_{1}^{n} f(x)\, dx.$$

As n becomes large, we have assumed that the integral approaches a limit. This means that

$$f(2) + f(3) + \cdots + f(n) \leqq \int_1^\infty f(x)\, dx.$$

Hence the partial sums are bounded, and hence by Theorem 2.1, they approach a limit. Therefore our series converges.

Conversely, assume that the partial sums

$$f(1) + \cdots + f(n)$$

approach a limit as n becomes large.

The area under the graph of f between 1 and n is less than or equal to the sum of the areas of the big rectangles. Thus

$$\int_1^2 f(x)\, dx \leqq f(1)(2 - 1) = f(1)$$

and

$$\int_2^3 f(x)\, dx \leqq f(2)(3 - 2) = f(2).$$

Proceeding stepwise, and taking the sum, we see that

$$\int_1^n f(x)\, dx \leqq f(1) + \cdots + f(n - 1).$$

The partial sums on the right are less than or equal to their limit. Call this limit L. Then for all positive integers n, we have

$$\int_1^n f(x)\, dx \leq L.$$

Given any number B, we can find an integer n such that $B \leqq n$. Then

$$\int_1^B f(x)\, dx \leqq \int_1^n f(x)\, dx \leqq L.$$

Hence the integral from 1 to B approaches a limit as B becomes large, and this limit is less than or equal to L. This proves our theorem.

Example. Prove that the series

$$\sum \frac{1}{n^2 + 1}$$

converges.

Let

$$f(x) = \frac{1}{x^2 + 1}.$$

Then f is decreasing, and

$$\int_1^B f(x)\,dx = \arctan B - \arctan 1 = \arctan B - \frac{\pi}{4}.$$

As B becomes large, $\arctan B$ approaches $\pi/2$ and therefore has a limit. Hence the integral converges. So does the series, by the theorem.

XIV, §4. EXERCISES

1. Show that the following series diverges: $\sum\limits_{n=2}^{\infty} 1/(n \log n)$.

2. Show that the following series converges: $\sum\limits_{n=1}^{\infty} (n + 1)/((n + 2)n!)$.

Test for convergence:

3. $\sum\limits_{n=1}^{\infty} n e^{-n^2}$

4. $\sum\limits_{n=2}^{\infty} \frac{1}{n(\log n)^3}$

5. $\sum\limits_{n=2}^{\infty} \frac{1}{n(\log n)^2}$

6. $\sum\limits_{n=1}^{\infty} \frac{n!}{n^n}$

7. $\sum\limits_{n=1}^{\infty} \frac{n}{e^n}$

8. $\sum\limits_{n=1}^{\infty} \frac{n+1}{n^3 + 2}$

9. $\sum\limits_{n=1}^{\infty} \frac{1}{n^2 + n - 1}$

10. $\sum\limits_{n=1}^{\infty} \frac{n}{n^3 - n + 5}$

11. Let ε be a number > 0. Show that the series $\sum\limits_{n=1}^{\infty} 1/n^{1+\varepsilon}$ converges.

12. Show that the following series converge.

(a) $\sum\limits_{n=1}^{\infty} \frac{\log n}{n^2}$

(b) $\sum\limits_{n=1}^{\infty} \frac{\log n}{n^{3/2}}$

(c) $\sum\limits_{n=1}^{\infty} \frac{\log n}{n^{1+\varepsilon}}$ if $\varepsilon > 0$.

(d) $\sum\limits_{n=1}^{\infty} \frac{(\log n)^2}{n^{3/2}}$

(e) $\sum\limits_{n=1}^{\infty} \frac{(\log n)^3}{n^2}$

13. If $\varepsilon > 0$ show that the series $\sum\limits_{n=2}^{\infty} 1/n(\log n)^{1+\varepsilon}$ converges.

XIV, §5. ABSOLUTE AND ALTERNATING CONVERGENCE

We consider a series $\sum\limits_{n=1}^{\infty} a_n$ in which we do not assume that the terms a_n are $\geqq 0$. We shall say that the series **converges absolutely** if the series

$$\sum_{n=1}^{\infty} |a_n|$$

formed with the absolute values of the terms a_n converges. This is now a series with terms $\geqq 0$, to which we can apply the tests for convergence given in the two preceding sections. This is important, because we have:

Theorem 5.1. *Let $\{a_n\}$ $(n = 1, 2,\ldots)$ be a sequence, and assume that the series*

$$\sum_{n=1}^{\infty} |a_n|$$

converges. Then so does the series $\sum\limits_{n=1}^{\infty} a_n$.

Proof. Let a_n^+ be equal to 0 if $a_n < 0$ and equal to a_n itself if $a_n \geqq 0$. Let a_n^- be equal to 0 if $a_n > 0$ and equal to $-a_n$ if $a_n \leqq 0$. Then both a_n^+ and a_n^- are $\geqq 0$. By assumption and comparison with $\sum |a_n|$, we see that each one of the series

$$\sum_{n=1}^{\infty} a_n^+ \qquad \text{and} \qquad \sum_{n=1}^{\infty} a_n^-$$

converges. Hence so does their difference

$$\sum_{n=1}^{\infty} a_n^+ - \sum_{n=1}^{\infty} a_n^-,$$

which is equal to

$$\sum_{n=1}^{\infty} (a_n^+ - a_n^-),$$

which is none other than $\sum\limits_{n=1}^{\infty} a_n$. This proves our theorem.

We shall use one more test for convergence of a series which may have positive and negative terms.

Theorem 5.2. *Let* $\sum\limits_{n=1}^{\infty} a_n$ *be a series such that*

$$\lim_{n \to \infty} a_n = 0,$$

such that the terms a_n *are alternately positive and negative, and such that* $|a_{n+1}| \leq |a_n|$ *for* $n \geq 1$. *Then the series is convergent.*

Proof. Let us write the series in the form

$$b_1 - c_1 + b_2 - c_2 + b_3 - c_3 + \cdots,$$

with $b_n, c_n \geq 0$. Let

$$s_n = b_1 - c_1 + b_2 - c_2 + \cdots + b_n,$$
$$t_n = b_1 - c_1 + b_2 - c_2 + \cdots + b_n - c_n.$$

Since the absolute values of the terms decrease, it follows that

$$s_1 \geq s_2 \geq s_3 \geq \cdots \qquad \text{and} \qquad t_1 \leq t_2 \leq t_3 \leq \cdots,$$

i.e. that the s_n are decreasing and the t_n are increasing. Indeed,

$$s_{n+1} = s_n - c_n + b_{n+1} \qquad \text{and} \qquad 0 \leq b_{n+1} \leq c_n.$$

Thus we subtract more from s_n by c_n than we add afterwards by b_{n+1}. Hence $s_n \geq s_{n+1}$. Furthermore, $s_n \geq t_n$. Hence we may visualize our sequences as follows:

$$s_n \geq s_{n+1} \geq \cdots \geq t_{n+1} \geq t_n.$$

Note that $s_n - t_n = c_n$, and that c_n approaches 0 as n becomes large. If we let L be the greatest lower bound for the sequence $\{s_n\}$, and M be the least upper bound for the sequence $\{t_n\}$, then

$$s_n \geq L \geq M \geq t_n$$

for all n. Since the difference $s_n - t_n$ becomes arbitrarily small, it follows that $L - M$ is arbitrarily small, and hence equal to 0. Thus $L = M$, and this proves that s_n and t_n approach L as a limit, whence our series $\sum a_n$ converges to L.

Example. The series

$$\sum_{n=1}^{\infty} (-1)^{n+1} \frac{1}{n} = 1 - \frac{1}{2} + \frac{1}{3} - \frac{1}{4} + \frac{1}{5} - \frac{1}{6} + \cdots$$

is convergent, but not absolutely convergent.

Remark. Not all series which are convergent are either absolutely convergent, or are of the above alternating type. However, these two kinds of series are the ones that arise most frequently in practice, so that we have laid emphasis on them.

XIV, §5. EXERCISES

Determine whether the following series converge absolutely:

1. $\sum \dfrac{\sin n}{n^3}$

2. $\sum \dfrac{1 + \cos \pi n}{n!}$

3. $\sum \dfrac{\sin \pi n + \cos 2\pi n}{n^{3/2}}$

4. $\sum \dfrac{(-1)^n}{n^2 + 1}$

5. $\sum \dfrac{(-1)^n + \cos 3n}{n^2 + n}$

Which of the following series converge and which do not?

6. $\sum \dfrac{(-1)^n}{n}$

7. $\sum \dfrac{(-1)^n}{n^2}$

8. $\sum (-1)^n \dfrac{1}{n + 1}$

9. $\sum \dfrac{(-1)^{n+1}}{\log (n + 2)}$

10. For each number x, show that the series $\sum (\sin n^2 x)/n^2$ converges absolutely. Let f be the function whose value at x is the above series. Show that f is continuous. Determine whether f is differentiable or not. (Remarkably enough, this was not known for a long time! Cf. J. P. Kahane, Bulletin of the American Mathematical Society, March 1964, p. 199. J. Gerver, a sophomore at Columbia College, showed that the series is differentiable at all points $m\pi/n$, where m, n are odd integers, that the derivative is $-\frac{1}{2}$, and that these are the only points where the function is differentiable. Cf. his articles in *Am. J. Math.*, 1970 and 1971.)

Determine whether the following series converge, and whether they converge absolutely.

11. $\sum \dfrac{(-1)^{n+1}}{\sqrt{n}}$

12. $\sum \dfrac{(-1)^{n+2}}{\log n}$

13. $\sum \dfrac{(-1)^n}{\sqrt[n]{n}}$

14. $\sum (-1)^n \dfrac{n^2}{n^2 + 1}$

15. $\sum (-1)^n \dfrac{n^2}{n^3 + 2}$

16. $\sum (-1)^{n+1} \dfrac{n^2 + 2}{n^3 + n - 1}$

17. $\sum (-1)^{n+1} \dfrac{\sqrt{n}}{n + 2}$

18. $\sum \dfrac{(-1)^n}{n^{5/2} + n}$

19. $\sum (-1)^n \dfrac{n}{n^2 + 1}$

20. $\sum \dfrac{(-1)^n}{\sqrt{\log n}}$

XIV, §6. POWER SERIES

Perhaps the most important series are power series. Let x be any number and let $\{a_n\}$ $(n = 0, 1, \ldots)$ be a sequence of numbers. Then we can form the series

$$\sum_{n=1}^{\infty} a_n x^n.$$

The partial sums are

$$a_0 + a_1 x + a_2 x^2 + \cdots + a_n x^n.$$

We have already met such sums when we discussed Taylor's formula.

Example. The power series

$$\sum_{n=1}^{\infty} \frac{x^n}{n!}$$

converges for all x, absolutely. Indeed, it will suffice to prove that for any number $R > 0$, the above series converges for $0 < x \leq R$. We use the ratio test. Let

$$b_n = \frac{x^n}{n!}.$$

Then

$$\frac{b_{n+1}}{b_n} = \frac{x^{n+1}}{(n+1)!} \frac{n!}{x^n} = \frac{x}{n+1} \leq \frac{R}{n+1}.$$

When n is sufficiently large, it follows that $R/(n+1)$ is small, and in particular is $< \frac{1}{2}$, so that we can apply the ratio test to prove our assertion.

Similarly, we could prove that the series

$$\sum_{n=1}^{\infty} (-1)^n \frac{x^{2n+1}}{(2n+1)!} \quad \text{and} \quad \sum_{n=1}^{\infty} (-1)^n \frac{x^{2n}}{(2n)!}$$

converge absolutely for all x, letting for instance

$$b_n = \frac{x^{2n+1}}{(2n+1)!}$$

for the first one. Then $b_{n+1}/b_n = x^2/[(2n+3)(2n+2)]$, and we can argue as before.

Theorem 6.1. *Assume that there is a number* $r \geq 0$ *such that the series*

$$\sum_{n=1}^{\infty} |a_n| r^n$$

converges. Then for all x *such that* $|x| \leq r$, *the series*

$$\sum_{n=1}^{\infty} a_n x^n$$

converges absolutely.

Proof. The absolute value of each term is

$$|a_n||x|^n \leq |a_n| r^n.$$

Our assertion follows from the comparison Theorem 2.2.

The least upper bound of all numbers r for which we have the convergence stated in the theorem is called the **radius of convergence** of the series. If there is no upper bound for the numbers r such that the power series above converges, then we say that the radius of convergence is **infinity**.

Suppose that there is an upper bound for the numbers r above, and thus let s be the radius of convergence of the series. Then if $x > s$, the series

$$\sum_{n=1}^{\infty} |a_n| x^n$$

does *not* converge. Thus the radius of convergence s is the number such that the series converges absolutely if $0 < x < s$ but does not converge absolutely if $x > s$.

Theorem 6.1 allows us to define a function f; namely, for all numbers x such that $|x| < r$, we define

$$f(x) = \lim_{n \to \infty} (a_0 + a_1 x + \cdots + a_n x^n).$$

Our proofs that the remainder term in Taylor's formula approaches 0 for various functions now allow us to say that these functions are given by their Taylor series. Thus

$$\sin x = x - \frac{x^3}{3!} + \frac{x^5}{5!} - \cdots,$$

$$e^x = 1 + x + \frac{x^2}{2!} + \cdots$$

for all x. Furthermore,

$$\log(1 + x) = x - \frac{x^2}{2} + \cdots$$

is valid for $-1 < x < 1$.

(Here we saw that the series converges for $x = 1$, but it does not converge absolutely, cf. §1.)

However, we can now define functions at random by means of a power series, provided we know the power series converges absolutely, for $|x| < r$.

The ratio test usually gives an easy way to determine when a power series converges, or when it diverges.

Example. Prove that the series

$$\sum_{n=2}^{\infty} \frac{\log n}{n^2} x^n$$

converges absolutely for $|x| < 1$, and diverges for $|x| > 1$.

Let $0 < c < 1$ and consider x such that $0 < x \leqq c$. Let

$$b_n = \frac{\log n}{n^2} x^n.$$

Then

$$\frac{b_{n+1}}{b_n} = \frac{\log(n + 1)}{(n + 1)^2} x^{n+1} \frac{n^2}{\log n} \frac{1}{x^n} = \frac{\log(n + 1)}{\log n} \left(\frac{n}{n + 1}\right)^2 x.$$

Since $\log(n + 1)/\log n$ and $(n/(n + 1))^2$ approach 1 when n becomes very large, it follows that if $c < c_1 < 1$, then for all n sufficiently large

$$\frac{b_{n+1}}{b_n} \leq c_1$$

and hence our series converges. This is true for every c such that $0 < c < 1$, and hence the series converges absolutely for $|x| < 1$.

Let $c > 1$. If $x \geq c$ then for all n sufficiently large, it follows that $b_{n+1}/b_n \geq 1$, whence the series does not converge. This is so for all $c > 1$, and hence the series does not converge if $x > 1$. Hence 1 is the radius of convergence.

If a power series converges absolutely only for $x = 0$, then we agree to say that its radius of convergence is 0. For example, the radius of convergence of the series

$$\sum_{n=1}^{\infty} n! x^n$$

is equal to 0, as one sees by using the ratio test in the divergent case.

Root test. *Let $\sum a_n x^n$ be a power series and assume that*

$$\lim |a_n|^{1/n} = s,$$

where s is a number. If $s \neq 0$ then the radius of convergence of the series is equal to $1/s$. If $s = 0$, then the radius of convergence is infinity. If $|a_n|^{1/n}$ becomes arbitrarily large as n becomes large, then the radius of convergence is 0.

Proof. Without loss of generality, we may assume that $a_n \geq 0$ for all n. Suppose first that s is a number $\neq 0$, and let $0 \leq r < 1/s$. Then $sr < 1$. The numbers $a_n^{1/n} r$ approach sr and hence there is some number $\varepsilon > 0$ such that

$$a_n^{1/n} r < 1 - \varepsilon$$

for all n sufficiently large. Hence the series $\sum a_n r^n$ converges by comparison with the geometric series. If on the other hand $r > 1/s$, then $a_n^{1/n} r$ approaches $sr > 1$, and hence we have

$$a_n^{1/n} r \geq 1 + \varepsilon$$

for all n sufficiently large. Comparison from below shows that the series $\sum a_n r^n$ diverges. We leave the cases $s = 0$ or $s = \infty$ to the reader.

Example. The series $\sum x^n/n^2$ has a radius of convergence equal to 1, because

$$\lim\left(\frac{1}{n^2}\right)^{1/n} = \lim \frac{1}{n^{2/n}} = 1,$$

by Corollary 5.5 of Chapter VIII.

To give other examples, we recall an inequality which you should have worked out in Chapter X, §1, namely

$$\boxed{(n-1)! \leq n^n e^{-n} e \leq n!.}$$

From it, we prove:

As n becomes large, the expression

$$\frac{(n!)^{1/n}}{n} = \left[\frac{n!}{n^n}\right]^{1/n}$$

approaches $1/e$.

Proof. Take the n-th root of the inequality $n^n e^{-n} e \leq n!$. We get

$$ne^{-1}e^{1/n} \leq (n!)^{1/n}.$$

Dividing by n yields

$$\frac{1}{e}\,e^{1/n} \leq \frac{(n!)^{1/n}}{n}.$$

On the other hand, multiply both sides of the inequality

$$(n-1)! \leq n^n e^{-n} e$$

by n. We get $n! \leq n^n e^{-n} en$. Take an n-th root:

$$(n!)^{1/n} \leq ne^{-1}e^{1/n}n^{1/n}.$$

Dividing by n yields

$$\frac{(n!)^{1/n}}{n} \leq \frac{1}{e}\,e^{1/n}n^{1/n}.$$

But we know that both $n^{1/n}$ and $e^{1/n}$ approach 1 as n becomes large. Thus our quotient is squeezed between two numbers approaching $1/e$, and must therefore approach $1/e$.

Example. We have

$$\lim_{n \to \infty} \left[\frac{(3n)!}{n^{3n}} \right]^{1/n} = \frac{27}{e^3}.$$

Proof. We write

$$\frac{(3n)!}{n^{3n}} = \frac{(3n)!}{(3n)^{3n}} 3^{3n}.$$

The $3n$-th root of this expression is

$$\left[\frac{(3n)!}{(3n)^{3n}} \right]^{1/3n} 3.$$

We have seen that

$$\left(\frac{m!}{m^m} \right)^{1/m} \qquad \text{approaches} \qquad \frac{1}{e}$$

as m becomes large. We use $m = 3n$. We conclude that

$$\left(\frac{(3n)!}{n^{3n}} \right)^{1/3n} \qquad \text{approaches} \qquad \frac{3}{e}.$$

Hence

$$\left(\frac{(3n)!}{n^{3n}} \right)^{1/n} \qquad \text{approaches} \qquad \frac{27}{e^3},$$

as desired.

XIV, §6. EXERCISES

1. Use the abbreviation $\lim_{n \to \infty}$ to mean: limit as n becomes very large. Prove that

   (a) $\lim_{n \to \infty} \left[\frac{(3n)!}{n^{3n}} \right]^{1/n} = \frac{27}{e^3}$ (b) $\lim_{n \to \infty} \left[\frac{(3n)!}{n! n^{2n}} \right]^{1/n} = \frac{27}{e^2}$

2. Find the limit:

   (a) $\lim_{n \to \infty} \left[\frac{(2n)!}{n^{2n}} \right]^{1/n}$ (b) $\lim_{n \to \infty} \left[\frac{(2n)!(5n)!}{n^{4n}(3n)!} \right]^{1/n}$

Find the radius of convergence of the following series:

3. (a) $\sum n^n x^n$ \qquad\qquad (b) $\sum \dfrac{x^n}{n^n}$

4. $\sum \dfrac{n}{n+5} x^n$ \qquad 5. $\sum (\log n) x^n$ \qquad 6. $\sum \dfrac{1}{\log n} x^n$

7. $\sum (\log n)^2 x^n$ \qquad 8. $\sum 2^n x^n$ \qquad 9. $\sum 2^{-n} x^n$

10. $\sum (1+n)^n x^n$ \qquad 11. $\sum \dfrac{x^n}{n}$ \qquad 12. $\sum \dfrac{x^n}{\sqrt{n}}$

13. $\sum (1+(-2)^n) x^n$ \qquad 14. $\sum (1+(-1)^n) x^n$ \qquad 15. $\sum_{n=1}^{\infty} \dfrac{(2n)!}{(n!)^2} x^n$

16. $\sum_{n=1}^{\infty} \dfrac{n^n}{n!} x^n$ \qquad 17. $\sum_{n=1}^{\infty} \dfrac{(n!)^3}{(3n)!} x^n$ \qquad 18. $\sum_{n=1}^{\infty} \dfrac{n^{5n}}{(2n)! n^{3n}} x^n$

19. $\sum_{n=1}^{\infty} \dfrac{(3n)!}{(n!)^2} x^n$ \qquad 20. $\sum_{n=1}^{\infty} \dfrac{\sin n\pi/2}{2^n} x^n$ \qquad 21. $\sum_{n=1}^{\infty} \dfrac{\log n}{2^n} x^n$

22. $\sum_{n=2}^{\infty} \dfrac{1+\cos 2\pi n}{3n} x^n$ \qquad 23. $\sum_{n=2}^{\infty} n x^n$ \qquad 24. $\sum_{n=1}^{\infty} \dfrac{\sin 2\pi n}{n!} x^n$

25. $\sum_{n=2}^{\infty} n^2 x^n$ \qquad 26. $\sum_{n=1}^{\infty} \dfrac{\cos n^2}{n^n} x^n$ \qquad 27. $\sum_{n=2}^{\infty} \dfrac{n}{\log n} x^n$

28. $\sum_{n=2}^{\infty} \dfrac{(-1)^n}{n!-1} x^n$ \qquad 29. $\sum_{n=1}^{\infty} \dfrac{n!}{n^n} x^n$ \qquad 30. $\sum_{n=1}^{\infty} \dfrac{(-1)^n+1}{n!} x^n$

Note: For some of the above radii of convergence, recall that

$$\lim_{n\to\infty} \left(1+\frac{1}{n}\right)^n = \lim_{n\to\infty} \left(\frac{1+n}{n}\right)^n = e.$$

XIV, §7. DIFFERENTIATION AND INTEGRATION OF POWER SERIES

If we have a polynomial

$$a_0 + a_1 x + \cdots + a_n x^n$$

with numbers $a_0, a_1, \ldots, a_n$ as coefficients, then we know how to find its derivative. It is $a_1 + 2a_2 x + \cdots + n a_n x^{n-1}$. We would like to say that the derivative of a series can be taken in the same way, and that the derivative converges whenever the series does.

Theorem 7.1. *Let r be a number > 0 and let $\sum a_n x^n$ be a series which converges absolutely for $|x| < r$. Then the series $\sum n a_n x^{n-1}$ also converges absolutely for $|x| < r$.*

Proof. Since we are interested in the absolute convergence, we may assume that $a_n \geq 0$ for all n. Let $0 < x < r$, and let c be a number such that $x < c < r$. Recall that

$$\lim_{n \to \infty} n^{1/n} = 1.$$

We may write

$$n a_n x^n = a_n (n^{1/n} x)^n.$$

Then for all n sufficiently large, we conclude that

$$n^{1/n} x < c$$

because $n^{1/n} x$ comes arbitrarily close to x. Hence for all n sufficiently large, we have

$$n a_n x^n < a_n c^n.$$

We can then compare the series $\sum n a_n x^n$ with $\sum a_n c^n$ to conclude that $\sum n a_n x^n$ converges. Since

$$\sum n a_n x^{n-1} = \frac{1}{x} \sum n a_n x^n$$

we have proved Theorem 7.1.

A similar result holds for integration, but trivially. Indeed, if we have a series $\sum_{n=1}^{\infty} a_n x^n$ which converges absolutely for $|x| < r$, then the series

$$\sum_{n=1}^{\infty} \frac{a_n}{n+1} x^{n+1} = x \sum_{n=1}^{\infty} \frac{a_n}{n+1} x^n$$

has terms whose absolute value is smaller than in the original series.

The preceding results can be expressed by saying that an absolutely convergent power series can be integrated and differentiated term by term and still yield an absolutely convergent power series.

It is natural to expect that if

$$f(x) = \sum_{n=1}^{\infty} a_n x^n,$$

then f is differentiable and its derivative is given by differentiating the series term by term. The next theorem proves this.

Theorem 7.2. *Let*

$$f(x) = \sum_{n=1}^{\infty} a_n x^n$$

be a power series, which converges absolutely for $|x| < r$. Then f is differentiable for $|x| < r$, and

$$f'(x) = \sum_{n=1}^{\infty} n a_n x^{n-1}.$$

Proof. Let $0 < b < r$. Let $\delta > 0$ be such that $b + \delta < r$. We consider values of x such that $|x| < b$ and values of h such that $|h| < \delta$. We have the numerator of the Newton quotient:

$$f(x + h) - f(x) = \sum_{n=1}^{\infty} a_n(x + h)^n - \sum_{n=1}^{\infty} a_n x^n = \sum_{n=1}^{\infty} a_n[(x + h)^n - x^n].$$

By mean value theorem, there exists a number x_n between x and $x + h$ such that

$$(x + h)^n - x^n = n x_n^{n-1} h$$

and consequently

$$f(x + h) - f(x) = \sum_{n=1}^{\infty} n a_n x_n^{n-1} h.$$

Therefore

$$\frac{f(x + h) - f(x)}{h} = \sum_{n=1}^{\infty} n a_n x_n^{n-1}.$$

We have to show that the Newton quotient above approaches the value of the series obtained by taking the derivative term by term. We have

$$\frac{f(x + h) - f(x)}{h} - \sum_{n=1}^{\infty} n a_n x^{n-1} = \sum_{n=1}^{\infty} n a_n x_n^{n-1} - \sum_{n=1}^{\infty} n a_n x^{n-1}$$

$$= \sum_{n=1}^{\infty} n a_n[x_n^{n-1} - x^{n-1}].$$

Using the mean value theorem again, there exists y_n between x_n and x such that the preceding expression is

$$\frac{f(x + h) - f(x)}{h} - \sum_{n=1}^{\infty} na_n x^{n-1} = \sum_{n=2}^{\infty} (n - 1)na_n y_n^{n-2}(x_n - x).$$

We have $|y_n| \leq b + \delta < r$, and $|x_n - x| \leq |h|$. Consequently,

$$\left| \frac{f(x + h) - f(x)}{h} - \sum_{n=1}^{\infty} na_n x^{n-1} \right| \leq \sum_{n=2}^{\infty} (n - 1)n|a_n||y_n|^{n-2}|h|$$

$$\leq |h| \sum_{n=2}^{\infty} (n - 1)n|a_n|(b + \delta)^{n-2}.$$

By Theorem 7.1 applied twice, we know that the series appearing on the right converges. It is equal to a fixed constant. As h approaches 0, it follows that the expression on the right approaches 0, so the expression on the left also approaches 0. This proves that f is differentiable at x, and that its derivative is equal to $\sum_{n=1}^{\infty} na_n x^{n-1}$, for all x such that $|x| < b$. This is true for all b, $0 < b < r$, and therefore concludes the proof of our theorem.

Theorem 7.3. *Let* $f(x) = \sum_{n=1}^{\infty} a_n x^n$ *be a power series, which converges absolutely for* $|x| < r$. *Then the relation*

$$\int f(x) \, dx = \sum_{n=1}^{\infty} \frac{a_n}{n + 1} x^{n+1}$$

is valid in the interval $|x| < r$.

Proof. We know that the series for f integrated term by term converges absolutely in the interval. By the preceding theorem, its derivative term by term is the series for the derivative of the function, thereby proving our assertion.

Example. If we had never heard of the exponential function, we could define a function

$$f(x) = 1 + x + \frac{x^2}{2!} + \frac{x^3}{3!} + \cdots.$$

Taking the derivative term by term, we see that

$$f'(x) = f(x).$$

Hence by what we know from Chapter VIII, §1, Exercise 8, we conclude that

$$f(x) = Ke^x$$

for some constant K. Letting $x = 0$ shows that

$$1 = f(0) = K.$$

Thus $K = 1$ and $f(x) = e^x$.

Similarly, if we had never heard of sine and cosine, we could **define** functions

$$S(x) = x - \frac{x^3}{3!} + \frac{x^5}{5!} - \cdots, \qquad C(x) = 1 - \frac{x^2}{2!} + \frac{x^4}{4!} - \cdots.$$

Differentiating term by term shows that

$$S'(x) = C(x), \qquad C'(x) = -S(x).$$

Furthermore, $S(0) = 0$ and $C(0) = 1$. It can then be shown easily that any pair of functions $S(x)$ and $C(x)$ satisfying these properties must be the sine and cosine.

XIV, §7. EXERCISES

1. Verify in detail that differentiating term by term the series for the sine and cosine given at the end of the section yields

$$S'(x) = C(x) \qquad \text{and} \qquad C'(x) = -S(x).$$

2. Let

$$f(x) = \sum_{n=0}^{\infty} \frac{x^{2n}}{(2n)!}.$$

   Prove that $f''(x) = f(x)$.

3. Let

$$f(x) = \sum_{n=0}^{\infty} \frac{x^{2n}}{(n!)^2}.$$

   Prove that

$$x^2 f''(x) + x f'(x) = 4x^2 f(x).$$

4. Let

$$f(x) = x - \frac{x^3}{3} + \frac{x^5}{5} - \frac{x^7}{7} + \cdots.$$

   Show that $f'(x) = 1/(1 + x^2)$.

5. Let

$$J(x) = \sum_{n=0}^{\infty} \frac{(-1)^n}{(n!)^2} \left(\frac{x}{2}\right)^{2n}.$$

Prove that

$$x^2 J''(x) + x J'(x) + x^2 J(x) = 0.$$

6. For any positive integer k, let

$$J_k(x) = \sum_{n=0}^{\infty} \frac{(-1)^n}{n!(n+k)!} \left(\frac{x}{2}\right)^{2n+k}.$$

Prove that

$$x^2 J_k''(x) + x J_k'(x) + (x^2 - k^2) J_k(x) = 0.$$

APPENDIX
To the First Four Parts

ε and δ

This appendix is intended to show how the notions of limits and the properties of limits can be explained and proved in terms of the notions and properties of numbers. We therefore assume the latter and carry out the proofs from there.

There remains the problem of showing how the real numbers can be defined in terms of the rational numbers, and the rational numbers in terms of integers. This takes too long to be included in this book.

Aside from the ordinary rules for addition, multiplication, subtraction, division (by non-zero numbers), ordering, positivity, and inequalities, there is one more basic property satisfied by the real numbers. This property is stated in §1. Our proofs then use only these properties.

Warning. The level of abstract understanding and use of language needed to master this appendix is considerably higher than for the rest of the book. We are involved in "proving" properties which are intuitively very clear. Hence you should not take this appendix too seriously unless you are theoretically inclined, or you wish to have an introduction to some essential tools of analysis, i.e. a first acquaintance, which will plant some ideas in your head for future use in higher courses, when the techniques described here become essential because more intricate estimates are needed when dealing with such higher analysis. It is useful to have seen the stuff previously, even though you may not have mastered it the first time. It is part of our psychology that we learn by approximation. Furthermore, knowledge at one level is fully mastered only when you use it at the next level of depth. Hence studying harder things, even though you have a limited understanding of them, makes it possible to understand fully the easier things.

Nevertheless, this appendix should be omitted under ordinary circumstances.

APP., §1. LEAST UPPER BOUND

We meet again the problem of where to jump into the theory. It would be long and tedious to jump in too early. Hence we assume known the contents of Chapter I, §1 and §2. These involve the ordinary operations of addition and multiplication, and the notion of ordering, positivity, negative numbers, and inequalities. Those who are interested in seeing the logical development of these notions are referred to books on analysis.

A collection of numbers will simply be called a **set** of numbers. This is shorter and is the usual terminology. If a set has at least one number in it, we say that it is **non-empty**. A set S' is called a **subset** of S if every element of S' is an element of S. In other words, if S' is part of S.

Let S be a non-empty set of numbers. We shall say that S is **bounded from above** if there exists a number B such that

$$x \leq B$$

for all x in our set S. We then call B an **upper bound** for S.

A **least upper bound** for S is an upper bound L such that any upper bound B for S satisfies the inequality $B \geq L$. If M is another least upper bound, then we have $M \geq L$ and $L \geq M$, whence $L = M$. Consequently, a least upper bound is unique.

Similarly, we define the notions of bounded from below, and of greatest lower bound. (Do it yourself.)

We shall now give examples, assuming that the reader has an intuitive notion of the real numbers. They make our meaning clearer, but we do not give proofs. Although this is the reverse of the logical order, it is the appropriate psychological order.

Example. The set of positive integers $\{1, 2, 3, ..\}$ is not bounded from above. It is bounded from below. The number 1 is a greatest lower bound.

Example. Let S be the set of numbers x such that $0 \leq x$ and $x^2 < 2$. This set is bounded from below, for instance by 0; it is also bounded from above, and 2 is certainly an upper bound. As a matter of fact, $\sqrt{2}$ is the least upper bound. Note that the least upper bound does not lie in the set S, that is, is not an element of S.

Example. Let T be the set of numbers x such that $0 \leq x$ and $x^2 \leq 2$. Again, $\sqrt{2}$ is the least upper bound of T, and is an element of T. It should be intuitively clear that T differs from S only in that it contains the additional element $\sqrt{2}$.

Example. Let U be the set of all numbers $1/n$, where n ranges over the positive integers. Thus U consists of $\{1, \frac{1}{2}, \frac{1}{3}, \ldots\}$. Then U is bounded. The number 1 is its least upper bound, and lies in U. The number 0 is the greatest lower bound, and does not lie in U.

The real numbers satisfy a property which is not satisfied by the set of rational numbers, namely:

Fundamental property. *Every non-empty set S of numbers which is bounded from above has a least upper bound. Every non-empty set of numbers S which is bounded from below has a greatest lower bound.*

Proposition 1.1. *Let a be a number such that*

$$0 \leq a < \frac{1}{n}$$

for every positive integer n. Then $a = 0$. There is no number b such that $b \geq n$ for every positive integer n.

Proof. Suppose there is a number $a \neq 0$ such that $a < 1/n$ for every positive integer n. Then $n < 1/a$ for every positive integer n. Thus to prove both our assertions, it will suffice to prove the second.

Suppose there is a number b such that $b \geq n$ for every positive integer n. Let S be the set of positive integers. Then S is bounded, and hence has a least upper bound. Let C be this least upper bound. No number strictly less than C can be an upper bound. Since $0 < 1$, we have $C < C + 1$, whence $C - 1 < C$. Hence there is a positive integer n such that

$$C - 1 < n.$$

This implies that $C < n + 1$ and $n + 1$ is a positive integer. We have contradicted our assumption that C is an upper bound for the set of postive integers, so no such upper bound can exist.

Observe that Proposition 1.1 proves that the set of positive integers is not bounded from above. It is reasonable to ask if this sort of obvious property really needs the least upper bound axiom to be proved, and the answer is *yes*. One can construct systems satisfying all the ordinary rules for addition, multiplication, division (by non-zero elements), inequalities, such that the least upper bound axiom is not satisfied, and such that there exists an element t in the system with the property that $n < t$ for all positive integers n. We don't want to make this appendix too long, and we won't go into the construction of such systems, but it is probably illuminating for you to have it confirmed that the least upper bound property was needed in the proof of Proposition 1.1.

APP., §1. EXERCISES

Determine in each case whether the set is bounded from above, from below, and describe the least upper bound and greatest lower bound if it exists, without giving proofs, just using your intuition of numbers.

1. (a) The set of all positive even integers.
   (b) The set of all positive odd integers.
   (c) The set of all rational numbers.

2. (a) The set of all numbers x such that $0 \leq x$ and $x^3 < 5$.
   (b) The set of all numbers x such that $0 \leq x$ and $x^3 \leq 5$.
   (c) The set of all numbers x such that $x^2 \leq 4$.
   (d) The set of all numbers x such that $2x - 7 < 4$.

3. Prove that there exists a positive integer N such that if n is an integer $\geq N$ then $3n > 150$.

4. Let B be a positive number. Prove that there exists a positive integer N such that if n is an integer $\geq N$ then $5n > B$.

5. Let S be the set of numbers x such that $0 \leq x$ and $x^2 \leq 2$. Prove that the least upper bound of S is a number b such that $b^2 = 2$. [*Hint*: Prove that $b^2 > 2$ and $b^2 < 2$ are impossible.]

APP., §2. LIMITS

Let S be a set of numbers and let f be a function defined for all numbers in S. Let x_0 be a number. We shall assume that S is **arbitrarily close to** x_0, i.e. given $\varepsilon > 0$ there exists an element x of S such that $|x - x_0| < \varepsilon$. Let L be a number. We shall say that $f(x)$ **approaches the limit** L **as** x **approaches** x_0 if the following condition is satisfied:

Given a number $\varepsilon > 0$, there exists a number $\delta > 0$ such that for all x in S satisfying

$$|x - x_0| < \delta$$

we have

$$|f(x) - L| < \varepsilon.$$

If that is the case, then we write

$$\lim_{x \to x_0} f(x) = L.$$

We could also rephrase this as follows. We write

$$\lim_{h \to 0} f(x_0 + h) = L$$

and say that **the limit of** $f(x_0 + h)$ **is** L **as** h **approaches** 0 if the following condition is satisfied:

Given $\varepsilon > 0$, there exists $\delta > 0$ such that whenever h is a number with $|h| < \delta$ and $x_0 + h$ in S, then

$$|f(x_0 + h) - L| < \varepsilon.$$

We note that our definition of limit depends on the set S on which f is defined. Thus we should say "limit with respect to S." The next proposition shows that this is really unnecessary.

Proposition 2.1. *Let S be a set of numbers arbitrarily close to x_0 and let S' be a subset of S, also arbitrarily close to x_0. Let f be a function defined on S. If*

$$\lim_{x \to x_0} f(x) = L \qquad \text{(with respect to S)},$$

$$\lim_{x \to x_0} f(x) = M \qquad \text{(with respect to S')},$$

then $L = M$. In particular, the limit is unique.

Proof. Given $\varepsilon > 0$, there exists $\delta_1 > 0$ such that whenever x is in S' and $|x - x_0| < \delta_1$ we have

$$|f(x) - L| < \frac{\varepsilon}{2},$$

and there exists $\delta_2 > 0$ such that whenever $|x - x_0| < \delta_2$ then

$$|f(x) - M| < \frac{\varepsilon}{2}.$$

Let $\delta = \min(\delta_1, \delta_2)$. If $|x - x_0| < \delta$ then

$$|L - M| \leq |L - f(x) + f(x) - M| < \frac{\varepsilon}{2} + \frac{\varepsilon}{2} = \varepsilon.$$

Hence $|L - M|$ is less than any $\varepsilon > 0$, and by Proposition 1.1, we must have $|L - M| = 0$, whence

$$L - M = 0$$

and

$$L = M.$$

In practice from now on, we omit to state that x is an element of S. The context makes it clear each time.

Furthermore, in many subsequent proofs, we shall need several simultaneous inequalities to be satisfied, just as in the preceding proof we had inequalities with δ_1 and δ_2. In each case, we use a similar trick, letting δ be the minimum of $\delta_1, \delta_2, \delta_3, \ldots$ needed to make each desired inequality valid. Thus in writing down the proofs, we omit the intermediate $\delta_1, \delta_2, \delta_3 \ldots$.

Remark. Suppose that $\lim\limits_{x \to x_0} f(x) = L$. Then there exists $\delta > 0$ such that whenever $|x - x_0| < \delta$ we have

$$|f(x)| < |L| + 1.$$

Indeed, given $1 > 0$ there exists δ such that whenever $|x - x_0| < \delta$ we have

$$|f(x) - L| < 1,$$

so that our assertion follows from standard properties of inequalities.

Also, note that we have trivially

$$\lim\limits_{x \to x_0} C = C$$

for any number C, viewed as a constant function on S. Indeed, given $\varepsilon > 0$,

$$|C - C| < \varepsilon.$$

Remark. We mention a word about limits "**when x becomes large.**" Let a be a number and f a function defined for all numbers $x \geq a$. Let L be a number. We shall say that $f(x)$ **approaches L as x becomes large,** and we write

$$\lim\limits_{x \to \infty} f(x) = L$$

if the following condition is satisfied. Given $\varepsilon > 0$ there exists a number A such that whenever $x > A$ we have

$$|f(x) - L| < \varepsilon.$$

In practice, instead of saying "when x becomes large," we sometimes say "when x approaches ∞." We leave it to you to define the analogous notion "when x becomes large negative," or "x approaches $-\infty$."

In the definition of $\lim\limits_{x \to \infty}$ we took a function f defined for $x \geq a$. If $a_1 > a$, and we restrict the function to all numbers $\geq a_1$, then the limit as x becomes very large will be the same.

Let us suppose that $a \geq 1$. Define a function g for values of x such that

$$0 < x \leq 1/a$$

by the condition

$$g(x) = f(1/x).$$

Then a second's thought will allow you to prove that

$$\lim_{x \to 0} g(x)$$

exists if and only if

$$\lim_{x \to \infty} f(x)$$

exists, and that they are equal.

Consequently all properties which we prove concerning limits as x approaches 0 (or a number) immediately give rise to similar properties concerning limits as x becomes very large. We leave their formulations to you.

An important case occurs when the function is defined for the positive integers. Then it is called a **sequence**. A sequence of numbers is usually denoted by

$$\{a_1, a_2, \ldots, a_n, \ldots\}$$

or simply $\{a_n\}$.

Example. Let $a_n = f(n) = (-1)^n$. Then

$$a_1 = -1, \qquad a_2 = 1, \qquad a_3 = -1, \qquad a_4 = 1,$$

and so forth. Observe that numbers of the sequence indexed by different integers, for instance a_2 and a_4, may be equal.

Example. Let $a_n = 2n$. This defines the sequence of positive even integers.

Example. Let $a_n = 2n + 1$. This defines the sequence of odd positive integers ≥ 3.

Example. We have $\lim_{n \to \infty} 1/nx = 0$ for any number $x \neq 0$. To prove this, say $x > 0$. Given ε, let N be a positive integer such that $1/N < \varepsilon x$. If $n \geq N$, then

$$\frac{1}{n} \leq \frac{1}{N} < \varepsilon x,$$

and therefore $1/nx < \varepsilon$. This proves our assertion when $x > 0$. The proof when $x < 0$ is similar. Carry it out completely.

The next theorems, concerning the basic properties of limits, describe limits of sums, products, quotients, inequalities, and composite functions.

Theorem 2.2. *Let S be a set of numbers, and f, g two functions defined for all numbers in S. Let x_0 be a number. If*

$$\lim_{x \to x_0} f(x) = L$$

and

$$\lim_{x \to x_0} g(x) = M,$$

then $\lim_{x \to x_0} (f + g)(x)$ exists and is equal to $L + M$.

Proof. Given $\varepsilon > 0$, there exists $\delta > 0$ such that, whenever $|x - x_0| < \delta$ (and x is in S), we have

$$|f(x) - L| < \frac{\varepsilon}{2},$$

$$|g(x) - M| < \frac{\varepsilon}{2}.$$

We observe that

$$|f(x) + g(x) - L - M| \leq |f(x) - L| + |g(x) - M| < \varepsilon.$$

This proves that $L + M$ is the limit of $(f + g)(x)$ as x approaches x_0.

Theorem 2.3. *Let S be a set of numbers, and f, g two functions defined for all numbers in S. Let x_0 be a number. If*

$$\lim_{x \to x_0} f(x) = L$$

and

$$\lim_{x \to x_0} g(x) = M,$$

then $\lim_{x \to x_0} f(x)g(x)$ *exists and is equal to LM.*

Proof. Given $\varepsilon > 0$ there exists $\delta > 0$ such that, whenever $|x - x_0| < \delta$, we have

$$|f(x) - L| < \frac{1}{2} \frac{\varepsilon}{|M| + 1},$$

$$|g(x) - M| < \frac{1}{2} \frac{\varepsilon}{|L| + 1},$$

$$|f(x)| < |L| + 1.$$

We have

$$
\begin{aligned}
|f(x)g(x) - LM| &= |f(x)g(x) - f(x)M + f(x)M - LM| \\
&\leq |f(x)g(x) - f(x)M| + |f(x)M - LM| \\
&\leq |f(x)||g(x) - M| + |f(x) - L||M| \\
&< (|L| + 1)\frac{1}{2}\frac{\varepsilon}{|L| + 1} + \frac{1}{2}\frac{\varepsilon}{|M| + 1}|M| \\
&\leq \frac{\varepsilon}{2} + \frac{\varepsilon}{2} \\
&\leq \varepsilon.
\end{aligned}
$$

Corollary 2.4. *Let C be a number and let the assumptions be as in the theorem. Then*

$$\lim_{x \to x_0} Cf(x) = CL.$$

Proof. Clear.

Corollary 2.5. *Let the notation be as in Theorem 2.3. Then*

$$\lim_{x \to x_0} [f(x) - g(x)] = L - M.$$

Proof. Clear.

Theorem 2.6. *Let S be a set of numbers, and f a function defined for all numbers in S. Let x_0 be a number. If*

$$\lim_{x \to x_0} f(x) = L$$

and $L \neq 0$, then the limit

$$\lim_{x \to x_0} \frac{1}{f(x)}$$

exists and is equal to $1/L$.

Proof. Given $\varepsilon > 0$, there exists $\delta > 0$ such that whenever $|x - x_0| < \delta$ we have

$$|f(x) - L| < \frac{|L|}{2}$$

and also

$$|f(x) - L| < \frac{\varepsilon |L|^2}{2}.$$

From the first inequality, we get

$$|f(x)| \geq |L| - \frac{|L|}{2} = \frac{|L|}{2}.$$

In particular, $f(x) \neq 0$ when $|x - x_0| < \delta$. For such x we get

$$\left| \frac{1}{f(x)} - \frac{1}{L} \right| = \frac{|L - f(x)|}{|f(x)L|}$$

$$\leq \frac{2}{|L|} \frac{|L - f(x)|}{|L|}$$

$$< \frac{2}{|L|^2} \frac{\varepsilon |L|^2}{2} = \varepsilon.$$

Corollary 2.7. *Let the hypotheses be as in Theorem 2.3, and assume that $L \neq 0$. Then*

$$\lim_{x \to x_0} \frac{g(x)}{f(x)}$$

exists and is equal to M/L.

Proof. Use Theorem 2.3 and Theorem 2.6.

Theorem 2.8. *Let S be a set of numbers, and f a function on S. Let x_0 be a number. Let g be a function on S such that $g(x) \leq f(x)$ for all x in S. Assume that*

$$\lim_{x \to x_0} f(x) = L \qquad and \qquad \lim_{x \to x_0} g(x) = M.$$

Then $M \leq L$.

Proof. Let $\phi(x) = f(x) - g(x)$. Then $\phi(x) \geq 0$ for all x is S. Also,

$$\lim_{x \to x_0} \phi(x) = L - M$$

by Corollary 2.5. Let K be this limit. We must show $K \geq 0$. Suppose $K < 0$. Then $-K > 0$ and $|K| = -K$. Given $\varepsilon > 0$ there exists $\delta > 0$ such that whenever $|x - x_0| < \delta$ we have

$$|\phi(x) - K| < \varepsilon,$$

whence

$$\phi(x) - K < \varepsilon.$$

Since $\phi(x) \geq 0$, we get $-K < \varepsilon$ for all $\varepsilon > 0$. In particular, for all positive integers n we get $-K < 1/n$. But $-K > 0$. This contradicts Proposition 1.1.

Theorem 2.9. *Let the notation be as in Theorem 2.8 and assume that $M = L$. Let ψ be a function on S such that*

$$g(x) \leq \psi(x) \leq f(x)$$

for all x in S. Then

$$\lim_{x \to x_0} \psi(x)$$

exists and is equal to L (or M).

Proof. Given $\varepsilon > 0$ there exists $\delta > 0$ such that whenever $|x - x_0| < \delta$ we have

$$|g(x) - L| < \frac{\varepsilon}{4},$$

$$|f(x) - L| < \frac{\varepsilon}{4}.$$

We also have

$$|f(x) - \psi(x)| \leq |f(x) - g(x)|$$
$$\leq |f(x) - L + L - g(x)|$$
$$\leq |f(x) - L| + |L - g(x)|$$
$$< \frac{\varepsilon}{2}.$$

But

$$|L - \psi(x)| \leq |L - f(x)| + |f(x) - \psi(x)|$$
$$< \frac{\varepsilon}{2} + \frac{\varepsilon}{2} = \varepsilon.$$

Theorem 2.10. *Let S, T be sets of numbers, and let f, g be functions defined on S and T respectively. Let x_0 be arbitrarily close to S. Assume that for all x in S we have $f(x)$ in T, so that $g \circ f$ is defined. Assume that*

$$\lim_{x \to x_0} f(x)$$

exists and is equal to a number y_0 arbitrarily close to T. Assume that

$$\lim_{y \to y_0} g(y)$$

exists and equals L. Then

$$\lim_{x \to x_0} g(f(x)) = L.$$

Proof. Give ε there exists δ such that whenever y is in T and

$$|y - y_0| < \delta$$

we have

$$|g(y) - L| < \varepsilon.$$

With the above δ being given, there exists δ_1 such that whenever x is in S and $|x - x_0| < \delta_1$ we have $|f(x) - y_0| < \delta$. From this it follows that

$$|g(f(x)) - L| < \varepsilon$$

whence proving our assertion.

[*Note*: Theorem 2.10 justifies the limit procedure used to prove the chain rule.]

This completely proves all the statements about limits we made in Chapter III.

APP., §2. EXERCISES

1. Let g be a bounded function on a set of numbers S. Let f be a function on S such that

$$\lim_{x \to x_0} f(x) = 0.$$

   Prove that $\lim_{x \to x_0} f(x)g(x) = 0$.

2. For an arbitrary number x, let

$$f(x) = \lim_{n \to \infty} \frac{1}{1 + nx}.$$

   Find the limit explicitly; prove all assertions you make.

3. (a) Let $b > 1$, and write $b = 1 + c$ with $c > 0$. Prove: Given a positive number B, there exists a positive integer N such that if $n \geq N$, then $b^n > B$.

   (b) Let $0 < x < 1$. Prove that

$$\lim_{n \to \infty} x^n = 0.$$

   (c) If $-1 < x < 0$, is the limit as in (b) still 0? If yes, give a proof. Look at what happens with an example, i.e. write down the values x^n when $x = -1/2$ and $n = 1, 2, 3, 4, 5, 6, 7$.

4. For which numbers x does the following limit exist:

$$f(x) = \lim_{n \to \infty} \frac{x^n}{1 + x^n}?$$

   Give explicitly the values $f(x)$ for the various x for which the limit exists.

5. For $x \neq -1$, prove that the following limit exists:

$$f(x) = \lim_{n \to \infty} \frac{x^n - 1}{x^n + 1}.$$

   (a) What is $f(1)$, $f(\frac{1}{2})$, $f(2)$?
   (b) What is $\lim_{x \to 1} f(x)$?
   (c) What is $\lim_{x \to -1} f(x)$?

6. Answer the same questions as in Exercise 5 if

$$f(x) = \lim_{n \to \infty} \left(\frac{x^n - 1}{x^n + 1}\right)^2,$$

and $x \neq -1$.

7. Find the following limits, as $n \to \infty$.

(a) $\dfrac{1}{\sqrt{n}}$ (b) $\dfrac{3}{n^2}$ (c) $\dfrac{5}{n^{1/4}}$ (d) $\dfrac{1}{\sqrt{n+1}}$

8. Find the following limits.

(a) $\sqrt{n+1} - \sqrt{n}$ (b) $\sqrt{n+2} - \sqrt{n}$ (c) $\sqrt{n-5} - \sqrt{n}$

[*Hint*: Rationalize the "numerator."]

APP., §3. POINTS OF ACCUMULATION

A **sequence** is a function defined on a set of integers ≥ 0. Usually, this set consists of all positive integers. In that case, a sequence amounts to giving numbers

$$a_1, a_2, a_3, \ldots$$

for each positive integer, and we denote the sequence by

$$\{a_n\} \ (n = 1, 2, \ldots).$$

If the set consists of all integers ≥ 0, then we denote the sequence by $\{a_n\} \ (n = 0, 1, 2, \ldots)$.

Let $\{a_n\} \ (n = 1, 2, \ldots)$ be a sequence. Let C be a number. We say that C is a **point of accumulation** of the sequence if given $\varepsilon > 0$ there exist infinitely many integers n such that

$$|a_n - C| < \varepsilon.$$

Let $\{a_n\} \ (n = 1, 2, \ldots)$ be a sequence, and L a number. We shall say that L is a **limit of the sequence** if given $\varepsilon > 0$ there exists an integer N such that for all $n > N$ we have

$$|a_n - L| < \varepsilon.$$

The limit is then unique (same type of proof as we had for limits of functions).

We shall say that the sequence $\{a_n\} \ (n = 1, 2, \ldots)$ is **increasing** if $a_n \leq a_{n+1}$ for all positive integers n.

Theorem 3.1. *Let $\{a_n\}$ $(n = 1, 2, \ldots)$ be an increasing sequence, and assume that it is bounded from above. Then the least upper bound L is a limit of the sequence.*

Proof. Given $\varepsilon > 0$ the number $L - (\varepsilon/2)$ is not an upper bound for the sequence. Hence there exists some number a_N such that

$$L - \frac{\varepsilon}{2} \leqq a_N.$$

This inequality is also satisfied for all $n > N$, since the sequence is increasing. But

$$a_n \leqq L$$

because L is an upper bound. Thus

$$|L - a_n| = L - a_n \leqq \frac{\varepsilon}{2} < \varepsilon$$

for all $n > N$, thereby proving our assertion.

Corollary 3.2. *Let $\{a_n\}$ $(n = 1, 2, \ldots)$ be a sequence, and let A, B be two numbers such that $A \leqq a_n \leqq B$ for all positive integers n. Then there exists a point of accumulation C of the sequence between A and B.*

Proof. For each integer n we let b_n be the greatest lower bound of the set of numbers $\{a_n, a_{n+1}, a_{n+2}, \ldots\}$. Then $b_n \leqq b_{n+1} \leqq \cdots$, i.e. $\{b_n\}$ $(n = 1, 2, \ldots)$ is an increasing sequence, and B is an upper bound. Let L be its limit, as in Theorem 3.1. We leave it to you as an exercise to prove that this limit is a point of accumulation.

One can reduce the notion of limit of a sequence to that of limits defined previously.

Let S be the set of numbers

$$1, \quad \frac{1}{2}, \quad \frac{1}{3}, \quad \ldots, \quad \frac{1}{n}, \quad \ldots,$$

i.e. the set of numbers which can be written as $1/n$, where n is a positive integer.

If $\{a_n\}$ $(n = 1, 2, \ldots)$ is a sequence, we let f be a function defined by S by the rule

$$f\!\left(\frac{1}{n}\right) = a_n.$$

Then you will verify immediately that

$$\lim_{n \to \infty} a_n$$

exists if and only if

$$\lim_{h \to 0} f(h)$$

exists, and in that case the two limits are equal. We say that a sequence $\{a_n\}$ **approaches a number L when n becomes large if**

$$L = \lim_{n \to \infty} a_n.$$

Thus properties concerning limits in the sense of §2 immediately give rise to properties concerning limits of sequences (for instance, limits of sums, products, quotients). We leave their translations to you.

APP., §3. EXERCISES

1. Let $\{I_n\}$ be a sequence of closed intervals, say $I_n = [a_n, b_n]$, where $[a, b]$ means the set of numbers x such that $a \leqq x \leqq b$. Suppose that the left-hand points of this sequence of intervals increase, that is

$$a_1 \leqq \cdots \leqq a_n \leqq a_{n+1} \leqq \cdots$$

and that the right-hand points decrease (that is $b_{n+1} \leqq b_n$ for all positive integers n). Let $L(I_n)$ be the length of the interval I_n, that is

$$L(I_n) = b_n - a_n.$$

If

$$\lim_{n \to \infty} L(I_n) = 0,$$

prove that there exists a point c in each interval I_n such that

$$\lim_{n \to \infty} a_n = \lim_{n \to \infty} b_n$$

$$= c.$$

2. Let c_n be an element of I_n in Exercise 1. Under the same hypothesis as in Exercise 1, prove that

$$\lim_{n \to \infty} c_n = c.$$

APP., §4. CONTINUOUS FUNCTIONS

Let f be a function defined on a set of numbers S. Let x_0 be a number in S. Then S is arbitrarily close to x_0. We say that f is **continuous** at x_0 if

$$\lim_{x \to x_0} f(x) = f(x_0).$$

Note that there may be two numbers a, b with $a < x_0 < b$ such that x_0 is the only point which is in the interval and lies also in S. (In this case, one could say that x_0 is an **isolated** point of S.)

If follows at once from our definition that if $\{a_n\}$ $(n = 1, 2, \ldots)$ is a sequence of numbers in S such that

$$\lim_{n \to \infty} a_n = x_0$$

and f is continuous at x_0, then

$$\lim_{n \to \infty} f(a_n) = f(x_0).$$

It is immediate that the sum, product, quotient of continuous functions are again continuous. (In the quotient, we have to assume that $f(x_0) \neq 0$, of course.) Every constant function is continuous. The function $f(x) = x$ is continuous for all x. This is trivially verified. From the quotient theorem, we see that the function

$$f(x) = \frac{1}{x}$$

(defined for $x \neq 0$) is continuous.

Theorem 4.1. *Let f and g be continuous functions such that the values of f are contained in the domain of definition of g. Then $g \circ f$ is continuous.*

Proof. Let x_0 be a number at which f is defined, and let

$$y_0 = f(x_0).$$

Given $\varepsilon > 0$, since g is continuous at y_0, there exists $\delta_1 > 0$ such that if $|y - y_0| < \delta_1$, then

$$|g(y) - g(y_0)| < \varepsilon.$$

Now with the above δ_1 being given, there exists $\delta > 0$ such that if $|x - x_0| < \delta$ then

$$|f(x) - f(x_0)| < \delta_1.$$

Hence

$$|g(f(x)) - g(f(x_0))| < \varepsilon,$$

thus proving our theorem.

Theorem 4.2. *Let f be a continuous function on a closed interval $a \leqq x \leqq b$. Then there exists a point c in the interval such that $f(c)$ is a maximum, and there exists a point d in the interval such that $f(d)$ is a minimum.*

Proof. We shall first prove that f is bounded, i.e. that there exists a number M such that $|f(x)| \leq M$ for all x in the interval.

If f is not bounded, then for every positive integer n we can find a number x_n in the interval such that $|f(x_n)| > n$. The sequence of such x_n has a point of accumulation C in the interval. We have

$$|f(x_n) - f(C)| \geqq |f(x_n)| - |f(C)|$$

$$\geqq n - f(C).$$

Given $\varepsilon > 0$, there exists a $\delta > 0$ such that, whenever

$$|x_n - C| < \delta$$

we have $|f(x_n) - f(C)| < \varepsilon$. This has to happen for infinitely many n, since C is an accumulation point. Our statements are contradictory, and we therefore conclude that the function is bounded.

Let β be the least upper bound of the set of values $f(x)$ for all x in the interval. Then given a positive integer n, we can find a number z_n in the interval such that

$$|f(z_n) - \beta| < \frac{1}{n}.$$

Let c be a point of accumulation of the sequence of numbers

$$\{z_n\} \qquad (n = 1, 2, \ldots).$$

Then $f(c) \leqq \beta$. We contend that

$$f(c) = \beta$$

(this will prove our theorem).

Given $\varepsilon > 0$, there exists $\delta > 0$ such that whenever $|z_n - c| < \delta$ we have

$$|f(z_n) - f(c)| < \varepsilon.$$

This happens for infinitely many n, since c is a point of accumulation of the sequence $\{z_n\}$. But

$$|f(c) - \beta| \leq |f(c) - f(z_n)| + |f(z_n) - \beta|$$

$$< \varepsilon + \frac{1}{n}.$$

This is true for every ε and infinitely many positive integers n. Hence $|f(c) - \beta| = 0$ and $f(c) = \beta$.

The proof for the minimum is similar and will be left as an exercise.

Theorem 4.3. *Let f be a continuous function on a closed interval $a \leq x \leq b$. Let $\alpha = f(a)$ and $\beta = f(b)$. Let γ be a number such that $\alpha < \gamma < \beta$. Then there exists a number c between a and b such that $f(c) = \gamma$.*

Proof. Let S be the set of numbers x in the interval such that $f(x) \leq \gamma$. Then S is not empty because a is in it, and b is an upper bound for S. Let c be its least upper bound. Then c is in our interval. We contend that $f(c) = \gamma$. If $f(c) < \gamma$, then $c \neq b$, and $f(x) < \gamma$ for all $x > c$ sufficiently close to c, because f is continuous at c. This contradicts the fact that c is an upper bound for S. If $f(c) > \gamma$, then $c \neq a$, and $f(x) > \gamma$ for all $x < c$ sufficiently close to c, again because f is continuous at c. This contradicts the fact that c is a least upper bound for S. We conclude that $f(c) = \gamma$, as was to be shown.

Part Five

Functions of
Several Variables

In the first chapter of this part, we consider vectors, which form the basic algebraic tool in investigating functions of several variables. The differentiation aspects of them which we take up are those which can be handled up to a point by "one variable" methods. The reason for this is that in higher dimensional space, we can join two points by a curve, and study a function by looking at its values only on this curve. This reduces many higher dimensional problems to problems of a one-dimensional situation.

CHAPTER XV

Vectors

The concept of a vector is basic for the study of functions of several variables. It provides geometric motivation for everything that follows. Hence the properties of vectors, both algebraic and geometric, will be discussed in full.

One significant feature of all the statements and proofs of this part is that they are neither easier nor harder to prove in 3-space than they are in 2-space.

XV, §1. DEFINITION OF POINTS IN SPACE

We know that a number can be used to represent a point on a line, once a unit length is selected.

A pair of numbers (i.e. a couple of numbers) (x, y) can be used to represent a point in the plane.

These can be pictured as follows:

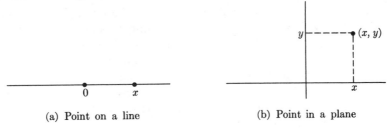

(a) Point on a line   (b) Point in a plane

Figure 1

We now observe that a triple of numbers (x, y, z) can be used to represent a point in space, that is 3-dimensional space, or 3-space. We simply introduce one more axis. Figure 2 illustrates this.

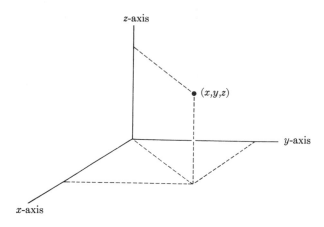

Figure 2

Instead of using x, y, z we could also use (x_1, x_2, x_3). The line could be called 1-space, and the plane could be called 2-space.

Thus we can say that a single number represents a point in 1-space. A couple represents a point in 2-space. A triple represents a point in 3-space.

Although we cannot draw a picture to go further, there is nothing to prevent us from considering a quadruple of numbers.

$$(x_1, x_2, x_3, x_4)$$

and decreeing that this is a point in 4-space. A quintuple would be a point in 5-space, then would come a sextuple, septuple, octuple,....

We let ourselves be carried away and **define a point in n-space** to be an n-tuple of numbers

$$(x_1, x_2, \ldots, x_n),$$

if n is a positive integer. We shall denote such an n-tuple by a capital letter X, and try to keep small letters for numbers and capital letters for points. We call the numbers $x_1, \ldots, x_n$ the **coordinates** of the point X. For example, in 3-space, 2 is the first coordinate of the point $(2, 3, -4)$, and -4 is its third coordinate. We denote n-space by $\mathbf{R}^n$.

Most of our examples will take place when $n = 2$ or $n = 3$. Thus the reader may visualize either of these two cases throughout the book. However, three comments must be made.

First, we have to handle $n = 2$ and $n = 3$, so that in order to avoid a lot of repetitions, it is useful to have a notation which covers both these cases simultaneously, even if we often repeat the formulation of certain results separately for both cases.

Second, no theorem or formula is simpler by making the assumption that $n = 2$ or 3.

Third, the case $n = 4$ does occur in physics.

Example 1. One classical example of 3-space is of course the space we live in. After we have selected an origin and a coordinate system, we can describe the position of a point (body, particle, etc.) by 3 coordinates. Furthermore, as was known long ago, it is convenient to extend this space to a 4-dimensional space, with the fourth coordinate as time, the time origin being selected, say, as the birth of Christ—although this is purely arbitrary (it might be more convenient to select the birth of the solar system, or the birth of the earth as the origin, if we could determine these accurately). Then a point with negative time coordinate is a BC point, and a point with positive time coordinate is an AD point.

Don't get the idea that "time is *the* fourth dimension", however. The above 4-dimensional space is only one possible example. In economics, for instance, one uses a very different space, taking for coordinates, say, the number of dollars expended in an industry. For instance, we could deal with a 7-dimensional space with coordinates corresponding to the following industries:

1. Steel	2. Auto	3. Farm products	4. Fish
5. Chemicals	6. Clothing	7. Transportation.	

We agree that a megabuck per year is the unit of measurement. Then a point

$$(1,000, 800, 550, 300, 700, 200, 900)$$

in this 7-space would mean that the steel industry spent one billion dollars in the given year, and that the chemical industry spent 700 million dollars in that year.

The idea of regarding time as a fourth dimension is an old one. Already in the *Encyclopédie* of Diderot, dating back to the eighteenth century, d'Alembert writes in his article on "dimension":

> Cette manière de considérer les quantités de plus de trois dimensions est aussi exacte que l'autre, car les lettres peuvent toujours être regardées comme représentant des nombres rationnels ou non. J'ai dit plus haut qu'il n'était pas possible de concevoir plus de trois dimensions. Un homme d'esprit de ma connaissance croit qu'on pourrait cependant regarder la durée comme une quatrième dimension, et que le produit temps par la solidité serait en quelque manière un produit de quatre dimensions; cette idée peut être contestée, mais elle a, ce me semble, quelque mérite, quand ce ne serait que celui de la nouveauté.

Encyclopédie, Vol. 4 (1754), p. 1010

Translated, this means:

> This way of considering quantities having more than three dimensions is just as right as the other, because algebraic letters can always be viewed as representing numbers, whether rational or not. I said above that it was not possible to conceive more than three dimensions. A clever gentleman with whom I am acquainted believes that nevertheless, one could view duration as a fourth dimension, and that the product time by solidity would be somehow a product of four dimensions. This idea may be challenged, but it has, it seems to me, some merit, were it only that of being new.

Observe how d'Alembert refers to a "clever gentleman" when he apparently means himself. He is being rather careful in proposing what must have been at the time a far out idea, which became more prevalent in the twentieth century.

D'Alembert also visualized clearly higher dimensional spaces as "products" of lower dimensional spaces. For instance, we can view 3-space as putting side by side the first two coordinates (x_1, x_2) and then the third x_3. Thus we write

$$\mathbf{R}^3 = \mathbf{R}^2 \times \mathbf{R}^1.$$

We use the product sign, which should not be confused with other "products", like the product of numbers. The word "product" is used in two contexts. Similarly, we can write

$$\mathbf{R}^4 = \mathbf{R}^3 \times \mathbf{R}^1.$$

There are other ways of expressing $\mathbf{R}^4$ as a product, namely

$$\mathbf{R}^4 = \mathbf{R}^2 \times \mathbf{R}^2.$$

This means that we view separately the first two coordinates (x_1, x_2) and the last two coordinates (x_3, x_4). We shall come back to such products later.

We shall now define how to add points. If A, B are two points, say in 3-space,

$$A = (a_1, a_2, a_3) \qquad \text{and} \qquad B = (b_1, b_2, b_3)$$

then we **define** $A + B$ to be the point whose coordinates are

$$A + B = (a_1 + b_1, a_2 + b_2, a_3 + b_3).$$

Example 2. In the plane, if $A = (1, 2)$ and $B = (-3, 5)$, then

$$A + B = (-2, 7).$$

In 3-space, if $A = (-1, \pi, 3)$ and $B = (\sqrt{2}, 7, -2)$, then

$$A + B = (\sqrt{2} - 1, \pi + 7, 1).$$

Using a neutral n to cover both the cases of 2-space and 3-space, the points would be written

$$A = (a_1, \ldots, a_n), \qquad B = (b_1, \ldots, b_n),$$

and we **define** $A + B$ to be the point whose coordinates are

$$(a_1 + b_1, \ldots, a_n + b_n).$$

We observe that the following rules are satisfied:

1. $(A + B) + C = A + (B + C)$.
2. $A + B = B + A$.
3. If we let

$$O = (0, 0, \ldots, 0)$$

be the point all of whose coordinates are 0, then

$$O + A = A + O = A$$

for all A.

4. Let $A = (a_1, \ldots, a_n)$ and let $-A = (-a_1, \ldots, -a_n)$. Then

$$A + (-A) = O.$$

All these properties are very simple, and are true because they are true for numbers, and addition of n-tuples is defined in terms of addition of their components, which are numbers.

Note. Do not confuse the number 0 and the n-tuple $(0, \ldots, 0)$. We usually denote this n-tuple by O, and also call it zero, because no difficulty can occur in practice.

We shall now interpret addition and multiplication by numbers geometrically in the plane (you can visualize simultaneously what happens in 3-space).

Example 3. Let $A = (2, 3)$ and $B = (-1, 1)$. Then

$$A + B = (1, 4).$$

The figure looks like a **parallelogram** (Fig. 3).

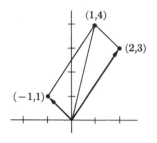

Figure 3

Example 4. Let $A = (3, 1)$ and $B = (1, 2)$. Then

$$A + B = (4, 3).$$

We see again that the geometric representation of our addition looks like a **parallelogram** (Fig. 4).

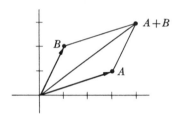

Figure 4

The reason why the figure looks like a **parallelogram** can be given in terms of plane geometry as follows. We obtain $B = (1, 2)$ by starting from the origin $O = (0, 0)$, and moving 1 unit to the right and 2 up. To get $A + B$, we start from A, and again move 1 unit to the right and 2 up. Thus the line segments between O and B, and between A and $A + B$ are the hypotenuses of right triangles whose corresponding legs are of the same length, and parallel. The above segments are therefore parallel and of the same length, as illustrated in Fig. 5.

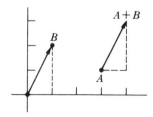

Figure 5

Example 5. If $A = (3, 1)$ again, then $-A = (-3, -1)$. If we plot this point, we see that $-A$ has opposite direction to A. We may view $-A$ as the reflection of A through the origin.

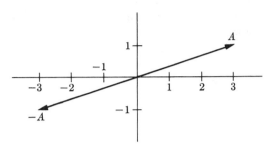

Figure 6

We shall now consider multiplication of A by a number. If c is any number, we **define** cA to be the point whose coordinates are

$$(ca_1, \ldots, ca_n).$$

Example 6. If $A = (2, -1, 5)$ and $c = 7$, then $cA = (14, -7, 35)$.

It is easy to verify the rules:

5. $c(A + B) = cA + cB$.
6. If c_1, c_2 are numbers, then

$$(c_1 + c_2)A = c_1 A + c_2 A \qquad \text{and} \qquad (c_1 c_2)A = c_1(c_2 A).$$

Also note that

$$(-1)A = -A.$$

What is the geometric representation of multiplication by a number?

Example 7. Let $A = (1, 2)$ and $c = 3$. Then

$$cA = (3, 6)$$

as in Fig. 7(a).

Multiplication by 3 amounts to stretching A by 3. Similarly, $\frac{1}{2}A$ amounts to stretching A by $\frac{1}{2}$, i.e. shrinking A to half its size. In general, if t is a number, $t > 0$, we interpret tA as a point in the same direction as A from the origin, but t times the distance. In fact, we define A and

B to have the **same direction** if there exists a number $c > 0$ such that $A = cB$. We emphasize that this means A and B have the same direction **with respect to the origin**. For simplicity of language, we omit the words "with respect to the origin".

Mulitiplication by a negative number reverses the direction. Thus $-3A$ would be represented as in Fig. 7(b).

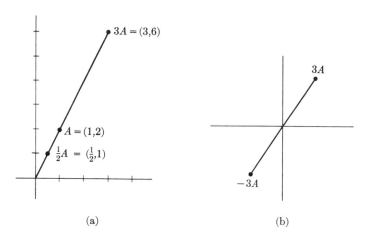

(a)   (b)

Figure 7

We define two vectors A, B (neither of which is zero) to have **opposite directions** if there is a number $c < 0$ such that $cA = B$. Thus when $B = -A$, then A, B have opposite direction.

XV, §1. EXERCISES

Find $A + B$, $A - B$, $3A$, $-2B$ in each of the following cases. Draw the points of Exercises 1 and 2 on a sheet of graph paper.

1. $A = (2, -1)$, $B = (-1, 1)$ 2. $A = (-1, 3)$, $B = (0, 4)$

3. $A = (2, -1, 5)$, $B = (-1, 1, 1)$ 4. $A = (-1, -2, 3)$, $B = (-1, 3, -4)$

5. $A = (\pi, 3, -1)$, $B = (2\pi, -3, 7)$ 6. $A = (15, -2, 4)$, $B = (\pi, 3, -1)$

7. Let $A = (1, 2)$ and $B = (3, 1)$. Draw $A + B$, $A + 2B$, $A + 3B$, $A - B$, $A - 2B$, $A - 3B$ on a sheet of graph paper.

8. Let A, B be as in Exercise 1. Draw the points $A + 2B$, $A + 3B$, $A - 2B$, $A - 3B$, $A + \frac{1}{2}B$ on a sheet of graph paper.

9. Let A and B be as drawn in Fig. 8. Draw the point $A - B$.

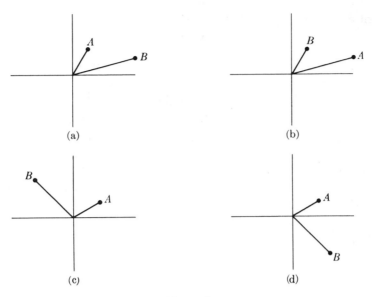

Figure 8

XV, §2. LOCATED VECTORS

We define a **located vector** to be an ordered pair of points which we write $\overrightarrow{AB}$. (This is *not* a product.) We visualize this as an arrow between A and B. We call A the **beginning point** and B the **end point** of the located vector (Fig. 9).

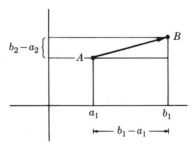

Figure 9

We observe that in the plane,

$$b_1 = a_1 + (b_1 - a_1).$$

Similarly,

$$b_2 = a_2 + (b_2 - a_2).$$

This means that

$$B = A + (B - A)$$

Let $\overrightarrow{AB}$ and $\overrightarrow{CD}$ be two located vectors. We shall say that they are **equivalent** if $B - A = D - C$. Every located vector $\overrightarrow{AB}$ is equivalent to one whose beginning point is the origin, because $\overrightarrow{AB}$ is equivalent to $\overrightarrow{O(B - A)}$. Clearly this is the only located vector whose beginning point is the origin and which is equivalent to $\overrightarrow{AB}$. If you visualize the parellogram law in the plane, then it is clear that equivalence of two located vectors can be interpreted geometrically by saying that the lengths of the line segments determined by the pair of points are equal, and that the "directions" in which they point are the same.

In the next figures, we have drawn the located vectors $\overrightarrow{O(B - A)}$, $\overrightarrow{AB}$, and $\overrightarrow{O(A - B)}$, $\overrightarrow{BA}$.

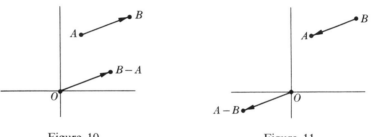

Figure 10   Figure 11

Example 1. Let $P = (1, -1, 3)$ and $Q = (2, 4, 1)$. Then $\overrightarrow{PQ}$ is equivalent to $\overrightarrow{OC}$, where $C = Q - P = (1, 5, -2)$. If

$$A = (4, -2, 5) \quad \text{and} \quad B = (5, 3, 3),$$

then $\overrightarrow{PQ}$ is equivalent to $\overrightarrow{AB}$ because

$$Q - P = B - A = (1, 5, -2).$$

Given a located vector $\overrightarrow{OC}$ whose beginning point is the origin, we shall say that it is **located at the origin.** Given any located vector $\overrightarrow{AB}$, we shall say that it is **located at** A.

A located vector at the origin is entirely determined by its end point. In view of this, we shall call an n-tuple either a point or a **vector,** depending on the interpretation which we have in mind.

Two located vectors $\overrightarrow{AB}$ and $\overrightarrow{PQ}$ are said to be **parallel** if there is a number $c \neq 0$ such that $B - A = c(Q - P)$. They are said to have the

same direction if there is a number $c > 0$ such that $B - A = c(Q - P)$, and have **opposite direction** if there is a number $c < 0$ such that

$$B - A = c(Q - P).$$

In the next pictures, we illustrate parallel located vectors.

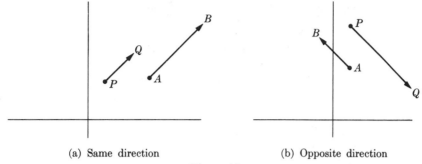

(a) Same direction   (b) Opposite direction

Figure 12

Example 2. Let

$$P = (3, 7) \quad \text{and} \quad Q = (-4, 2).$$

Let

$$A = (5, 1) \quad \text{and} \quad B = (-16, \ -14).$$

Then

$$Q - P = (-7, -5) \quad \text{and} \quad B - A = (-21, -15).$$

Hence $\overrightarrow{PQ}$ is parallel to $\overrightarrow{AB}$, because $B - A = 3(Q - P)$. Since $3 > 0$, we even see that $\overrightarrow{PQ}$ and $\overrightarrow{AB}$ have the same direction.

In a similar manner, any definition made concerning n-tuples can be carried over to located vectors. For instance, in the next section, we shall define what it means for n-tuples to be perpendicular.

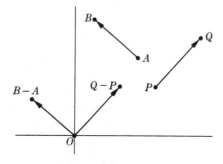

Figure 13

Then we can say that two located vectors $\overrightarrow{AB}$ and $\overrightarrow{PQ}$ are **perpendicular** if $B - A$ is perpendicular to $Q - P$. In Fig. 13, we have drawn a picture of such vectors in the plane.

XV, §2. EXERCISES

In each case, determine which located vectors $\overrightarrow{PQ}$ and $\overrightarrow{AB}$ are equivalent.

1. $P = (1, -1)$, $Q = (4, 3)$, $A = (-1, 5)$, $B = (5, 2)$.

2. $P = (1, 4)$, $Q = (-3, 5)$, $A = (5, 7)$, $B = (1, 8)$.

3. $P = (1, -1, 5)$, $Q = (-2, 3, -4)$, $A = (3, 1, 1)$, $B = (0, 5, 10)$.

4. $P = (2, 3, -4)$, $Q = (-1, 3, 5)$, $A = (-2, 3, -1)$, $B = (-5, 3, 8)$.

In each case, determine which located vectors $\overrightarrow{PQ}$ and $\overrightarrow{AB}$ are parallel.

5. $P = (1, -1)$, $Q = (4, 3)$, $A = (-1, 5)$, $B = (7, 1)$.

6. $P = (1, 4)$, $Q = (-3, 5)$, $A = (5, 7)$, $B = (9, 6)$.

7. $P = (1, -1, 5)$, $Q = (-2, 3, -4)$, $A = (3, 1, 1)$, $B = (-3, 9, -17)$.

8. $P = (2, 3, -4)$, $Q = (-1, 3, 5)$, $A = (-2, 3, -1)$, $B = (-11, 3, -28)$.

9. Draw the located vectors of Exercises 1, 2, 5, and 6 on a sheet of paper to illustrate these exercises. Also draw the located vectors $\overrightarrow{QP}$ and $\overrightarrow{BA}$. Draw the points $Q - P$, $B - A$, $P - Q$, and $A - B$.

XV, §3. SCALAR PRODUCT

It is understood that throughout a discussion we select vectors always in the same n-dimensional space. You may think of the cases $n = 2$ and $n = 3$ only.

In 2-space, let $A = (a_1, a_2)$ and $B = (b_1, b_2)$. We define their **scalar product** to be

$$A \cdot B = a_1 b_1 + a_2 b_2.$$

In 3-space, let $A = (a_1, a_2, a_3)$ and $B = (b_1, b_2, b_3)$. We define their **scalar product** to be

$$A \cdot B = a_1 b_1 + a_2 b_2 + a_3 b_3.$$

In n-space, covering both cases with one notation, let $A = (a_1, \ldots, a_n)$ and $B = (b_1, \ldots, b_n)$ be two vectors. We define their **scalar** or **dot product** $A \cdot B$ to be

$$a_1 b_1 + \cdots + a_n b_n.$$

This product is a **number**. For instance, if

$$A = (1, 3, -2) \qquad \text{and} \qquad B = (-1, 4, -3),$$

then

$$A \cdot B = -1 + 12 + 6 = 17.$$

For the moment, we do not give a geometric interpretation to this scalar product. We shall do this later. We derive first some important properties. The basic ones are:

SP 1. *We have* $A \cdot B = B \cdot A$.

SP 2. *If* A, B, C *are three vectors, then*

$$A \cdot (B + C) = A \cdot B + A \cdot C = (B + C) \cdot A.$$

SP 3. *If* x *is a number, then*

$$(xA) \cdot B = x(A \cdot B) \qquad \text{and} \qquad A \cdot (xB) = x(A \cdot B).$$

SP 4. *If* $A = O$ *is the zero vector, then* $A \cdot A = 0$, *and otherwise*

$$A \cdot A > 0.$$

We shall now prove these properties.
Concerning the first, we have

$$a_1 b_1 + \cdots + a_n b_n = b_1 a_1 + \cdots + b_n a_n,$$

because for any two numbers a, b, we have $ab = ba$. This proves the first property.

For **SP 2**, let $C = (c_1, \ldots, c_n)$. Then

$$B + C = (b_1 + c_1, \ldots, b_n + c_n)$$

and

$$
\begin{aligned}
A \cdot (B + C) &= a_1(b_1 + c_1) + \cdots + a_n(b_n + c_n) \\
&= a_1 b_1 + a_1 c_1 + \cdots + a_n b_n + a_n c_n.
\end{aligned}
$$

Reordering the terms yields

$$a_1 b_1 + \cdots + a_n b_n + a_1 c_1 + \cdots + a_n c_n.$$

which is none other than $A \cdot B + A \cdot C$. This proves what we wanted. We leave property **SP 3** as an exercise.

Finally, for **SP 4**, we observe that if one coordinate a_i of A is not equal to 0, then there is a term $a_i^2 \neq 0$ and $a_i^2 > 0$ in the scalar product

$$A \cdot A = a_1^2 + \cdots + a_n^2.$$

Since every term is $\geqq 0$, it follows that the sum is > 0, as was to be shown.

In much of the work which we shall do concerning vectors, we shall use only the ordinary properties of addition, multiplication by numbers, and the four properties of the scalar product. We shall give a formal discussion of these later. For the moment, observe that there are other objects with which you are familiar and which can be added, subtracted, and multiplied by numbers, for instance the continuous functions on an interval $[a, b]$ (cf. Exercise 6).

Instead of writing $A \cdot A$ for the scalar product of a vector with itself, it will be convenient to write also A^2. (This is the only instance when we allow ourselves such a notation. Thus A^3 has no meaning.) As an exercise, verify the following identities:

$$(A + B)^2 = A^2 + 2A \cdot B + B^2,$$

$$(A - B)^2 = A^2 - 2A \cdot B + B^2.$$

A dot product $A \cdot B$ may very well be equal to 0 without either A or B being the zero vector. For instance, let

$$A = (1, 2, 3) \qquad \text{and} \qquad B = (2, 1, -\tfrac{4}{3}).$$

Then

$$A \cdot B = 0$$

We define two vectors A, B to be **perpendicular** (or as we shall also say, **orthogonal**), if $A \cdot B = 0$. For the moment, it is not clear that in the plane, this definition coincides with our intuitive geometric notion of perpendicularity. We shall convince you that it does in the next section. Here we merely note an example. Say in $\mathbf{R}^3$, let

$$E_1 = (1, 0, 0), \qquad E_2 = (0, 1, 0), \qquad E_3 = (0, 0, 1)$$

be the three unit vectors, as shown on the diagram (Fig. 14).

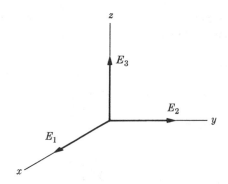

Figure 14

Then we see that $E_1 \cdot E_2 = 0$, and similarly $E_i \cdot E_j = 0$ if $i \neq j$. And these vectors look perpendicular. If $A = (a_1, a_2, a_3)$, then we observe that the i-th component of A, namely

$$a_i = A \cdot E_i$$

is the dot product of A with the i-th unit vector. We see that A is perpendicular to E_i (according to our definition of perpendicularity with the dot product) if and only if its i-th component is equal to 0.

XV, §3. EXERCISES

1. Find $A \cdot A$ for each of the following n-tuples.
   (a) $A = (2, -1)$, $B = (-1, 1)$ (b) $A = (-1, 3)$, $B = (0, 4)$
   (c) $A = (2, -1, 5)$, $B = (-1, 1, 1)$ (d) $A = (-1, -2, 3)$, $B = (-1, 3, -4)$
   (e) $A = (\pi, 3, -1)$, $B = (2\pi, -3, 7)$ (f) $A = (15, -2, 4)$, $B = (\pi, 3, -1)$

2. Find $A \cdot B$ for each of the above n-tuples.

3. Using only the four properties of the scalar product, verify in detail the identities given in the text for $(A + B)^2$ and $(A - B)^2$.

4. Which of the following pairs of vectors are perpendicular?
   (a) $(1, -1, 1)$ and $(2, 1, 5)$ (b) $(1, -1, 1)$ and $(2, 3, 1)$
   (c) $(-5, 2, 7)$ and $(3, -1, 2)$ (d) $(\pi, 2, 1)$ and $(2, -\pi, 0)$

5. Let A be a vector perpendicular to every vector X. Show that $A = O$.

XV, §4. THE NORM OF A VECTOR

We define the **norm** of a vector A, and denote by $\|A\|$, the number

$$\|A\| = \sqrt{A \cdot A}.$$

Since $A \cdot A \geq 0$, we can take the square root. The norm is also some-times called the **magnitude** of A.

When $n = 2$ and $A = (a, b)$, then

$$\|A\| = \sqrt{a^2 + b^2},$$

as in the following picture (Fig. 15).

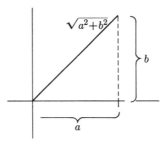

Figure 15

Example 1. If $A = (1, 2)$, then

$$\|A\| = \sqrt{1 + 4} = \sqrt{5}.$$

When $n = 3$ and $A = (a_1, a_2, a_3)$, then

$$\|A\| = \sqrt{a_1^2 + a_2^2 + a_3^2}.$$

Example 2. If $A = (-1, 2, 3)$, then

$$\|A\| = \sqrt{1 + 4 + 9} = \sqrt{14}.$$

If $n = 3$, then the picture looks like Fig. 16, with $A = (x, y, z)$.

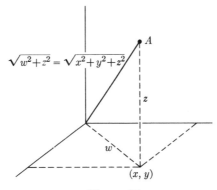

Figure 16

If we first look at the two components (x, y), then the length of the segment between $(0, 0)$ and (x, y) is equal to $w = \sqrt{x^2 + y^2}$, as indicated. Then again the norm of A by the Pythagoras theorem would be

$$\sqrt{w^2 + z^2} = \sqrt{x^2 + y^2 + z^2}.$$

Thus when $n = 3$, our definition of norm is compatible with the geometry of the Pythagoras theorem.

In terms of coordinates, $A = (a_1, \ldots, a_n)$ we see that

$$\|A\| = \sqrt{a_1^2 + \cdots + a_n^2}.$$

If $A \neq O$, then $\|A\| \neq 0$ because some coordinate $a_i \neq 0$, so that $a_i^2 > 0$, and hence $a_1^2 + \cdots + a_n^2 > 0$, so $\|A\| \neq 0$.

Observe that for any vector A we have

$$\boxed{\|A\| = \|-A\|.}$$

This is due to the fact that

$$(-a_1)^2 + \cdots + (-a_n)^2 = a_1^2 + \cdots + a_n^2,$$

because $(-1)^2 = 1$. Of course, this is as it should be from the picture:

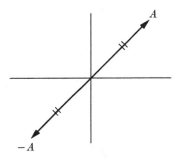

Figure 17

Recall that A and $-A$ are said to have **opposite direction**. However, they have the same norm (magnitude, as is sometimes said when speaking of vectors).

Let A, B be two points. We define the **distance** between A and B to be

$$\|A - B\| = \sqrt{(A - B) \cdot (A - B)}.$$

This definition coincides with our geometric intuition when A, B are points in the plane (Fig. 18). It is the same thing as the length of the located vector $\overrightarrow{AB}$ or the located vector $\overrightarrow{BA}$.

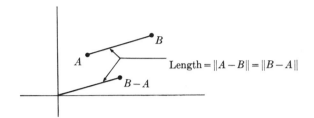

Figure 18

Example 3. Let $A = (-1, 2)$ and $B = (3, 4)$. Then the length of the located vector $\overrightarrow{AB}$ is $\|B - A\|$. But $B - A = (4, 2)$. Thus

$$\|B - A\| = \sqrt{16 + 4} = \sqrt{20}.$$

In the picture, we see that the horizontal side has length 4 and the vertical side has length 2. Thus our definitions reflect our geometric intuition derived from Pythagoras.

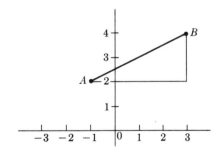

Figure 19

Let P be a point in the plane, and let a be a number > 0. The set of points X such that

$$\|X - P\| < a$$

will be called the **open disc** of radius a centered at P. The set of points X such that

$$\|X - P\| \leqq a$$

will be called the **closed disc** of radius a and center P. The set of points X such that

$$\|X - P\| = a$$

is called the circle of radius a and center P. These are illustrated in Fig. 20.

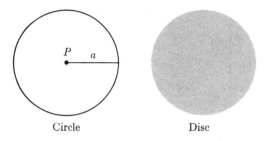

Circle   Disc

Figure 20

In 3-dimensional space, the set of points X such that

$$\|X - P\| < a$$

will be called the **open ball** of radius a and center P. The set of points X such that

$$\|X - P\| \leqq a$$

will be called the **closed ball** of radius a and center P. The set of points X such that

$$\|X - P\| = a$$

will be called the **sphere** of radius a and center P. In higher dimensional space, one uses this same terminology of ball and sphere.

Figure 21 illustrates a sphere and a ball in 3-space.

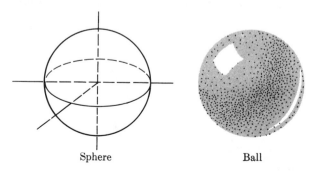

Sphere   Ball

Figure 21

The sphere is the outer shell, and the ball consists of the region inside the shell. The open ball consists of the region inside the shell excluding the shell itself. The closed ball consists of the region inside the shell *and* the shell itself.

From the geometry of the situation, it is also reasonable to expect that if $c > 0$, then $\|cA\| = c\|A\|$, i.e. if we stretch a vector A by multiplying by a positive number c, then the length stretches also by that amount. We verify this formally using our definition of the length.

Theorem 4.1 *Let x be a number. Then*

$$\|xA\| = |x|\ \|A\|$$

(absolute value of x times the norm of A).

Proof. By definition, we have

$$\|xA\|^2 = (xA) \cdot (xA),$$

which is equal to

$$x^2(A \cdot A)$$

by the properties of the scalar product. Taking the square root now yields what we want.

Let S_1 be the sphere of radius 1, centered at the origin. Let a be a number > 0. If X is a point of the sphere S_1, then aX is a point of the sphere of radius a, because

$$\|aX\| = a\|X\| = a.$$

In this manner, we get all points of the sphere of radius a. (Proof?) Thus the sphere of radius a is obtained by stretching the sphere of radius 1, through multiplication by a.

A similar remark applies to the open and closed balls of radius a, they being obtained from the open and closed balls of radius 1 through multiplication by a.

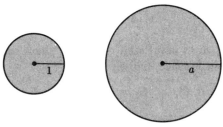

Disc of radius 1   Disc of radius a

Figure 22

We shall say that a vector E is a **unit** vector if $\|E\| = 1$. Given any vector A, let $a = \|A\|$. If $a \neq 0$, then

$$\frac{1}{a} A$$

is a unit vector, because

$$\left\| \frac{1}{a} A \right\| = \frac{1}{a} a = 1.$$

We say that two vectors A, B (neither of which is O) have the **same direction** if there is a number $c > 0$ such that $cA = B$. In view of this definition, we see that the vector

$$\frac{1}{\|A\|} A$$

is a unit vector in the direction of A (provided $A \neq O$).

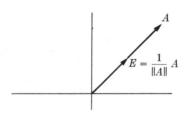

Figure 23

If E is the unit vector in the direction of A, and $\|A\| = a$, then

$$A = aE.$$

Example 4. Let $A = (1, 2, -3)$. Then $\|A\| = \sqrt{14}$. Hence the unit vector in the direction of A is the vector

$$E = \left(\frac{1}{\sqrt{14}}, \frac{2}{\sqrt{14}}, \frac{-3}{\sqrt{14}} \right).$$

Warning. There are as many unit vectors as there are directions. The three **standard unit vectors** in 3-space, namely

$$E_1 = (1, 0, 0), \qquad E_2 = (0, 1, 0), \qquad E_3 = (0, 0, 1)$$

are merely the three unit vectors in the directions of the coordinate axes.

We are also in the position to justify our definition of perpendicularity. Given A, B in the plane, the condition that

$$\|A + B\| = \|A - B\|$$

(illustrated in Fig. 24(b)) coincides with the geometric property that A should be perpendicular to B.

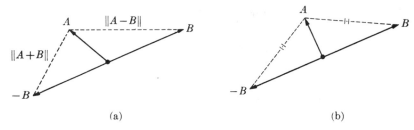

(a)   (b)

Figure 24

We shall prove:

$$\boxed{\|A + B\| = \|A - B\| \text{ if and only if } A \cdot B = 0.}$$

Let $\Leftrightarrow$ denote "if and only if". Then

$$\|A + B\| = \|A - B\| \;\Leftrightarrow\; \|A + B\|^2 = \|A - B\|^2$$
$$\Leftrightarrow\; A^2 + 2A \cdot B + B^2 = A^2 - 2A \cdot B + B^2$$
$$\Leftrightarrow\; 4A \cdot B = 0$$
$$\Leftrightarrow\; A \cdot B = 0.$$

This proves what we wanted.

General Pythagoras theorem. *If A and B are perpendicular, then*

$$\|A + B\|^2 = \|A\|^2 + \|B\|^2.$$

The theorem is illustrated on Fig. 25.

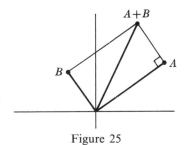

Figure 25

To prove this, we use the definitions, namely

$$\|A + B\|^2 = (A + B)\cdot(A + B) = A^2 + 2A\cdot B + B^2$$
$$= \|A\|^2 + \|B\|^2,$$

because $A\cdot B = 0$, and $A\cdot A = \|A\|^2$, $B\cdot B = \|B\|^2$ by definition.

Remark. If A is perpendicular to B, and x is any number, then A is also perpendicular to xB because

$$A\cdot xB = xA\cdot B = 0.$$

We shall now use the notion of perpendicularity to derive the notion of **projection**. Let A, B be two vectors and $B \neq 0$. Let P be the point on the line through $\overrightarrow{OB}$ such that $\overrightarrow{PA}$ is perpendicular to $\overrightarrow{OB}$, as shown on Fig. 26(a).

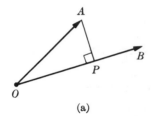
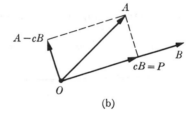

(a)   (b)

Figure 26

We can write

$$P = cB$$

for some number c. We want to find this number c explicitly in terms of A and B. The condition $\overrightarrow{PA} \perp \overrightarrow{OB}$ means that

$$A - P \text{ is perpendicular to } B,$$

and since $P = cB$ this means that

$$(A - cB)\cdot B = 0,$$

in other words,

$$A\cdot B - cB\cdot B = 0.$$

We can solve for c, and we find $A\cdot B = cB\cdot B$, so that

$$\boxed{c = \frac{A\cdot B}{B\cdot B}.}$$

Conversely, if we take this value for c, and then use distributivity, dotting $A - cB$ with B yields 0, so that $A - cB$ is perpendicular to B. Hence we have seen that there is a unique number c such that $A - cB$ is perpendicular to B, and c is given by the above formula.

Definition. The **component** of A along B is the number $c = \dfrac{A \cdot B}{B \cdot B}$.

The **projection** of A along B is the vector $cB = \dfrac{A \cdot B}{B \cdot B} B$.

Example 5. Suppose

$$B = E_i = (0, \ldots, 0, 1, 0, \ldots, 0)$$

is the i-th unit vector, with 1 in the i-th component and 0 in all other components.

$$\text{If } A = (a_1, \ldots, a_n), \text{ then } A \cdot E_i = a_i.$$

Thus $A \cdot E_i$ is the ordinary i-th component of A.

More generally, if B is a unit vector, not necessarily one of the E_i, then we have simply

$$c = A \cdot B$$

because $B \cdot B = 1$ by definition of a unit vector.

Example 6. Let $A = (1, 2, -3)$ and $B = (1, 1, 2)$. Then the component of A along B is the **number**

$$c = \frac{A \cdot B}{B \cdot B} = \frac{-3}{6} = -\frac{1}{2}.$$

Hence the projection of A along B is the **vector**

$$cB = (-\tfrac{1}{2}, \; -\tfrac{1}{2}, \; -1).$$

Our construction gives an immediate geometric interpretation for the scalar product. Namely, assume $A \neq O$ and look at the angle θ between A and B (Fig. 27). Then from plane geometry we see that

$$\cos \theta = \frac{c \|B\|}{\|A\|},$$

or substituting the value for c obtained above,

$$\boxed{A \cdot B = \|A\| \; \|B\| \cos \theta} \qquad \text{and} \qquad \boxed{\cos \theta = \frac{A \cdot B}{\|A\| \; \|B\|}.}$$

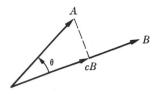

Figure 27

In some treatments of vectors, one takes the relation

$$A \cdot B = \|A\| \ \|B\| \cos \theta$$

as definition of the scalar product. This is subject to the following disadvantages, not to say objections:

(a) The four properties of the scalar product **SP 1** through **SP 4** are then by no means obvious.
(b) Even in 3-space, one has to rely on geometric intuition to obtain the cosine of the angle between A and B, and this intuition is less clear than in the plane. In higher dimensional space, it fails even more.
(c) It is extremely hard to work with such a definition to obtain further properties of the scalar product.

Thus we prefer to lay obvious algebraic foundations, and then recover very simply all the properties. We used plane geometry to see the expression

$$A \cdot B = \|A\| \ \|B\| \cos \theta.$$

After working out some examples, we shall prove the inequality which allows us to justify this in n-space.

Example 7. Let $A = (1, 2, -3)$ and $B = (2, 1, 5)$. Find the cosine of the angle θ between A and B.
By definition,

$$\cos \theta = \frac{A \cdot B}{\|A\| \ \|B\|} = \frac{2 + 2 - 15}{\sqrt{14} \sqrt{30}} = \frac{-11}{\sqrt{420}}.$$

Example 8. Find the cosine of the angle between the two located vectors $\overrightarrow{PQ}$ and $\overrightarrow{PR}$ where

$$P = (1, 2, -3), \qquad Q = (-2, 1, 5), \qquad R = (1, 1, -4).$$

The picture looks like this:

Figure 28

We let

$$A = Q - P = (-3, -1, 8) \quad \text{and} \quad B = R - P = (0, -1, -1).$$

Then the angle between $\overrightarrow{PQ}$ and $\overrightarrow{PR}$ is the same as that between A and B. Hence its cosine is equal to

$$\cos \theta = \frac{A \cdot B}{\|A\| \, \|B\|} = \frac{0 + 1 - 8}{\sqrt{74} \, \sqrt{2}} = \frac{-7}{\sqrt{74} \, \sqrt{2}}.$$

We shall prove further properties of the norm and scalar product using our results on perpendicularity. First note a special case. If

$$E_i = (0, \dots, 0, 1, 0, \dots, 0)$$

is the i-th unit vector of $\mathbf{R}^n$, and

$$A = (a_1, \dots, a_n),$$

then

$$A \cdot E_i = a_i$$

is the i-th component of A, i.e. the component of A along E_i. We have

$$|a_i| = \sqrt{a_i^2} \leqq \sqrt{a_1^2 + \cdots + a_n^2} = \|A\|,$$

so that the absolute value of each component of A is at most equal to the length of A.

We don't have to deal only with the special unit vector as above. Let E be any unit vector, that is a vector of norm 1. Let c be the component of A along E. We saw that

$$c = A \cdot E.$$

Then $A - cE$ is perpendicular to E, and

$$A = A - cE + cE.$$

Then $A - cE$ is also perpendicular to cE, and by the Pythagoras theorem, we find

$$\|A\|^2 = \|A - cE\|^2 + \|cE\|^2 = \|A - cE\|^2 + c^2.$$

Thus we have the inequality $c^2 \leq \|A\|^2$, and $|c| \leq \|A\|$.

In the next theorem, we generalize this inequality to a dot product $A \cdot B$ when B is not necessarily a unit vector.

Theorem 4.2. *Let A, B be two vectors in $\mathbf{R}^n$. Then*

$$|A \cdot B| \leq \|A\| \, \|B\|.$$

Proof. If $B = O$, then both sides of the inequality are equal to 0, and so our assertion is obvious. Suppose that $B \neq O$. Let c be the component of A along B, so $c = (A \cdot B)/(B \cdot B)$. We write

$$A = A - cB + cB.$$

By Pythagoras,

$$\|A\|^2 = \|A - cB\|^2 + \|cB\|^2 = \|A - cB\|^2 + c^2 \|B\|^2.$$

Hence $c^2 \|B\|^2 \leq \|A\|^2$. But

$$c^2 \|B\|^2 = \frac{(A \cdot B)^2}{(B \cdot B)^2} \|B\|^2 = \frac{|A \cdot B|^2}{\|B\|^4} \|B\|^2 = \frac{|A \cdot B|^2}{\|B\|^2}.$$

Therefore

$$\frac{|A \cdot B|^2}{\|B\|^2} \leq \|A\|^2.$$

Multiply by $\|B\|^2$ and take the square root to conclude the proof.

In view of Theorem 4.2, we see that for vectors A, B in n-space, the number

$$\frac{A \cdot B}{\|A\| \, \|B\|}$$

has absolute value ≤ 1. Consequently,

$$-1 \leq \frac{A \cdot B}{\|A\| \, \|B\|} \leq 1,$$

and there exists a unique angle θ such that $0 \le \theta \le \pi$, and such that

$$\cos \theta = \frac{A \cdot B}{\|A\| \, \|B\|}.$$

We define this angle to be the **angle between** A **and** B.

The inequality of Theorem 4.2 is known as the **Schwarz inequality**.

Theorem 4.3. *Let A, B be vectors. Then*

$$\|A + B\| \le \|A\| + \|B\|.$$

Proof. Both sides of this inequality are positive or 0. Hence it will suffice to prove that their squares satisfy the desired inequality, in other words,

$$(A + B) \cdot (A + B) \le (\|A\| + \|B\|)^2.$$

To do this, we consider

$$(A + B) \cdot (A + B) = A \cdot A + 2A \cdot B + B \cdot B.$$

In view of our previous result, this satisfies the inequality

$$\le \|A\|^2 + 2\|A\| \, \|B\| + \|B\|^2,$$

and the right-hand side is none other than

$$(\|A\| + \|B\|)^2.$$

Our theorem is proved.

Theorem 4.3 is known as the **triangle inequality**. The reason for this is that if we draw a triangle as in Fig. 29, then Theorem 4.3 expresses the fact that the length of one side is $\le$ the sum of the lengths of the other two sides.

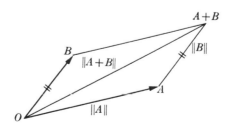

Figure 29

Remark. All the proofs do not use coordinates, only properties **SP 1** through **SP 4** of the dot product. In n-space, they give us inequalities which are by no means obvious when expressed in terms of coordinates. For instance, the Schwarz inequality reads, in terms of coordinates:

$$|a_1 b_1 + \cdots + a_n b_n| \leq (a_1^2 + \cdots + a_n^2)^{1/2}(b_1^2 + \cdots + b_n^2)^{1/2}.$$

Just try to prove this directly, without the "geometric" intuition of Pythagoras, and see how far you get.

XV, §4. EXERCISES

1. Find the norm of the vector A in the following cases.
   (a) $A = (2, -1)$, $B = (-1, 1)$
   (b) $A = (-1, 3)$, $B = (0, 4)$
   (c) $A = (2, -1, 5)$, $B = (-1, 1, 1)$
   (d) $A = (-1, -2, 3)$, $B = (-1, 3, -4)$
   (e) $A = (\pi, 3, -1)$, $B = (2\pi, -3, 7)$
   (f) $A = (15, -2, 4)$, $B = (\pi, 3, -1)$

2. Find the norm of vector B in the above cases.

3. Find the projection of A along B in the above cases.

4. Find the projection of B along A in the above cases.

5. Find the cosine between the following vectors A and B.
   (a) $A = (1, -2)$ and $B = (5, 3)$
   (b) $A = (-3, 4)$ and $B = (2, -1)$
   (c) $A = (1, -2, 3)$ and $B = (-3, 1, 5)$
   (d) $A = (-2, 1, 4)$ and $B = (-1, -1, 3)$
   (e) $A = (-1, 1, 0)$ and $B = (2, 1, -1)$

6. Determine the cosine of the angles of the triangle whose vertices are
   (a) $(2, -1, 1)$, $(1, -3, -5)$, $(3, -4, -4)$.
   (b) $(3, 1, 1)$, $(-1, 2, 1)$, $(2, -2, 5)$.

7. Let $A_1, \ldots, A_r$ be non-zero vectors which are mutually perpendicular, in other words $A_i \cdot A_j = 0$ if $i \neq j$. Let $c_1, \ldots, c_r$ be numbers such that

$$c_1 A_1 + \cdots + c_r A_r = O.$$

   Show that all $c_i = 0$.

8. For any vectors A, B, prove the following relations:
   (a) $\|A + B\|^2 + \|A - B\|^2 = 2\|A\|^2 + 2\|B\|^2$.
   (b) $\|A + B\|^2 = \|A\|^2 + \|B\|^2 + 2A \cdot B$.
   (c) $\|A + B\|^2 - \|A - B\|^2 = 4A \cdot B$.
   Interpret (a) as a "parallelogram law".

9. Show that if θ is the angle between A and B, then

$$\|A - B\|^2 = \|A\|^2 + \|B\|^2 - 2\|A\|\,\|B\|\,\cos\theta.$$

10. Let A, B, C be three non-zero vectors. If $A \cdot B = A \cdot C$, show by an example that we do not necessarily have $B = C$.

XV, §5. PARAMETRIC LINES

We define the **parametric equation** or **parametric representation** of a straight line passing through a point P in the direction of a vector $A \neq O$ to be

$$X = P + tA,$$

where t runs through all numbers (Fig. 30).

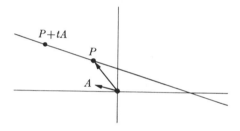

Figure 30

When we give such a parametric representation, we may think of a bug starting from a point P at time $t = 0$, and moving in the direction of A. At time t, the bug is at the position $P + tA$. Thus we may interpret physically the parametric representation as a description of motion, in which A is interpreted as the velocity of the bug. At a given time t, the bug is at the point.

$$X(t) = P + tA,$$

which is called the **position** of the bug at time t.

This parametric representation is also useful to describe the set of points lying on the line segment between two given points. Let P, Q be two points. Then the **segment** between P and Q consists of all the points

$$S(t) = P + t(Q - P) \qquad \text{with} \qquad 0 \leq t \leq 1.$$

Indeed, $\overrightarrow{O(Q - P)}$ is a vector having the same direction as $\overrightarrow{PQ}$, as shown on Fig. 31.

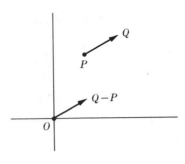

Figure 31

When $t = 0$, we have $S(0) = P$, so at time $t = 0$ the bug is at P. When $t = 1$, we have

$$S(1) = P + (Q - P) = Q,$$

so when $t = 1$ the bug is at Q. As t goes from 0 to 1, the bug goes from P to Q.

Example 1. Let $P = (1, -3, 4)$ and $Q = (5, 1, -2)$. Find the coordinates of the point which lies one third of the distance from P to Q.

Let $S(t)$ as above be the parametric representation of the segment from P to Q. The desired point is $S(1/3)$, that is:

$$S\left(\frac{1}{3}\right) = P + \frac{1}{3}(Q - P) = (1, -3, 4) + \frac{1}{3}(4, 4, -6)$$

$$= \left(\frac{7}{3}, \frac{-5}{3}, 2\right).$$

Warning. The desired point in the above example is *not* given by

$$\frac{P + Q}{3}.$$

Example 2. Find a parametric representation for the line passing through the two points $P = (1, -3, 1)$ and $Q = (-2, 4, 5)$.

We first have to find a vector in the direction of the line. We let

$$A = P - Q,$$

so

$$A = (3, -7, -4).$$

The parametric representation of the line is therefore

$$X(t) = P + tA = (1, -3, 1) + t(3, -7, -4).$$

Remark. It would be equally correct to give a parametric representation of the line as

$$Y(t) = P + tB \qquad \text{where} \qquad B = Q - P.$$

Interpreted in terms of the moving bug, however, one parametrization gives the position of a bug moving in one direction along the line, starting from P at time $t = 0$, while the other parametrization gives the position of another bug moving in the **opposite** direction along the line, also starting from P at time $t = 0$.

We shall now discuss the relation between a parametric representation and the ordinary equation of a line in the plane.

Suppose that we work in the plane, and write the coordinates of a point X as (x, y). Let $P = (p, q)$ and $A = (a, b)$. Then in terms of the coordinates, we can write

$$x = p + ta, \qquad y = q + tb.$$

We can then eliminate t and obtain the usual equation relating x and y.

Example 3. Let $P = (2, 1)$ and $A = (-1, 5)$. Then the parametric representation of the line through P in the direction of A gives us

(*) $x = 2 - t, \qquad y = 1 + 5t.$

Multiplying the first equation by 5 and adding yields

(**) $5x + y = 11,$

which is the familiar equation of a line.

This elimination of t shows that every pair (x, y) which satisfies the parametric representation (*) for some value of t also satisfies equation (**). Conversely, suppose we have a pair of numbers (x, y) satisfying (**). Let $t = 2 - x$. Then

$$y = 11 - 5x = 11 - 5(2 - t) = 1 + 5t.$$

Hence there exists some value of t which satisfies equation (∗). Thus we have proved that the pairs (x, y) which are solutions of (∗∗) are exactly the same pairs of numbers as those obtained by giving arbitrary values for t in (∗). Thus the straight line can be described parametrically as in (∗) or in terms of its usual equation (∗∗). Starting with the ordinary equation

$$5x + y = 11,$$

we let $t = 2 - x$ in order to recover the specific parametrization of (∗).

When we parametrize a straight line in the form

$$X = P + tA,$$

we have of course infinitely many choices for P on the line, and also infinitely many choices for A, differing by a scalar multiple. We can always select at least one. Namely, given an equation

$$ax + by = c$$

with numbers a, b, c, suppose that $a \neq 0$. We use y as parameter, and let

$$y = t.$$

Then we can solve for x, namely

$$x = \frac{c}{a} - \frac{b}{a} t.$$

Let $P = (c/a, 0)$ and $A = (-b/a, 1)$. We see that an arbitrary point (x, y) satisfying the equation

$$ax + by = c$$

can be expressed parametrically, namely

$$(x, y) = P + tA.$$

In higher dimensions, starting with a parametric representation

$$X = P + tA,$$

we cannot eliminate t, and thus the parametric representation is the only one available to describe a straight line.

XV, §5. EXERCISES

1. Find a parametric representation for the line passing through the following
   pairs of points.
   (a) $P_1 = (1, 3, -1)$ and $P_2 = (-4, 1, 2)$
   (b) $P_1 = (-1, 5, 3)$ and $P_2 = (-2, 4, 7)$

Find a parametric representation for the line passing through the following
points.

2. $(1, 1, -1)$ and $(-2, 1, 3)$ 3. $(-1, 5, 2)$ and $(3, -4, 1)$

4. Let $P = (1, 3, -1)$ and $Q = (-4, 5, 2)$. Determine the coordinates of the following points:
   (a) The midpoint of the line segment between P and Q.
   (b) The two points on this line segment lying one-third and two-thirds of the
       way from P to Q.
   (c) The point lying one-fifth of the way from P to Q.
   (d) The point lying two-fifths of the way from P to Q.

5. If P, Q are two arbitrary points in n-space, give the general formula for the
   midpoint of the line segment between P and Q.

XV, §6. PLANES

We can describe planes in 3-space by an equation analogous to the
single equation of the line. We proceed as follows.

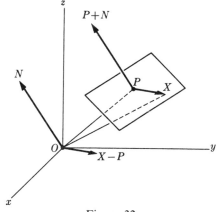

Figure 32

Let P be a point in 3-space and consider a located vector $\overrightarrow{ON}$. We
define the **plane passing through P perpendicular to** $\overrightarrow{ON}$ to be the collection of all points X such that the located vector $\overrightarrow{PX}$ is perpendicular
to $\overrightarrow{ON}$. According to our definitions, this amounts to the condition

$$(X - P) \cdot N = 0,$$

which can also be written as

$$\boxed{X \cdot N = P \cdot N.}$$

We shall also say that this plane is the one perpendicular to N, and consists of all vectors X such that $X - P$ is perpendicular to N. We have drawn a typical situation in 3-spaces in Fig. 32.

Instead of saying that N is **perpendicular** to the plane, one also says that N is **normal** to the plane.

Let t be a number $\neq 0$. Then the set points X such that

$$(X - P) \cdot N = 0$$

coincides with the set of points X such that

$$(X - P) \cdot tN = 0.$$

Thus we may say that our plane is the plane passing through P and perpendicular to the **line** in the direction of N. To find the equation of the plane, we could use any vector tN (with $t \neq 0$) instead of N.

Example 1. Let

$$P = (2, 1, -1) \qquad \text{and} \qquad N = (-1, 1, 3).$$

Let $X = (x, y, z)$. Then

$$X \cdot N = (-1)x + y + 3z.$$

Therefore the equation of the plane passing through P and perpendicular to N is

$$-x + y + 3z = -2 + 1 - 3$$

or

$$-x + y + 3z = -4.$$

Observe that in 2-space, with $X = (x, y)$, the formulas lead to the equation of the line in the ordinary sense.

Example 2. The equation of the line in the (x, y)-plane, passing through $(4, -3)$ and perpendicular to $(-5, 2)$ is

$$-5x + 2y = -20 - 6 = -26.$$

We are now in position to interpret the coefficients $(-5, 2)$ of x and y in this equation. They give rise to a vector perpendicular to the line. **In any equation**

$$ax + by = c$$

the vector (a, b) **is perpendicular to the line determined by the equation.** Similarly, in 3-space, the vector (a, b, c) is perpendicular to the plane determined by the equation

$$ax + by + cz = d.$$

Example 3. The plane determined by the equation

$$2x - y + 3z = 5$$

is perpendicular to the vector $(2, -1, 3)$. If we want to find a point in that plane, we of course have many choices. We can give arbitrary values to x and y, and then solve for z. To get a concrete point, let $x = 1$, $y = 1$. Then we solve for z, namely

$$3z = 5 - 2 + 1 = 4,$$

so that $z = \frac{4}{3}$. Thus

$$(1, 1, \tfrac{4}{3})$$

is a point in the plane.

In n-space, the equation $X \cdot N = P \cdot N$ is said to be the equation of a **hyperplane**. For example,

$$3x - y + z + 2w = 5$$

is the equation of a hyperplane in 4-space, perpendicular to $(3, -1, 1, 2)$.

Two vectors A, B are said to be parallel if there exists a number $c \neq 0$ such that $cA = B$. Two lines are said to be **parallel** if, given two distinct points P_1, Q_1 on the first line and P_2, Q_2 on the second, the vectors

$$P_1 - Q_1$$

and

$$P_2 - Q_2$$

are parallel.

Two planes are said to be **parallel** (in 3-space) if their normal vectors are parallel. They are said to be **perpendicular** if their normal vectors are perpendicular. The **angle** between two planes is defined to be the angle between their normal vectors.

Example 4. Find the cosine of the angle θ between the planes.

$$2x - y + z = 0,$$
$$x + 2y - z = 1.$$

This cosine is the cosine of the angle between the vectors.

$$A = (2, -1, 1) \quad \text{and} \quad B = (1, 2, -1).$$

Therefore

$$\cos \theta = \frac{A \cdot B}{\|A\| \, \|B\|} = -\frac{1}{6}.$$

Example 5. Let

$$Q = (1, 1, 1) \quad \text{and} \quad P = (1, -1, 2).$$

Let

$$N = (1, 2, 3)$$

Find the point of intersection of the line through P in the direction of N, and the plane through Q perpendicular to N.

The parametric representation of the line through P in the direction of N is

(1) $$X = P + tN.$$

The equation of the plane through Q perpendicular to N is

(2) $$(X - Q) \cdot N = 0.$$

We visualize the line and plane as follows:

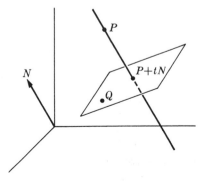

Figure 33

We must find the value of t such that the vector X in (1) also satisfies (2), that is

$$(P + tN - Q) \cdot N = 0,$$

or after using the rules of the dot product,

$$(P - Q) \cdot N + tN \cdot N = 0.$$

Solving for t yields

$$t = \frac{(Q - P) \cdot N}{N \cdot N} = \frac{1}{14}.$$

Thus the desired point of intersection is

$$P + tN = (1, -1, 2) + \tfrac{1}{14}(1, 2, 3) = (\tfrac{15}{14}, -\tfrac{12}{14}, \tfrac{31}{14}).$$

Example 6. Find the equation of the plane passing through the three points

$$P_1 = (1, 2, -1). \qquad P_2 = (-1, 1, 4), \qquad P_3 = (1, 3, -2).$$

We visualize schematically the three points as follows:

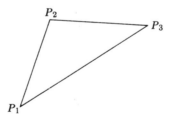

Figure 34

Then we find a vector N perpendicular to $\overrightarrow{P_1P_2}$ and $\overrightarrow{P_1P_3}$, or in other words, perpendicular to $P_2 - P_1$ and $P_3 - P_1$. We have

$$P_2 - P_1 = (-2, -1, +5),$$
$$P_3 - P_1 = (0, 1, -1).$$

Let $N = (a, b, c)$. We must solve

$$N \cdot (P_2 - P_1) = 0 \qquad \text{and} \qquad N \cdot (P_3 - P_1) = 0,$$

in other words,

$$- 2a - b + 5c = 0,$$
$$b - \;\; c = 0.$$

We take $b = c = 1$ and solve for $a = 2$. Then

$$N = (2, 1, 1)$$

satisfies our requirements. The plane perpendicular to N, passing through P_1 is the desired plane. Its equation is therefore $X \cdot N = P_1 \cdot N$, that is

$$2x + y + z = 2 + 2 - 1 = 3.$$

Distance between a point and a plane. Consider a plane defined by the equation

$$(X - P) \cdot N = 0,$$

and let Q be an arbitrary point. We wish to find a formula for the distance between Q and the plane. By this we mean the length of the segment from Q to the point of intersection of the perpendicular line to the plane through Q, as on the figure. We let Q' be this point of intersection.

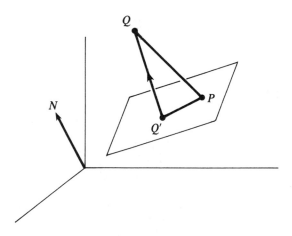

Figure 35

From the geometry, we have:

length of the segment $\overline{QQ'}$ = length of the projection of $\overline{QP}$ on $\overline{QQ'}$.

We can express the length of this projection in terms of the dot product as follows. A unit vector in the direction of N, which is perpendicular to the plane, is given by $N/\|N\|$. Then

length of the projection of $\overline{QP}$ on $\overline{QQ'}$

$$= \text{norm of the projection of } Q - P \text{ on } N/\|N\|$$

$$= \left| (Q - P) \cdot \frac{N}{\|N\|} \right|.$$

This can also be written in the form:

$$\text{distance between } Q \text{ and the plane} = \frac{|(Q - P) \cdot N|}{\|N\|}.$$

Example 7. Let

$$Q = (1, 3, 5), \qquad P = (-1, 1, 7) \qquad \text{and} \qquad N = (-1, 1, -1).$$

The equation of the plane is

$$-x + y - z = -5.$$

We find $\|N\| = \sqrt{3}$,

$$Q - P = (2, 2, -2) \qquad \text{and} \qquad (Q - P) \cdot N = -2 + 2 + 2 = 2.$$

Hence the distance between Q and the plane is $2/\sqrt{3}$.

XV, §6. EXERCISES

1. Show that the lines $2x + 3y = 1$ and $5x - 5y = 7$ are not perpendicular.

2. Let $y = mx + b$ and $y = m'x + c$ be the equations of two lines in the plane. Write down vectors perpendicular to these lines. Show that these vectors are perpendicular to each other if and only if $mm' = -1$.

Find the equation of the line in 2-space, perpendicular to N and passing through P, for the following values of N and P.

3. $N = (1, -1)$, $P = (-5, 3)$ 4. $N = (-5, 4)$, $P = (3, 2)$

5. Show that the lines

$$3x - 5y = 1, \qquad 2x + 3y = 5$$

are not perpendicular.

6. Which of the following pairs of lines are perpendicular?
   (a) $3x - 5y = 1$ and $2x + y = 2$
   (b) $2x + 7y = 1$ and $x - y = 5$
   (c) $3x - 5y = 1$ and $5x + 3y = 7$
   (d) $-x + y = 2$ and $x + y = 9$

7. Find the equation of the plane perpendicular to the given vector N and passing through the given point P.
   (a) $N = (1, -1, 3)$, $P = (4, 2, -1)$
   (b) $N = (-3, -2, 4)$, $P = (2, \pi, -5)$
   (c) $N = (-1, 0, 5)$, $P = (2, 3, 7)$

8. Find the equation of the plane passing through the following three points.
   (a) $(2, 1, 1)$, $(3, -1, 1)$, $(4, 1, -1)$
   (b) $(-2, 3, -1)$, $(2, 2, 3)$, $(-4, -1, 1)$
   (c) $(-5, -1, 2)$, $(1, 2, -1)$, $(3, -1, 2)$

9. Find a vector perpendicular to $(1, 2, -3)$ and $(2, -1, 3)$, and another vector perpendicular to $(-1, 3, 2)$ and $(2, 1, 1)$.

10. Find a vector parallel to the line of intersection of the two planes

$$2x - y + z = 1, \qquad 3x + y + z = 2.$$

11. Same question for the planes,

$$2x + y + 5z = 2, \qquad 3x - 2y + z = 3.$$

12. Find a parametric representation for the line of intersection of the planes of Exercises 10 and 11.

13. Find the cosine of the angle between the following planes:
   (a) $x + y + z = 1$ (b) $2x + 3y - z = 2$
       $x - y - z = 5$ $x - y + z = 1$
   (c) $x + 2y - z = 1$ (d) $2x + y + z = 3$
       $-x + 3y + z = 2$ $-x - y + z = \pi$

14. (a) Let $P = (1, 3, 5)$ and $A = (-2, 1, 1)$. Find the intersection of the line through P in the direction of A, and the plane $2x + 3y - z = 1$.
    (b) Let $P = (1, 2, -1)$. Find the point of intersection of the plane

$$3x - 4y + z = 2,$$

with the line through P, perpendicular to that plane.

15. Let $Q = (1, -1, 2)$, $P = (1, 3, -2)$, and $N = (1, 2, 2)$. Find the point of the intersection of the line through P in the direction of N, and the plane through Q perpendicular to N.

16. Find the distance between the indicated point and plane.
   (a) $(1, 1, 2)$ and $3x + y - 5z = 2$
   (b) $(-1, 3, 2)$ and $2x - 4y + z = 1$
   (c) $(3, -2, 1)$ and the yz-plane
   (d) $(-3, -2, 1)$ and the yz-plane

17. Draw the triangle with vertices $A = (1, 1)$, $B = (2, 3)$, and $C = (3, -1)$. Draw the point P such that $\overrightarrow{AP} \perp \overrightarrow{BC}$ and P belongs to the line passing through the points B and C.
    (a) Find the cosine of the angle of the triangle whose vertex is at A.
    (b) What are the coordinates of P?

18. (a) Find the equation of the plane M passing through the point $P = (1, 1, 1)$ and perpendicular to the vector $\overrightarrow{ON}$, where $N = (1, 2, 0)$.
    (b) Find a parametric representation of the line L passing through

$$Q = (1, 4, 0)$$

    and perpendicular to the plane M.
    (c) What is the distance from Q to the plane M?

19. Find the cosine of the angle between the planes

$$2x + 4y - z = 5 \quad \text{and} \quad x - 3y + 2z = 0.$$

CHAPTER XVI

Differentiation of Vectors

XVI, §1. DERIVATIVE

Consider a bug moving along some curve in 3-dimensional space. The position of the bug at time t is given by the three coordinates

$$(x(t),\ y(t),\ z(t)),$$

which depend on t. We abbreviate these by $X(t)$. For instance, the position of a bug moving along a straight line was seen in the preceding chapter to be given by

$$X(t) = P + tA,$$

where P is the starting point, and A gives the direction of the bug. However, we can give examples when the bug does not move on a straight line. First we look at an example in the plane.

Example 1. Let $X(\theta) = (\cos \theta,\ \sin \theta)$. Then the bug moves around a circle of radius 1 in counterclockwise direction.

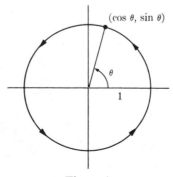

Figure 1

Here we used θ as the variable, corresponding to the angle as shown on the figure. Let ω be the angular speed of the bug, and assume ω constant. Thus $d\theta/dt = \omega$ and

$$\theta = \omega t + \text{a constant.}$$

For simplicity, assume that the constant is 0. Then we can write the position of the bug as

$$X(\theta) = X(\omega t) = (\cos \omega t, \sin \omega t).$$

If the angular speed is 1, then we have simply the representation

$$X(t) = (\cos t, \sin t).$$

Example 2. If the bug moves around a circle of radius 2 with angular speed equal to 1, then its position at time t is given by

$$X(t) = (2 \cos t, 2 \sin t).$$

More generally, if the bug moves around a circle of radius r, then the position is given by

$$X(t) = (r \cos t, r \sin t).$$

In these examples, we assume of course that at time $t = 0$ the bug starts at the point $(r, 0)$, that is

$$X(0) = (r, 0),$$

where r is the radius of the circle.

Example 3. Suppose the position of the bug is given in 3-space by

$$X(t) = (\cos t, \sin t, t).$$

Then the bug moves along a spiral. Its coordinates are given as functions of t by

$$x(t) = \cos t,$$

$$y(t) = \sin t,$$

$$z(t) = t.$$

The position at time t is obtained by plugging in the special value of t. Thus:

$$X(\pi) = (\cos \pi, \sin \pi, \pi) = (-1, 0, \pi)$$

$$X(1) = (\cos 1, \sin 1, 1).$$

We may now give the definition of a curve in general.

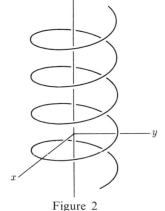

Figure 2

Definition. Let I be an interval. A **parametrized curve** (defined on this interval) is an association which to each point of I associates a vector. If X denotes a curve defined on I, and t is a point of I, then $X(t)$ denotes the vector associated to t by X. We often write the association $t \mapsto X(t)$ as an arrow

$$X : I \to \mathbf{R}^n.$$

We also call this association the **parametrization** of a curve. We call $X(t)$ the **position vector** at time t. It can be written in terms of coordinates,

$$X(t) = (x_1(t), \ldots, x_n(t)),$$

each $x_i(t)$ being a function of t. We say that this curve is **differentiable** if each function $x_i(t)$ is a differentiable function of t.

Remark. We take the intervals of definition for our curves to be open, closed, or also half-open or half-closed. When we define the derivative of a curve, it is understood that the interval of definition contains more than one point. In that case, at an end point the usual limit of

$$\frac{f(a + h) - f(a)}{h}$$

is taken for those h such that the quotient makes sense, i.e. $a + h$ lies in the interval. If a is a left end point, the quotient is considered only for $h > 0$. If a is a right end point the quotient is considered only for $h < 0$. Then the usual rules for differentiation of functions are true in this greater generality, and thus Rules 1 through 4 below, and the chain rule of §2 remain true also. [An example of a statement which is not always true for curves defined over closed intervals is given in Exercise 11(b).]

Let us try to differentiate curves. We consider the Newton quotient

$$\frac{X(t + h) - X(t)}{h} .$$

Its numerator is illustrated in Fig. 3.

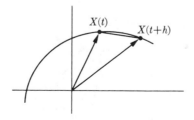

Figure 3

As h approaches 0, we see geometrically that

$$\frac{X(t + h) - X(t)}{h}$$

should approach a vector pointing in the direction of the curve. We can write the Newton quotient in terms of coordinates,

$$\frac{X(t + h) - X(t)}{h} = \left(\frac{x_1(t + h) - x_1(t)}{h}, \ldots, \frac{x_n(t + h) - x_n(t)}{h}\right)$$

and see that each component is a Newton quotient for the corresponding coordinate. We assume that each $x_i(t)$ is differentiable. Then each quotient

$$\frac{x_i(t + h) - x_i(t)}{h}$$

approaches the derivatives dx_i/dt. For this reason, we define the **derivative** dX/dt to be

$$\frac{dX}{dt} = \left(\frac{dx_1}{dt}, \ldots, \frac{dx_n}{dt}\right).$$

In fact, we could also say that the vector

$$\left(\frac{dx_1}{dt}, \ldots, \frac{dx_n}{dt}\right)$$

is the limit of the Newton quotient

$$\frac{X(t + h) - X(t)}{h}$$

as h approaches 0. Indeed, as h approaches 0, each component

$$\frac{x_i(t + h) - x_i(t)}{h}$$

approaches dx_i/dt. Hence the Newton quotient approaches the vector

$$\left(\frac{dx_1}{dt}, \ldots, \frac{dx_n}{dt}\right).$$

Example 4. If $X(t) = (\cos t, \sin t, t)$ then

$$\frac{dX}{dt} = (-\sin t, \cos t, 1).$$

Physicists often denote dX/dt by $\dot{X}$; thus in the previous example, we could also write

$$\dot{X}(t) = (-\sin t, \cos t, 1) = X'(t).$$

We define the **velocity vector** of the curve at time t to be the velocity $X'(t)$.

Example 5. When $X(t) = (\cos t, \sin t, t)$, then

$$X'(t) = (-\sin t, \cos t, 1);$$

the velocity vector at $t = \pi$ is

$$X'(\pi) = (0, -1, 1),$$

and for $t = \pi/4$ we get

$$X'(\pi/4) = (-1/\sqrt{2}, 1/\sqrt{2}, 1).$$

The velocity vector is located at the origin, but when we translate it to the point $X(t)$, then we visualize it as tangent to the curve, as in the next figure.

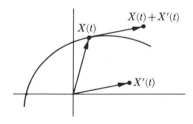

Figure 4

We define the **tangent line** to a curve X at time t to be the line passing through $X(t)$ in the direction of $X'(t)$, provided that $X'(t) \neq O$. Otherwise, we don't define a tangent line. We have therefore given two interpretations for $X'(t)$:

> $X'(t)$ *is the velocity at time t;*
>
> $X'(t)$ *is parallel to a tangent vector at time t.*

By abuse of language, we sometimes call $X'(t)$ a tangent vector, although strictly speaking, we should refer to the located vector $\overrightarrow{X(t)(X(t) + X'(t))}$ as the tangent vector. However, to write down this located vector each time is cumbersome.

Example 6. Find a parametric equation of the tangent line to the curve $X(t) = (\sin t, \cos t)$ at $t = \pi/3$.

We have $X'(t) = (\cos t, -\sin t)$, so that at $t = \dfrac{\pi}{3}$ we get

$$X'\left(\frac{\pi}{3}\right) = \left(\frac{1}{2}, -\frac{\sqrt{3}}{2}\right) \quad \text{and} \quad X\left(\frac{\pi}{3}\right) = \left(\frac{\sqrt{3}}{2}, \frac{1}{2}\right).$$

Let $P = X(\pi/3)$ and $A = X'(\pi/3)$. Then a parametric equation of the tangent line at the required point is

$$L(t) = P + tA = \left(\frac{\sqrt{3}}{2}, \frac{1}{2}\right) + \left(\frac{1}{2}, \frac{\sqrt{3}}{2}\right)t.$$

(We use another letter L because X is already occupied.) In terms of the coordinates $L(t) = (x(t), y(t))$, we can write the tangent line as

$$x(t) = \frac{\sqrt{3}}{2} + \frac{1}{2}t,$$

$$y(t) = \frac{1}{2} - \frac{\sqrt{3}}{2}t.$$

Example 7. Find the equation of the plane perpendicular to the spiral

$$X(t) = (\cos t, \sin t, t)$$

when $t = \pi/3$.

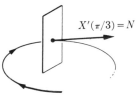

$X'(\pi/3) = N$

Figure 5

Let the given point be

$$P = X\left(\frac{\pi}{3}\right) = \left(\cos\frac{\pi}{3}, \sin\frac{\pi}{3}, \frac{\pi}{3}\right),$$

so that more simply,

$$P = \left(\frac{1}{2}, \frac{\sqrt{3}}{2}, \frac{\pi}{3}\right).$$

We must then find a vector N perpendicular to the plane at the given point P.

We have $X'(t) = (-\sin t, \cos t, 1)$, so

$$X'\left(\frac{\pi}{3}\right) = \left(-\frac{\sqrt{3}}{2}, \frac{1}{2}, 1\right) = N.$$

The equation of the plane through P perpendicular to N is

$$X \cdot N = P \cdot N,$$

so the equation of the desired plane is

$$-\frac{\sqrt{3}}{2}x + \frac{1}{2}y + z = -\frac{\sqrt{3}}{4} + \frac{\sqrt{3}}{4} + \frac{\pi}{3}$$

$$= \frac{\pi}{3}.$$

We define the **speed** of the curve $X(t)$ to be the norm of the velocity vector. If we denote the speed by $v(t)$, then by definition we have

$$\boxed{v(t) = \|X'(t)\|,}$$

and thus

$$\boxed{v(t)^2 = X'(t)^2 = X'(t) \cdot X'(t).}$$

We can also omit the t from the notation, and write

$$v^2 = X' \cdot X' = X'^2.$$

Example 8. The speed of the bug moving on the circle

$$X(t) = (\cos t, \sin t)$$

is the norm of the velocity $X'(t) = (-\sin t, \cos t)$, and so is

$$v(t) = \sqrt{(-\sin t)^2 + (\cos^2 t)} = 1.$$

Example 9. The speed of the bug moving on the spiral

$$X(t) = (\cos t, \sin t, t)$$

is the norm of the velocity $X'(t) = (-\sin t, \cos t, 1)$, and so is

$$v(t) = \sqrt{(-\sin t)^2 + (\cos^2 t) + 1}$$
$$= \sqrt{2}.$$

We define the **acceleration vector** to be the derivative

$$\frac{dX'(t)}{dt} = X''(t),$$

provided of course that X' is differentiable. We shall also denote the acceleration vector by $X''(t)$ as above.

We shall now discuss acceleration. There are two possible definitions for a **scalar acceleration**:

First there is the *rate of change of the speed*, that is

$$\frac{dv}{dt} = v'(t).$$

Second, there is the *norm of the acceleration vector*, that is

$$\|X''(t)\|.$$

Warning. These two are usually not equal. Almost any example will show this.

Example 10. Let

$$X(t) = (\cos t, \sin t).$$

Then:

$$v(t) = \|X'(t)\| = 1 \qquad \text{so} \qquad dv/dt = 0.$$
$$X''(t) = (-\cos t, -\sin t) \qquad \text{so} \qquad \|X''(t)\| = 1.$$

Thus if and when we need to refer to scalar acceleration, we must always say which one we mean. One could use the notation $a(t)$ for scalar acceleration, but one must specify which of the two possibilities $a(t)$ denotes.

The fact that the above two quantities are not equal reflects the physical interpretation. A bug moving around a circle at uniform speed has

$dv/dt = 0$. However, the acceleration vector is not O, because the velocity vector is constantly changing. Hence the norm of the acceleration vector is not equal to 0.

We shall list the rules for differentiation. These will concern sums, products, and the chain rule which is postponed to the next section.

The derivative of a curve is defined componentwise. Thus the rules for the derivative will be very similar to the rules for differentiating functions.

Rule 1. *Let $X(t)$ and $Y(t)$ be two differentiable curves (defined for the same values of t). Then the sum $X(t) + Y(t)$ is differentiable, and*

$$\frac{d(X(t) + Y(t))}{dt} = \frac{dX}{dt} + \frac{dY}{dt}.$$

Rule 2. *Let c be a number, and let $X(t)$ be differentiable. Then $cX(t)$ is differentiable, and*

$$\frac{d(cX(t))}{dt} = c\,\frac{dX}{dt}.$$

Rule 3. *Let $X(t)$ and $Y(t)$ be two differentiable curves (defined for the same values of t). Then $X(t) \cdot Y(t)$ is a differentiable function whose derivative is*

$$\frac{d}{dt}\,[X(t) \cdot Y(t)] = X(t) \cdot Y'(t) + X'(t) \cdot Y(t).$$

(This is formally analogous to the derivative of a product of functions, namely **the first times the derivative of the second plus the second times the derivative of the first,** except that the product is now a scalar product.)

As an example of the proofs we shall give the third one in detail, and leave the others to you as exercises.

Let for simplicity

$$X(t) = (x_1(t),\ x_2(t)) \qquad \text{and} \qquad Y(t) = (y_1(t),\ y_2(t)).$$

Then

$$\frac{d}{dt}\,X(t) \cdot Y(t) = \frac{d}{dt}\,[x_1(t)y_1(t) + x_2(t)y_2(t)]$$

$$= x_1(t)\,\frac{dy_1(t)}{dt} + \frac{dx_1}{dt}\,y_1(t) + x_2(t)\,\frac{dy_2}{dt} + \frac{dx_2}{dt}\,y_2(t)$$

$$= X(t) \cdot Y'(t) + X'(t) \cdot Y(t),$$

by combining the appropriate terms.

The proof for 3-space or n-space is obtained by replacing 2 by 3 or n, and inserting ... in the middle to take into account the other coordinates.

Example 11. The square $X(t)^2 = X(t) \cdot X(t)$ comes up frequently in applications, for instance because it can be interpreted as the square of the distance of $X(t)$ from the origin. Using the rule for the derivative of a product, we find the formula

$$\frac{d}{dt} X(t)^2 = 2X(t) \cdot X'(t).$$

You should memorize this formula by repeating it out loud.

Suppose that $\|X(t)\|$ is constant. This means that $X(t)$ lies on a sphere of constant radius k. Taking the square yields

$$X(t)^2 = k^2$$

that is, $X(t)^2$ is also constant. Differentiate both sides with respect to t. Then we obtain

$$2X(t) \cdot X'(t) = 0 \qquad \text{and therefore} \qquad X(t) \cdot X'(t) = 0$$

Interpretation. *Suppose a bug moves along a curve $X(t)$ which remains at constant distance from the origin, i.e. $\|X(t)\| = k$ is constant. Then the position vector $X(t)$ is perpendicular to the velocity $X'(t)$.*

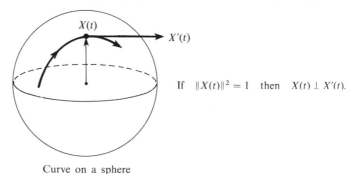

If $\|X(t)\|^2 = 1$ then $X(t) \perp X'(t)$.

Curve on a sphere

If $X(t)$ is a curve and $f(t)$ is a function, defined for the same values of t, then we may also form the product $f(t)X(t)$ of the number $f(t)$ by the vector $X(t)$.

Example 12. Let $X(t) = (\cos t, \sin t, t)$ and $f(t) = e^t$, then

$$f(t)X(t) = (e^t \cos t, e^t \sin t, e^t t),$$

and

$$f(\pi)X(\pi) = (e^\pi(-1), e^\pi(0), e^\pi \pi) = (-e^\pi, 0, e^\pi \pi).$$

If $X(t) = (x(t),\ y(t),\ z(t))$, then

$$f(t)X(t) = (f(t)x(t),\ f(t)y(t),\ f(t)z(t)).$$

We have a rule for such differentiation analogous to Rule 3.

Rule 4. *If both $f(t)$ and $X(t)$ are defined over the same interval, and are differentiable, then so is $f(t)X(t)$, and*

$$\frac{d}{dt}\ f(t)X(t) = f(t)X'(t) + f'(t)X(t).$$

The proof is just the same as for Rule 3.

Example 13. Let A be a fixed vector, and let f be an ordinary differentiable function of one variable. Let $F(t) = f(t)A$. Then $F'(t) = f'(t)A$. For instance, if $F(t) = (\cos t)A$ and $A = (a,\ b)$ where $a,\ b$ are fixed numbers, then

$$F(t) = (a \cos t,\ b \cos t)$$

and thus

$$F'(t) = (-a \sin t,\ -b \sin t) = (-\sin t)A.$$

Similarly, if A, B are fixed vectors, and

$$G(t) = (\cos t)A + (\sin t)B,$$

then

$$G'(t) = (-\sin t)A + (\cos t)B.$$

XVI, §1. EXERCISES

Find the velocity of the following curves.

1. $(e^t,\ \cos t,\ \sin t)$ 2. $(\sin 2t,\ \log(1 + t),\ t)$

3. $(\cos t,\ \sin t)$ 4. $(\cos 3t,\ \sin 3t)$

5. (a) In Exercises 3 and 4, show that the velocity vector is perpendicular to the position vector. Is this also the case in Exercises 1 and 2?

   (b) In Exercises 3 and 4, show that the acceleration vector is in the opposite direction from the position vector.

6. Let A, B be two constant vectors. What is the velocity vector of the curve

$$X = A + tB?$$

7. Let $X(t)$ be a differentiable curve. A plane or line which is perpendicular to the velocity vector $X'(t)$ at the point $X(t)$ is said to be **normal** to the curve at the point t or also at the point $X(t)$. Find the equation of a line normal to the curves of Exercises 3 and 4 at the point $\pi/3$.

8. (a) Find the equation of a plane normal to the curve

$$(e^t,\ t,\ t^2)$$

   at the point $t = 1$.
   (b) Same question at the point $t = 0$.

9. Let P be the point $(1, 2, 3, 4)$ and Q the point $(4, 3, 2, 1)$. Let A be the vector $(1, 1, 1, 1)$. Let L be the line passing through P and parallel to A.
   (a) Given a point X on the line L, compute the distance between Q and X (as a function of the parameter t).
   (b) Show that there is precisely one point X_0 on the line such that this distance achieves a minimum, and that this minimum is $2\sqrt{5}$.
   (c) Show that $X_0 - Q$ is perpendicular to the line.

10. Let P be the point $(1, -1, 3, 1)$ and Q the point $(1, 1, -1, 2)$. Let A be the vector $(1, -3, 2, 1)$. Solve the same questions as in the preceding problem, except that in this case the minimum distance is $\sqrt{146/15}$.

11. Let $X(t)$ be a differentiable curve defined on an open interval. Let Q be a point which is not on the curve.
   (a) Write down the formula for the distance between Q and an arbitrary point on the curve.
   (b) If t_0 is a value of t such that the distance between Q and $X(t_0)$ is at a minimum, show that the vector $Q - X(t_0)$ is normal to the curve, at the point $X(t_0)$. [*Hint*: Investigate the minimum of the square of the distance.]
   (c) If $X(t)$ is the parametric representation of a straight line, show that there exists a unique value t_0 such that the distance between Q and $X(t_0)$ is a minimum.

12. Let N be a non-zero vector, c a number, and Q a point. Let P_0 be the point of intersection of the line passing through Q, in the direction of N, and the plane $X \cdot N = c$. Show that for all points P of the plane, we have

$$\|Q - P_0\| \leqq \|Q - P\|.$$

13. Prove that if the speed is constant, then the acceleration is perpendicular to the velocity.

14. Prove that if the acceleration of a curve is always perpendicular to its velocity, then its speed is constant.

15. Let B be a non-zero vector, and let $X(t)$ be such that $X(t) \cdot B = t$ for all t. Assume also that the angle between $X'(t)$ and B is constant. Show that $X''(t)$ is perpendicular to $X'(t)$.

16. Write a parametric representation for the tangent line to the given curve at the given point in each of the following cases.
    (a) $(\cos 4t, \sin 4t, t)$ at the point $t = \pi/8$
    (b) $(t, 2t, t^2)$ at the point $(1, 2, 1)$
    (c) $(e^{3t}, e^{-3t}, 3\sqrt{2}\, t)$ at $t = 1$
    (d) (t, t^3, t^4) at the point $(1, 1, 1)$

17. Let A, B be fixed non-zero vectors. Let

$$X(t) = e^{2t}A + e^{-2t}B.$$

Show that $X''(t)$ has the same direction as $X(t)$.

18. Show that the two curves $(e^t, e^{2t}, 1 - e^{-t})$ and $(1 - \theta, \cos \theta, \sin \theta)$ intersect at the point $(1, 1, 0)$. What is the angle between their tangents at that point?

19. At what points does the curve $(2t^2, 1 - t, 3 + t^2)$ intersect the plane

$$3x - 14y + z - 10 = 0?$$

20. Let $X(t)$ be a differentiable curve.
    (a) Suppose that $X'(t) = O$ for all t throughout its interval of definition I. What can you say about the curve?
    (b) Suppose $X'(t) \neq O$ but $X''(t) = O$ for all t in the interval. What can you say about the curve?

21. Let $X(t) = (a \cos t, a \sin t, bt)$, where a, b are constant. Let $\theta(t)$ be the angle which the tangent line at a given point of the curve makes with the z-axis. Show that $\cos \theta(t)$ is the constant $b/\sqrt{a^2 + b^2}$.

22. Show that the velocity and acceleration vectors of the curve in Exercise 21 have constant norms (magnitudes).

23. Let B be a fixed unit vector, and let $X(t)$ be a curve such that $X(t) \cdot B = e^{2t}$ for all t. Assume also that the velocity vector of the curve has a constant angle θ with the vector B, with $0 < \theta < \pi/2$.
    (a) Show that the speed is $2e^{2t}/\cos \theta$.
    (b) Determine the dot product $X'(t) \cdot X''(t)$ in terms of t and θ.

24. Let

$$X(t) = \left(\frac{2t}{1 + t^2}, \frac{1 - t^2}{1 + t^2}, 1 \right).$$

Show that the cosine of the angle between $X(t)$ and $X'(t)$ is constant.

25. Suppose that a bug moves along a differentiable curve $B(t) = (x(t), y(t), z(t))$, lying in the surface $z^2 = 1 + x^2 + y^2$. (This means that the coordinates (x, y, z) of the curve satisfy this equation.)

(a) Show that

$$2x(t)x'(t) = B(t) \cdot B'(t).$$

(b) Assume that the cosine of the angle between the vector $B(t)$ and the velocity vector $B'(t)$ is always positive. Show that the distance of the bug to the yz-plane increases whenever its x-coordinate is positive.

26. A bug is moving in space on a curve given by

$$X(t) = (t, \ t^2, \ \tfrac{2}{3}t^3),$$

(a) Find a parametric representation of the tangent line at $t = 1$.
(b) Write the equation of the normal plane to the curve at $t = 1$.

27. Let a particle move in the plane so that its position at time t is

$$C(t) = (e^t \cos t, \ e^t \sin t).$$

Show that the tangent vector to the curve makes a constant angle of $\pi/4$ with the position vector.

XVI, §2. LENGTH OF CURVES

Suppose a bug travels along a curve $X(t)$. The rate of change of the distance traveled is equal to the speed, so we may write the equation

$$\frac{ds(t)}{dt} = v(t).$$

Consequently it is reasonable to make the following definition.

We define the **length** of a curve X between two values a, b of t $(a \leq b)$ in the interval of definition of the curve to be the integral of the speed:

$$\int_b^a v(t) \ dt = \int_a^b \|X'(t)\| \ dt.$$

By definition, we can rewrite this integral in the form

$$\int_a^b \sqrt{\left(\frac{dx}{dt}\right)^2 + \left(\frac{dy}{dt}\right)^2} \ dt \qquad \text{when} \qquad X(t) = (x(t), y(t)),$$

$$\int_a^b \sqrt{\left(\frac{dx}{dt}\right)^2 + \left(\frac{dy}{dt}\right)^2 + \left(\frac{dz}{dt}\right)^2} \ dt \qquad \text{when} \qquad X(t) = (x(t), y(t), z(t)),$$

$$\int_a^b \sqrt{\left(\frac{dx_1}{dt}\right)^2 + \cdots + \left(\frac{dx_n}{dt}\right)^2} \ dt \qquad \text{when} \qquad X(t) = (x_1(t), \ldots x_n(t)).$$

Example 1. Let the curve be defined by

$$X(t) = (\sin t, \ \cos t).$$

Then $X'(t) = (\cos t, \ -\sin t)$ and $v(t) = \sqrt{\cos^2 t + \sin^2 t} = 1$. Hence the length of the curve between $t = 0$ and $t = 1$ is

$$\int_0^1 v(t) \ dt = t \Big|_0^1 = 1.$$

In this case, of course, the integral is easy to evaluate. There is no reason why this should always be the case.

Example 2. Set up the integral for the length of the curve

$$X(t) = (e^t, \ \sin t, \ t)$$

between $t = 1$ and $t = \pi$.

We have $X'(t) = (e^t, \ \cos t, \ 1)$. Hence the desired integral is

$$\int_1^\pi \sqrt{e^{2t} + \cos^2 t + 1} \ dt.$$

In this case, there is no easy formula for the integral. In the exercises, however, the functions are adjusted in such a way that the integral can be evaluated by elementary techniques of integration. Don't expect this to be the case in real life, though. The presence of the square root sign usually makes it impossible to evaluate the length integral by elementary functions.

XVI, §2. EXERCISES

1. Find the length of the spiral $(\cos t, \sin t, t)$ between $t = 0$ and $t = 1$.

2. Find the length of the spirals.
   (a) $(\cos 2t, \sin 2t, 3t)$ between $t = 1$ and $t = 3$.
   (b) $(\cos 4t, \sin 4t, t)$ between $t = 0$ and $t = \pi/8$.

3. Find the length of the indicated curve for the given interval:
   (a) $(t, 2t, t^2)$ between $t = 1$ and $t = 3$. [*Hint*: You will get at some point the integral $\int \sqrt{1 + u^2} \, du$. The easiest way of handling that is to let

$$u = \frac{e^t - e^{-t}}{2} = \sinh t, \quad \text{so} \quad 1 + \sinh^2 t = \cosh^2 t,$$

where

$$\cosh t = \frac{e^t + e^{-t}}{2}.$$

This makes the expression under the square root sign into a perfect square. This method will in fact prove the general formula

$$\int \sqrt{a^2 + x^2} \, dx = \frac{1}{2} \left[x\sqrt{a^2 + x^2} + a^2 \log(x + \sqrt{a^2 + x^2}) \right].$$

Of course, you can check the formula by differentiating the right-hand side, and just use it for the exercise.
   (b) $(e^{3t}, e^{-3t}, 3\sqrt{2} \, t)$ between $t = 0$ and $t = \frac{1}{3}$.

   [*Hint*: At some point you will meet a square root.

$$\sqrt{e^{6t} + e^{-6t} + 2}.$$

   The expression under the square root is a perfect square. Try squaring $(e^{3t} + e^{-3t})$. What do you get?]

4. Find the length of the curve defined by

$$X(t) = (t - \sin t, \, 1 - \cos t)$$

between (a) $t = 0$ and $t = 2\pi$, (b) $t = 0$ and $t = \pi/2$.

[*Hint*: Remember the identity

$$\sin^2 \theta = \frac{1 - \cos 2\theta}{2}.$$

Therefore letting $t = 2\theta$ gives

$$1 - \cos t = 2 \sin^2(t/2).$$

The expression under the integral sign will then be a perfect square.]

5. Find the length of the curve $X(t) = (t, \log t)$ between:
   (a) $t = 1$ and $t = 2$, (b) $t = 3$ and $t = 5$. [*Hint*: Substitute $u^2 = 1 + t^2$ to evaluate the integral. Use partial fractions.]

6. Find the length of the curve defined by $X(t) = (t, \log \cos t)$ between $t = 0$ and $t = \pi/4$.

7. Let $X(t) = (t, t^2, \frac{2}{3}t^3)$.
   (a) Find the speed of this curve.
   (b) Find the length of the curve between $t = 0$ and $t = 1$.

8. Let $X(t) = (6t, 2t^3, 3\sqrt{2}\ t^2)$. Find the length of the curve between $t = 0$ and $t = 1$.

CHAPTER XVII

Functions of Several Variables

We view functions of several variables as functions of points in space. This appeals to our geometric intuition, and also relates such functions more easily with the theory of vectors. The gradient will appear as a natural generalization of the derivative. In this chapter we are mainly concerned with basic definitions and notions. We postpone the important theorems to the next chapter.

XVII, §1. GRAPHS AND LEVEL CURVES

In order to conform with usual terminology, and for the sake of brevity, a collection of objects will simply be called a **set**. In this chapter, we are mostly concerned with sets of points in space.

Let S be a set of points in n-space. A **function** (defined on S) is an association which to each element of S associates a **number**. For instance, if to each point we associate the numerical value of the temperature at that point, we have the temperature function.

Remark. In the previous chapter, we considered parametrized curves, associating a vector to a point. We do **not** call these functions. Only when the values of the association are **numbers** do we use the word **function**. We find this to be the most useful convention for this course.

In practice, we sometimes omit mentioning explicitly the set S, since the context usually makes it clear for which points the function is defined.

Example 1. In 2-space (the plane) we can define a function f by the rule

$$f(x, y) = x^2 + y^2.$$

It is defined for all points (x, y) and can be interpreted geometrically as the square of the distance between the origin and the point.

Example 2. Again in 2-space, we can define a function f by the formula

$$f(x, y) = \frac{x^2 - y^2}{x^2 + y^2} \qquad \text{for all} \quad (x, y) \neq (0, 0).$$

We do not define f at $(0, 0)$ (also written O).

Example 3. In 3-space, we can define a function f by the rule

$$f(x, y, z) = x^2 - \sin(xyz) + yz^3.$$

Since a point and a vector are represented by the same thing (namely an n-tuple), we can think of a function such as the above also as a function of vectors. When we do not want to write the coordinates, we write $f(X)$ instead of $f(x_1, \ldots, x_n)$. As with numbers, we call $f(X)$ the **value** of f at the point (or vector) X.

Just as with functions of one variable, we define the **graph** of a function f of n variables $x_1, \ldots, x_n$ to be the set of points in $(n + 1)$-space of the form

$$(x_1, \ldots, x_n, f(x_1, \ldots, x_n)),$$

the $(x_1, \ldots, x_n)$ being in the domain of definition of f.

When $n = 1$, the graph of a function f is a set of points $(x, f(x))$. Thus the graph itself is in 2-space.

When $n = 2$, the graph of a function f is the set of points

$$(x, y, f(x, y)).$$

When $n = 2$, it is already difficult to draw the graph since it involves a figure in 3-space. The graph of a function of two variables may look like this:

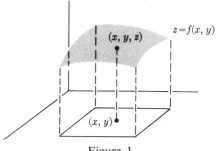

Figure 1

For each number c, the equation $f(x, y) = c$ is the equation of a curve in the plane. We have considerable experience in drawing the graphs of such curves, and we may therefore assume that we know how to draw this graph in principle. This curve is called the **level curve** of f at c. It gives us the set of points (x, y) where f takes on the value c. By drawing a number of such level curves, we can get a good description of the function.

Example 4. Let $f(x, y) = x^2 + y^2$. The level curves are described by equations

$$x^2 + y^2 = c.$$

These have a solution only when $c \geq 0$. In that case, they are circles (unless $c = 0$ in which case the circle of radius 0 is simply the origin). In Fig. 2, we have drawn the level curves for $c = 1$ and 4.

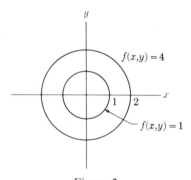

Figure 2

The graph of the function $z = f(x, y) = x^2 + y^2$ is then a figure in 3-space, which we may represent as follows.

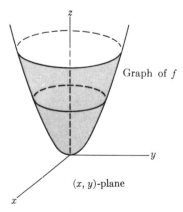

Figure 3

Example 5. Let the elevation of a mountain in meters be given by the formula

$$f(x, y) = 4{,}000 - 2x^2 - 3y^4.$$

We see that $f(0, 0) = 4{,}000$ is the highest point of the mountain. As x, y increase, the altitude decreases. The mountain and its level curves might look like this.

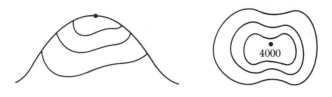

Figure 4

In this case, the highest point is at the origin, and the level curves indicate decreasing altitude as they move away from the origin.

If we deal with a function of three variables, say $f(x, y, z)$, then $(x, y, z) = X$ is a point in 3-space. In that case, the set of points satisfying the equation

$$f(x, y, z) = c$$

for some constant c is a surface. The notion analogous to that of level curve is that of **level surface**.

Example 6. Let $f(x, y, z) = x^2 + y^2 + z^2$. Then f is the square of the distance from the origin. The equation

$$x^2 + y^2 + z^2 = c$$

is the equation of a sphere for $c > 0$, and the radius is of course $\sqrt{c}$. If $c = 0$ this is the equation of a point, namely the origin itself. If $c < 0$ there is no solution. Thus the level surfaces for the function f are spheres.

Example 7. Let $f(x, y, z) = 3x^2 + 2y^2 + z^2$. Then the level surfaces for f are defined by the equations

$$3x^2 + 2y^2 + z^2 = c.$$

They have the same shape as ellipses, and called ellipsoids, for $c > 0$.

It is harder to draw figures in 3 dimensions than in 2 dimensions, so we restrict ourselves to drawing level curves.

The graph of a function of three variables is the set of points

$$(x, y, z, f(x, y, z))$$

in 4-dimensional space. Not only is this graph hard to draw, it is impossible to draw. It is, however, possible to define it as we have done by writing down coordinates of points.

In physics, a function f might be a potential function, giving the value of the potential energy at each point of space. The level surfaces are then sometimes called surfaces of **equipotential**. The function f might also give a temperature distribution (i.e. its value at a point X is the temperature at X). In that case, the level surfaces are called **isothermal** surfaces.

XVII, §1. EXERCISES

Sketch the level curves for the functions $z = f(x, y)$, where $f(x, y)$ is given by the following expressions.

1. $x^2 + 2y^2$ 2. $y - x^2$ 3. $y - 3x^2$

4. $x - y^2$ 5. $3x^2 + 3y^2$ 6. xy

7. $(x - 1)(y - 2)$ 8. $(x + 1)(y + 3)$ 9. $\dfrac{x^2}{4} + \dfrac{y^2}{16}$

10. $2x - 3y$ 11. $\sqrt{x^2 + y^2}$ 12. $x^2 - y^2$

13. $y^2 - x^2$ 14. $(x - 1)^2 + (y + 3)^2$ 15. $(x + 1)^2 + y^2$

XVII, §2. PARTIAL DERIVATIVES

In this section and the next, we discuss the notion of differentiability for functions of several variables. When we discussed the derivative of functions of one variable, we assumed that such a function was defined on an interval. We shall have to make a similar assumption in the case of several variables, and for this we need to introduce a new notion.

Let U be a set in the plane. We shall say that U is an **open set** if the following condition is satisfied. Given a point P in U, there exists an open disc D of radius $a > 0$ which is centered at P and such that D is contained in U.

Let U be a set in space. We shall say that U is an **open set** in space if given a point P in U, there exists an open ball B of radius $a > 0$ which is centered at P and such that B is contained in U.

A similar definition is given of an open set in n-space.

Given a point P in an open set, we can go in all directions from P by a small distance and still stay within the open set.

Example 1. In the plane, the set consisting of the first quadrant, excluding the x- and y-axes, is an open set.

The x-axis is not open in the plane (i.e. in 2-space). Given a point on the x-axis, we cannot find an open disc centered at the point and contained in the x-axis.

Example 2. Let U be the open ball of radius $a > 0$ centered at the origin. Then U is an open set. This is illustrated on Fig. 5.

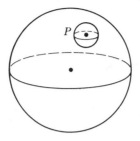

Figure 5

In the next picture we have drawn an open set in the plane, consisting of the region inside the curve, but not containing any point of the boundary. We have also drawn a point P in U, and a ball (disc) around P contained in U.

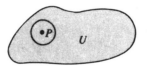

Figure 6

When we defined the derivative as a limit of

$$\frac{f(x + h) - f(x)}{h},$$

we needed the function f to be defined in some open interval around the point x.

Now let f be a function of n variables, defined on an open set U. Then for any point X in U, the function f is also defined at all points which are close to X, namely all points which are contained in an open ball centered at X and contained in U. We shall obtain the partial derivative of f by keeping all but one variable fixed, and taking the ordinary derivative with respect to the one variable.

Let us start with two variables. Given a function $f(x, y)$ of two variables x, y, let us keep y constant and differentiate with respect to x. We are then led to consider the limit as h approaches 0 of

$$\frac{f(x + h, y) - f(x, y)}{h}$$

Definition. If this limit exists, we call it the **derivative** of f **with respect to the first variable**, or also the **first partial derivative** of f, and denote it by

$$(D_1 f)(x, y).$$

This notation allows us to use any letters to denote the variables. For instance,

$$\lim_{h \to 0} \frac{f(u + h, v) - f(u, v)}{h} = D_1 f(u, v).$$

Note that $D_1 f$ is a single function. We often omit the parentheses, writing

$$D_1 f(u, v) = (D_1 f)(u, v)$$

for simplicity.

Also, if the variables x, y are agreed upon, then we write

$$D_1 f(x, y) = \frac{\partial f}{\partial x}.$$

Similarly, we define

$$D_2 f(x, y) = \lim_{h \to 0} \frac{f(x, y + k) - f(x, y)}{h}$$

and also write

$$D_2 f(x, y) = \frac{\partial f}{\partial y}.$$

Example 3. Let $f(x, y) = x^2 y^3$. Then

$$\frac{\partial f}{\partial x} = 2xy^3 \qquad \text{and} \qquad \frac{\partial f}{\partial y} = 3x^2 y^2.$$

We observe that the partial derivatives are themselves functions. This is the reason why the notation $D_i f$ is sometimes more useful than the notation $\partial f / \partial x_i$. It allows us to write $D_i f(P)$ for any point P in the set where the partial is defined. There cannot be any ambiguity or confusion with a (meaningless) symbol $D_i (f(P))$, since $f(P)$ is a number. Thus $D_i f(P)$ means $(D_i f)(P)$. It is the value of the function $D_i f$ at P.

Example 4. Let $f(x, y) = \sin xy$. To find $D_2 f(1, \pi)$, we first find $\partial f / \partial y$, or $D_2 f(x, y)$, which is simply

$$D_2 f(x, y) = (\cos xy)x.$$

Hence

$$D_2 f(1, \pi) = (\cos \pi) \cdot 1 = -1.$$

Also,

$$D_2 f\left(3, \frac{\pi}{4}\right) = \left(\cos \frac{3\pi}{4}\right) \cdot 3 = -\frac{1}{\sqrt{2}} \cdot 3 = -\frac{3}{\sqrt{2}}.$$

A similar definition of the partial derivatives is given in 3-space. Let f be a function of three variables (x, y, z), defined on an open set U in 3-space. We define, for instance,

$$(D_3 f)(x, y, z) = \frac{\partial f}{\partial z} = \lim_{h \to 0} \frac{f(x, y, z + h) - f(x, y, z)}{h},$$

and similarly for the other variables.

Example 5. Let $f(x, y, z) = x^2 y \sin(yz)$. Then

$$D_3 f(x, y, z) = \frac{\partial f}{\partial z} = x^2 y \cos(yz)y = x^2 y^2 \cos(yz).$$

Let $X = (x, y, z)$ for abbreviation. Let

$$E_1 = (1, 0, 0), \qquad E_2 = (0, 1, 0), \qquad E_3 = (0, 0, 1)$$

be the three standard unit vectors in the directions of the coordinate axes. Then we can abbreviate the Newton quotient for the partial derivatives by writing

$$D_i f(X) = \frac{\partial f}{\partial x_i} = \lim_{h \to 0} \frac{f(X + hE_i) - f(X)}{h}.$$

Indeed, observe that

$$hE_1 = (h, 0, 0) \qquad \text{so} \qquad f(X + hE_1) = f(x + h, y, z),$$

and similarly for the other two variables.

In a similar fashion we can define the partial derivatives in n-space, by a definition which applies simultaneously to 2-space and 3-space. Let f be a function defined on an open set U in n-space. Let the variables be $(x_1, \ldots, x_n)$.

For small values of h, the point

$$(x_1 + h, x_2, \ldots, x_n)$$

is contained in U. Hence the function is defined at that point, and we may form the quotient

$$\frac{f(x_1 + h, x_2, \ldots, x_n) - f(x_1, \ldots, x_n)}{h}.$$

If the limit exists as h tends to 0, then we call it the **first partial derivative** of f and denote it by

$$D_1 f(x_1, \ldots, x_n), \quad \text{or} \quad D_1 f(X), \quad \text{or also by} \quad \frac{\partial f}{\partial x_1}.$$

Similarly, we let

$$D_i f(X) = \frac{\partial f}{\partial x_i}$$

$$= \lim_{h \to 0} \frac{f(x_1, \ldots, x_i + h, \ldots, x_n) - f(x_1, \ldots, x_n)}{h}$$

if it exists, and call it the i-th **partial derivative**.

Let

$$E_i = (0, \ldots, 0, 1, 0, \ldots, 0)$$

be the i-th vector in the direction of the i-th coordinate axis, having components equal to 0 except for the i-th component which is 1. Then we have

$$(D_i f)(X) = \lim_{h \to 0} \frac{f(X + hE_i) - f(X)}{h}.$$

This is a very useful brief notation which applies simultaneously to 2-space, 3-space, or n-space.

Definition. Let f be a function of two variables (x, y). We define the **gradient** of f, written **grad** f, to be the vector

$$\operatorname{grad} f(x, y) = \left(\frac{\partial f}{\partial x}, \frac{\partial f}{\partial y}\right).$$

Example 6. Let $f(x, y) = x^2 y^3$. Then

$$\operatorname{grad} f(x, y) = (2xy^3, 3x^2y^2),$$

so that in this case,

$$\operatorname{grad} f(1, 2) = (16, 12).$$

Thus the gradient of a function f associates a **vector** to a point X.

If f is a function of three variables (x, y, z), then we define the gradient to be

$$\operatorname{grad} f(x, y, z) = \left(\frac{\partial f}{\partial x}, \frac{\partial f}{\partial y}, \frac{\partial f}{\partial z}\right).$$

Example 7. Let $f(x, y, z) = x^2 y \sin(yz)$. Find grad $f(1, 1, \pi)$. First we find the three partial derivatives, which are:

$$\frac{\partial f}{\partial x} = 2xy \sin(yz),$$

$$\frac{\partial f}{\partial y} = x^2[y \cos(yz)z + \sin(yz)],$$

$$\frac{\partial f}{\partial z} = x^2 y \cos(yz)y = x^2 y^2 \cos(yz).$$

We then substitute $(1, 1, \pi)$ for (x, y, z) in these partials, and get

$$\operatorname{grad} f(1, 1, \pi) = (0, -\pi, -1).$$

Let f be defined in an open set U in n-space, and assume that the partial derivatives of f exist at each point X of U. We define the **gradient of** f **at** X to be the vector

$$\operatorname{grad} f(X) = \left(\frac{\partial f}{\partial x_1}, \ldots, \frac{\partial f}{\partial x_n}\right) = (D_1 f(X), \ldots, D_n f(X)),$$

whose components are the partial derivatives. One must read this

$$(\mathrm{grad}\, f)(X),$$

but we shall usually omit the parentheses around grad f. Sometimes one also writes ∇f instead of grad f. Thus in 2-space we also write

$$\nabla f(x, y) = (\nabla f)(x, y) = (D_1 f(x, y), D_2 f(x, y)),$$

and similarly in 3-space,

$$\nabla f(x, y, z) = (\nabla f)(x, y, z) = (D_1 f(x, y, z), D_2 f(x, y, z), D_3 f(x, y, z)).$$

So far, we defined the gradient only by a formula with partial derivatives. We shall give a geometric interpretation for the gradient in Chapter XVIII, §3. There we shall see that it gives the direction of maximal increase of the function, and that its magnitude is the rate of increase in that direction.

Using the formula for the derivative of a sum of two functions, and the derivative of a constant times a function, we conclude at once that the gradient satisfies the following properties:

Theorem 2.1. *Let f, g be two functions defined on an open set U, and assume that their partial derivatives exist at every point of U. Let c be a number. Then*

$$\mathrm{grad}(f + g) = \mathrm{grad}\, f + \mathrm{grad}\, g,$$

$$\mathrm{grad}(cf) = c\, \mathrm{grad}\, f.$$

We shall give later several geometric and physical interpretations for the gradient.

XVII, §2. EXERCISES

Find the partial derivatives

$$\frac{\partial f}{\partial x}, \quad \frac{\partial f}{\partial y}, \quad \text{and} \quad \frac{\partial f}{\partial z},$$

for the following functions $f(x, y)$ or $f(x, y, z)$.

1. $xy + z$ 2. $x^2 y^5 + 1$ 3. $\sin(xy) + \cos z$

4. $\cos(xy)$ 5. $\sin(xyz)$ 6. e^{xyz}

7. $x^2 \sin(yz)$ 8. xyz 9. $xz + yz + xy$

10. $x \cos(y - 3z) + \arcsin(xy)$

11. Find grad $f(P)$ if P is the point $(1, 2, 3)$ in Exercises 1, 2, 6, 8, and 9.

12. Find grad $f(P)$ if P is the point $(1, \pi, \pi)$ in Exercises 4, 5, 7.

13. Find grad $f(P)$ if

$$f(x, y, z) = \log(z + \sin(y^2 - x))$$

and

$$P = (1, -1, 1).$$

14. Find the partial derivatives of x^y. [*Hint:* $x^y = e^{y \log x}$.]

Find the gradient of the following functions at the given point.

15. $f(x, y, z) = e^{-2x} \cos(yz)$ at $(1, \pi, \pi)$

16. $f(x, y, z) = e^{3x+y} \sin(5z)$ at $(0, 0, \pi/6)$

XVII, §3. DIFFERENTIABILITY AND GRADIENT

Let f be a function defined on an open set U. Let X be a point of U. For all vectors H such that $\|H\|$ is small (and $H \neq O$), the point $X + H$ also lies in the open set. However, we **cannot** form a quotient

$$\frac{f(X + H) - f(x)}{H}$$

because it is **meaningless to divide by a vector**. In order to define what we mean for a function f to be differentiable, we must therefore find a way which does not involve dividing by H.

We reconsider the case of functions of one variable. Let us fix a number x. We had defined the derivative to be

$$f'(x) = \lim_{h \to 0} \frac{f(x + h) - f(x)}{h}.$$

Let

$$\varphi(h) = \frac{f(x + h) - f(x)}{h} - f'(x).$$

Then $\varphi(h)$ is not defined when $h = 0$, but

$$\lim_{h \to 0} \varphi(h) = 0.$$

We can write

$$f(x + h) - f(x) = f'(x)h + h\varphi(h).$$

This relation has meaning so far only when $h \neq 0$. However, we observe that if we define $\varphi(0)$ to be 0, then the preceding relation is obviously true when $h = 0$ (because we just get $0 = 0$).

Let

$$g(h) = \varphi(h) \qquad \text{if} \quad h > 0,$$

$$g(h) = -\varphi(h) \quad \text{if} \quad h < 0.$$

Then we have shown that if f is differentiable, there exists a function g such that

(1) $$f(x + h) - f(x) = f'(x)h + |h|g(h),$$

and

$$\lim_{h \to 0} g(h) = 0.$$

Conversely, suppose that there exists a number a and a function $g(h)$ such that

(1a) $$f(x + h) - f(x) = ah + |h|g(h).$$

and

$$\lim_{h \to 0} g(h) = 0.$$

We find for $h \neq 0$,

$$\frac{f(x + h) - f(x)}{h} = a + \frac{|h|}{h} g(h).$$

Taking the limit as h approaches 0, we observe that

$$\lim_{h \to 0} \frac{|h|}{h} g(h) = 0.$$

Hence the limit of the Newton quotient exists and is equal to a. Hence f is differentiable, and its derivative $f'(x)$ is equal to a.

Therefore, the existence of a number a and a function g satisfying (1a) above could have been used as the definition of differentiability in the case of functions of one variable. The great advantage of (1) is that no h appears in the denominator. It is this relation which will suggest to us how to define differentiability for functions of several variables, and how to prove the chain rule for them.

Let us begin with two variables. We let

$$X = (x, y) \qquad \text{and} \qquad H = (h, k).$$

Then the notion corresponding to $x + h$ in one variable is here

$$X + H = (x + h, y + k).$$

We wish to compare the values of a function f at X and $X + H$, i.e. we wish to investigate the difference

$$f(X + H) - f(X) = f(x + h, y + k) - f(x, y).$$

Definition. We say that f is **differentiable** at X if the partial derivatives

$$\frac{\partial f}{\partial x} \qquad \text{and} \qquad \frac{\partial f}{\partial y}$$

exist, and if there exists a function g (defined for small H) such that

$$\lim_{H \to O} g(H) = 0$$

and

(2)
$$f(x + h, y + k) - f(x, y) = \frac{\partial f}{\partial x} h + \frac{\partial f}{\partial y} k + \|H\| g(H).$$

We view the term

$$\frac{\partial f}{\partial x} h + \frac{\partial f}{\partial y} k$$

as an approximation to $f(X + H) - f(X)$, depending in a particularly simple way on h and k.

If we use the abbreviation

$$\operatorname{grad} f = \nabla f,$$

then formula (2) can be written

$$f(X + H) - f(X) = \nabla f(x) \cdot H + \|H\| g(H).$$

As with grad f, one must read $(\nabla f)(X)$ and not the meaningless $\nabla(f(X))$ since $f(X)$ is a number for each value of X, and thus it makes no sense to apply ∇ to a number. The symbol ∇ is applied to the function f, and $(\nabla f)(X)$ is the value of ∇f at X.

We now consider a function of n variables.

Let f be a function defined on an open set U. Let X be a point of U. If $H = (h_1,\ldots,h_n)$ is a vector such that $\|H\|$ is small enough, then $X + H$ will also be a point of U and so $f(X + H)$ is defined. Note that

$$X + H = (x_1 + h_1,\ldots,x_n + h_n).$$

This is the generalization of the $x + h$ with which we dealt previously in one variable, or the $(x + h, y + k)$ in two variables. For three variables, we already run out of convenient letters, so we may as well write n instead of 3.

Definition. We say that f is **differentiable** at X if the partial derivatives $D_1 f(X),\ldots,D_n f(X)$ exist, and if there exists a function g (defined for small H) such that

$$\lim_{H \to O} g(H) = 0 \quad \left(\text{also written} \quad \lim_{\|H\| \to 0} g(H) = 0 \right)$$

and

$$f(X + H) - f(X) = D_1 f(X)h_1 + \cdots + D_n f(x)h_n + \|H\|g(H).$$

With the other notation for partial derivatives, this last relation reads:

$$f(X + H) - f(X) = \frac{\partial f}{\partial x_1} h_1 + \cdots + \frac{\partial f}{\partial x_n} h_n + \|H\|g(H).$$

We say that f is **differentiable** in the open set U if it is differentiable at every point of U, so that the above relation holds for every point X in U.

In view of the definition of the gradient in §2, we can rewrite our fundamental relation in the form

(3) $$\boxed{f(X + H) - f(X) = (\operatorname{grad} f(X)) \cdot H + \|H\|g(H).}$$

The term $\|H\|g(H)$ has an order of magnitude smaller than the previous term involving the dot product. This is one advantage of the present notation. We know how to handle the formalism of dot products and are accustomed to it, and its geometric interpretation. This will help us later in interpreting the gradient geometrically.

Example 1. Suppose that we consider values for H pointing only in the direction of the standard unit vectors. In the case of two variables, consider for instance $H = (h, 0)$. Then for such H, the condition for differentiability reads:

$$f(X + H) = f(x + h, y) = f(x, y) + \frac{\partial f}{\partial x} h + |h| g(H).$$

In higher dimensional space, let $E_i = (0, \ldots, 0, 1, 0, \ldots, 0)$ be the i-th unit vector. Let $H = hE_i$ for some number h, so that

$$H = (0, \ldots, 0, h, 0, \ldots, 0).$$

Then for such H,

$$f(X + H) = f(X + hE_i) = f(X) + \frac{\partial f}{\partial x_i} h + |h| g(H),$$

and therefore if $h \neq 0$, we obtain

$$\frac{f(X + H) - f(x)}{h} = D_i f(X) + \frac{|h|}{h} g(H).$$

Because of the special choice of H, we can divide by the *number* h, but we are *not* dividing by the vector H.

The functions which we meet in practice are differentiable. The next theorem gives a criterion which shows that this is true. A function $\varphi(X)$ is said to be **continuous** if

$$\lim_{H \to 0} \varphi(X + H) = \varphi(X),$$

for all X in the domain of definition of the function.

Theorem 3.1. *Let f be a function defined on some open set U. Assume that its partial derivatives exist for every point in this open set, and that they are continuous. Then f is differentiable.*

We shall omit the proof. Observe that in practice, the partial derivatives of a function are given by formulas from which it is clear that they are continuous.

XVII, §3. EXERCISES

1. Let $f(x, y) = 2x - 3y$. What is $\partial f/\partial x$ and $\partial f/\partial y$?

2. Let $A = (a, b)$ and let f be the function on $\mathbf{R}^2$ such that $f(X) = A \cdot X$. Let $X = (x, y)$. In terms of the coordinates of A, determine $\partial f/\partial x$ and $\partial f/\partial y$.

3. Let $A = (a, b, c)$ and let f be the function on $\mathbf{R}^3$ such that $f(X) = A \cdot X$. Let $X = (x, y, z)$. In terms of the coordinates of A, determine $\partial f/\partial x$, $\partial f/\partial y$, and $\partial f/\partial z$.

4. Generalize the above two exercises to n-space.

5. Let f be defined on an open set U. Let X be a point of U. Let A be a vector, and let g be a function defined for small H, such that

$$\lim_{H \to O} g(H) = 0.$$

Assume that

$$f(X + H) - f(X) = A \cdot H + \|H\| g(H).$$

Prove that $A = \operatorname{grad} f(X)$. You may do this exercise in 2 variables first and then in 3 variables, and let it go at that. Use coordinates, e.g. let $A = (a, b)$ and $X = (x, y)$. Use special values of H, as in Example 1.

CHAPTER XVIII

The Chain Rule and the Gradient

In this chapter, we prove the chain rule for functions of several variables and give a number of applications. Among them will be several interpretations for the gradient. These form one of the central points of our theory. They show how powerful the tools we have accumulated turn out to be.

XVIII, §1. THE CHAIN RULE

Let f be a function defined on some open set U. Let $C(t)$ be a curve such that the values $C(t)$ are contained in U. Then we can form the composite function $f \circ C$, which is a function of t, given by

$$(f \circ C)(t) = f(C(t)).$$

Example 1. Take $f(x, y) = e^x \sin(xy)$. Let $C(t) = (t^2, t^3)$. Then

$$f(C(t)) = e^{t^2} \sin(t^5).$$

The expression on the right is obtained by substituting t^2 for x and t^3 for y in $f(x, y)$. This is a function of t in the old sense of functions of one variable. If we interpret f as the temperature, then $f(C(t))$ is the temperature of a bug traveling along the curve $C(t)$ at time t.

The chain rule tells us how to find the derivative of this function, provided we know the gradient of f and the derivative C'. Its statement is as follows.

Chain rule. *Let f be a function which is defined and differentiable on an open set U. Let C be a differentiable curve (defined for some interval of numbers t) such that the values C(t) lie in the open set U. Then the function*

$$f(C(t))$$

is differentiable (as a function of t), and

$$\frac{df(C(t))}{dt} = (\operatorname{grad} f(C(t))) \cdot C'(t).$$

Memorize this formula by repeating it out loud.

In the notation dC/dt, this also reads

$$\frac{df(C(t))}{dt} = (\operatorname{grad} f)(C(t)) \cdot \frac{dC}{dt}.$$

Proof of the Chain Rule. By definition, we must investigate the quotient

$$\frac{f(C(t+h)) - f(C(t))}{h}.$$

Let

$$K = K(t, h) = C(t+h) - C(t).$$

Then our quotient can be rewritten in the form

$$\frac{f(C(t) + K) - f(C(t))}{h}.$$

Using the definition of differentiability for f, we have

$$f(X + K) - f(X) = (\operatorname{grad} f)(X) \cdot K + \|K\| g(K)$$

and

$$\lim_{\|K\| \to 0} g(K) = 0.$$

Replacing K by what it stands for, namely $C(t+h) - C(t)$, and dividing by h, we obtain:

$$\frac{f(C(t+h)) - f(C(t))}{h} = (\operatorname{grad} f)(C(t)) \cdot \frac{C(t+h) - C(t)}{h}$$

$$\pm \left\| \frac{C(t+h) - C(t)}{h} \right\| g(K).$$

As h approaches 0, the first term of the sum approaches what we want, namely

$$(\operatorname{grad} f)(C(t)) \cdot C'(t).$$

The second term approaches

$$\pm \|C'(t)\| \lim_{h \to 0} g(K),$$

and when h approaches 0, so does $K = C(t + h) - C(t)$. Hence the second term of the sum approaches 0. This proves our chain rule.

To use the chain rule for certain computations, it is convenient to reformulate it in terms of components, and in terms of the two notations we have used for partial derivatives

$$\frac{\partial f}{\partial x} = D_1 f(x, y), \qquad \frac{\partial f}{\partial y} = D_2 f(x, y)$$

when the variables are x, y.

Suppose $C(t)$ is given in terms of coordinates by

$$C(t) = (x_1(t), \ldots, x_n(t)),$$

then

$$\boxed{\frac{d(f(C(t)))}{dt} = \frac{\partial f}{\partial x_1} \frac{dx_1}{dt} + \cdots + \frac{\partial f}{\partial x_n} \frac{dx_n}{dt}.}$$

If f is a function of two variables (x, y) then

$$\boxed{\frac{df(C(t))}{dt} = \frac{\partial f}{\partial x} \frac{dx}{dt} + \frac{\partial f}{\partial y} \frac{dy}{dt}.}$$

In the D_1, D_2 notation, we can write this formula in the form

$$\boxed{\frac{d}{dt}(f(x(t), y(t))) = (D_1 f)(x, y) \frac{dx}{dt} + (D_2 f)(x, y) \frac{dy}{dt},}$$

and similarly for several variables. For simplicity we usually omit the parentheses around $D_1 f$ and $D_2 f$. Also on the right-hand side we have abbreviated $x(t)$, $y(t)$ to x, y, respectively. Without any abbreviation, the formula reads:

$$\frac{d}{dt}(f(x(t), y(t))) = D_1 f(x(t), y(t)) \frac{dx}{dt} + D_2 f(x(t), y(t)) \frac{dy}{dt}.$$

Example 2. Let $C(t) = (e^t, t, t^2)$ and let $f(x, y, z) = x^2 yz$. Then putting

$$x = e^t, \qquad y = t, \qquad z = t^2$$

we get:

$$\frac{d}{dt} f(C(t)) = \frac{\partial f}{\partial x} \frac{dx}{dt} + \frac{\partial f}{\partial y} \frac{dy}{dt} + \frac{\partial f}{\partial z} \frac{dz}{dt}$$

$$= 2xyze^t + x^2 z + x^2 y 2t.$$

If we want this function entirely in terms of t, we substitute back the values for x, y, z in terms of t, and get

$$\frac{d}{dt} f(C(t)) = 2e^t t t^2 e^t + e^{2t} t^2 + e^{2t} t 2t$$

$$= 2t^3 e^{2t} + t^2 e^{2t} + 2t^2 e^{2t}.$$

In some cases, as in the next example, one does not use the chain rule in several variables, just the old one from one-variable calculus.

Example 3. Let

$$f(x, y, z) = \sin(x^2 - 3zy + xz).$$

Then keeping y and z constant, and differentiating with respect to x, we find

$$\frac{\partial f}{\partial x} = \cos(x^2 - 3zy + xz) \cdot (2x + z).$$

More generally, let

$$f(x, y, z) = g(x^2 - 3zy + xz),$$

where g is a differentiable function of one variable. [In the special case above, we have $g(u) = \sin u$.] Then the chain rule gives

$$\frac{\partial f}{\partial x} = g'(x^2 - 3zy + xz)(2x + z).$$

We denote the derivative of g by g' as usual. We do *not* write it as dg/dx, because x is a letter which is already occupied for other purposes. We could let

$$u = x^2 - 3zy + xz,$$

in which case it would be all right to write

$$\boxed{\frac{\partial f}{\partial x} = \frac{dg}{du}\frac{\partial u}{\partial x},}$$

and we would get the same answer as above.

XVIII, §1. EXERCISES

1. Let P, A be constant vectors. If $g(t) = f(P + tA)$, show that

$$g'(t) = (\operatorname{grad} f)(P + tA) \cdot A.$$

2. Suppose that f is a function such that

$$\operatorname{grad} f(1, 1, 1) = (5, 2, 1).$$

Let $C(t) = (t^2, t^{-3}, t)$. Find

$$\frac{d}{dt}(f(C(t))) \qquad \text{at} \quad t = 1.$$

3. Let $f(x, y) = e^{9x + 2y}$ and $g(x, y) = \sin(4x + y)$. Let C be a curve such that $C(0) = (0, 0)$. Given:

$$\left.\frac{d}{dt}f(C(t))\right|_{t=0} = 2 \qquad \text{and} \qquad \left.\frac{d}{dt}g(C(t))\right|_{t=0} = 1,$$

Find $C'(0)$.

4. (a) Let P be a constant vector. Let $g(t) = f(tP)$, where f is some differentiable function. What is $g'(t)$?
   (b) Let f be a differentiable function defined on all of space. Assume that $f(tP) = tf(P)$ for all numbers t and all points P. Show that for all P we have

$$f(P) = \operatorname{grad} f(O) \cdot P.$$

5. Let f be a differentiable function of two variables and assume that there is an integer $m \geq 1$ such that

$$f(tx, ty) = t^m f(x, y)$$

for all numbers t and all x, y. Prove **Euler's relation**

$$x\frac{\partial f}{\partial x} + y\frac{\partial f}{\partial y} = mf(x, y).$$

[*Hint*: Let $C(t) = (tx, ty)$. Differentiate both sides of the given equation with respect to t, keeping x and y **constant**. Then put $t = 1$.]

6. Generalize Exercise 5 to n variables, namely let f be a differentiable function of n variables and assume that there exists an integer $m \geq 1$ such that $f(tX) = t^m f(X)$ for all numbers t and all points X in $\mathbf{R}^n$. Show that

$$x_1\frac{\partial f}{\partial x_1} + \cdots + x_n\frac{\partial f}{\partial x_n} = mf(X),$$

which can also be written $X \cdot \operatorname{grad} f(X) = mf(X)$.

7. (a) Let $f(x, y) = (x^2 + y^2)^{1/2}$. Find $\partial f/\partial x$ and $\partial f/\partial y$.
   (b) Let $f(x, y, z) = (x^2 + y^2 + z^2)^{1/2}$. Find $\partial f/\partial x$, $\partial f/\partial y$, $\partial f/\partial z$.

8. Let $r = (x_1^2 + \cdots + x_n^2)^{1/2}$. What is $\partial r/\partial x_i$?

9. Find the derivatives with respect to x and y of the following functions.
   (a) $\sin(x^3y + 2x^2)$ (b) $\cos(3x^2y - 4x)$
   (c) $\log(x^2y + 5y)$ (d) $(x^2y + 4x)^{1/2}$

XVIII, §2. TANGENT PLANE

We begin by an example analyzing a function along a curve where the values of the function are constant. This gives rise to a very important principle of perpendicularity.

Example 1. Let f be a function on $\mathbf{R}^3$. Let us interpret f as giving the temperature, so that at any point X in $\mathbf{R}^3$, the value of the function $f(X)$ is the temperature at X. Suppose that a bug moves in space along a differentiable curve, which we may denote in parametric form by

$$B(t).$$

Thus $B(t) = (x(t), y(t), z(t))$ is the position of the bug at time t. Let us assume that the bug starts from a point where it feels that the temperature is comfortable, and therefore that the temperature is constant along the path on which it moves. In other words, f is constant along the curve $B(t)$. This means that for all values of t, we have

$$f(B(t)) = k,$$

where k is constant. Differentiating with respect to t, and using the chain rule, we find that

$$\operatorname{grad} f(B(t)) \cdot B'(t) = 0.$$

This means that the gradient of f is perpendicular to the velocity vector at every point of the curve.

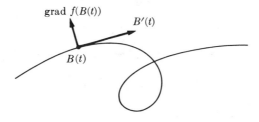

Figure 1

Let f be a differentiable function defined on an open set U in 3-space, and let k be a number. The set of points X such that

$$f(X) = k \quad \text{and} \quad \operatorname{grad} f(X) \neq O$$

is called a **surface**. It is the level surface of level k, for the function f. For the applications we have in mind, we impose the additional condition that $\operatorname{grad} f(X) \neq O$. It can be shown that this eliminates the points where the surface is not smooth.

Let $C(t)$ be a differentiable curve. We shall say that the curve **lies on the surface** if, for all t, we have

$$f(C(t)) = k.$$

This simply means that all the points of the curve satisfy the equation of the surface. For instance, let the surface be defined by the equation

$$x^2 + y^2 + z^2 = 1.$$

The surface is the sphere of radius 1, centered at the origin, and here we have $f(x, y, z) = x^2 + y^2 + z^2$. Let

$$C(t) = (x(t), y(t), z(t))$$

be a curve, defined for t in some interval. Then $C(t)$ lies on the surface means that

$$x(t)^2 + y(t)^2 + z(t)^2 = 1 \quad \text{for all } t \text{ in the interval.}$$

In other words,

$$f(C(t)) = 1, \quad \text{or also} \quad C(t)^2 = 1.$$

For theoretical purposes, it is neater to write $f(C(t)) = 1$. For computational purposes, we have to go back to coordinates if we want specific numerical values in a given problem.

Now suppose that a curve $C(t)$ lies on a surface $f(X) = k$. Thus we have

$$f(C(t)) = k \quad \text{for all } t.$$

If we differentiate this relation, we get from the chain rule:

$$\operatorname{grad} f(C(t)) \cdot C'(t) = 0.$$

Let P be a point of the surface, and let $C(t)$ be a curve on the surface passing through P. This means that there is a number t_0 such that $C(t_0) = P$. For this value t_0, we obtain

$$\operatorname{grad} f(P) \cdot C'(t_0) = 0.$$

Thus the gradient of f at P is perpendicular to the tangent vector of the curve at P. [We assume that $C'(t_0) \neq O$.] This is true for **every** differentiable curve on the surface passing through P. It is therefore very reasonable to make the following

Definition. The **tangent plane** to the surface $f(X) = k$ at the point P is the plane through P, perpendicular to $\operatorname{grad} f(P)$.

We know from Chapter XV how to find such a plane. The definition applies only when $\operatorname{grad} f(P) \neq O$. If

$$\operatorname{grad} f(P) = O,$$

then we do not define the notion of tangent plane.

The fact that $\operatorname{grad} f(P)$ is perpendicular to every curve passing through P on the surface also gives us an interpretation of the gradient as being perpendicular to the surface

$$f(X) = k.$$

which is one of the level surfaces for the function f (Fig. 2).

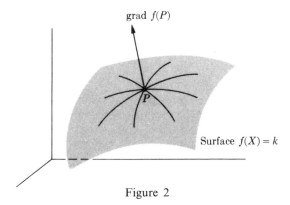

grad $f(P)$

P

Surface $f(X) = k$

Figure 2

Example 2. Find the tangent plane to the surface

$$x^2 + y^2 + z^2 = 3$$

at the point (1, 1, 1).

Let $f(X) = x^2 + y^2 + z^2$. Then at the point $P = (1, 1, 1)$,

$$\operatorname{grad} f(P) = (2, 2, 2).$$

The equation of a plane passing through P and perpendicular to a vector N is

$$X \cdot N = P \cdot N.$$

In the present case, this yields

$$2x + 2y + 2z = 2 + 2 + 2 = 6.$$

Observe that our arguments also give us a means of finding a vector perpendicular to a curve in 2-space at a given point, simply by applying the preceding discussion to the plane instead of 3-space. A curve is defined by an equation $f(x, y) = k$, and in this case, $\operatorname{grad} f(x_0, y_0)$ is perpendicular to the curve at the point (x_0, y_0) on the curve.

Example 3. Find the tangent line to the curve

$$x^2 y + y^3 = 10$$

at the point $P = (1, 2)$, and find a vector perpendicular to the curve at that point.

Let $f(x, y) = x^2 y + y^3$. Then

$$\operatorname{grad} f(x, y) = (2xy, x^2 + 3y^2),$$

and so

$$\operatorname{grad} f(P) = \operatorname{grad} f(1, 2) = (4, 13).$$

Let $N = (4, 13)$. Then N is perpendicular to the curve at the given point. The tangent line is given by $X \cdot N = P \cdot N$, and thus its equation is

$$4x + 13y = 4 + 26 = 30.$$

Example 4. A surface may also be given in the form $z = g(x, y)$ where g is some function of two variables. In this case, the tangent plane is determined by viewing the surface as expressed by the equation

$$g(x, y) - z = 0.$$

For instance, suppose the surface is given by $z = x^2 + y^2$. We wish to determine the tangent plane at $(1, 2, 5)$. Let $f(x, y, z) = x^2 + y^2 - z$. Then

$$\text{grad } f(x, y, z) = (2x, 2y, -1) \quad \text{and} \quad \text{grad } f(1, 2, 5) = (2, 4, -1).$$

The equation of the tangent plane at $P = (1, 2, 5)$ perpendicular to

$$N = (2, 4, -1)$$

is

$$2x + 4y - z = P \cdot N = 5.$$

This is the desired equation.

Example 5. Find a parametric equation for the tangent line to the curve of intersection of the two surfaces

$$x^2 + y^2 + z^2 = 6 \quad \text{and} \quad x^3 - y^2 + z = 2,$$

at the point $P = (1, 1, 2)$.

The tangent line to the curve is the line in common with the tangent planes of the two surfaces at the point P. We know how to find these tangent planes, and in Chapter XV, we learned how to find the parametric representation of the line common to two planes, so we know how to do this problem. We carry out the numerical computation in full.

The first surface is defined by the equation $f(x, y, z) = 6$. A vector N_1 perpendicular to this first surface at P is given by

$$N_1 = \text{grad } f(P), \quad \text{where} \quad \text{grad } f(x, y, z) = (2x, 2y, 2z).$$

Thus for $P = (1, 1, 2)$ we find

$$N_1 = (2, 2, 4).$$

The second surface is given by the equation $g(x, y, z) = 2$, and

$$\text{grad } g(x, y, z) = (3x^2, -2y, 1).$$

Thus a vector N_2 perpendicular to the second surface at P is

$$N_2 = \text{grad } g(1, 1, 2) = (3, -2, 1).$$

A vector $A = (a, b, c)$ in the direction of the line of intersection is perpendicular to both N_1 and N_2. To find A, we therefore have to solve the equations

$$A \cdot N_1 = 0 \quad \text{and} \quad A \cdot N_2 = 0.$$

This amounts to solving

$$2a + 2b + 4c = 0,$$

$$3a - 2b + c = 0.$$

Let, for instance, $a = 1$. Solving for b and c yields

$$a = 1, \quad b = 1, \quad c = -1.$$

Thus $A = (1, 1, -1)$. Finally, the parametric representation of the desired line is

$$P + tA = (1, 1, 2) + t(1, 1, -1).$$

XVIII, §2. EXERCISES

1. Find the equation of the tangent plane and normal line to each of the following surfaces at the specific point.
   (a) $x^2 + y^2 + z^2 = 49$ at $(6, 2, 3)$
   (b) $xy + yz + zx - 1 = 0$ at $(1, 1, 0)$
   (c) $x^2 + xy^2 + y^3 + z + 1 = 0$ at $(2, -3, 4)$
   (d) $2y - z^3 - 3xz = 0$ at $(1, 7, 2)$
   (e) $x^2y^2 + xz - 2y^3 = 10$ at $(2, 1, 4)$
   (f) $\sin xy + \sin yz + \sin xz = 1$ at $(1, \pi/2, 0)$

2. Let $f(x, y, z) = z - e^x \sin y$, and $P = (\log 3, 3\pi/2, -3)$. Find:
   (a) grad $f(P)$,
   (b) the normal line at P to the level surface for f which passes through P,
   (c) the tangent plane to this surface at P.

3. Find a parametric representation of the tangent line to the curve of intersection of the following surfaces at the indicated point.
   (a) $x^2 + y^2 + z^2 = 49$ and $x^2 + y^2 = 13$ at $(3, 2, -6)$
   (b) $xy + z = 0$ and $x^2 + y^2 + z^2 = 9$ at $(2, 1, -2)$
   (c) $x^2 - y^2 - z^2 = 1$ and $x^2 - y^2 + z^2 = 9$ at $(3, 2, 2)$
   [*Note*: The tangent line above may be defined to be the line of intersection of the tangent planes of the given point.]

4. Let $f(X) = 0$ be a differentiable surface. Let Q be a point which does not lie on the surface. Given a differentiable curve $C(t)$ on the surface, defined on an open interval, give the formula for the distance between Q and a point $C(t)$. Assume that this distance reaches a minimum for $t = t_0$. Let $P = C(t_0)$. Show that the line joining Q to P is perpendicular to the curve at P.

5. Find the equation of the tangent plane to the surface $z = f(x, y)$ at the given point P when f is the following function:
   (a) $f(x, y) = x^2 + y^2$, $P = (3, 4, 25)$
   (b) $f(x, y) = x/(x^2 + y^2)^{1/2}$, $P = (3, -4, \frac{3}{5})$
   (c) $f(x, y) = \sin(xy)$ at $P = (1, \pi, 0)$

6. Find the equation of the tangent plane to the surface $x = e^{2y-z}$ at $(1, 1, 2)$.

7. Let $f(x, y, z) = xy + yz + zx$. (a) Write down the equation of the level surface for f through the point $P = (1, 1, 0)$. (b) Find the equation of the tangent plane to this surface at P.

8. Find the equation of the tangent plane to the surface

$$3x^2 - 2y + z^3 = 9$$

at the point $(1, 1, 2)$

9. Find the equation of the tangent plane to the surface

$$z = \sin(x + y)$$

at the point where $x = 1$ and $y = 2$.

10. Find the tangent plane to the surface $x^2 + y^2 - z^2 = 18$ at the point $(3, 5, -4)$.

11. (a) Find a unit vector perpendicular to the surface

$$x^3 + xz = 1$$

at the point $(1, 2, -1)$.
(b) Find the equation of the tangent plane at that point.

12. Find the cosine of the angle between the surfaces

$$x^2 + y^2 + z^2 = 3 \quad \text{and} \quad x - z^2 - y^2 = -3$$

at the point $(-1, 1, -1)$. (This angle is the angle between the normal vectors at the point.)

13. (a) A differentiable curve $C(t)$ lies on the surface

$$x^2 + 4y^2 + 9z^2 = 14,$$

and is so parametrized that $C(0) = (1, 1, 1)$. Let

$$f(x, y, z) = x^2 + 4y^2 + 9z^2$$

and let $h(t) = f(C(t))$. Find $h'(0)$.
(b) Let $g(x, y, z) = x^2 + y^2 + z^2$ and let $k(t) = g(C(t))$. Suppose in addition that $C'(0) = (4, -1, 0)$, find $k'(0)$.

14. Find the equation of the tangent plane to the level surface

$$(x + y + z)e^{xyz} = 3e$$

at the point $(1, 1, 1)$.

XVIII, §3. DIRECTIONAL DERIVATIVE

Let f be defined on an open set and assume that f is differentiable. Let P be a point of the open set, and let A be a **unit vector** (i.e. $\|A\| = 1$). Then $P + tA$ is the parametric representation of a straight line in the direction of A and passing through P. We observe that

$$\frac{d(P + tA)}{dt} = A.$$

For instance, if $n = 2$ and $P = (p, q)$, $A = (a, b)$, then

$$P + tA = (p + ta, q + tb),$$

or in terms of coordinates,

$$x = p + ta, \qquad y = q + tb.$$

Hence

$$\frac{dx}{dt} = a \qquad \text{and} \qquad \frac{dy}{dt} = b$$

so that

$$\frac{d(P + tA)}{dt} = (a, b) = A.$$

The same argument works in higher dimensions.

We wish to consider the rate of change of f in the direction of A. It is natural to consider the values of f on the line $P + tA$, that is to consider the values

$$f(P + tA).$$

The rate of change of f along this line will then be given by taking the derivative of this expression, which we know how to do. We illustrate the line $P + tA$ in the figure.

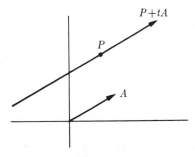

Figure 3

If f represents a temperature at the point P, we look at the variation of temperature in the direction of A, starting from the point P. The value $f(P + tA)$ gives the temperature at the point $P + tA$. This is a function of t, say

$$g(t) = f(P + tA).$$

The rate of change of this temperature function is $g'(t)$, the derivative with respect to t, and $g'(0)$ is the rate of change at time $t = 0$, i.e. the rate of change of f at the point P, in the direction of A.

By the chain rule, if we take the derivative of the function

$$g(t) = f(P + tA),$$

which is defined for small values of t, we obtain

$$\frac{df(P + tA)}{dt} = \operatorname{grad} f(P + tA) \cdot A.$$

When t is equal to 0, this derivative is equal to

$$\operatorname{grad} f(P) \cdot A.$$

For obvious reasons, we now make the

Definition. Let A be a unit vector. The **directional derivative of f in the direction of A at P** is the number

$$D_A f(P) = \operatorname{grad} f(P) \cdot A.$$

We interpret this directional derivative as the rate of change of f along the straight line in the direction of A, at the point P. Thus if we agree on the notation $D_A f(P)$ for the directional derivative of f at P in the direction of the unit vector A, then we have

$$\boxed{D_A f(P) = \left. \frac{df(P + tA)}{dt} \right|_{t=0} = \operatorname{grad} f(P) \cdot A.}$$

In using this formula, the reader should remember that A is taken to be a **unit vector.** When a direction is given in terms of a vector whose norm is not 1, then one must first divide this vector by its norm before applying the formula.

Example 1. Let $f(x, y) = x^2 + y^3$ and let $B = (1, 2)$. Find the directional derivative of f in the direction of B, at the point $(-1, 3)$.

We note that B is not a unit vector. Its norm is $\sqrt{5}$. Let

$$A = \frac{1}{\sqrt{5}} B.$$

Then A is a unit vector having the same direction as B. Let

$$P = (-1, 3).$$

Then grad $f(P) = (-2, 27)$. Hence by our formula, the directional derivative is equal to:

$$\operatorname{grad} f(P) \cdot A = \frac{1}{\sqrt{5}} (-2 + 54) = \frac{52}{\sqrt{5}}.$$

Consider again a differentiable function f on an open set U.

Let P be a point of U. *Let us assume that* grad $f(P) \neq O$, and let A be a unit vector. We know that

$$D_A f(P) = \operatorname{grad} f(P) \cdot A = \|\operatorname{grad} f(P)\| \|A\| \cos \theta,$$

where θ is the angle between grad $f(P)$ and A. Since $\|A\| = 1$, we see that the directional derivative is equal to

$$\boxed{D_A f(P) = \|\operatorname{grad} f(P)\| \cos \theta.}$$

We remind the reader that this formula holds only when A is a unit vector.

The value of $\cos \theta$ varies between -1 and $+1$ when we select all possible unit vectors A.

The maximal value of $\cos \theta$ is obtained when we select A such that $\theta = 0$, i.e. when we select A to have the same direction as grad $f(P)$. In that case, the directional derivative is equal to the norm of the gradient.

Thus we have obtained another interpretation for the gradient:

> *The direction of the gradient is that of maximal increase of the function.*
>
> *The norm of the gradient is the rate of increase of the function in that direction (i.e. in the direction of maximal increase).*

The directional derivative in the direction of A is at a minimum when $\cos \theta = -1$. This is the case when we select A to have opposite direction to grad $f(P)$. That direction is therefore the direction of maximal decrease of the function.

For example, f might represent a temperature distribution in space. At any point P, a particle which feels cold and wants to become warmer fastest should move in the direction of grad $f(P)$. Another particle which is warm and wants to cool down fastest should move in the direction of $-$grad $f(P)$.

Example 2. Let $f(x, y) = x^2 + y^3$ again, and let $P = (-1, 3)$. Find the directional derivative of f at P, in the direction of maximal increase of f.

We have found previously that grad $f(P) = (-2, 27)$. The directional derivative of f in the direction of maximal increase is precisely the norm of the gradient, and so is equal to

$$\|\text{grad} f(P)\| = \|(-2, 27)\| = \sqrt{4 + 27^2} = \sqrt{733}.$$

XVIII, §3. EXERCISES

1. Let $f(x, y, z) = z - e^x \sin y$, and $P = (\log 3, 3\pi/2, -3)$. Find:
   (a) the directional derivative of f at P in the direction of $(1, 2, 2)$,
   (b) the maximum and minimum values for the directional derivative of f at P.

2. Find the directional derivatives of the following functions at the specified points in the specified directions.
   (a) $\log(x^2 + y^2)^{1/2}$ at $(1, 1)$, direction $(2, 1)$
   (b) $xy + yz + zx$ at $(-1, 1, 7)$, direction $(3, 4, -12)$
   (c) $4x^2 + 9y^2$ at $(2, 1)$ in the direction of maximum directional derivative

3. A temperature distribution in space is given by the function

$$f(x, y) = 10 + 6 \cos x \cos y + 3 \cos 2x + 4 \cos 3y.$$

   At the point $(\pi/3, \pi/3)$, find the direction of greatest increase of temperature, and the direction of greatest decrease of temperature.

4. In what direction are the following functions of X increasing most rapidly at the given point?
   (a) $x/\|X\|^{3/2}$ at $(1, -1, 2)$ $(X = (x, y, z))$
   (b) $\|X\|^5$ at $(1, 2, -1, 1)$ $(X = (x, y, z, w))$

5. (a) Find the directional derivative of the function

$$f(x, y) = 4xy + 3y^2$$

   in the direction of $(2, -1)$, at the point $(1, 1)$.

(b) Find the directional derivative in the direction of maximal increase of the function.

6. Let $f(x, y, z) = (x + y)^2 + (y + z)^2 + (z + x)^2$. What is the direction of greatest increase of the function at the point $(2, -1, 2)$? What is the directional derivative of f in this direction at that point?

7. Let $f(x, y) = x^2 + xy + y^2$. What is the direction in which f is increasing most rapidly at the point $(-1, 1)$? Find the directional derivative of f in this direction.

8. Suppose the temperature in (x, y, z)-space is given by

$$f(x, y, z) = x^2 y + yz - e^{xy}.$$

Compute the rate of change of temperature at the point $P = (1, 1, 1)$ in the direction of $\overrightarrow{PO}$.

9. (a) Find the directional derivative of the function

$$f(x, y, z) = \sin(xyz)$$

at the point $P = (\pi, 1, 1)$ in the direction of $\overrightarrow{OA}$ where A is the unit vector $(0, 1/\sqrt{2}, -1/\sqrt{2})$.

(b) Let U be a unit vector whose direction is *opposite* to that of

$$(\operatorname{grad} f)\,(P).$$

What is the value of the directional derivative of f at P in the direction of U?

10. Let f be a differentiable function defined on an open set U. Suppose that P is a point of U such that $f(P)$ is a maximum, i.e. suppose we have

$$f(P) \geqq f(X) \qquad \text{for all} \quad X \text{ in } U.$$

Show that grad $f(P) = O$.

XVIII, §4. FUNCTIONS DEPENDING ONLY ON THE DISTANCE FROM THE ORIGIN

The first such function which comes to mind is the distance function. In 2-space, it is given by

$$r = \sqrt{x^2 + y^2}.$$

In 3-space, it is given by

$$r = \sqrt{x^2 + y^2 + z^2}.$$

In n-space, it is given by

$$r = \sqrt{x_1^2 + x_2^2 + \cdots + x_n^2}.$$

Let us find its gradient. For instance, in 2-space,

$$\frac{\partial r}{\partial x} = \frac{1}{2}(x^2 + y^2)^{-1/2}2x$$

$$= \frac{x}{\sqrt{x^2 + y^2}} = \frac{x}{r}.$$

Differentiating with respect to y instead of x you will find

$$\frac{\partial r}{\partial y} = \frac{y}{r}.$$

Hence

$$\boxed{\operatorname{grad} r = \left(\frac{x}{r}, \frac{y}{r}\right).}$$

This can also be written

$$\boxed{\operatorname{grad} r = \frac{X}{r}.}$$

Thus the gradient of r is the unit vector in the direction of the position vector. It points outward from the origin.

If we are dealing with functions on 3-space, so

$$r = \sqrt{x^2 + y^2 + z^2}$$

then the chain rule again gives

$$\boxed{\frac{\partial r}{\partial x} = \frac{x}{r}, \quad \frac{\partial r}{\partial y} = \frac{y}{r}, \quad \text{and} \quad \frac{\partial r}{\partial z} = \frac{z}{r}}$$

so again

$$\boxed{\operatorname{grad} r = \frac{X}{r}.}$$

Warning: Do **not** write $\partial r / \partial X$. This suggests dividing by a vector X and is therefore bad notation. The notation $\partial r / \partial x$ was correct and good notation since we differentiate only with respect to the single variable x. Information coming from differentiating with respect to all the variables is correctly expressed by the formula grad $r = X/r$ in the box.

In n-space, let

$$r = \sqrt{x_1^2 + \cdots + x_n^2}.$$

Then

$$\frac{\partial r}{\partial x_i} = \tfrac{1}{2}(x_1^2 + \cdots + x_n^2)^{-1/2} 2x_i$$

so

$$\boxed{\frac{\partial r}{\partial x_i} = \frac{x_i}{r}.}$$

By definition of the gradient, it follows that

$$\boxed{\text{grad } r = \frac{X}{r}.}$$

We now come to other functions depending on the distance. Such functions arise frequently. For instance, a temperature function may be inversely proportional to the distance from the source of heat. A potential function may be inversely proportional to the square of the distance from a certain point. The gradient of such functions has special properties which we discuss further.

Example 1. Let

$$f(x, y) = \sin r = \sin \sqrt{x^2 + y^2}.$$

Then $f(x, y)$ depends only on the distance r of (x, y) from the origin. By the chain rule,

$$\frac{\partial f}{\partial x} = \frac{d \sin r}{dr} \cdot \frac{\partial r}{\partial x}$$

$$= (\cos r)\tfrac{1}{2}(x^2 + y^2)^{-1/2} 2x$$

$$= (\cos r) \frac{x}{r}.$$

Similarly, $\partial f / \partial y = (\cos r) y / r$. Consequently

$$\operatorname{grad} f(x, y) = \left((\cos r) \frac{x}{r}, (\cos r) \frac{y}{r} \right)$$

$$= \frac{\cos r}{r} (x, y)$$

$$= \frac{\cos r}{r} X.$$

The same use of the chain rule as in the special case

$$f(x, y) = \sin r$$

which we worked out in Example 1 shows:

Let g be a differentiable function of one variable, and let $f(X) = g(r)$. Then

$$\boxed{\operatorname{grad} f(X) = \frac{g'(r)}{r} X.}$$

Work out all the examples given in Exercise 2. You should memorize and keep in mind this simple expression for the gradient of a function which depends only on the distance. Such dependence is expressed by the function g.

Exercises 9 and 10 give important information concerning functions which depend only on the distance from the origin, and should be seen as essential complements of this section. They will prove the following result.

A differentiable function $f(X)$ depends only on the distance of X from the origin if and only if $\operatorname{grad} f(X)$ *is parallel to X, or O.*

In this situation, the gradient $\operatorname{grad} f(X)$ may point towards the origin, or away from the origin, depending on whether the function is decreasing or increasing as the point moves away from the origin.

Example 2. Suppose a heater is located at the origin, and the temperature at a point decreases as a function of the distance from the origin, say is inversely proportional to the square of the distance from the origin. Then temperature is given as

$$h(X) = g(r) = k / r^2$$

for some constant $k > 0$. Then the gradient of temperature is

$$\operatorname{grad} h(X) = -2k\frac{1}{r^3}\frac{X}{r} = -\frac{2k}{r^4}X.$$

The factor $2k/r^4$ is positive, and we see that grad $h(X)$ points in the direction of $-X$. Each circle centered at the origin is a level curve for temperature. Thus the gradient may be drawn as on the following figure. The gradient is parallel to X but in opposite direction. A bug traveling along the circle will stay at constant temperature. If it wants to get warmer fastest, it must move toward the origin.

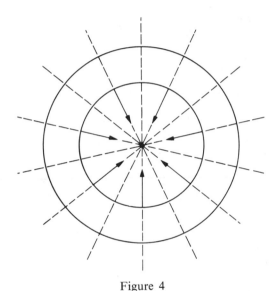

Figure 4

The dotted lines indicate the path of the bug when moving in the direction of maximal increase of the function. These lines are perpendicular to the circles of constant temperature.

XVIII, §4. EXERCISES

1. Let g be a function of r, let $r = \|X\|$, and $X = (x, y, z)$. Let $f(X) = g(r)$. Show that

$$\left(\frac{dg}{dr}\right)^2 = \left(\frac{\partial f}{\partial x}\right)^2 + \left(\frac{\partial f}{\partial y}\right)^2 + \left(\frac{\partial f}{\partial z}\right)^2.$$

2. Let g be a function of r, and $r = \|X\|$. Let $f(X) = g(r)$. Find grad $f(X)$ for the following functions.

(a) $g(r) = 1/r$ (b) $g(r) = r^2$ (c) $g(r) = 1/r^3$
(d) $g(r) = e^{-r^2}$ (e) $g(r) = \log 1/r$ (f) $g(r) = 4/r^m$
(g) $g(r) = \cos r$

You may either work out each exercise separately, writing

$$r = \sqrt{x_1^2 + \cdots + x_n^2},$$

and use the chain rule, finding $\partial f / \partial x_i$ in each case, or you may apply the general formula obtained in Example 1, that if $f(X) = g(r)$, we have

$$\operatorname{grad} f(X) = \frac{g'(r)}{r} X.$$

Probably you should do both for a while to get used to the various notations and situations which may rise.

The next five exercises concern certain parametrizations, and some of the results from them will be used in Exercise 9.

3. Let A. B be two unit vectors such that $A \cdot B = 0$. Let

$$F(t) = (\cos t)A + (\sin t)B.$$

Show that $F(t)$ lies on the sphere of radius 1 centered at the origin, for each value of t. [*Hint*: What is $F(t) \cdot F(t)$?]

4. Let P, Q be two points on the sphere of radius 1, centered at the origin. Let $L(t) = P + t(Q - P)$, with $0 \le t \le 1$. If there exists a value of t in $[0, 1]$ such that $L(t) = O$, show that $t = \frac{1}{2}$, and that $P = -Q$.

5. Let P, Q be two points on the sphere of radius 1. Assume that $P \ne -Q$. Show that there exists a curve joining P and Q on the sphere of radius 1, centered at the origin. By this we mean there exists a curve $C(t)$ such that $C(t)^2 = 1$, or if you wish $\|C(t)\| = 1$ for all t, and there are two numbers t_1 and t_2 such that $C(t_1) = P$ and $C(t_2) = Q$. [*Hint*: Divide $L(t)$ in Exercise 4 by its norm.]

6. If P, Q are two unit vectors such that $P = -Q$, show that there exists a differentiable curve joining P and Q on the sphere of radius 1, centered at the origin. You may assume that there exists a unit vector A which is perpendicular to P. Then use Exercise 3.

7. Parametrize the ellipse $(x^2/a^2) + (y^2/b^2) = 1$ by a differentiable curve.

8. Let f be a differentiable function (in two variables) such that $\operatorname{grad} f(X) = cX$ for some constant c and all X in 2-space. Show that f is constant on any circle of radius $a > 0$, centered at the origin. [*Hint*: Put $x = a \cos t$ and $y = a \sin t$ and find df/dt.]

Exercise 8 is a special case of a general phenomenon, stated in Exercise 9.

9. Let f be a differentiable function in n variables, and assume that there exists a function h such that $\operatorname{grad} f(X) = h(X)X$. Show that f is constant on the sphere of radius $a > 0$ centered at the origin.

[That f is constant on the sphere of radius a means that given any two points P, Q on this sphere, we must have $f(P) = f(Q)$. To prove this, use the fact proved in Exercises 5 and 6 that given two such points, there exists a curve $C(t)$ joining the two points, i.e. $C(t_1) = P$, $C(t_2) = Q$, and $C(t)$ lies on the sphere for all t in the interval of definition, so

$$C(t) \cdot C(t) = a^2.$$

The hypothesis that grad $f(X)$ can be written in the form $h(X)X$ for some function h means that grad $f(X)$ is *parallel* to X (or O). Indeed, we know that grad $f(X)$ parallel to X means that grad $f(X)$ is equal to a scalar multiple of X, and this scalar may depend on X, so we have to write it as a function $h(X)$.]

10. Let $r = \|X\|$. Let g be a differentiable function of one variable whose derivative is never equal to 0. Let $f(X) = g(r)$. Show that grad $f(X)$ is parallel to X for $X \neq O$.

[This statement is the converse of Exercise 9. The proof is quite easy, cf. Example 1. The function $h(X)$ of Exercise 9 is then seen to be equal to $g'(r)/r$.]

XVIII, §5. CONSERVATION LAW

Definition. Let U be an open set. By a **vector field** on U we mean an association which to every point of U associates a vector of the same dimension.

If F is a vector field on U, and X a point of U, then we denote by $F(X)$ the vector associated to X by F and call it the **value of F at X**, as usual.

Example 1. Let $F(x, y) = (x^2 y, \sin xy)$. Then F is a vector field which to the point (x, y) associates $(x^2 y, \sin xy)$, having the same number of coordinates, namely two of them in this case.

A vector field in physics is often interpreted as a field of forces. A vector field may be visualized as a field of arrows, which to each point associates an arrow as shown on the figure.

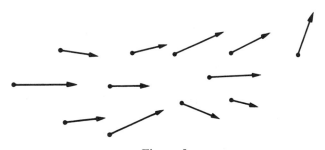

Figure 5

Each arrow points in the direction of the force, and the length of the arrow represents the magnitude of the force.

If f is a differentiable function on U, then we observe that grad f is a vector field, which associates the vector grad $f(P)$ to the point P of U.

If F is a vector field, and if there exists a differentiable function f such that $F =$ grad f, then the vector field is called **conservative**. Since

$$-\operatorname{grad} f = \operatorname{grad}(-f)$$

it does not matter whether we use f or $-f$ in the definition of conservative.

Let us assume that F is a conservative field on U, and let ψ be a differentiable function such that for all points X in U we have

$$F(X) = -\operatorname{grad} \psi.$$

In physics, one interprets ψ as the **potential energy**. Suppose that a particle of mass m moves on a differentiable curve $C(t)$ in U. **Newton's law** states that

$$\boxed{F(C(t)) = mC''(t)}$$

for all t where $C(t)$ is defined. Newton's law says that force equals mass times acceleration.

Physicists define the **kinetic energy** to be

$$\tfrac{1}{2}mC'(t)^2 = \tfrac{1}{2}mv(t)^2.$$

Conservation Law. *Assume the vector field F is conservative, that is $F = -\operatorname{grad} \psi$, where ψ is the potential energy. Assume that a particle moves on a curve satisfying Newton's law. Then the sum of the potential energy and kinetic energy is constant.*

Proof. We have to prove that

$$\psi(C(t)) + \tfrac{1}{2}mC'(t)^2$$

is constant. To see this, we differentiate the sum. By the chain rule, we see that the derivative is equal to

$$\operatorname{grad} \psi(C(t)) \cdot C'(t) + mC'(t) \cdot C''(t).$$

By Newton's law, $mC''(t) = F(C(t)) = -\text{grad }\psi(C(t))$. Hence this derivative is equal to

$$\text{grad }\psi(C(t)) \cdot C'(t) - \text{grad }\psi(C(t)) \cdot C'(t) = 0.$$

This proves what we wanted.

It is not true that all vector fields are conservative. We shall discuss the problem of determining which ones are conservative in the next book.

The fields of classical physics are for the most part conservative.

Example 2. Consider a force $F(X)$ which is inversely proportional to the square of the distance from the point X to the origin, and in the direction of X. Then there is a constant k such that for $X \neq O$ we have

$$F(X) = k \frac{1}{\|X\|^2} \frac{X}{\|X\|},$$

because $X/\|X\|$ is the unit vector in the direction of X. Thus

$$F(X) = k \frac{1}{r^3} X,$$

where $r = \|X\|$. A potential energy for F is given by

$$\psi(X) = \frac{k}{r}.$$

This is immediately verified by taking the partial derivatives of this function.

If there exists a function $\varphi(X)$ such that

$$F(X) = (\text{grad }\varphi)(X), \qquad \text{that is} \quad F = \text{grad }\varphi,$$

then we shall call such a function φ a **potential function** for F. Our conventions are such that a potential function is equal to *minus* the potential energy.

XVIII, §5. EXERCISES

1. Find a potential function for a force field $F(X)$ that is inversely proportional to the distance from the point X to the origin and is in the direction of X.

2. Same question, replacing "distance" with "cube of the distance."

3. Let k be an integer ≥ 1. Find a potential function for the vector field F given by

$$F(X) = \frac{1}{r^k} X, \qquad \text{where} \quad r = \|X\|.$$

[*Hint*: Recall the formula that if $\varphi(X) = g(r)$, then

$$\text{grad } \varphi(X) = \frac{g'(r)}{r} X.$$

Set $F(X)$ equal to the right-hand side and solve for g.]

Answers to Exercises

I am much indebted to Anthony Petrello for some of the answers to the exercises.

I, §2, p. 13

1. $-3 < x < 3$ 2. $-1 \leqq x \leqq 0$ 3. $-\sqrt{3} \leqq x \leqq -1$ or $1 \leqq x \leqq \sqrt{3}$
4. $x < 3$ or $x > 7$ 5. $-1 < x < 2$ 6. $x < -1$ or $x > 1$ 7. $-5 < x < 5$
8. $-1 \leqq x \leqq 0$ 9. $x \geqq 1$ or $x = 0$ 10. $x \leqq -10$ or $x = 5$
11. $x \leqq -10$ or $x = 5$ 12. $x \geqq 1$ or $x = -\frac{1}{2}$ 13. $x < -4$
14. $-5 < x < -3$ 15. $-3 < x < -2$ and $-2 < x < -1$ 16. $-2 < x < 2$
17. $-2 < x < 8$ 18. $2 < x < 4$ 19. $-4 < x < 10$ 20. $x < -4$ and $x > 10$
21. $x < -10$ and $x > 4$

I, §3, p. 17

1. $-\frac{3}{2}$ 2. $\dfrac{1}{(2x + 1)}$ 3. 0, 2, 108 4. $2z - z^2$, $2w - w^2$

5. $x \neq \sqrt{2}$ or $-\sqrt{2}$. $f(5) = \frac{1}{23}$ 6. All x. $f(27) = 3$
7. (a) 1 (b) 1 (c) -1 (d) -1 8. (a) 1 (b) 4 (c) 0 (d) 0
9. (a) -2 (b) -6 (c) $x^2 + 4x - 2$ 10. $x \geqq 0$, 2
11. (a) odd (b) even (c) odd (d) odd

I, §4, p. 20

1. 8 and 9 2. $\frac{1}{5}$ and -1 3. $\frac{1}{16}$ and 2 4. $\frac{1}{9}$ and $2^{1/3}$ 5. $\frac{1}{16}$ and $\frac{1}{2}$ 6. 9 and 8
7. $-\frac{1}{3}$ and -1 8. $\frac{1}{4}$ and $\frac{1}{4}$ 9. 1 and $-\frac{1}{4}$ 10. $-\frac{1}{512}$ and $\frac{1}{3}$

11. Yes. Suppose a is negative, so write $a = -b$ where b is positive. Let c be a positive number such that $c^n = b$. Then $(-c)^n = a$ because $(-1)^n = -1$ since n is odd.

II, §1, p. 24

3. x negative, y positive   **4.** x negative, y negative

II, §3, p. 33

5. $y = -\frac{8}{5}x + \frac{5}{3}$ **6.** $y = -\frac{3}{2}x + 5$ **7.** $x = \sqrt{2}$

8. $y = \dfrac{9}{\sqrt{3}+3}\,x + 4 - \dfrac{9\sqrt{3}}{\sqrt{3}+3}$ **9.** $y = 4x - 3$ **10.** $y = -2x + 2$

11. $y = -\frac{1}{2}x + 3 + \dfrac{\sqrt{2}}{2}$ **12.** $y = \sqrt{3}x + 5 + \sqrt{3}$ **19.** $-\frac{1}{4}$ **20.** -8

21. $2 + \sqrt{2}$ **22.** $\frac{1}{6}(3 + \sqrt{3})$ **23.** $y = (x - \pi)\left(\dfrac{2}{\sqrt{2} - \pi}\right) + 1$

24. $y = (x - \sqrt{2})\left(\dfrac{\pi - 2}{1 - \sqrt{2}}\right) + 2$ **25.** $y = -(x + 1)\left(\dfrac{3}{\sqrt{2} + 1}\right) + 2$

26. $y = (x + 1)(3 + \sqrt{2}) + \sqrt{2}$ **29.** (a) $x = -4$, $y = -7$ (b) $x = \frac{3}{2}$, $y = \frac{5}{2}$
(c) $x = -\frac{1}{3}$, $y = \frac{7}{3}$ (d) $x = -6$, $y = -5$

II, §4, p. 35

1. $\sqrt{97}$ **2.** $\sqrt{2}$ **3.** $\sqrt{52}$ **4.** $\sqrt{13}$ **5.** $\frac{1}{2}\sqrt{5}$ **6.** $(4, -3)$ **7.** 5 and 5 **8.** $(-2, 5)$
9. 5 and 7

II, §8, p. 51

5. $(x - 2)^2 + (y + 1)^2 = 25$ **6.** $x^2 + (y - 1)^2 = 9$ **7.** $(x + 1)^2 + y^2 = 3$
8. $y + \frac{25}{8} = 2(x + \frac{1}{4})^2$ **9.** $y - 1 = (x + 2)^2$ **10.** $y + 4 = (x - 1)^2$
11. $(x + 1)^2 + (y - 2)^2 = 2$ **12.** $(x - 2)^2 + (y - 1)^2 = 2$
13. $x + \frac{25}{8} = 2(y + \frac{1}{4})^2$ **14.** $x - 1 = (y + 2)^2$

III, §1, p. 61

1. 4 **2.** -2 **3.** 2 **4.** $\frac{3}{4}$ **5.** $-\frac{1}{4}$ **6.** 0 **7.** 4 **8.** 6 **9.** 3 **10.** 12 **11.** 2
12. 3 **13.** a

III, §2, p. 70

	Tangent line at $x = 2$	Slope at $x = 2$
1. $2x$	$y = 4x - 3$	4
2. $3x^2$	$y = 12x - 16$	12
3. $6x^2$	$y = 24x - 32$	24
4. $6x$	$y = 12x - 12$	12
5. $2x$	$y = 4x - 9$	4
6. $4x + 1$	$y = 9x - 8$	9
7. $4x - 3$	$y = 5x - 8$	5
8. $\dfrac{3x^2}{2} + 2$	$y = 8x - 8$	8
9. $-\dfrac{1}{(x + 1)^2}$	$y = -\frac{1}{9}x + \frac{5}{9}$	$-\frac{1}{9}$
10. $-\dfrac{2}{(x + 1)^2}$	$y = -\frac{2}{9}x + \frac{10}{9}$	$-\frac{2}{9}$

III, §3, p. 75

1. $4x + 3$ **2.** $-\dfrac{2}{(2x + 1)^2}$ **3.** $\dfrac{1}{(x + 1)^2}$ **4.** $2x + 1$ **5.** $-\dfrac{1}{(2x - 1)^2}$ **6.** $9x^2$

7. $4x^3$ **8.** $5x^4$ **9.** $6x^2$ **10.** $\dfrac{3x^2}{2} + 1$ **11.** $-2/x^2$ **12.** $-3/x^2$

13. $-2/(2x - 3)^2$ **14.** $-3/(3x + 1)^2$ **15.** $-1/(x + 5)^2$ **16.** $-1/(x - 2)^2$
17. $-2x^{-3}$ **18.** $-2(x + 1)^{-3}$

III, §4, p. 78

1. $x^4 + 4x^3h + 6x^2h^2 + 4xh^3 + h^4$ **2.** $4x^3$
3. (a) $\frac{2}{3}x^{-1/3}$ (b) $-\frac{3}{2}x^{-5/2}$ (c) $\frac{7}{6}x^{1/6}$ **4.** $y = 9x - 8$ **5.** $y = \frac{1}{3}x + \frac{4}{3}$, slope $\frac{1}{3}$

6. $y = \dfrac{-3}{2^9}x + \dfrac{7}{32}$, slope $\dfrac{-3}{2^9}$ **7.** $y = \dfrac{1}{2\sqrt{3}}x + \dfrac{\sqrt{3}}{2}$, slope $\dfrac{1}{2\sqrt{3}}$

8. (a) $\frac{1}{4}5^{-3/4}$ (b) $-\frac{1}{4}7^{-5/4}$ (c) $\sqrt{2}(10^{\sqrt{2}-1})$ (d) $\pi 7^{\pi-1}$

III, §5, p. 89

1. (a) $\frac{2}{3}x^{-2/3}$ (b) $\frac{9}{4}x^{-1/4}$ (c) x (d) $\frac{9}{4}x^2$
2. (a) $55x^{10}$ (b) $-8x^{-3}$ (c) $\frac{4}{3}x^3 - 15x^2 + 2x$
3. (a) $-\frac{3}{8}x^{-7/4}$ (b) $3 - 6x^2$ (c) $20x^4 - 21x^2 + 2$
4. (a) $21x^2 + 8x$ (b) $\frac{8}{3}x^{-1/3} + 20x^3 - 3x^2 + 3$
5. (a) $-25x^{-2} + 6x^{-1/2}$ (b) $6x^2 + 35x^6$ (c) $16x^3 - 21x^2 + 1$
6. (a) $\frac{6}{5}x - 16x^7$ (b) $12x^3 - 4x + 1$ (c) $7\pi x^6 - 40x^4 + 1$
7. $(x^3 + x) + (3x^2 + 1)(x - 1)$ **8.** $(2x^2 - 1)4x^3 + 4x(x^4 + 1)$
9. $(x + 1)(2x + \frac{15}{2}x^{1/2}) + (x^2 + 5x^{3/2})$
10. $(2x - 5)(12x^3 + 5) + 2(3x^4 + 5x + 2)$

11. $(x^{-2/3} + x^2)\left(3x^2 - \dfrac{1}{x^2}\right) + (-\tfrac{2}{3}x^{-5/3} + 2x)\left(x^3 + \dfrac{1}{x}\right)$

12. $(2x + 3)(-2x^{-3} - x^{-2}) + 2(x^{-2} + x^{-1})$ **13.** $\dfrac{9}{(x + 5)^2}$ **14.** $\dfrac{(-2x^2 + 2)}{(x^2 + 3x + 1)^2}$

15. $\dfrac{(t + 1)(t - 1)(2t + 2) - (t^2 + 2t - 1)2t}{(t^2 - 1)^2}$

16. $\dfrac{(t^2 + t - 1)(-5/4)t^{-9/4} - t^{-5/4}(2t + 1)}{(t^2 + t - 1)^2}$

17. $\tfrac{5}{49}$, $y = \tfrac{5}{49}t + \tfrac{4}{49}$ **18.** $\tfrac{1}{2}$, $y = \tfrac{1}{2}t$

III, §5, Supplementary Exercises, p. 89

1. $9x^2 - 4$ **3.** $2x + 1$ **5.** $\tfrac{5}{2}x^{3/2} - \tfrac{5}{2}x^{-7/2}$ **7.** $x^2 - 1 + (x + 5)(2x)$

9. $(\tfrac{3}{2}x^{1/2} + 2x)(x^4 - 99) + (x^{3/2} + x^2)(4x^3)$

11. $(4x)\left(\dfrac{1}{x^2} + 4x + 8\right) + (2x^2 + 1)\left(\dfrac{-2}{x^3} + 4\right)$

13. $(x + 2)(x + 3) + (x + 1)(x + 3) + (x + 1)(x + 2)$

15. $3x^2(x^2 + 1)(x + 1) + x^3(2x)(x + 1) + (x^3)(x^2 + 1)$

17. $\dfrac{-2}{(2x + 3)^2}$ **19.** $\dfrac{5(3x^2 + 4x)}{(x^3 + 2x^2)^2}$ **21.** $\dfrac{-2(x + 1) + 2x}{(x + 1)^2}$

23. $\dfrac{(x + 1)(x - 1)3(\tfrac{1}{2}x^{-1/2}) - 3x^{1/2}[(x - 1) + (x + 1)]}{(x + 1)^2(x - 1)^2}$

25. $\dfrac{(x^2 + 1)(x + 7)(5x^4) - (x^5 + 1)((x^2 + 1) + (2x)(x + 7))}{(x^2 + 1)^2(x + 7)^2}$

27. $\dfrac{(1 - x^2)(3x^2) - x^3(-2x)}{(1 - x^2)^2}$ **29.** $\dfrac{(x^2 + 1)(2x - 1) - (x^2 - x)(2x)}{(x^2 + 1)^2}$

31. $\dfrac{(x^2 + x - 4)(2) - (2x + 1)(2x + 1)}{(x^2 + x - 4)^2}$

33. $\dfrac{(x^2 + 2)(4 - 3x^2) - (4x - x^3)(2x)}{(x^2 + 2)^2}$ **35.** $\dfrac{-5x - (1 - 5x)}{x^2}$

37. $\dfrac{(x + 1)(x - 2)(2x) - x^2((x - 2) + (x + 1))}{(x + 1)^2(x - 2)^2}$

39. $\dfrac{(4x^3 - x^5 + 1)(12x^3 + \tfrac{5}{4}x^{1/4}) - (3x^4 + x^{5/4})(12x^2 - 5x^4)}{(4x^3 - x^5 + 1)^2}$

41. $(y - 18) = \tfrac{25}{32}(x - 16)$ **43.** $(y + 12) = 19x$ **45.** $(y - 10) = 14(x - 1)$

47. $y - \dfrac{4}{9} = \dfrac{-12}{81}(x - 2)$ **49.** $y - \dfrac{4}{3} = \dfrac{-4}{9}(x - 2)$

51. Point of tangency: $(3, -3)$. Both curves intersect here and have slope -1.

53. Both curves have the point $(1, 3)$ in common and have slope 6 at this point.

55. Tangent line $(y - 7) = 16x$ at $(0, 7)$; tangent line $(y - 19) = 16(x - 1)$ at $(1, 19)$; tangent line $(y + 13) = 16(x + 1)$ at $(-1, -13)$.

III, §6, p. 99

1. $8(x + 1)^7$ **2.** $\frac{1}{2}(2x - 5)^{-1/2} \cdot 2$ **3.** $3(\sin x)^2 \cos x$ **4.** $5(\log x)^4 \left(\frac{1}{x}\right)$

5. $(\cos 2x)2$ **6.** $\frac{1}{x^2 + 1}(2x)$ **7.** $e^{\cos x}(-\sin x)$ **8.** $\frac{1}{e^x + \sin x}(e^x + \cos x)$

9. $\cos\left[\log x + \frac{1}{x}\right]\left(\frac{1}{x} - \frac{1}{x^2}\right)$ **10.** $\dfrac{\sin 2x - (x + 1)(\cos 2x)2}{(\sin 2x)^2}$

11. $3(2x^2 + 3)^2(4x)$ **12.** $-[\sin(\sin 5x)](\cos 5x)5$ **13.** $\dfrac{1}{\cos 2x}(-\sin 2x)2$

14. $[\cos(2x + 5)^2](2(2x + 5))(2).$ **15.** $[\cos(\cos(x + 1))](-\sin(x + 1))$

16. $(\cos e^x)e^x$ **17.** $-\dfrac{1}{(3x - 1)^8}[4(3x - 1)^3] \cdot 3$ **18.** $-\dfrac{1}{(4x)^6} \cdot 3(4x)^2 \cdot 4$

19. $-\dfrac{1}{(\sin 2x)^4} 2(\sin 2x)(\cos 2x) \cdot 2$ **20.** $-\dfrac{1}{(\cos 2x)^4} 2(\cos 2x)(-\sin 2x)2$

21. $-\dfrac{1}{(\sin 3x)^2}(\cos 3x) \cdot 3$ **22.** $-\sin^2 x + \cos^2 x$ **23.** $(x^2 + 1)e^x + 2xe^x$

24. $(x^3 + 2x)(\cos 3x) \cdot 3 + (3x^2 + 2)\sin 3x$

25. $-\dfrac{1}{(\sin x + \cos x)^2}(\cos x - \sin x)$ **26.** $\dfrac{2e^x \cos 2x - (\sin 2x)e^x}{e^{2x}}$

27. $\dfrac{(x^2 + 3)/x - (\log x)(2x)}{(x^2 + 3)^2}$ **28.** $\dfrac{\cos 2x - (x + 1)(-\sin 2x) \cdot 2}{\cos^2 2x}$

29. $(2x - 3)(e^x + 1) + 2(e^x + x)$ **30.** $(x^3 - 1)(e^{3x} \cdot 3 + 5) + 3x^2(e^{3x} + 5x)$

31. $\dfrac{(x - 1)3x^2 - (x^3 + 1)}{(x - 1)^2}$ **32.** $\dfrac{(2x + 3)2x - (x^2 - 1)2}{(2x + 3)^2}$

33. $2(x^{4/3} - e^x) + (\frac{4}{3}x^{1/3} - e^x)(2x + 1)$

34. $(\sin 3x)\frac{1}{4}x^{-3/4} + 3(\cos 3x)(x^{1/4} - 1)$ **35.** $[\cos(x^2 + 5x)](2x + 5)$

36. $e^{3x^2 + 8}(6x)$ **37.** $\dfrac{-1}{[\log(x^4 + 1)]^2} \cdot \dfrac{1}{x^4 + 1} \cdot 4x^3$

38. $\dfrac{-1}{[\log(x^{1/2} + 2x)]^2} \dfrac{1}{(x^{1/2} + 2x)}(\frac{1}{2}x^{-1/2} + 2)$ **39.** $\dfrac{2e^x - 2xe^x}{e^{2x}}$

40. $\dfrac{2x}{1 + x^6}; \frac{4}{65}$

III, §6, Supplementary Exercises, p. 100

1. $2(2x + 1)2$ **3.** $7(5x + 3)^6 5$ **5.** $3(2x^2 + x - 5)^2(4x + 1)$

7. $\frac{1}{2}(3x + 1)^{-1/2}(3)$ **9.** $-2(x^2 + x - 1)^{-3}(2x + 1)$ **11.** $-\frac{5}{3}(x + 5)^{-8/3}$

13. $(x - 1)3(x - 5)^2 + (x - 5)^3$ **15.** $4(x^3 + x^2 - 2x - 1)^3(3x^2 + 2x - 2)$

17. $\dfrac{(x - 1)^{1/2}(\frac{3}{4})(x + 1)^{-1/4} - (x + 1)^{3/4}(\frac{1}{2})(x - 1)^{-1/2}}{x - 1}$

19. $\dfrac{(3x + 2)^9(\frac{5}{2})(2x^2 + x - 1)^{3/2}(4x + 1) - (2x^2 + x - 1)^{5/2}(9)(3x + 2)^8(3)}{(3x + 2)^{18}}$

21. $\frac{1}{2}(2x + 1)^{-1/2}(2)$ **23.** $\frac{1}{2}(x^2 + x + 5)^{-1/2}(2x + 1)$

25. $3x^2 \cos(x^3 + 1)$ **27.** $(e^{x^3+1})(3x^2)$ **29.** $(\cos(\cos x))(-\sin x)$

31. $(e^{\sin(x^3+1)})(3x^2 \cos(x^3 + 1))$

33. $[\cos((x + 1)(x^2 + 2))][(x + 1)(2x) + (x^2 + 2)]$

35. $(e^{(x+1)(x-3)})((x + 1) + (x - 3))$ **37.** $2\cos(2x + 5)$

39. $\dfrac{2}{2x + 1}$ **41.** $\left(\cos\dfrac{x - 5}{2x + 4}\right)\left(\dfrac{(2x + 4) - (x - 5)2}{(2x + 4)^2}\right)$

43. $(e^{2x^2+3x+1})(4x + 3)$ **45.** $\dfrac{1}{2x + 1}[\cos(\log 2x + 1)]2$

47. $-(6x - 2)\sin(3x^2 - 2x + 1)$ **49.** $80(2x + 1)^{79}(2)$

51. $49(\log x)^{48}(x^{-1})$ **53.** $5(e^{2x+1} - x)^4(2e^{2x+1} - 1)$

55. $\frac{1}{2}(3\log(x^2 + 1) - x^3)^{-1/2}\left(\dfrac{3}{x^2 + 1}(2x) - 3x^2\right)$

57. $\dfrac{2(\cos 3x)(\cos 2x) - 3(\sin 2x)(-\sin 3x)}{(\cos 3x)^2}$

59. $\dfrac{(\sin x^3)(1/2x^2)4x - (\log 2x^2)(\cos x^3)3x^2}{(\sin x^3)^2}$

61. $\dfrac{(\cos 2x)(4x^3) - 2(x^4 + 4)(-\sin 2x)}{(\cos 2x)^2}$

63. $\dfrac{(\cos x^3)(4)(2x^2 + 1)^3(4x) + (2x^2 + 1)^4(\sin x^3)(3x^2)}{\cos^2 x^3}$

65. $-3e^{-3x}$ **67.** $(e^{-4x^2+x})(-8x + 1)$

69. $\dfrac{e^{-x}[2x/(x^2 + 2)] - [\log(x^2 + 2)](e^{-x})(-1)}{e^{-2x}}$

III, §7, p. 103

1. $18x$ **2.** $5(x^2 + 1)^4 \cdot 2 + 20(x^2 + 1)^3 4x^2$ **3.** 0 **4.** $7!$ **5.** 0 **6.** 6

7. $-\cos x$ **8.** $\cos x$ **9.** $-\sin x$ **10.** $-\cos x$ **11.** $\sin x$ **12.** $\cos x$

In Problems 7 through 12, there is a pattern. Note that the derivatives of $\sin x$ are:

$$f(x) = \sin x;$$
$$f^{(1)}(x) = \cos x;$$
$$f^{(2)}(x) = -\sin x;$$
$$f^{(3)}(x) = -\cos x;$$
$$f^{(4)}(x) = \sin x.$$

Then the derivatives repeat. Thus every fourth derivative is the same. Hence to find the n-th derivative, we just divide n by 4, and if r is the remainder, so

$n = 4q + r$, then

$$f^{(n)}(x) = f^{(r)}(x).$$

13. (a) 5! (b) 7! (c) 13! **14.** (a) $k!$ (b) $k!$ (c) 0 (d) 0

III, §8, p. 106

1. $-(2x + y)/x$ **2.** $\dfrac{3 - x}{y + 1}$ **3.** $\dfrac{y - 3x^2}{3y^2 - x}$ **4.** $\dfrac{6x^2}{3y^2 + 1}$ **5.** $\dfrac{1 - 2y}{2x + 2y - 1}$

6. $-y^2/x^2$ **7.** $-\dfrac{1 + 4xy}{2(y + x^2)}$ **8.** $\dfrac{x(y^2 - 1)}{y(1 - x^2)}$ **9.** $(y - 3) = 3(x + 1)$

10. $y + 1 = 4(x - 3)$ **11.** $(y - 2) = \frac{3}{4}(x - 6)$ **12.** $y + 2 = -\frac{3}{4}(x - 1)$
13. $(y + 4) = \frac{3}{4}(x - 3)$ **14.** $y - 2 = -\frac{1}{8}(x + 4)$ **15.** $y - 3 = \frac{7}{4}(x - 2)$

III, §9, p. 114

1. (a) 1/6 (b) 0 (c) Impossible **2.** 0 **3.** 320 ft/sec^2 **4.** 0 **5.** 240 m^3/sec

6. 36π cm^2/sec **7.** $2\pi r, \dfrac{\pi d}{2}, \dfrac{c}{2\pi}$ **8.** $\frac{3}{16}$ units/sec

9. (a) $\frac{4}{3}$ ft/sec (b) 1 ft/sec **10.** (a) $\frac{5}{3}$ ft/sec (b) 2 ft/sec
11. $(\frac{1}{2}, \frac{1}{4})$
12. The picture is as follows.

We are given $dx/dt = -1$ and $dy/dt = 2$. The area is $A = \frac{1}{2}xy$, so

$$\frac{dA}{dt} = \frac{1}{2}\left[x\,\frac{dy}{dt} + y\,\frac{dx}{dt}\right] = \frac{1}{2}[2x - y].$$

We are given that at some time, $x = 8$. Since the speed is uniform toward the origin, after 2 min we find $x = 8 - 2 = 6$. Also after 2 min we find $y = 6 + 4 = 10$. Hence after 2 min we get

$$\frac{dA}{dt} = \frac{1}{2}[12 - 10] = 1 \text{ cm}^2/\text{min}.$$

13. 90 cm^2/sec **14.** $-8/5$ ft/sec **15.** 3/200 ft/min **16.** $5/4\pi$ ft/min
17. $t = \frac{1}{4}$, acc $= 4$

18. Both x and y are functions of time t. Differentiating each side of

$$y = x^2 - 6x$$

with respect to t, we find

$$\frac{dy}{dt} = 2x\,\frac{dx}{dt} - 6\,\frac{dx}{dt}.$$

When $dy/dt = 4\,dx/dt$ this yields

$$4\,\frac{dx}{dt} = 2x\,\frac{dx}{dt} - 6\,\frac{dx}{dt}.$$

Canceling dx/dt yields $4 = 2x - 6$, so $x = 5$, and $y = -5$.

19. $4/75\pi$ ft/min. Draw the picture

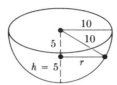

When $h = 5$, then the distance from the top of the water to the top of the hemisphere is also 5, so by Pythagoras,

$$5^2 + r^2 = 10^2.$$

You can then solve for r. Use that $dV/dt = 4$ to find dh/dt.

20. $\frac{1}{2}(50^2 \cdot 7 + 60^2 \cdot 3)(50^2 \cdot 3.5^2 + 60^2 \cdot 1.5^2)^{-1/2}$

21. $1/12\pi$ ft/min

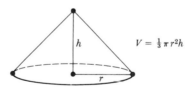

The assumption about the diameter implies that $r = 3h/2$ so $V = \frac{3}{4}\pi h^3$. Then

$$\frac{dV}{dt} = \frac{9}{4}\pi h^2\,\frac{dh}{dt} = 3.$$

When $h = 4$ this gives the answer.

22. $-3/400$ cm/min, $-6\pi/5$ cm^2/min   **23.** $1/32\pi$ m/min   **24.** 100π ft^3/sec

IV, §1, p. 131

1. $\sqrt{2}/2$ **2.** $\sqrt{3}/2$ **3.** $\sqrt{3}/2$ **4.** $\frac{1}{2}$ **5.** $-\sqrt{3}/2$ **6.** $-\frac{1}{2}$ **7.** $\sqrt{3}/2$ **8.** $-\sqrt{2}/2$
9. 1 **10.** $\sqrt{3}$ **11.** 1 **12.** -1 **13.** $-\frac{1}{2}$ **14.** $-\sqrt{3}/2$ **15.** $-1/2$ **16.** $\sqrt{3}/2$
17. $-\sqrt{3}/2$ **18.** $\frac{1}{2}$

IV, §2, p. 135

2.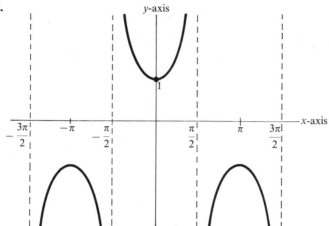

3.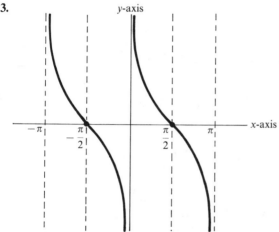

4. (a) (b)

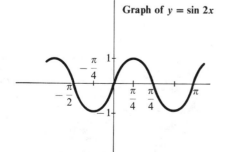

Graph of $y = \sin 2x$

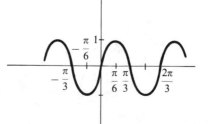

Graph of $y = \sin 3x$

5. (a)

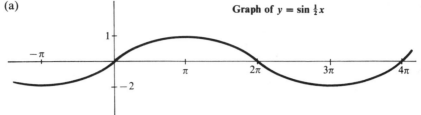

Graph of $y = \sin \frac{1}{2}x$

(b)

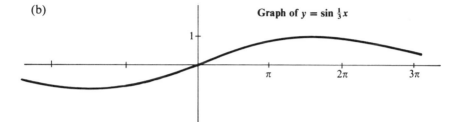

Graph of $y = \sin \frac{1}{3}x$

6. (a)

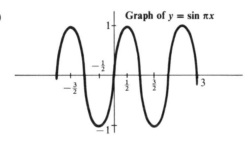

Graph of $y = \sin \pi x$

(c) $y = \sin 2\pi x$

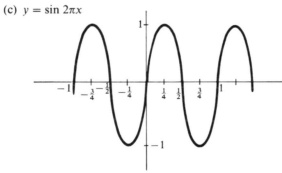

7. (a) $y = |\sin x|$

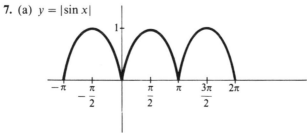

IV, §3, p. 140

1. $\dfrac{\sqrt{2}}{4}(\sqrt{3}+1)$ 2. $\dfrac{-\sqrt{2}}{4}(\sqrt{3}-1)$

3. (a) $\dfrac{\sqrt{2}}{4}(\sqrt{3}-1)$ (b) $\dfrac{\sqrt{2}}{4}(\sqrt{3}+1)$ (c) $\dfrac{\sqrt{2}}{4}(\sqrt{3}+1)$ (d) $\dfrac{\sqrt{2}}{4}(\sqrt{3}-1)$

(e) $\dfrac{\sqrt{2}}{4}(\sqrt{3}-1)$ (f) $\dfrac{-\sqrt{2}}{4}(\sqrt{3}+1)$ (g) $\frac{1}{2}$ (h) $-\sqrt{3}/2$

5. $\sin 3x = 3\sin x - 4\sin^3 x,\ \cos 3x = 4\cos^3 x - 3\cos x$

IV, §4, p. 145

1. $-\csc^2 x$ 2. $3\cos 3x$ 3. $-5\sin 5x$
4. $(8x+1)\cos(4x^2+x)$ 5. $(3x^2)\sec^2(x^3-5)$ 6. $(4x^3-3x^2)\sec^2(x^4-x^3)$
7. $\cos x \sec^2(\sin x)$ 8. $\sec^2 x \cos(\tan x)$ 9. $-\sec^2 x \sin(\tan x)$ 10. -1
11. 0 12. $\sqrt{3}/2$ 13. $-\sqrt{2}$ 14. 2 15. $-2\sqrt{3}$

16. (a) $y = 1$ (b) $\left(y - \dfrac{\sqrt{3}}{2}\right) = \dfrac{-1}{2}\left(x - \dfrac{\pi}{6}\right)$

(c) $y = 1$ (d) $(y+1) = 6\left(x - \dfrac{\pi}{4}\right)$

(e) $y = 1$ (f) $(y - \sqrt{2}) = -\sqrt{2}\left(x - \dfrac{\pi}{4}\right)$

(g) $(y-1) = -2\left(x - \dfrac{\pi}{4}\right)$ (h) $y - 1 = x - \dfrac{\pi}{2}$

(i) $(y - \frac{1}{2}) = \dfrac{\sqrt{3}}{4}\left(x - \dfrac{\pi}{3}\right)$ (j) $(y - \frac{1}{2}) = \dfrac{-\pi\sqrt{3}}{6}(x-1)$

(k) $y = 1$ (l) $\left(y - \dfrac{1}{\sqrt{3}}\right) = \dfrac{4\pi}{3}\left(x - \dfrac{1}{6}\right)$

17. (a) -1.6 (b) $\frac{2}{3}$ (c) $\sqrt{3}/30$

18.

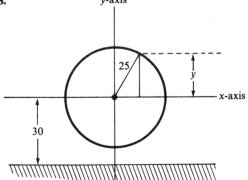

One revolution every two minutes is half a revolution per minute. Hence $d\theta/dt = \pi$ (in radians per minute). But

$$y = 25 \sin \theta$$

so

$$\frac{dy}{dt} = 25 \, \frac{d \sin \theta}{d\theta} \, \frac{d\theta}{dt} = 25(\cos \theta)\pi.$$

When the height of a point on the wheel is 42.5 then $y = 42.5 - 30 = 12.5$. Therefore $\sin \theta = 12.5/25 = \frac{1}{2}$ and $\theta = \pi/6$. Hence

$$\frac{dy}{dt} \Big|_{\theta = \pi/6} = 25\left(\cos \frac{\pi}{6}\right)\pi = \tfrac{25}{2} \sqrt{3}\pi \text{ ft/min.}$$

19. (a) 180 ft/sec   (b) 360 ft/sec   (c) 2250 ft/sec   (d) 9000/91 ft/sec
      (e) 1530 ft/sec   **20.** 25 rad/hr

21.

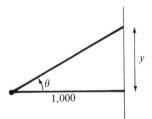

We are given $d\theta/dt = 4\pi$ (two revolutions $= 4\pi$ radians). Then

$$\tan \theta = \frac{y}{1000} \qquad \text{so} \quad y = 1000 \tan \theta.$$

Using the chain rule,

$$\frac{dy}{dt} = 1000(1 + \tan^2 \theta) \, \frac{d\theta}{dt}.$$

(a) The point on the wall nearest to the light is when $\theta = 0$. Then $\tan 0 = 0$, so

$$\frac{dy}{dt} \Big|_{\theta = 0} = 1000 \cdot 4\pi = 4{,}000\pi \text{ ft/min.}$$

(b) When $y = 500$ then $\tan \theta = \frac{1}{2}$, so we substitute $\frac{1}{2}$ for $\tan \theta$ and get

$$\frac{dy}{dt} \Big|_{y = 500} = 1000(\tfrac{5}{4})4\pi = 5000\pi \text{ ft/min.}$$

22. $8000\pi/27$ ft/sec

23. Let s be the length of the shadow.

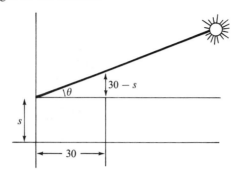

We are given $d\theta/dt = \pi/10$. We also have

$$\tan \theta = \frac{30 - s}{30} = 1 - \frac{s}{30}.$$

Differentiating with respect to t and using the chain rule gives:

$$(1 + \tan^2 \theta) \frac{d\theta}{dt} = -\frac{1}{30} \frac{ds}{dt}.$$

If $\theta = \pi/6$ then $\tan(\pi/6) = 1/\sqrt{3}$. Substituting yields:

$$\left(1 + \frac{1}{3}\right) \frac{\pi}{10} = -\frac{1}{30} \frac{ds}{dt} \Big|_{\theta = \pi/6}.$$

We can solve for ds/dt, namely

$$\frac{ds}{dt}\Big|_{\theta = \pi/6} = -30\left(\frac{4}{3}\right)\frac{\pi}{10} = -4\pi \text{ ft/hr.}$$

24. $54/5\pi$ deg/min

25. We are given $dx/dt = 3$. Find $d\theta/dt$ when $x = 15$.

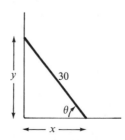

We have

$$\cos \theta = \frac{x}{30}.$$

Hence by the chain rule,

$$-\sin\theta\,\frac{d\theta}{dt} = \frac{1}{30}\frac{dx}{dt}.$$

When $x = 15$ we have $\cos\theta = 1/2$ and so $\theta = \pi/3$. Hence $\sin\theta = \sqrt{3}/2$. Then

$$\frac{d\theta}{dt}\bigg|_{x=15} = -\frac{1}{\sqrt{3}/2}\frac{1}{30}\,3$$

$$= \frac{-1}{5\sqrt{3}}\text{ rad/sec.}$$

26. $9/2\pi$ deg/sec

27.

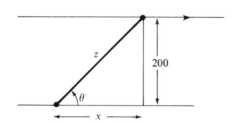

We are given $dx/dt = 20$. We want to find $d\theta/dt$ when $z = 400$. We have

$$\tan\theta = \frac{200}{x},$$

so differentiating with respect to t and using the chain rule,

$$(1 + \tan^2\theta)\frac{d\theta}{dt} = 200\left(\frac{-1}{x^2}\right)\frac{dx}{dt}.$$

When $z = 400$, then $\sin\theta = 200/400 = 1/2$. Hence $\theta = \pi/6$ and therefore $\tan\theta = 1/\sqrt{3}$. Also $x = 200\sqrt{3}$. This gives

$$\left(1 + \frac{1}{3}\right)\frac{d\theta}{dt}\bigg|_{z=400} = 200\cdot\frac{-1}{200^2\cdot 3}\,20$$

$$= \frac{-1}{30}.$$

Hence finally

$$\frac{d\theta}{dt}\bigg|_{z=400} = -\frac{1}{30}\frac{3}{4} = -\frac{1}{40}\text{ ft/sec.}$$

28. $-1/25$ rad/sec

IV, §6, p. 157

3. (a) $(\sqrt{2}, \pi/4)$ (b) $(\sqrt{2}, 5\pi/4)$ (c) $(6, \pi/3)$ (d) $(1, \pi)$
4. (a) $(y-1)^2 + x^2 = 1$ (b) $(x-\frac{3}{2})^2 + y^2 = \frac{9}{4}$
5. (a) $\left(y - \dfrac{a}{2}\right)^2 + x^2 = \left(\dfrac{a}{2}\right)^2$ $\qquad\qquad$ (b) $\left(x - \dfrac{a}{2}\right)^2 + y^2 = \left(\dfrac{a}{2}\right)^2$
$\quad$ (c) $x^2 + (y-a)^2 = a^2$ $\qquad\qquad$ (d) $(x-a)^2 + y^2 = a^2$
6. $r^2 = \cos\theta$. This is equivalent with $r = \sqrt{\cos\theta}$. Only values of θ such that $\cos\theta \geqq 0$ will give a contribution to r. Also, since $\cos(-\theta) = \cos\theta$ the curve is symmetric with respect to the x-axis. We make a table.

θ	r
0 to $\pi/2$	dec. 1 to 0
$-\pi/$ to 0	inc. 0 to 1

In these intervals, we have $0 \leqq \cos\theta \leqq 1$, and hence

$$\sqrt{\cos\theta} \geqq \cos\theta, \qquad \text{(watch out!)}$$

with equality only at the end points. Since $r = \cos\theta$ is the equation of a circle (similarly to $r = \sin\theta$, see Problem 5), the graph looks like this.

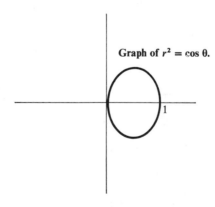

Graph of $r^2 = \cos\theta$.

8. (a) $r = \sin^2\theta$. The right-hand side is always $\geqq 0$ so there is a value of r for each value of θ. Since

$$-1 \leqq \sin\theta \leqq 1,$$

it follows that $\sin^2 \theta \leq |\sin \theta|$. Also the regions of increase and decrease are over intervals of length $\pi/2$. You should make a table of these, and then see that the graph looks like this.

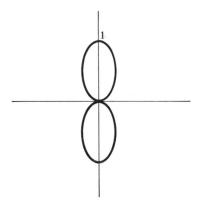

The ovals are thinner than circles, contrary to Problem 6, where they were more expanded than circles.

9. $r = 4 \sin^2 \theta$. Note that the right-hand side is always ≥ 0, and so there is a value of r for each value of θ. Also

$$\sin(-\theta) = -\sin \theta \quad \text{and} \quad \sin^2(-\theta) = \sin^2 \theta$$

so the graph is symmetric with respect to the x-axis. We make a table.

θ	r
0 to $\pi/2$	inc. 0 to 4
$\pi/2$ to π	dec. 4 to 0.

Also observe that $\sin^2 \theta \leq |\sin \theta|$ because $|\sin \theta| \leq 1$. Hence the graph is something like this:

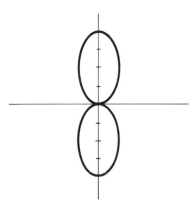

10. $x^2 + y^2 = 25$ (circle)   **11.** $x^2 + y^2 = 16$ (circle)

12. (a) $r\cos\theta = 1$ is equivalent with $x = 1$, which is a vertical line!

13. $r = 3/\cos\theta$. This is defined for $\cos\theta \neq 0$. In this case, the equation is equivalent with $r\cos\theta = 3$, and $x = r\cos\theta$, so the equation in rectangular coordinates is $x = 3$, which is a vertical line.

16. $(x^2 + x + y^2)^2 = x^2 + y^2$ **17.** $(x^2 + y^2 + 2y)^2 = x^2 + y^2$ **27.** $y^2 = 2x + 1$

28. $r = \dfrac{2}{2 - \cos\theta}$. We make a table:

θ	$\cos\theta$	$2 - \cos\theta$	r
inc. 0 to $\pi/2$	dec. 1 to 0	inc. 1 to 2	dec. 2 to 1
inc. $\pi/2$ to π	dec. 0 to -1	inc. 2 to 3	dec. 1 to 2/3
inc. π to $3\pi/2$	inc. -1 to 0	dec. 3 to 2	inc. 2/3 to 1
inc. $5\pi/2$ to 2π	inc. 0 to 1	dec. 2 to 1	inc. 1 to 2

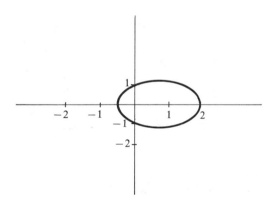

One can see that this equation is an ellipse by converting to (x, y)-coordinates. The equation is equivalent with

$$r(2 - \cos\theta) = 2, \qquad \text{that is} \qquad 2\sqrt{x^2 + y^2} - x = 2.$$

By algebra, this is equivalent to $4(x^2 + y^2) = (x + 2)^2$, that is

$$3x^2 - 4x + 4y^2 = 4.$$

By completing the square this is the equation of an ellipse

$$3\left(x - \frac{2}{3}\right)^2 + 4y^2 = \frac{16}{3}.$$

29. $\sqrt{x^2 + y^2} = 4 - 2x$. Since $r \geq 0$ by assumption, we must have $4 - 2x \geq 0$, or equivalently $x \leq 2$. Conversely, for $x \leq 2$ the relation is equivalent with what we obtain when we square both sides, and the equation becomes

$$x^2 + y^2 = 16 - 16x + 4x^2,$$

or equivalently

$$3x^2 - 16x - y^2 = -16.$$

This is the equation of a hyperbola. Thus the equation in polar coordinates is equivalent with the equation of a hyperbola, together with the additional condition $x \le 2$.

30. $r = \tan \theta = \sin \theta / \cos \theta$. Multiply both sides by $\cos \theta$ to see that this equation is equivalent to $r \cos \theta = \sin \theta$, that is

$$x = \sin \theta.$$

Of course, the function is not defined when $\cos \theta = 0$. Since

$$-1 < \sin \theta < 1,$$

it follows that $-1 < x < 1$. We make a small table:

θ	x
0	0
$\pi/4$	$1/\sqrt{2}$
$\pi/3$	$\sqrt{3}/2$
inc. 0 to $\pi/2$	inc. 0 to 1

As θ approaches $\pi/2$, x approaches 1. But $\cos \theta$ approaches 0 and so $r = \tan \theta$ becomes very large positive. Hence the graph looks as in the following figure for $0 \le x < 1$.

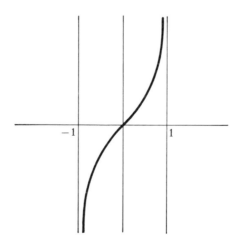

The graph also has a symmetry. Since $r = y/x$ and $r \ge 0$, both y and x must have the same sign, that is both $x, y > 0$ or both $x, y < 0$, unless $x = 0$.

The next interval of θ for which $\tan \theta$ is positive is then

$$\pi \le \theta < \frac{3\pi}{2}.$$

Either by making a table again, or by symmetry, using

$$\tan(\theta + \pi) = \tan \theta$$

you see that the graph is as shown for $-1 < x \le 0$.

31. $r = 5 + 2 \sin \theta$. We make a table:

θ	$\sin \theta$	r
0 to $\pi/2$	inc. 0 to 1	inc. 5 to 7
$\pi/2$ to π	dec. 1 to 0	dec. 7 to 5
π to $3\pi/2$	dec. 0 to -1	dec. 5 to 3
$3\pi/2$ to 2π	inc. -1 to 0	inc. 3 to 5

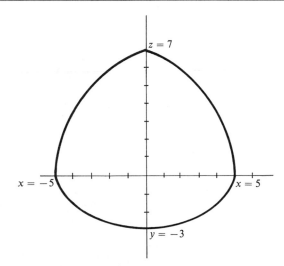

32. $r = |1 + 2 \cos \theta|$. Again we make a table. The absolute value sign makes the right-hand side positive, so we get a value of r for every value of θ. However, we want to choose intervals which take into account changes of sign of $1 + 2 \cos \theta$, this is when $\cos \theta = -1/2$. We make the table accordingly, when $\theta = 2\pi/3$ or $\theta = 4\pi/3$.

θ	$\cos \theta$	r
0 to $\pi/2$	dec. 1 to 0	dec. 3 to 1
$\pi/2$ to $2\pi/3$	dec. 0 to $-1/2$	dec. 1 to 0
$2\pi/3$ to π	dec. $-1/2$ to -1	inc. 0 to 1

Since $\cos(-\theta) = \cos\theta$, the graph is symmetric with respect to the x-axis, and looks like this:

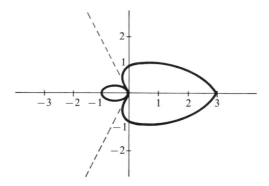

33. (a) $r = 2 + \sin 2\theta$. Since $\sin 2\theta$ lies between -1 and $+1$, it follows that the right-hand side is positive for all θ and so there is an r corresponding to every value of θ. We make a table, choosing the intervals to reflect the regions of increase of $\sin 2\theta$, so by intervals of length $\pi/4$.

θ	2θ	$\sin 2\theta$	r
0 to $\pi/4$	0 to $\pi/2$	inc. 0 to 1	inc. 2 to 3
$\pi/4$ to $\pi/2$	$\pi/2$ to π	dec. 1 to 0	dec. 3 to 2
$\pi/2$ to $3\pi/4$	π to $3\pi/2$	dec. 0 to -1	dec. 2 to 1
$3\pi/4$ to π	$3\pi/2$ to 2π	inc. -1 to 0	inc. 1 to 2

The graph looks like this.

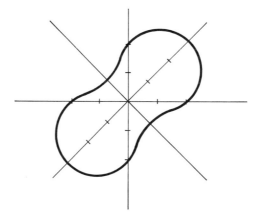

33. (b) Make a table. We use intervals of length $\pi/4$ because $\sin\theta$ changes its behavior on intervals of length $\pi/2$, so $\sin 2\theta$ changes its behavior on intervals of length $\pi/4$.

θ	2θ	$\sin 2\theta$	$r = 2 - \sin 2\theta$
inc. 0 to $\pi/4$	inc. 0 to $\pi/2$	inc. 0 to 1	dec. 2 to 1
inc. $\pi/4$ to $\pi/2$	inc. $\pi/2$ to π	dec. 1 to 0	inc. 1 to 2
inc. $\pi/2$ to $3\pi/4$	inc. π to $3\pi/2$	dec. 0 to -1	inc. 2 to 3
inc. $3\pi/4$ to π	inc. $3\pi/2$ to 2π	inc. -1 to 0	dec. 3 to 2

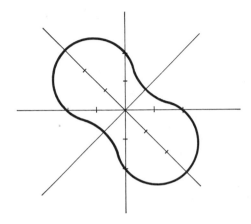

The part for $\pi \leq \theta \leq 2\pi$ is obtained similarly, or by symmetry.

34. $x \leq 0$, $y = 0$ (negative x-axis)

35. $x = 0$, $y \geq 0$ (positive y-axis)

36. $x = 0$, $y \leq 0$ (negative y-axis)

37.

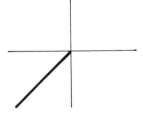

38.

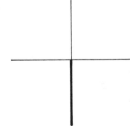

39.

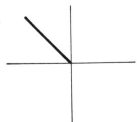

V, §1, p. 165

1. 1   **2.** $\frac{3}{4}$ **3.** $\frac{1}{6}$ **4.** 1   **5.** $\frac{3}{4}$ **6.** 0   **7.** ± 1

8. $\frac{\pi}{4} + 2n\pi$ and $\frac{5\pi}{4} + 2n\pi$, $n =$ integer.   **9.** $n\pi$, $n =$ interger

10. $\frac{\pi}{2} + n\pi$, $n =$ integer

V, §2, p. 175

1. Increasing for all x.
2. Decreasing for $x \leq \frac{1}{2}$, increasing for $x \geq \frac{1}{2}$.
3. Increasing all x.
4. Decreasing $x \leq -\sqrt{2/3}$ and $x \geq \sqrt{2/3}$. Increasing $-\sqrt{2/3} \leq x \leq \sqrt{2/3}$.
5. Increasing all x.
6. Decreasing for $x \leq 0$. Increasing for $x \geq 0$.
17. Min: 1; max: 4
19. Max: -1; min: 4
21. Min: -1; max: -2, 1
24. Let $f(x) = \tan x - x$. Then $f'(x) = 1 + \tan^2 x - 1 = \tan^2 x$. But $\tan^2 x$ is a square, and so is > 0 for $0 < x < \pi/2$. Hence f is strictly increasing. Since $f(0) = 0$ it follows that $f(x) > 0$ for all x with $0 < x < \pi/2$.
26. Base $= \sqrt{C/3}$, height $= \sqrt{C/12}$ **27.** Radius $= \sqrt{C/3\pi}$, height $= \sqrt{C/3\pi}$
28. Base $= \sqrt{C/6}$, height $= \sqrt{C/6}$; Radius $= \sqrt{C/6\pi}$, height $= 2\sqrt{C/6\pi}$
30. (a) $f(t) = -3t + C$ (b) $f(t) = 2t + C$ **31.** $f(t) = -3t + 1$
32. $f(t) = 2t - 5$ **33.** $x(t) = 7t - 61$ **34.** $H(t) = -2t + 30$

VI, §1, p. 187

1. 0, 0   **2.** 0, 0   **3.** 0, 0   **4.** $\frac{1}{\pi}, \frac{1}{\pi}$ **5.** 0, 0   **6.** ∞, $-\infty$ **7.** $-\infty$, ∞
8. $-\frac{1}{2}, -\frac{1}{2}$ **9.** $-\infty$, $+\infty$ **10.** 0, 0   **11.** ∞, $-\infty$ **12.** $-\infty$, ∞
13. ∞, ∞ **14.** $-\infty$, $-\infty$ **15.** ∞, $-\infty$ **16.** $-\infty$, ∞
17. ∞, ∞ **18.** $-\infty$, $-\infty$

19.

n	a_n	$x \to \infty$	$x \to -\infty$
Odd	>0	$f(x) \to \infty$	$f(x) \to -\infty$
Odd	<0	$f(x) \to -\infty$	$f(x) \to \infty$
Even	>0	$f(x) \to \infty$	$f(x) \to \infty$
Even	<0	$f(x) \to -\infty$	$f(x) \to -\infty$

20. Suppose a polynomial has odd degree, say

$$f(x) = ax^n + \text{lower terms},$$

and $a \neq 0$. Suppose first $a > 0$. If $x \to \infty$ then $f(x) \to \infty$ and in particular, $f(x) > 0$ for some x. If $x \to -\infty$ then $f(x) \to -\infty$, and in particular, $f(x) < 0$ for some x. By the intermediate value theorem, there is some number c such that $f(c) = 0$. The same argument works if $a < 0$.

VI, §2, p. 191

1. For $\sin x$: $0, \pi$, and add $n\pi$ with any integer n.

2. For $\cos x$: $\dfrac{\pi}{2}, \dfrac{3\pi}{2}$, and add $n\pi$.

3. Let $f(x) = \tan x$. Then $f'(x) = 1 + \tan^2 x$, and

$$f''(x) = \frac{d}{dx} (1 + \tan^2 x) = 2(\tan x)(1 + \tan^2 x).$$

The expression $1 + \tan^2 x$ is always > 0, and $f''(x) = 0$ if and only if $x = 0$ (in the given interval). If $x > 0$ then $\tan x > 0$ and if $x < 0$ then $\tan x < 0$. Hence $x = 0$ is the inflection point.

4. Sketch of graphs of $\sin^2 x$ and $|\sin x|$.

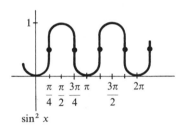

$\sin^2 x$

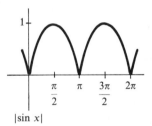

$|\sin x|$

Observe that the function $\sin^2 x$ is differentiable, and its derivative is 0 at $x = 0$, $\pi/2$, π, etc., so the curve is flat at these points. On the other hand, $|\sin x|$ is not differentiable at 0, π, 2π, etc., where it is "pointed." For instance, the graph of $|\sin x|$ for $\pi \leq x \leq 2\pi$ is obtained by reflecting the graph of $\sin x$ across the x-axis.

Let $f(x) = \sin^2 x$. Then $f'(x) = 2 \sin x \cos x = \sin 2x$. Also $f''(x) = 2 \cos 2x$. These allow you to find easily the regions of increase, decrease, and the inflection points, when $f''(x) = 0$. The inflection points are when $\cos 2x = 0$, that is $2x = \pi/2 + n\pi$ with an integer n, so $x = \pi/4 + n\pi/2$ with an integer n.

6. Bending up for $x > 0$; down for $x < 0$.

7. Bending up for $x > \sqrt{3}$, $-\sqrt{3} < x \leq 0$. Down for $x < -\sqrt{3}$, $0 \leq x < \sqrt{3}$.

8. Bending up for $x > 1$, $-1 < x \leq 0$. Down for $x < -1$ and $0 \leq x < 1$.

9. Max at $x = \pi/4$. Min at $x = 5\pi/4$.
Strictly increasing for $0 \leq x \leq \pi/4$, $5\pi/4 \leq x \leq 8\pi/4$. Decreasing for

$$\pi/4 \leq x \leq 5\pi/4.$$

Inflection points: $3\pi/4$ and $7\pi/4$.

Sketch of curve $f(x) = \sin x + \cos x$.

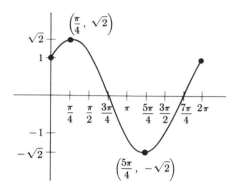

VI, §3, p. 196

1. Let $f(x) = ax^3 + bx^2 + cx + d$. Then

$$f''(x) = 6ax + 2b.$$

There is a unique solution $x = -b/3a$. Furthermore, if $a > 0$:

$$f''(x) = 6ax + 2b > 0 \quad \Leftrightarrow \quad x > -b/3a,$$
$$f''(x) = 6ax + 2b < 0 \quad \Leftrightarrow \quad x < -b/3a.$$

Hence $x = -b/3a$ is an inflection point, and is the only one. If $a < 0$, dividing an inequality by a changes the direction of the inequality, but the argument is the same.

17. (a), (e)   **18.** (c)

19. $f''(x) = 12x^2 + 18x - 2$. So $\frac{1}{2}f''(x) = 6x^2 + 9x - 1$, and $f''(x) = 0$ if and only if

$$x = \frac{-9 - \sqrt{105}}{12} \quad \text{or} \quad x = \frac{-9 + \sqrt{105}}{12}$$

Furthermore the graph of f'' is a parabola bending up (because the coefficient 12 of x^2 is positive) and so

$$f''(x) < 0 \quad \text{if and only if} \quad \frac{-9 - \sqrt{105}}{12} < x < x = \frac{-9 + \sqrt{105}}{12}.$$

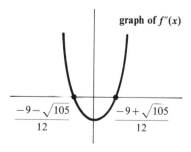

graph of $f''(x)$

$$\frac{-9 - \sqrt{105}}{12} \qquad \frac{-9 + \sqrt{105}}{12}$$

Therefore $f''(x)$ changes sign at the two roots of $f''(x)$, whence these two roots are inflection points of f.

We have $f'(x) = 4x^3 + 9x^2 - 2x = x(4x^2 + 9x - 2)$. The critical points of f are the roots of f', that is

$$x_1 = 0, \qquad x_2 = \frac{-9 - \sqrt{113}}{8}, \qquad x_3 = \frac{-9 + \sqrt{113}}{8}.$$

Note that $f(-2) < 0$ (by direct calculation, so $f(x)$ is negative at some $x < 0$. Also

$$\text{if} \quad x \to -\infty \quad \text{then} \quad f(x) \to \infty,$$

$$\text{if} \quad x \to \infty \quad \text{then} \quad f(x) \to \infty.$$

If $x \geq 0$ then $f(x) > 0$. Indeed, if $0 < x \leq 1$ then $-x^2 + 5 > 0$ and the other two terms $x^4 + 3x^3$ are both positive, so $f(x) > 0$. If $x \geq 1$, then

$$x^4 - x^2 > 0,$$

and $3x^3 + 5 > 0$ so again $f(x) > 0$.

We can now sketch the graph of f.

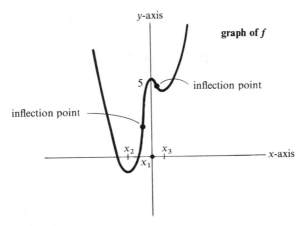

21. We have $f'(x) = 6x^5 - 6x^3 + \frac{9}{8}x = 3x(2x^4 - 2x^2 + \frac{3}{8})$. Let $u = x^2$. Then the roots of $f'(x)$ (which are the critical points of f) are $x = 0$, and those values coming from the quadratic formula applied to u, namely

$$2u^2 - 2u + \tfrac{3}{8} = 0.$$

This yields $u = 1/4$ or $u = 3/4$, or in terms of x,

$$x = \pm 1/2 \qquad \text{and} \qquad x = \pm\sqrt{3}/2.$$

Observe that $f(x)$ can also be written in terms of $u = x^2$, namely

$$f(x) = u^3 - \tfrac{3}{2}u^2 + \tfrac{9}{16}u - \tfrac{1}{32}.$$

Then you will find that the values of f at the critical points are all equal to 1/32 or $-1/32$. (Neat?) The graph looks like this.

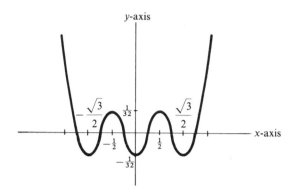

The inflection points can also be found easily. We have

$$f''(x) = 30x^4 - 18x^2 + \tfrac{9}{8} = 30u^2 - 18u + \tfrac{9}{8}.$$

Hence $f''(x) = 0$ if and only if

$$u = \frac{-18 \pm \sqrt{189}}{60}.$$

This gives two values for u, and then $x = \pm\sqrt{u}$ are the inflection points.

VI, §4, p. 201

Also see graphs below.

	c.p.	Increasing	Decreasing
1.	$3 \pm \sqrt{11}$	$x \leq 3 - \sqrt{11}$ and $x \geq 3 + \sqrt{11}$	$3 - \sqrt{11} \leq x < 3$ and $3 < x \leq 3 + \sqrt{11}$
2.	$3 \pm \sqrt{10}$	$3 - \sqrt{10} \leq x \leq 3 + \sqrt{10}$	$3 + \sqrt{10} \leq x$ and $x \leq 3 - \sqrt{10}$
3.	$-1 \pm \sqrt{2}$	$-1 - \sqrt{2} \leq x \leq -1 + \sqrt{2}$	$x \leq -1 - \sqrt{2}$ and $x \geq -1 + \sqrt{2}$
	For **4** and **5**, see graphs below.		
6.	0	$x \leq 0$	$\sqrt{2} < x$ and $0 \leq x < \sqrt{2}$
7.	None	$x < -\tfrac{1}{3},\ x > -\tfrac{1}{3}$	

	Rel. Max.	Rel. Min.	Increasing	Decreasing
9.	$-\sqrt{3}$	$\sqrt{3}$	$x < -\sqrt{3},\ x > \sqrt{3}$	$-\sqrt{3} < x < 0,$ $0 < x < \sqrt{3}$
12.	None	None	Nowhere	$x < 5/3,\ x > 5/3$
14.	None	0	$x > 0$	$-1 < x < 0$
16.	None	None	Nowhere	$x < -\sqrt{5},$ $-\sqrt{5} < x < \sqrt{5},$ $x > \sqrt{5}$
18.	0	None	$x < -2,$ $-2 < x < 0$	$0 < x < 2,\ x > 2$

4. $y = f(x) = x - 1/x$; no critical point. Function strictly increasing on every interval where defined.

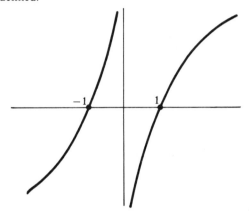

5. $y = f(x) = x/(x^3 - 1)$; $f'(x) = -(2x^3 + 1)/(x^3 - 1)^2$

$$f''(x) = 6x^2(x^3 - 1)(x^3 + 2)/(x^3 - 1)^4$$

$$= ((x^3 + 2)/(x^3 - 1)) \cdot \text{square.}$$

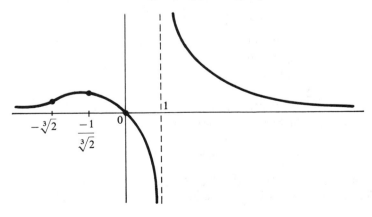

Critical point at $x = -1/\sqrt[3]{2}$. Inflection point at $x = -\sqrt[3]{2}$.
Local max at critical point.

VI, §5, p. 211

1. $r\sqrt{2}$ by $\frac{1}{2}r\sqrt{2}$ where r is the radius of the semicircle.
2. Let x be the side of the base and y the other side, as shown on the figure.

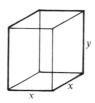

Let $A(x)$ be the combined area. Then

$$A(x) = x^2 + 4xy = 48, \qquad \text{whence} \qquad y = \frac{48 - x^2}{4x}.$$

If the volume is V, then

$$V = x^2 y = 12x - \frac{x^3}{4}.$$

Then $V'(x) = 12 - 3x^2/4$ and $V'(x) = 0$ if and only if $x = 4$, and $y = 2$ by substituting back in the expression for y in terms of x. We have for $x > 0$:

$$V'(x) > 0 \quad \Leftrightarrow \quad 12 - 3x^2/4 > 0 \quad \Leftrightarrow \quad x < 4,$$
$$V'(x) < 0 \quad \Leftrightarrow \quad 12 - 3x^2/4 < 0 \quad \Leftrightarrow \quad x > 4.$$

Hence $x = 4$ is a maximum point.
4. The total cost $f(x)$ is given by

$$f(x) = \frac{300}{x}\left(2 + \frac{x^2}{600}\right)\frac{30}{100} + D\,\frac{300}{x} = \frac{300D + 180}{x} + \frac{3x}{20}.$$

Then $f'(x) = 0$ if and only if $x^2 = 20(300D + 180)/3$. Answers: (a) $x = 20\sqrt{3}$ (b) $x = 40\sqrt{2}$ (c), (d), (e) $x = 60$ because the critical points are larger than 60, namely $20\sqrt{13}$, $60\sqrt{2}$ and $20\sqrt{23}$ in the respective cases.
5. $4\sqrt{2}$ by $8\sqrt{2}$
6. Answer: $x = 8/3$. For $x, y \geq 0$ we require $x + y = 4$ and we want to minimize $x^2 + y^3$. Since $y = 4 - x$ we must minimize

$$f(x) = x^2 + (4 - x)^3.$$

Then $f'(x) = 2x - 3(4 - x)^2$, and $f'(x) = 0$ if and only if $x = (26 \pm 10)/6$. The solution $36/6$ is beyond the range $0 \leq x \leq 4$. The critical point in this range is therefore $x = 16/6 = 8/3$. By a direction computation you can see that $f(8/3)$ is smaller than $f(0)$ or $f(4)$. Since there is only one critical point in the given interval, it must be the required minimum. You could also graph $f'(x)$ (parabola) to see that $f'(x) < 0$ if $0 < x < 8/3$ and $f'(x) > 0$ if $8/3 < x < 4$, so $f(x)$ is decreasing to the left of the critical point, and increasing to the right of the critical point, whence the critical point is a minimum in the given interval.

7. Minimum: use $24\pi/(4 + \pi)$ cm for circle; $96/(4 + \pi)$ cm for square.
   Maximum: use whole wire for circle.
   We show how to set up Exercise 7. Let x be the side of the square. Then $4x$ is the perimeter of the square, and $0 \le 4x \le 24$, so $0 \le x \le 6$. Also, $24 - 4x$ is the length of the circle, i.e. the circumference. But

$$24 - 4\pi = 2\pi r, \quad \text{where } r \text{ is the radius,}$$

and πr^2 is the area of the circle. The sum of the areas is

$$\pi r^2 + x^2.$$

This is expressed in terms of two variables, x and r, but we have the relation

$$r = \frac{24 - 4x}{2\pi},$$

so that the sum of the areas can be expressed in terms of one variable,

$$f(x) = \pi \left(\frac{24 - 4x}{2\pi}\right)^2 + x^2.$$

You can now minimize or maximize by finding first the critical points, and then investigate if they are maxima, minima, or whether such extrema occur when $4x = 24$ or $4x = 0$. Note that the graph of f is a parabola, bending up. What is $f(0)$? What is $f(6)$? The possible values for x are in the interval $0 \le x \le 6$. The maximum or minimum of f in this interval is either $f(0)$, $f(6)$, or at the critical point x such that $f'(x) = 0$. Draw the graph of f to get the idea of what happens.

8. Answer: $(1, 2)$. The square of the distance of (x, y) to $(2, 1)$ is

$$(x - 2)^2 + (y - 1)^2.$$

Since $y^2 = 4x$, we get $x = y^2/4$ and so we have to minimize

$$f(y) = \left(\frac{y^2}{4} - 2\right)^2 + (y - 1)^2.$$

But $f'(y) = (y^3/4) - 2$ and $f'(y) = 0$ if and only if $y = 2$, so $x = 1$. You can verify for yourself that $f'(y) > 0$ if $y > 2$ and $f'(y) < 0$ if $y < 2$, so $y = 2$ is a minimum for $f(y)$.

9. $(\sqrt{5/2}, \frac{1}{2})$, $(-\sqrt{5/2}, \frac{1}{2})$

10. The square of the distance between $(11, 1)$ and (x, y) is

$$(x - 11)^2 + (y - 1)^2 = (x - 11)^2 + (x^3 - 3x - 1)^2 = f(x),$$

which is a function of x alone. The picture is as shown roughly.

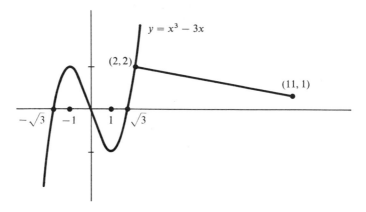

We have to minimize $f(x)$. As x becomes large positive or negative, $f(x)$ becomes large positive because

$$f(x) = x^6 + \text{lower degree terms.}$$

Hence a minimum occurs at a critical point. We have

$$f'(x) = 2(x - 11) + 2(x^3 - 3x - 1)(3x^2 - 3)$$

so

$$\tfrac{1}{2} f'(x) = 3x^5 - 12x^3 - 3x^2 + 10x - 8.$$

Plugging 2 directly shows that $f'(2) = 0$, and therefore 2 is a critical point. If we could show that 2 is the only critical point of f, then we would be done. Let us get more information on $f'(x)$. By long division, we obtain a factoring

$$\tfrac{1}{2} f'(x) = (x - 2)g(x) \qquad \text{where} \qquad g(x) = 3x^4 + 6x^3 - 3x + 4.$$

The function $g(x)$ is still complicated. If $g(x) \neq 0$ for all x we would be done, but it looks hard, even if true. Let us avoid technical complications and let us simplify our original problem by inspection. The picture suggests that all points (x, y) on the curve $y = x^3 - 3x$ such that $x \leq 1$ will be at further distance from $(11, 1)$ than $(2, 2)$. This is actually easily proved, for $f(2) = 82$, and if $x \leq 1$ then

$$f(x) \geq (x - 11)^2 \geq 100 > 82.$$

Therefore it suffices to prove that $f(2)$ is a minimum for $f(x)$ when $x > 1$, and it suffices to prove that 2 is the only critical point of $f(x)$ for $x > 1$. Thus it suffices to prove that $g(x) \neq 0$ for $x > 1$. This is easy because $3x^4 + 4$ is positive, and

$$6x^3 - 3x = 3(2x^3 - 1) > 0 \qquad \text{for} \quad x > 1.$$

This proves that $g(x) > 0$ for $x > 1$. Therefore $x = 2$ is the only critical point of f for $x > 1$, and finally we have proved that the minimum of $f(x)$ is at $x = 2$.

11. $(5, 3), (-5, 3)$ **12.** $(-1/2, 1\sqrt{2})$ **13.** $(-1, 0)$

14. Answer $(1, 2)$. The square of the distance between (x, y) and $(9, 0)$ is $(x - 9)^2 + y^2$. Since $y = 2x^2$, we have to minimize

$$f(x) = (x - 9)^2 + (2x^2)^2 = (x - 9)^2 + 4x^4.$$

Then $f'(x) = 16x^3 + 2x - 18$, and $f'(1) = 0$. You can graph $f'(x)$ as usual. Since $f''(x) = 48x^2 + 2 > 0$ for all x, we see that $f'(x)$ is strictly increasing, so $f'(x) = 0$ only for $x = 1$, which is the only critical point of f. But $f(x)$ becomes large when x becomes large positive or negative, and so $f(x)$ has a minimum. Since there is only one critical point for f, the minimum is equal to the critical point, thus giving the answer.

15. $F = 2\sqrt{3}Q/9b^2$

16. Answer: $y = 2h/3$. We have $F'(y) = y\frac{1}{2}(h - y)^{-1/2}(-1) + (h - y)^{1/2}$, so

$$F'(y) = \frac{2h - 3y}{2(h - y)^{1/2}}.$$

Thus $F'(y) = 0$ if and only if $y = 2h/3$. But $F(0) = F(h) = 0$ and $F(y) > 0$ for all y in the interval $0 < y < h$. Hence the critical point must be the maximum.

17. $(2, 0)$

18. max $x = 0$ (no triangle); min $x = \dfrac{L}{2\pi\left(\dfrac{1}{6\sqrt{3}} + \dfrac{1}{2\pi}\right)}$. Cut the wire with x de-

voted to the triangle and $L - x$ to the circle. Let s be the side of the triangle so $3s = x$; let h be the height of the triangle, and r the radius of the circle. Then

$$L = x + 2\pi r \quad \text{and} \quad A = \tfrac{1}{2}sh + \pi r^2.$$

We need other relations to make A a function of one variable x, namely

$$h = s\sqrt{3}/2 = x\sqrt{3}/6 \quad \text{and} \quad r = (L - x)/2\pi.$$

Then $A(x) = x^2\sqrt{3}/36 + ((L - x)/2\pi)^2$ and $A'(x) = x\sqrt{3}/18 - (L - x)/2\pi$. Thus the critical point is as given in the answer. Since the graph of $A(x)$ is a parabola bending up, the critical point is a minimum. The maximum occurs at the end point of the interval $0 \le x \le L$. To find out which end point, evaluate $A(0)$ and $A(L)$ and compare the two values to see that $A(0)$ is bigger.

19. $4\left[1 + \left(\dfrac{13.5}{4}\right)^{2/3}\right]^{3/2} = \dfrac{13\sqrt{13}}{2}$

20. We just set it up. Let r be the radius of the base of the cylinder, and h the height. Then the total volume, which is constant, is

$$V = \pi r^2 h + \tfrac{4}{3}\pi r^3.$$

This allows to solve for h, namely

$$h = \frac{V - 4\pi r^3/3}{\pi r^2}.$$

The cost of material is a constant times:

(Area of cylinder) + (twice area of sphere).

We can express this cost as a function of r and h, namely

$$2\pi r h + 2 \cdot 4\pi r^2.$$

Since h is expressed as a function of r above, we get the total cost expressed as a function of r only, namely

$$f(r) = C\left[2\pi r\left(\frac{V - 4\pi r^3/3}{\pi r^2}\right) + 8\pi r^2 \right] = C\left[\frac{2V}{r} + \frac{16}{3}\pi r^2\right].$$

From here on, you can find $f'(r)$ and proceed in the usual way to find when $f'(r) = 0$. There is only one critical point, and $f(r)$ becomes large when r approaches 0 or r becomes large. Hence the critical point is a minimum.

21. The picture is as follows:

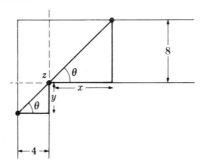

The longest rod is that which will fit the minimal distance labeled z in the figure. We let x, y be as shown in the figure. We have

$$z = \sqrt{4^2 + y^2} + \sqrt{x^2 + 8^2}.$$

This depends on the two variables x, y. But we can find a relation between them by using similar triangles, namely

$$\frac{8}{x} = \frac{y}{4},$$

so that

$$y = \frac{32}{x}.$$

Hence

$$z = f(x) = \sqrt{16 + \left(\frac{32}{x}\right)^2} + \sqrt{x^2 + 64}.$$

You then have to minimize $f(x)$. The answer comes out

$$x = 4\sqrt[3]{4}, \qquad z = \sqrt{4^2 + \left(\frac{32}{4\sqrt[3]{4}}\right)^2} + \sqrt{(4\sqrt[3]{4})^2 + 8^2}$$

22. From the figure in the text, we have $0 \leq x \leq a$, and

$$\text{dist}(P, R) = \sqrt{x^2 + y_1^2},$$

$$\text{dist}(R, Q) = \sqrt{(a - x)^2 + y_2^2}.$$

Hence the sum of the distances is

$$f(x) = \sqrt{x^2 + y_1^2} + \sqrt{(a - x)^2 + y_2^2}.$$

Then

$$f'(x) = \frac{x}{\sqrt{x^2 + y_1^2}} - \frac{(a - x)}{\sqrt{(a - x)^2 + y_2^2}}.$$

We have $f'(x) = 0$ if and only if

$$\frac{x}{\sqrt{x^2 + y_1^2}} = \frac{a - x}{\sqrt{(a - x)^2 + y_2^2}},$$

which is the desired cosine relation.

In particular, we see that there is only one critical point. Hence the minimum of f is either at the end points $x = 0$ or $x = a$, or at the critical point. We shall now prove that the critical point is the minimum. We are trying to prove that the graph of f looks something like this.

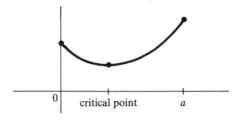

0   critical point   a

When x is near 0 and $x > 0$ then the first term in $f'(x)$ is small and the second term is near

$$-\frac{a}{\sqrt{a^2 + y_2^2}},$$

which is negative. Hence $f'(x) < 0$ when x is near 0, and the function is decreasing when x is near 0.

On the other hand, suppose x is near a. Then the second term in $f'(x)$ is near 0 and the first term is near

$$\frac{a}{\sqrt{a^2 + y_1^2}} > 0.$$

Consequently $f'(x) > 0$ when x is near a. Therefore f is increasing when x is near a. Since $f(x)$ is decreasing for x near 0 and increasing for x near a it follows that the minimum cannot be at the end points, and hence is somewhere in the middle. Hence the minimum is a critical point, and we have seen that there is only one critical point. Hence the minimum is at the critical point. This proves what we wanted.

23. Let x and $a - x$ be as on the figure. Then $0 \leq x \leq a$.

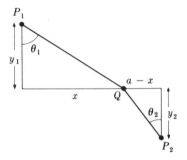

Let t_1 be the time needed to travel from P_1 to Q, and t_2 the time needed to travel from Q to P_2. Then

$$t_1 = \frac{\text{dist}(P_1, Q)}{v_1} \quad \text{and} \quad t_2 = \frac{\text{dist}(Q, P_2)}{v_2}.$$

Then

$$t_1 + t_2 = f(x) = \frac{1}{v_1} \sqrt{x^2 + y_1^2} + \frac{1}{v_2} \sqrt{(a - x)^2 + y_2^2}.$$

Both v_1, v_2 are given as constant. Hence again we have to take $f'(x)$ and set it equal to 0. This is similar to Exercise 22, and we find exactly the relation that is to be proved.

24. $p = s/n$. Since $L(0) = L(1) = 0$ and $L(p) > 0$ for $0 < p < 1$, it follows that the maximum is at a critical point. But

$$L'(p) = p^s(n - s)(1 - p)^{n-s-1}(-1) + sp^{s-1}(1 - p)^{n-s}$$

$$= p^{s-1}(1 - p)^{n-s-1}[p(s - n) + s(1 - p)]$$

$$= p^{s-1}(1 - p)^{n-s-1}(-np + s).$$

For $0 < p < 1$, the factor $p^{s-1}(1 - p)^{n-s-1}$ is not 0, so $L'(p) = 0$ if and only if $-np + s = 0$, that is $p = s/n$. Hence there is only one critical point, so the maximum is at the critical point.

25. (a)

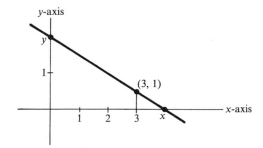

Let x, y be the intercepts of the line with the axes. Then the area of the triangle is equal to

$$A = \tfrac{1}{2}xy.$$

We want to minimize the area. By similar triangles, we know that

$$\frac{y}{x} = \frac{1}{x-3} \qquad \text{so} \qquad y = \frac{x}{x-3}.$$

Then the area is given by

$$A(x) = \tfrac{1}{2}x\,\frac{x}{x-3} = \frac{1}{2}\frac{x^2}{x-3}.$$

From the physical considerations, we are limited to the interval $x > 3$. As x approaches 3 and $x > 3$ the denominator approaches 0 and is positive. Since x^2 approaches 9, it follows that $A(x)$ becomes large positive. Also as $x \to \infty$, $A(x) \to \infty$. Hence the minimum of A will occur at a critical point.

   We find the critical points. We have

$$2A'(x) = \frac{(x-3)2x - x^2}{(x-3)^2} = \frac{x^2 - 6x}{(x-3)^2}.$$

Then $A'(x) = 0$ if and only if $x = 6$ ($x = 0$ is excluded because $x > 3$). Hence the desired line passes through the point $(6, 0)$. The equation is then

$$y - 0 = \frac{1-0}{3-6}\,(x-6) = -\tfrac{1}{3}(x-6).$$

or also

$$y - 2 = -\tfrac{1}{3}x.$$

25. (b) $3y = -2x + 12$

26. $x = (a_1 + \cdots + a_n)/n$. We are given

$$f(x) = (x - a_1)^2 + \cdots + (x - a_n)^2.$$

Then

$$f'(x) = 2(x - a_1) + \cdots + 2(x - a_n)$$
$$= 2x - 2a_1 + 2x - 2a_2 + \cdots + 2x - 2a_n$$
$$= n2x - 2(a_1 + \cdots + a_n).$$

So $f'(x) = 0$ if and only if $nx = a_1 + \cdots + a_n$. Divide by n to get the critical point of f. Since $f(x) \to \infty$ as $x \to \pm\infty$ because for instance just one square term $(x - a_1)^2$ becomes large when x becomes large positive or negative, it follows that the minimum must be at a critical point, and there is only one critical point. Hence the critical point is the minimum.

27. $25/\sqrt{2}$

28. Answer: $\theta = \pi/2$. Let h be the height of the triangle as shown on the picture.

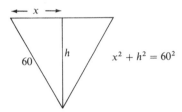

It suffices to maximize the area given by

$$A(x) = \tfrac{1}{2}xh = \tfrac{1}{2}x\sqrt{60^2 - x^2} \qquad \text{for} \quad 0 < x < 60.$$

Then $A'(x) = (-2x^2 + 60^2)/2(60^2 - x^2)^{1/2}$ and $A'(x) = 0$ if and only if $x = 60/\sqrt{2}$. But $A(x) \geq 0$ and $A(0) = A(60) = 0$. Hence the critical point is the maximum. The above value for x implies that $\theta = \pi/2$, because the triangle is similar to the triangle with sides 1, 1, $\sqrt{2}$.

29. $\pi/3$. We work it out. Let the depth be y. Then $\sin \theta = y/100$. Maximum capacity occurs when the area of the cross section is maximum. This area is equal to

$$A = 100y + 2(\tfrac{1}{2}y \ 100 \cos \theta).$$

You have a choice whether to express A entirely in terms of y or entirely in terms of θ. Suppose we do it in terms of θ. Then

$$A(\theta) = 10^4 \sin \theta + 10^4 \sin \theta \cos \theta = 10^4[\sin \theta + \tfrac{1}{2} \sin 2\theta].$$

So $A'(\theta) = 10^4[\cos \theta + \cos 2\theta]$, and $0 \leq \theta \leq \pi/2$. But in this interval, $\cos \theta$ and $\cos 2\theta$ are strictly decreasing so $A'(\theta)$ is strictly decreasing. We have $A'(\theta) = 0$ precisely when $\cos \theta + \cos 2\theta = 0$, which occurs when $\theta = \pi/3$. Thus $A'(\theta) > 0$ if $0 < \theta < \pi/3$ and $A'(\theta) < 0$ if $\pi/3 < \theta < \pi/2$. Hence $A(\theta)$ is increasing for $0 \leq \theta \leq \pi/3$ and decreasing for $\pi/3 \leq \theta \leq \pi/2$. Hence the maximum occurs when $\theta = \pi/3$.

30. (a) $a = 16$ (b) $a = -54$. To see this, note that

$$f'(x) = 2x - a/x^2,$$

and $f'(x) = 0$ if and only if $a = 2x^3$. For $x = 2$ and $x = -3$, this gives the desired value for a. You can check that this is a minimum directly by determining when $f'(x) > 0$ or $f'(x) < 0$. As for part (c), this is one of the rare cases when taking the second derivative is useful. The second derivative is $f''(x) = 2 + 2a/x^3$, and the critical point has been determined to be when $a = 2x^3$; so if x is the critical point we get $f''(x) = 6 > 0$. Hence the critical point must be a local minimum.

31. $\dfrac{c}{1 + \sqrt[3]{a/b}}$ away from b

32. base $= 24/(4 + \pi)$ and height $= 12/(4 + \pi y)$. The picture is as follows.

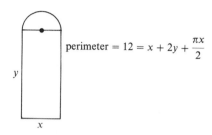

perimeter $= 12 = x + 2y + \dfrac{\pi x}{2}$

Area $= xy + \frac{1}{2}\pi r^2 = xy + \frac{1}{2}\pi(x/2)^2$. You can solve for y in terms of x by using the perimeter equation, so the area is given as a function of x, $A(x)$. Finding $A'(x) = 0$ yields the desired value for x. It is a critical point, the only critical point, and $A(x)$ is a parabola which bends down, so the critical point is a maximum.

33. (a) Find the radius and angle of a circular sector of maximum area if the perimeter is 20 cm.

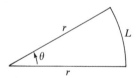

Let r be the radius of the sector, and L the length of the circular arc. Then $L = (\theta/2\pi)2\pi r = \theta r$. The perimeter is

Graph of $A(r)$

$$P = 2r + L = 2r + \theta r = 20.$$

Hence we can solve for θ in terms of r, that is

$$\theta = \frac{20}{r} - 2.$$

The area of the sector is

$$A = \frac{\theta}{2\pi}\,\pi r^2 = \frac{\theta r^2}{2}$$

5   r-axis

so in terms of r alone:

$$A(r) = \left(\frac{20}{r} - 2\right)\frac{r^2}{2} = 10r - r^2.$$

The graph of $A(r)$ is a parabola bending down, so the maximum is at the critical point. But

$$A'(r) = 10 - 2r$$

so the maximum is when $2r = 10$ so $r = 5$. Then

$$\theta = \frac{20}{5} - 2 = 2.$$

33. (b) radius $= 4$ cm, angle $= 2$ radians

34. Don't make things more complicated than they need to be. Observe that $2\sin\theta\cos\theta = \sin 2\theta$. This is a maximum when $\theta = \pi/4$.

35. $2(1 + \sqrt[3]{36})^{3/2}$

36. $r = (V/\pi)^{1/3} = y$. Note that $V = \pi r^2 y$ so $y = V/\pi r^2$.

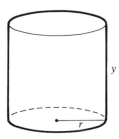

Let $S =$ surface area so $S = \pi r^2 + 2\pi ry$. Then

$$S(r) = \pi r^2 + 2V/r.$$

We have $S'(r) = 2\pi r - 2V/r^2 = 0$ if and only if $r = (V/\pi)^{1/3}$. There is only one critical point. But $S(r)$ becomes large when r approaches 0 or also when r becomes large. There is a minimum since $S(r) > 0$ for $r > 0$, and so the minimum is equal to the single critical point.

37. $P = 2r + L = 2r + r\theta.$

From $A = \theta r^2/2$ we get $\theta = 2A/r^2$ so P can be expressed in terms of r only by

$$P(r) = 2r + 2A/r.$$

We have $P'(r) = 2 - 2A/r^2$, and $P'(r) = 0$ if and only if $r = A^{1/2}$. So P has only one critical point, and $P(r) \to \infty$ as $r \to \infty$ and as $r \to 0$. Hence P has a minimum and that minimum is at the critical point. This is a minimum for all values of $r > 0$. In part (a), we have $\theta \leq \pi$ so $r^2 \geq 2A/\pi$, and the data limits us to the interval

$$\sqrt{2A/\pi} \leq r.$$

Hence in part (a) the minimum is at the critical point. In part (b), we have $\theta \leq \pi/2$ so $r^2 \geq 4A/\pi$, which limits us to the interval

$$\sqrt{4A/\pi} \leq r.$$

Since $\sqrt{4A/\pi} > \sqrt{A}$, the minimum is at the end point $r = \sqrt{4A/\pi}$.

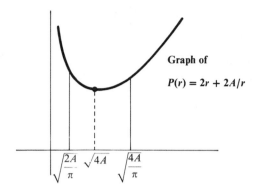

Graph of

$$P(r) = 2r + 2A/r$$

$\sqrt{\dfrac{2A}{\pi}} \quad \sqrt{4A} \quad \sqrt{\dfrac{4A}{\pi}}$

38. $x = 20$. The profits are given by

$$P(x) = 50x - f(x).$$

Then $P'(x) = -3x^2 + 90x - 600$. By the quadratic formula, $P'(x) = 0$ if and only if $x = 20$ or $x = 10$. But the graph of $P(x)$ is that of a cubic, which you should know how to do. Also $P(10)$ is negative, and $P(0)$ is also negative. So the maximum is at $x = 20$.

39. 18 units, daily profit $266

40. $(50 + 5\sqrt{94})/3$. Same method as Problems 38 and 39.

41. 30 units; $8900

42. 20. Let $g(x)$ be the profits. Since $p = 1000 - 10x$, we get

$$g(x) = x(1000 - 10x) - f(x) = -20x^2 + 800x - 6000.$$

The graph of $g(x)$ is a parabola bending down, and $g'(x) = 0$ if and only if $x = 20$, which gives the maximum for $g(x)$.

VII, §1, p. 221

1. Yes; all real numbers   **3.** Yes; all real numbers   **5.** Yes; for $y < 1$
7. Yes; for $y \geq 1$ **9.** Yes; for $y \leq -1$ **11.** Yes; for $y \geq 2$
13. Yes; for $-1 \leq y \leq 1$

VII, §2, p. 224

0. Let $f(x) = -x^3 + 2x + 1$. Then

$$f'(x) = -3x^2 + 2$$

$$= 0 \quad \text{if and only if} \quad 3x^2 = 2$$
$$\text{if and only if} \quad x = \sqrt{2/3} \quad \text{and} \quad x = -\sqrt{2/3}.$$

These are the critical points of f. Also

$$f'(x) > 0 \quad \Leftrightarrow \quad -3x^2 + 2 > 0$$
$$\Leftrightarrow \quad 3x^2 < 2$$
$$\Leftrightarrow \quad |x| < \sqrt{2/3},$$
$$f'(x) < 0 \quad \Leftrightarrow \quad x^2 > 2/3,$$
$$\Leftrightarrow \quad x > \sqrt{2/3} \quad \text{and} \quad x < -\sqrt{2/3}.$$

There are three maximal intervals where an inverse function of f could be defined (excluding the end points):

$$x < -\sqrt{2/3}, \qquad -\sqrt{2/3} < x < \sqrt{2/3}, \qquad x > \sqrt{2/3}.$$

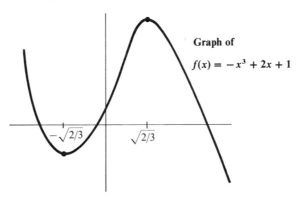

Graph of

$$f(x) = -x^3 + 2x + 1$$

Over each such interval, there is an inverse function of f whose value at 2 is some point in the interval. We work out two out of the three cases, but you only had to pick one of them.

Case 1. Observe that $f(1) = 2$, and $1 > \sqrt{2/3}$. Therefore in this case, if g is the inverse function, then

$$g'(2) = \frac{1}{f'(1)} = \frac{1}{-3 + 2} = -1.$$

Case 2. Take the interval $-\sqrt{2/3} < x < \sqrt{2/3}$. We want to solve $f(x) = 2$, that is

$$-x^3 + 2x + 1 = 2, \qquad \text{or} \qquad x^3 - 2x + 1 = 0.$$

Factoring, this is the same as

$$(x - 1)(x^2 + x - 1) = 0.$$

In the given interval, $x = 1$ is not a solution. There are two other possible solutions:

$$x = \frac{-1 + \sqrt{5}}{2} \qquad \text{and} \qquad x = \frac{-1 - \sqrt{5}}{2}.$$

But $(-1 - \sqrt{5})/2$ is not in the present interval of definition. Hence there is only one possible x in this case, namely $x_1 = (-1 + \sqrt{5})/2$. If g is the inverse function for the given interval, then

$$g'(2) = \frac{1}{f'(x_1)} = \frac{1}{-3x_1^2 + 2}.$$

(Alternative answers depend on the choice of intervals.)

1. $\frac{1}{3}$ 2. $\frac{1}{11}$ 3. $\frac{1}{3}$ or $-\frac{1}{3}$ 4. 1 or -1 5. 1 or -1 6. $\pm\frac{1}{2}$ or $\pm\dfrac{1}{2\sqrt{2}}$

7. $\frac{1}{4}$ 8. -1 or $\frac{1}{2} \pm \frac{3}{10}\sqrt{5}$ 9. $\frac{1}{24}$ 10. $\dfrac{1}{10\sqrt{2}}$ or $\dfrac{-1}{10\sqrt{2}}$

11. $g'(y) = \dfrac{1}{f'(x)} = \dfrac{1}{f'(g(y))}$,

$$g''(y) = \frac{-1}{f'(g(y))^2} f''(g(y))g'(y) = \frac{-1}{f'(x)^2} f''(x)g'(y).$$

If $f'(x) > 0$ then $g'(y) > 0$ by the first formula, and $g''(y) < 0$ by the second.

VII, §3, p. 229

1 and 2. View the cosine as defined on the interval

$$0 \leqq x \leqq \pi.$$

On this interval, the cosine is strictly decreasing, and for $0 < x < \pi$ we have

$$\frac{d \cos x}{dx} = -\sin x < 0.$$

Hence the inverse function $x = g(y)$ exists, and is called the **arccosine**. Since $y = \cos x$ decreases from 1 to -1, the arccosine is defined on the interval $[-1, 1]$. Its derivative is given by $g'(y) = 1/f'(x)$, so that

$$\frac{d \arccos y}{dy} = g'(y) = \frac{1}{-\sin x}.$$

But we have the relationship

$$\sin^2 x = 1 - \cos^2 x,$$

and for $0 < x < \pi$ we have $\sin x > 0$ so that

$$\sin x = \sqrt{1 - \cos^2 x} = \sqrt{1 - y^2}.$$

Consequently

$$g'(y) = \frac{-1}{\sqrt{1 - y^2}}.$$

The graph of arccos looks like this.

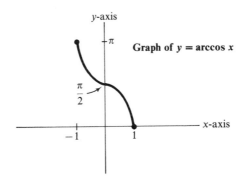

Graph of $y = \arccos x$

3. (a) $2/\sqrt{3}$ (b) $\sqrt{2}$ (c) $\pi/6$ (d) $\pi/4$ (e) 2 (f) $\pi/3$

4. $-2/\sqrt{3}$, $-\sqrt{2}$, $\pi/3$, $\pi/4$

5. Let $y = \sec x$ on interval $0 < x < \pi/2$. Then $x = \text{arcsec } y$ is defined on $1 < y$,

and $dx/dy = \dfrac{1}{y\sqrt{y^2 - 1}}.$

6. $-\pi/2$ **7.** 0   **8.** $\pi/2$ **9.** $\pi/2$ **10.** $-\pi/4$ **11.** $\dfrac{1}{\sqrt{1 - (x^2 - 1)^2}} 2x$

12. $\dfrac{-1}{\sqrt{-(x^2 + 5x + 6)}}$ **13.** $\dfrac{-1}{(\arcsin x)^2 \sqrt{1 - x^2}}$ **14.** $\dfrac{4}{\sqrt{1 - 4x^2}(\arccos 2x)^2}$

VII, §4, p. 233

1. $\pi/4$, $\pi/6$, $-\pi/4$, $\pi/3$ **2.** $\frac{1}{2}$, $\frac{3}{4}$, $\frac{1}{2}$, $\frac{1}{4}$ **3.** $1/(1 + y^2)$

4. (a) $-\pi/4$ (b) 0 (c) $-\pi/6$ (d) $\pi/6$ **5.** $\dfrac{3}{1 + 9x^2}$ **7.** 0   **9.** $\dfrac{2 \cos 2x}{1 + \sin^2 2x}$

11. $\dfrac{(\cos x)(\arcsin x) - (\sin x)/\sqrt{1 - x^2}}{(\arcsin x)^2}$ **13.** $\dfrac{-1}{1 + x^2}$

15. $\dfrac{9}{\sqrt{1 - 9x^2}} (1 + \arcsin 3x)^2$ **17.** $\left(y - \dfrac{\pi}{4}\right) = \sqrt{2}\left(x - \dfrac{1}{\sqrt{2}}\right)$

19. $\left(y - \dfrac{\pi}{3}\right) = \dfrac{1}{2}\left(x - \dfrac{\sqrt{3}}{2}\right)$ **21.** $\left(y + \dfrac{\pi}{6}\right) = \dfrac{2}{\sqrt{5}}\left(x + \dfrac{1}{2}\right)$ **22.** $2/25$

23. 440 ft/sec   **24.** 3/26   **25.** 0.02 rad/sec   **27.** $\dfrac{d\theta}{dt} = \frac{1}{6}\sin\theta\tan\theta$ **28.** $\dfrac{2}{25\sqrt{21}}$

29. $\frac{1}{82}$ rad/sec   **31.** (a) $\dfrac{1500}{(400)^2 + (\frac{225}{4})^2}$ rad/sec   (b) $\dfrac{1500}{(600)^2 + (\frac{375}{4})^2}$ rad/sec

33. Picture:

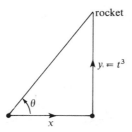

We are given $dx/dt = -50$. When $t = 0$ we have $x(0) = 300$, so in general

$$x(t) = 300 - 50t.$$

Furthermore, $dy/dt = 3t^2$. We want to find $d\theta/dt$. We have

$$\tan \theta = y/x \qquad \text{so} \qquad \theta = \arctan y/x.$$

Then

$$\frac{d\theta}{dt} = \frac{1}{1 + (y/x)^2} \frac{x\, dy/dt - y\, dx/dt}{x^2}.$$

But $y(5) = 125$ and $x(5) = 300 - 250 = 50$. Hence

$$\frac{d\theta}{dt}\bigg|_{t=5} = \frac{1}{1 + \left(\dfrac{125}{50}\right)^2} \frac{50 \cdot 3 \cdot 25 - 125 \cdot (-50)}{50^2}$$

$$= 16/29 \text{ rad/sec.}$$

VIII, §1, p. 244

1. (a) $y = 2e^2x - e^2$ (b) $y = 2e^{-4}x + 5e^{-4}$ (c) $y = 2x + 1$
2. (a) $y = \frac{1}{2}e^{-2}x + 3e^{-2}$ (b) $y = \frac{1}{2}e^{1/2}x + \frac{1}{2}e^{1/2}$ (c) $y = \frac{1}{2}x + 1$
3. $y = 3e^2x - 4e^2$ **4.** (a) $e^{\sin 3x}(\cos 3x)3$ (b) $\cos(e^x + \sin x)(e^x + \cos x)$
  (c) $\cos(e^{x+2})e^{x+2}$ (d) $4 \cos(e^{4x-5})e^{4x-5}$

5. (a) $\dfrac{1}{1 + e^{2x}} e^x$ (b) $e^x(-\sin(3x + 5))3 + e^x \cos(3x + 5)$

  (c) $2(\cos 2x)e^{\sin 2x}$ (d) $-\dfrac{1}{\sqrt{1 - x^2}} e^{\arccos x}$ (e) $-e^{-x}$

  (h) $e^{-\arcsin x}\left(\dfrac{-1}{\sqrt{1 - x^2}}\right)$ (i) $e^x \sec^2 e^x$ (j) $\dfrac{1}{1 + e^{4x}} 2e^{2x}$ (k) $\dfrac{-1}{\sin^2 e^x} (\cos e^x)e^x$

  (l) $\dfrac{1}{\sqrt{1 - (e^x + x)^2}} (e^x + 1)$ (m) $\sec^2 (xe^{\tan x})$ (n) $(1 + \tan^2 e^x)e^x$

6. (c) Let $f(x) = xe^x$. Suppose you have already proved that

$$f^{(n)}(x) = (x + n)e^x.$$

Then

$$f^{(n+1)}(x) = \frac{d}{dx}[(x + n)e^x]$$

$$= (x + n)e^x + e^x \qquad \text{(derivative of a product)}$$

$$= (x + n + 1)e^x.$$

This proves the formula for the $(n + 1)$-th derivative.

8. (c) Differentiate $f(x)/e^{h(x)}$ by the rule for quotients and the chain rule. We get:

$$\frac{d}{dx}\left(\frac{f(x)}{e^{h(x)}}\right) = \frac{e^{h(x)}f'(x) - f(x)e^{h(x)}h'(x)}{e^{2h(x)}}$$

$$= \frac{e^{h(x)}h'(x)f(x) - f(x)e^{h(x)}h'(x)}{e^{2h(x)}}$$

$$= 0.$$

Hence $f(x)/e^{h(x)}$ is constant, so there is a constant C such that

$$\frac{f(x)}{e^{h(x)}} = C.$$

Now cross multiply to get $f(x) = Ce^{h(x)}$.

9. $(y - e^2) = 2e^2(x - 1)$ **10.** $y - 2e^2 = 3e^2(x - 2)$ **11.** $y - 5e^5 = 6e^5(x - 5)$

12. $y = x$ **13.** $(y - 1) = -x$ **14.** $y - e^{-1} = e^{-1}(x - 1)$

15. Let $f(x) = e^x + x$. Then $f'(x) = e^x + 1 > 0$ for all x. Hence f is strictly increasing. We have $f(0) = 1$, and $f(-1) = 1/e - 1 < 0$ because $1/e < 1$. By the intermediate value theorem, there exists some x such that $f(x) = 0$, and this value of x is unique because f is strictly increasing.

16. See the proof of Theorem 5.1.

17. If $x = 1$ in Exercise 16(b) we get $2 < e$. If $x = 1$ in Exercise 16(c) we get $2.5 < e$.

18. See Theorem 5.2.

19. (a) Let $f_1(x) = e^{-x} - (1 - x)$. Then $f'_1(x) = -e^{-x} + 1$ and since $e^x > 1$ for $x > 0$ we get

$$f'_1(x) = -\frac{1}{e^x} + 1 > 0 \qquad \text{for} \quad x > 0.$$

Hence f is strictly increasing for $x \geq 0$. Since $f_1(0) = 0$ we conclude $f_1(x) > 0$ for $x > 0$, in other words

$$e^{-x} - (1 - x) > 0 \qquad \text{for} \quad x > 0,$$

and therefore $e^{-x} > 1 - x$ for $x > 0$, as desired.

(b) Let $f_2(x) = 1 - x + x^2/2 - e^{-x}$. Then

$$f'_2(x) = -1 + x + e^{-x} = f_1(x).$$

By part (a), we know that $f_1(x) > 0$ for $x > 0$. Hence f_2 is strictly increasing for $x \geq 0$. Since $f_2(0) = 0$ we conclude that $f_2(x) > 0$ for $x > 0$, whence (b) follows at once.

(c) Let $f_3(x) = e^{-x} - \left(1 - x + \dfrac{x^2}{2} - \dfrac{x^3}{3 \cdot 2} \right)$. Then $f'_3(x) = f_2(x)$. Use part (b) and similar arguments as before to conclude $f_3(x) > 0$ for $x > 0$, whence (c) follows.

(d) Left to you.

20. If we put $x = 1/2$ in Exercise 19(a), then we find $\frac{1}{2} < e^{-1/2}$, or in other words, $\frac{1}{2} < 1/e^{1/2}$. Hence $e^{1/2} < 2$ and $e < 4$. If we put $x = 1$ in Exercise 19(c), then we find

$$\frac{1}{2} - \frac{1}{6} < e^{-1} = \frac{1}{e}, \qquad \text{that is} \qquad \frac{1}{3} < \frac{1}{e},$$

whence $e < 3$.

21. $\cosh^2(t) = \frac{1}{4}(e^t + e^{-t})^2 = \frac{1}{4}(e^{2t} + 2e^t e^{-t} + e^{-2t})$

$$= \frac{1}{4}(e^{2t} + 2 + e^{-2t}).$$

Similarly,

$$\sinh^2(t) = \frac{1}{4}(e^{2t} - 2 + e^{-2t}).$$

Subtracting yields $\cosh^2 - \sinh^2 = 1$.

As for the derivative, $\cosh'(t) = \frac{1}{2}(e^t - e^{-t})$; don't forget how to use the chain rule: let $u = -t$. Then

$$\frac{de^{-t}}{dt} = \frac{de^u}{du} \frac{du}{dt} = -e^{-t}.$$

22.

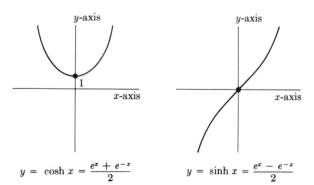

$$y = \cosh x = \frac{e^x + e^{-x}}{2} \qquad\qquad y = \sinh x = \frac{e^x - e^{-x}}{2}$$

VIII, §2, p. 255

1. (a) $(y - \log 2) = \frac{1}{2}(x - 2)$
   (b) $(y - \log 5) = \frac{1}{5}(x - 5)$
   (c) $(y - \log \frac{1}{2}) = 2(x - \frac{1}{2})$

2. (a) $(y - \log 2) = -(x + 1)$
   (b) $(y - \log 5) = \frac{4}{5}(x - 2)$

   (c) $(y - \log 10) = \dfrac{-3}{5} (x + 3)$

3. (a) $\dfrac{\cos x}{\sin x}$ (b) $\cos(\log(2x + 3)) \dfrac{1}{2x + 3} \cdot 2$ (c) $\dfrac{1}{x^2 + 5} \cdot 2x$

   (d) $\dfrac{(\sin x)/x - (\log 2x) \cos x}{\sin^2 x}$

4. $(y - \log 4) = \frac{1}{4}(x - 3)$ 5. $(y - \log 3) = \frac{2}{3}(x - 4)$ 7. $(y - 1) = \dfrac{1}{e} (x - e)$

8. $(y - e) = 2(x - e)$ 9. $(y - 2\log 2) = (1 + \log 2)(x - 2)$

10. $(y - 3) = \dfrac{3}{e} (x - e)$ 11. $(y - 1) = \dfrac{-1}{e} (x - e)$

12. $\left(y - \dfrac{1}{\log 2} \right) = \dfrac{-1}{2(\log 2)^2} (x - 2)$

14. $\dfrac{2x}{x^2 + 3}$ 15. $\dfrac{-1}{x(\log x)^2}$ 16. $\dfrac{\log x - 1}{(\log x)^2}$ 17. $\frac{1}{3}(\log x)^{-2/3} + (\log x)^{1/3}$

18. $\dfrac{-x}{1 - x^2}$

19. Let $f(x) = x + \log x$. Then $f'(x) = 1 + 1/x > 0$ for $x > 0$. Hence f is strictly increasing. Also $f''(x) = -1/x^2$, so f is bending down. Note that $f(1) = 1$. If $x \to \infty$ then $f(x) \to \infty$ because both x and $\log x$ become large. In fact $f(x)$ lies at a distance $\log x$ above the line $y = x$. As $x \to 0$, $\log x \to -\infty$ (think of $x = 1/e^z = e^{-z}$ where $z \to \infty$). Hence the graph looks like this:

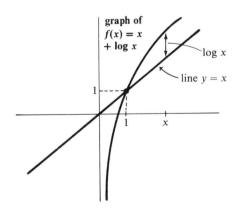

VIII, §3, p. 261

1. $10^x \log 10$, $7^x \log 7$ 2. $3^x \log 3$, $\pi^x \log \pi$
5. $(y - 1) = (\log 10)x$ 6. $y - \pi^2 = \pi^2 \log \pi (x - 2)$
7. (a) $e^{x \log x}[\log x + 1]$ (b) $x^{(x^x)}[x^{x-1} + (\log x)x^x(1 + \log x)]$ In (b), we write

$$x^{(x^x)} = e^{(x^x) \log x}$$

and use the chain rule. The derivative of $(x^x) \log x$ is found by the rule for differentiating a product, and we have

$$\frac{d}{dx} e^{(x^x) \log x} = e^{(x^x) \log x}\left[x^x \cdot \frac{1}{x} + \frac{d(x^x)}{dx} \log x\right].$$

The derivative of x^x was found in (a), and the answer drops out.
8. (a) $y - 1 = x - 1$ (b) $y - 4 = 2(1 + \log 2)(x - 2)$
   (c) $y - 27 = 27(1 + \log 3)(x - 3)$
9. (a) Let $f(x) = x^{x^{1/2}} = e^{x^{1/2} \log x}$. Then

$$f'(x) = x^{\sqrt{x}}\left[\frac{1}{\sqrt{x}} + \frac{1}{2\sqrt{x}} \log x\right]$$

so

$$f'(2) = 2^{\sqrt{2}}\left[\frac{1}{\sqrt{2}} + \frac{1}{2\sqrt{2}} \log 2\right].$$

The tangent line at $x = 2$ is

$$y - 2^{\sqrt{2}} = 2^{\sqrt{2}}\left[\frac{1}{\sqrt{2}} + \frac{1}{2\sqrt{2}} \log 2\right](x - 2).$$

(b) $y - 5^{\sqrt{5}} = 5^{\sqrt{5}}\left[\frac{1}{\sqrt{5}} + \frac{1}{2\sqrt{5}} \log 5\right](x - 5)$

10. Let $f(x) = x^{\sqrt[3]{x}} = e^{x^{1/3} \log x}$. Then

$$f'(x) = x^{\sqrt[3]{x}}\left[x^{1/3}\frac{1}{x} + \tfrac{1}{3}x^{-2/3} \log x\right].$$

(a) Tangent line at $x = 2$ is

$$y - 2^{2^{1/3}} = 2^{2^{1/3}}[2^{-2/3} + \tfrac{1}{3}2^{-2/3} \log 2](x - 2).$$

(b) Tangent line at $x = 5$ is

$$y - 5^{5^{1/3}} = 5^{5^{1/3}}[5^{-2/3} + \tfrac{1}{3}5^{-2/3} \log 5](x - 5).$$

11. Let $f(x) = x^a - 1 - a(x - 1)$. Then $f(1) = 0$, and

$$f'(x) = ax^{a-1} - a.$$

If $x > 1$, then $f'(x) > 0$ so $f(x)$ is strictly increasing. If $x < 1$ then $f'(x) < 0$ so $f(x)$ is strictly decreasing. Hence $f(1)$ is as minimum value, so that for all $x > 0$ and $x \neq 1$ we get $f(x) > 0$.

12. $x = 0$ and $x = -2/\log a$

13. We have $\left(1 + \dfrac{r}{x}\right)^x = \left(1 + \dfrac{1}{y}\right)^{yr}$. If $y \to \infty$, then $\left(1 + \dfrac{1}{y}\right)^y$ approaches e, by Limit 3, so its r-th power approaches e^r. This uses the fact that the r-th power function is continuous. If z approaches z_0, then z^r approaches z_0^r. Here

$$z = \left(1 + \frac{1}{y}\right)^y$$

and z approaches e by Limit 3.

14. (a) We can write $\dfrac{a^h - 1}{h} = \dfrac{f(h) - f(0)}{h}$ where $f(x) = a^x$. Hence the limit is equal to $f'(0)$. Since $f'(x) = a^x \log a$ we find $f'(0) = \log a$.

(b) Putting $h = 1/n$ we have $n(a^{1/n} - 1) = \dfrac{a^h - 1}{h}$ so the limit comes from part (a).

VIII, §4, p. 266

1. $-\log 25$ **2.** $5e^{-4}$ **3.** $e^{-(\log 10)10^{-6t}}$ **4.** $20/e$ **5.** $-(\log 2)/K$ **6.** $(\log 3)/4$

7. $12 \log 10/\log 2$ **8.** $\dfrac{-3 \log 2}{\log 9 - \log 10}$ **10.** 1984: $(50{,}000)2^{84/50}$; 2000: $2 = 10^5$

11. $30[\frac{4}{5}]^{5/3}$ **12.** $4\left[\dfrac{\log \frac{1}{20}}{\log \frac{1}{3}}\right]$ **13.** $2\left[\dfrac{\log \frac{5}{8}}{\log \frac{1}{2}}\right]$

14. (a) $40\left[\dfrac{\log \frac{7}{10}}{\log \frac{2}{5}}\right]$ (b) $40\left[\dfrac{\log \frac{4}{25}}{\log \frac{2}{5}}\right]$ (c) $100[\frac{2}{5}]^{1/2}$ **15.** $\log 2$

16. (a) $\frac{1}{5568}\log \frac{1}{2}$ (b) $5568(\log 4/5)/(\log 1/2)$

VIII, §5, p. 274

2.   **3.**

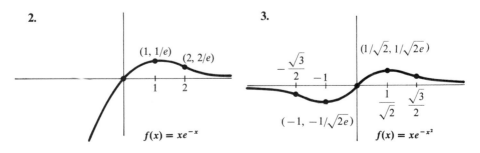

$f(x) = xe^{-x}$

$f(x) = xe^{-x^2}$

4.

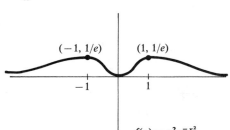

$(-1, 1/e)$ $(1, 1/e)$

-1 1

$f(x) = x^2 e^{-x^2}$

5.

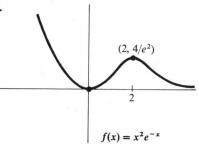

$(2, 4/e^2)$

2

$f(x) = x^2 e^{-x}$

6.

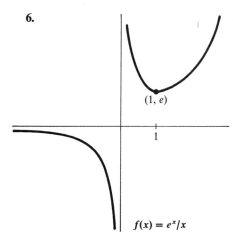

$(1, e)$

1

$f(x) = e^x/x$

7.

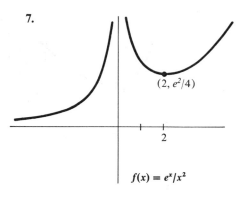

$(2, e^2/4)$

2

$f(x) = e^x/x^2$

8.

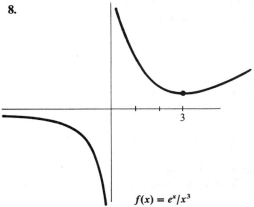

3

$f(x) = e^x/x^3$

9.   **10.**

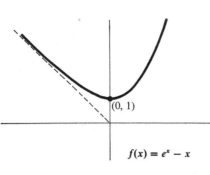

$f(x) = e^x - x$

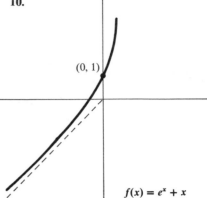

(0, 1)

$f(x) = e^x + x$

11.   **12.**

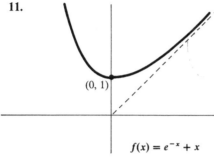

(0, 1)

$f(x) = e^{-x} + x$

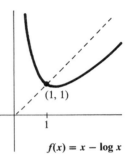

(1, 1)

1

$f(x) = x - \log x$

13. Suppose first $a < 0$. Let $f(x) = e^x - ax$. When x is large positive, then

$$f(x) = e^x\left(1 - \frac{ax}{e^x}\right)$$

is large positive. When x is large negative, then e^x is close to 0 and $-ax$ is large negative, so $f(x)$ is negative. By the intermediate value theorem, the equation $f(x) = 0$ has a solution.

Suppose next that $a \geqq e$. Then $f(1) = e - a \leqq 0$, and again $f(x)$ is large when x is large positive. The intermediate value theorem again provides a solution.

14. (a) $-\dfrac{n}{2^n} \log 2$

(b) Limit is 0 in both cases. For instance, let $x = e^{-y}$. As x approaches 0, then y becomes large, and

$$x \log x = -ye^{-y} = -\frac{y}{e^y},$$

which approaches 0 by Theorem 5.1. Also $x^2 \log x = x(x \log x)$, and the product of the limits is the limit of the product of x and $x \log x$, so is equal to 0.

15. Let $x = e^{-y}$. Then $\log x = -y$ and

$$x(\log x)^n = e^{-y}y^n = \frac{y^n}{e^y}.$$

As $x \to 0$, $y \to \infty$ so $x(\log x)^n \to 0$ by Theorem 5.1.

16. Let $x = e^y$. As $x \to \infty$, $y \to \infty$ so

$$\frac{(\log x)^n}{x} = \frac{y^n}{e^y} \to 0 \qquad \text{by Theorem 5.1.}$$

17. (a)

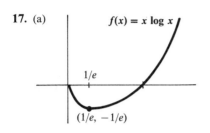

(b)

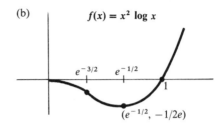

(c)

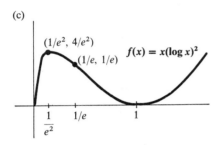

(d)

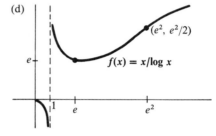

17. (a) $f(x) = x \log x$, defined for $x > 0$. Then:

$$f'(x) = x \cdot \frac{1}{x} + \log x = 1 + \log x.$$

We have:

$$f'(x) = 0 \quad \Leftrightarrow \quad \log x = -1 \quad \Leftrightarrow \quad x = e^{-1},$$
$$f'(x) > 0 \quad \Leftrightarrow \quad \log x > -1 \quad \Leftrightarrow \quad x > e^{-1},$$
$$f'(x) < 0 \quad \Leftrightarrow \quad \log x < -1 \quad \Leftrightarrow \quad x < e^{-1}.$$

So there is only one critical point, and the regions of increase and decrease are given by the regions where $f' > 0$ and $f' < 0$.

We also get $f''(x) = 1/x > 0$ for all $x > 0$, so f is bending up.

If $x \to \infty$ then $\log x \to \infty$ also, so $f(x) \to \infty$.

If $x \to 0$ then $f(x) \to 0$ by Exercise 14(b).

This justifies all the items in the graph.

17. (b) We carry out the details of the graph for $f(x) = x^2 \log x$. We have

$$f'(x) = x + 2x \log x = x(1 + 2 \log x).$$

Since $x > 0$, it follows that $f'(x) > 0$ if and only if $1 + 2 \log x > 0$, and

$$1 + 2 \log x > 0 \quad \text{if and only if} \quad \log x > -1/2$$
$$\text{if and only if} \quad x > e^{-1/2}.$$

Thus f is strictly increasing for $x \geqq e^{-1/2}$ and is strictly decreasing for

$$0 < x \leqq e^{-1/2}.$$

From Exercise 14 we know that $x \log x$ approaches 0 as x approaches 0. Hence $f'(x)$ approaches 0 as x approaches 0, which means the curve looks flat near 0. We have

$$f''(x) = (1 + 2 \log x) + 2 = 3 + 2 \log x.$$

Then $f''(x) = 0$ if and only if $3 + 2 \log x = 0$, or, in other words, $\log x = -3/2$ and $x = e^{-3/2}$. Thus the inflection point occurs for $x = e^{-3/2}$. This explains all the indicated features of the graph.

17. (c) $f(x) = x(\log x)^2$ for $x > 0$. Then

$$f'(x) = x \cdot 2(\log x)\frac{1}{x} + (\log x)^2$$

$$= (\log x)(2 + \log x).$$

The signs of the two factors $\log x$ and $2 + \log x$ will vary according to the intervals when either factor is 0. We have

$$f'(x) = 0 \quad \Leftrightarrow \quad \log x = 0 \quad \text{or} \quad 2 + \log x = 0$$
$$\Leftrightarrow \quad x = 1 \quad \text{or} \quad \log x = -2$$
$$\Leftrightarrow \quad x = 1 \quad \text{or} \quad x = e^{-2}.$$

So there are two critical points at $x = 1$ and $x = e^{-2}$. We make a table of regions of increase and decrease corresponding to the intervals between critical points.

interval	$\log x$	$2 + \log x$	$f'(x)$	f
$0 < x < e^{-2}$	neg.	neg.	pos.	s.i.
$e^{-2} < x < 1$	neg.	pos.	neg.	s.d.
$1 < x$	pos.	pos.	pos.	s.i.

The second derivative is not too bad:

$$f''(x) = (\log x) \frac{1}{x} + \frac{1}{x} (2 + \log x) = \frac{2}{x} (1 + \log x).$$

For $x > 0$ we have $2/x > 0$ so $f''(x) = 0$ if and only if $x = e^{-1}$. Since

$$f''(x) > 0 \quad \Leftrightarrow \quad \log x > -1 \quad \Leftrightarrow \quad x > e^{-1},$$
$$f''(x) < 0 \quad \Leftrightarrow \quad \log x < -1 \quad \Leftrightarrow \quad x < e^{-1},$$

it follows that $x = e^{-1}$ is an inflection point. The graph bends up if $x > e^{-1}$ and bends down if $x < e^{-1}$.

If $x \to \infty$ then $\log x \to \infty$ so $f(x) \to \infty$.

If $x \to 0$ then $f(x) \to 0$ by Exercise 15.

This justifies all features of the graph as drawn.

17. (d) We carry out the details. Let $f(x) = x/\log x$ for $x > 0$, and $x \neq 1$. Then

$$f'(x) = \left(\log x - x \cdot \frac{1}{x} \right) / (\log x)^2 = \frac{\log x - 1}{(\log x)^2}.$$

For $x \neq 1$ the denominator is positive (being a square). Hence

$$f'(x) = 0 \quad \Leftrightarrow \quad \log x = 1 \quad \Leftrightarrow \quad x = e,$$
$$f'(x) > 0 \quad \Leftrightarrow \quad \log x > 1 \quad \Leftrightarrow \quad x > e,$$
$$f'(x) < 0 \quad \Leftrightarrow \quad \log x < 1 \quad \Leftrightarrow \quad x < e.$$

Next we list the behavior as x becomes large positive, x approaches 0, and x approaches 1 (since the denominator is not defined at $x = 1$).

If $x \to \infty$ then $x/\log x \to \infty$ by Theorem 4.3.

If $x \to 0$, then $x/\log x \to 0$.

This is because $\log x$ becomes large negative, but is in the denominator, so dividing by a large negative number contributes to the fraction approaching 0.

If $x \to 1$ and $x > 1$ then $x/\log x \to \infty$. *Proof:* The numerator x approaches 1. The denominator $\log x$ approaches 0, and is positive for $x > 1$. So $x/\log x \to \infty$.

If $x \to 1$ and $x < 1$ then $x/\log x \to -\infty$. *Proof:* Again the numerator x approaches 1, and the denominator $\log x$ approaches 0 but is negative for $x < 1$, so $x/\log x \to -\infty$.

This already justifies the graph as drawn in so far as regions of increase and decrease are concerned, and for the critical point (there is only one critical point). Let us now look at the regions of bending up and down. We write the first derivative in the form

$$f'(x) = \frac{1}{\log x} - \frac{1}{(\log x)^2}.$$

Then

$$f''(x) = \frac{-1}{(\log x)^2} \frac{1}{x} - (-2)(\log x)^{-3} \frac{1}{x}$$

$$= \frac{-1}{(\log x)^3} \frac{1}{x} (\log x - 2).$$

Therefore:

$$f''(x) = 0 \quad \Leftrightarrow \quad \log x = 2 \quad \Leftrightarrow \quad x = e^2.$$

We shall now analyze the sign of $f''(x)$ in various intervals, taken between the points 0, 1, and e^2 which are the points where the factors of $f''(x)$ change sign. Note that the sign of $f''(x)$ (plus or minus) will be determined by the signs of $\log x$, x, and $\log x - 2$, together with the minus sign in front.

If $x > e^2$ then $f''(x) < 0$ and the graph bends down, because $\log x - 2 > 0$, both $\log x$ and x are positive, and the minus sign in front makes $f''(x)$ negative.

If $1 < x < e^2$ then $f''(x) > 0$ because $\log x - 2 < 0$, both $\log x$ and x are positive, and the minus sign in front together with the fact that $\log x - 2$ is negative make $f''(x)$ positive. Hence the graph bends up for $1 < x < e^2$.

If $0 < x < 1$ then $f''(x) < 0$ because x is positive, $\log x$ is negative, $\log x - 2$ is negative, and the minus sign in front combines with the other signs to make $f''(x)$ negative. Hence the graph bends down for $0 < x < 1$.

18. Let $f(x) = x^x = e^{x \log x}$. Then

$$f'(x) = e^{x \log x}\left(x \cdot \frac{1}{x} + \log x\right) = e^{x \log x}(1 + \log x).$$

Since $e^u > 0$ for all numbers u, we have $e^{x \log x} > 0$ for all $x > 0$, so

$$f'(x) = 0 \quad \Leftrightarrow \quad \log x = -1 \quad \Leftrightarrow \quad x = e^{-1},$$

$$f'(x) > 0 \quad \Leftrightarrow \quad \log x > -1 \quad \Leftrightarrow \quad x > e^{-1}.$$

This already takes care of Exercise 18.

19. Note that $f(e^{-1}) = (e^{-1})^{1/e} = e^{-1/e} = 1/e^{1/e}$. Also

$$f'(x) < 0 \quad \Leftrightarrow \quad \log x < -1 \quad \Leftrightarrow \quad x < e^{-1}.$$

Finally $f''(x) = e^{x \log x}[1/x + (1 + \log x)^2] > 0$ for all $x > 0$ so the graph bends up.

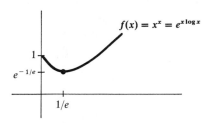

20. Note that $x^{-x} = 1/x^x$. But you should go through the rigamarole about taking the derivative etc. to see the graph as follows.

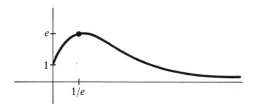

21. Let $f(x) = 2^x x^x = e^{x \log 2} e^{x \log x} = e^{x(\log 2 + \log x)}$. Then

$$f'(x) = e^{x(\log 2 + \log x)}\left[x \cdot \frac{1}{x} + \log 2 + \log x\right]$$

$$= 2^x x^x(1 + \log 2 + \log x).$$

Since $2^x x^x > 0$ for all $x > 0$, we get

$$f'(x) > 0 \quad \Leftrightarrow \quad 1 + \log 2 + \log x > 0$$
$$\Leftrightarrow \quad \log x > -1 - \log 2$$
$$\Leftrightarrow \quad x > e^{-1 - \log 2}$$

and $e^{-1 - \log 2} = e^{-1} e^{-\log 2} = 1/2e$, as was to be shown.

22. (a) 1 (b) 1 (c) $1/e$ (d) 1

IX, §1, p. 291

1. $-(\cos 2x)/2$ **2.** $\dfrac{\sin 3x}{3}$ **3.** $\log(x + 1)$, $x > -1$ **4.** $\log(x + 2)$, $x > -2$

IX, §3, p. 296

1. 156 **2.** 2 **3.** 2 **4.** $\log 2$ **5.** $\log 3$ **6.** $\frac{2}{5}$ **7.** $e - 1$

IX, §4, p. 307

1. (a) $U_1^2 = \frac{9}{4}(\frac{3}{2} - 1) + 4(2 - \frac{3}{2})$

$L_1^2 = 1(\frac{3}{2} - 1) + \frac{9}{4}(2 - \frac{3}{2})$

(b) $U_1^2 = \frac{16}{9}(\frac{4}{3} - 1) + \frac{25}{9}(\frac{5}{3} - \frac{4}{3}) + 4(2 - \frac{5}{3})$

$L_1^2 = 1(\frac{4}{3} - 1) + \frac{16}{9}(\frac{5}{3} - \frac{4}{3}) + \frac{25}{9}(2 - \frac{5}{3})$

(c) $U_1^2 = \frac{25}{16}(\frac{5}{4} - 1) + \frac{9}{4}(\frac{3}{2} - \frac{5}{4}) + \frac{49}{16}(\frac{7}{4} - \frac{3}{2}) + 4(2 - \frac{7}{4})$

$L_1^2 = 1(\frac{5}{4} - 1) + \frac{25}{16}(\frac{3}{2} - \frac{5}{4}) + \frac{9}{4}(\frac{7}{4} - \frac{3}{2}) + \frac{49}{16}(2 - \frac{7}{4})$

(d) $U_1^2 = \frac{1}{n} \cdot \left[\left(1 + \frac{1}{n}\right)^2 + \left(1 + \frac{2}{n}\right)^2 + \cdots + \left(1 + \frac{n}{n}\right)^2 \right]$

$L_1^2 = \frac{1}{n} \left[1 + \left(1 + \frac{1}{n}\right)^2 + \cdots + \left(1 + \frac{n-1}{n}\right)^2 \right]$

2. (a) $U_1^3 = 1(\frac{3}{2} - 1) + \frac{2}{3}(2 - \frac{3}{2}) + \frac{1}{2}(\frac{5}{2} - 2) + \frac{2}{5}(3 - \frac{5}{2})$

$L_1^3 = \frac{2}{3}(\frac{3}{2} - 1) + \frac{1}{2}(2 - \frac{3}{2}) + \frac{2}{5}(\frac{5}{2} - 2) + \frac{1}{3}(3 - \frac{5}{2})$

(b) $U_1^3 = 1(\frac{4}{3} - 1) + \frac{3}{4}(\frac{5}{3} - \frac{4}{3}) + \frac{3}{5}(\frac{6}{3} - \frac{5}{3}) + \frac{3}{6}(\frac{7}{3} - \frac{6}{3}) + \frac{3}{7}(\frac{8}{3} - \frac{7}{3})$
$\qquad + \frac{3}{8}(\frac{9}{3} - \frac{8}{3})$

$L_1^3 = \frac{3}{4}(\frac{4}{3} - 1) + \frac{3}{5}(\frac{5}{3} - \frac{4}{3}) + \frac{3}{6}(\frac{6}{3} - \frac{5}{3}) + \frac{3}{7}(\frac{7}{3} - \frac{6}{3}) + \frac{3}{8}(\frac{8}{3} - \frac{7}{3}) + \frac{3}{9}(\frac{9}{3} - \frac{8}{3})$

(c) $U_1^3 = 1(\frac{5}{4} - \frac{4}{4}) + \frac{4}{5}(\frac{6}{4} - \frac{5}{4}) + \frac{4}{6}(\frac{7}{4} - \frac{6}{4}) + \frac{4}{7}(\frac{8}{4} - \frac{7}{4})$
$\qquad + \frac{4}{8}(\frac{9}{4} - \frac{8}{4}) + \frac{4}{9}(\frac{10}{4} - \frac{9}{4}) + \frac{4}{10}(\frac{11}{4} - \frac{10}{4}) + \frac{4}{11}(\frac{12}{4} - \frac{11}{4})$

$L_1^3 = \frac{4}{5}(\frac{5}{4} - \frac{4}{4}) + \frac{4}{6}(\frac{6}{4} - \frac{5}{4}) + \frac{4}{7}(\frac{7}{4} - \frac{6}{4}) + \frac{4}{8}(\frac{8}{4} - \frac{7}{4})$
$\qquad + \frac{4}{9}(\frac{9}{4} - \frac{8}{4}) + \frac{4}{10}(\frac{10}{4} - \frac{9}{4}) + \frac{4}{11}(\frac{11}{4} - \frac{10}{4}) + \frac{4}{12}(\frac{12}{4} - \frac{11}{4})$

(d) $U_1^3 = \frac{1}{n} \left[1 + \dfrac{1}{\left(1 + \dfrac{1}{n}\right)} + \cdots + \dfrac{1}{\left(1 + \dfrac{2n-1}{n}\right)} \right]$

$\qquad = \left[\frac{1}{n} + \frac{1}{n+1} + \cdots + \frac{1}{3n-1} \right]$

$L_1^3 = \frac{1}{n} \left[\dfrac{1}{\left(1 + \dfrac{1}{n}\right)} + \cdots + \dfrac{1}{\left(1 + \dfrac{2n}{n}\right)} \right] = \left[\frac{1}{n+1} + \cdots + \frac{1}{3n} \right]$

3. (a) $U_0^2 = \frac{1}{2}(\frac{1}{2} - 0) + 1(1 - \frac{1}{2}) + \frac{3}{2}(\frac{3}{2} - 1) + 2(2 - \frac{3}{2})$

$L_0^2 = 0(\frac{1}{2} - 0) + \frac{1}{2}(1 - \frac{1}{2}) + 1(\frac{3}{2} - 1) + \frac{3}{2}(2 - \frac{3}{2})$

(b) $U_0^2 = \frac{1}{3}(\frac{1}{3} - 0) + \frac{2}{3}(\frac{2}{3} - \frac{1}{3}) + 1(\frac{3}{3} - \frac{2}{3}) + \frac{4}{3}(\frac{4}{3} - 1)$
$\qquad + \frac{5}{3}(\frac{5}{3} - \frac{4}{3}) + 2(\frac{6}{3} - \frac{5}{3})$

$L_0^2 = 0(\frac{1}{3} - 0) + \frac{1}{3}(\frac{2}{3} - \frac{1}{3}) + \frac{2}{3}(\frac{3}{3} - \frac{2}{3}) + 1(\frac{4}{3} - 1)$
$\qquad + \frac{4}{3}(\frac{5}{3} - \frac{4}{3}) + \frac{5}{3}(\frac{6}{3} - \frac{5}{3})$

(c) $U_0^2 = \frac{1}{4}(\frac{1}{4} - 0) + \frac{1}{2}(\frac{1}{2} - \frac{1}{4}) + \frac{3}{4}(\frac{3}{4} - \frac{1}{2}) + 1(1 - \frac{3}{4})$
$\qquad + \frac{5}{4}(\frac{5}{4} - 1) + \frac{3}{2}(\frac{3}{2} - \frac{5}{4}) + \frac{7}{4}(\frac{7}{4} - \frac{3}{2}) + 2(2 - \frac{7}{4})$

$L_0^2 = 0(\frac{1}{4} - 0) + \frac{1}{4}(\frac{1}{2} - \frac{1}{4}) + \frac{1}{2}(\frac{3}{4} - \frac{1}{2}) + \frac{3}{4}(1 - \frac{3}{4})$
$\qquad + 1(\frac{5}{4} - 1) + \frac{5}{4}(\frac{3}{2} - \frac{5}{4}) + \frac{3}{2}(\frac{7}{4} - \frac{3}{2}) + \frac{7}{4}(2 - \frac{7}{4})$

(d) $U_0^2 = \frac{1}{n} \left[\frac{1}{n} + \frac{2}{n} + \cdots + \frac{2n}{n} \right] \quad L_0^2 = \frac{1}{n} \left[0 + \frac{1}{n} + \cdots + \frac{2n-1}{n} \right]$

4. (a) $U_0^2 = \frac{1}{4}(\frac{1}{2} - 0) + 1(1 - \frac{1}{2}) + \frac{9}{4}(\frac{3}{2} - 1) + 4(2 - \frac{3}{2})$

$L_0^2 = 0(\frac{1}{2} - 0) + \frac{1}{4}(1 - \frac{1}{2}) + 1(\frac{3}{2} - 1) + \frac{9}{4}(2 - \frac{3}{2})$

(b) $U_0^2 = \frac{1}{9}(\frac{1}{3} - 0) + \frac{4}{9}(\frac{2}{3} - \frac{1}{3}) + 1(1 - \frac{2}{3}) + \frac{16}{9}(\frac{4}{3} - 1)$
$\qquad + \frac{25}{9}(\frac{5}{3} - \frac{4}{3}) + 4(2 - \frac{5}{3})$

$L_0^2 = 0(\frac{1}{3} - 0) + \frac{1}{9}(\frac{2}{3} - \frac{1}{3}) + \frac{4}{9}(1 - \frac{2}{3}) + 1(\frac{4}{3} - 1)$
$\qquad + \frac{16}{9}(\frac{5}{3} - \frac{4}{3}) + \frac{25}{9}(2 - \frac{5}{3})$

(c) $U_0^2 = \frac{1}{16}(\frac{1}{4} - 0) + \frac{1}{4}(\frac{1}{2} - \frac{1}{4}) + \frac{9}{16}(\frac{3}{4} - \frac{1}{2}) + 1(1 - \frac{3}{4})$
$\qquad + \frac{25}{16}(\frac{5}{4} - 1) + \frac{9}{4}(\frac{3}{2} - \frac{5}{4}) + \frac{49}{16}(\frac{7}{4} - \frac{3}{2}) + 4(2 - \frac{7}{4})$

$L_0^2 = 0(\frac{1}{4} - 0) + \frac{1}{16}(\frac{1}{2} - \frac{1}{4}) + \frac{1}{4}(\frac{3}{4} - \frac{1}{2}) + \frac{9}{16}(1 - \frac{3}{4})$
$\qquad + 1(\frac{5}{4} - 1) + \frac{25}{16}(\frac{3}{2} - \frac{5}{4}) + \frac{9}{4}(\frac{7}{4} - \frac{3}{2}) + \frac{49}{16}(2 - \frac{7}{4})$

(d) $U_0^2 = \frac{1}{n}\left[\frac{1}{n^2} + \left(\frac{2}{n}\right)^2 + \cdots + \left(\frac{2n}{n}\right)^2\right]$

$L_0^2 = \frac{1}{n}\left[0 + \frac{1}{n^2} + \cdots + \left(\frac{2n-1}{n}\right)^2\right]$

5. $U_1^2 = \frac{1}{n}\left[1 + \dfrac{1}{\left(1 + \dfrac{1}{n}\right)} + \cdots + \dfrac{1}{\left(1 + \dfrac{n-1}{n}\right)}\right]$

$\quad = \left[\dfrac{1}{n} + \dfrac{1}{n+1} + \cdots + \dfrac{1}{2n-1}\right]$

$L_1^2 = \frac{1}{n}\left[\dfrac{1}{\left(1 + \dfrac{1}{n}\right)} + \dfrac{1}{\left(1 + \dfrac{2}{n}\right)} + \cdots + \dfrac{1}{\left(1 + \dfrac{n}{n}\right)}\right]$

$\quad = \left[\dfrac{1}{n+1} + \dfrac{1}{n+2} + \cdots + \dfrac{1}{2n}\right]$

6. The area under the curve $y = 1/x$ between $x = 1$ and $x = 2$ is $\log 2 - \log 1 = \log 2$. Write down that this area is less than an upper sum and greater than a lower sum, and use Exercise 5 to get the desired inequalities.

7. $U_1^n = [\log 2 + \log 3 + \cdots + \log n] = \log n!$
$L_1^n = [\log 1 + \log 2 + \cdots + \log(n - 1)] = \log(n - 1)!$

X, §1, p. 317

1. $\frac{63}{6}$ **2.** 0 **3.** 0 **4.** 0

5. (b) and (c) Let $f(x) = x^{-1/2}$, and let the partition be $P = \{1, 2, 3, \ldots, n\}$ for the interval $[1, n]$. Then compare the integral

$$\int_1^n f(x)\, dx = \frac{x^{1/2}}{1/2}\bigg|_1^n = 2(\sqrt{n} - 1)$$

with the upper and lower sums.

6. (a) Use $f(x) = x^2$ and $P = \{0, 2, \ldots, n\}$, with interval $[0, n]$. The lower sum could start with 0, but this 0 may be omitted since $0 + A = A$ for all numbers A.
(b) Let $f(x) = x^3$, interval $[0, n]$, partition $P = \{0, \ldots, n\}$.
(c) Let $f(x) = x^{1/4}$, interval $[0, n]$, partition $P = \{0, \ldots, n\}$.

7. (d) Use $f(x) = 1/x^4$ over the interval $[1, n]$ with the partition $\{1, .., n\}$ consisting of the positive integers from 1 to n. Then

$$\int_1^n f(x)\, dx = \frac{x^{-3}}{-3}\Big|_1^n = -\frac{1}{3}\left(\frac{1}{n^3} - 1\right) = \frac{1}{3} - \frac{1}{3n^3}.$$

Comparing with the lower and upper sum yields

$$\frac{1}{2^4} + \frac{1}{3^4} + \cdots + \frac{1}{n^4} \leqq \frac{1}{3} - \frac{1}{3n^3} \leqq 1 + \frac{1}{2^4} + \cdots + \frac{1}{(n-1)^4}.$$

8. Let $f(x) = \dfrac{1}{1 + x^2}$. Then $\displaystyle\int_0^1 f(x)\, dx = \arctan x \Big|_0^1 = \pi/4$. On the interval

$[0, 1]$ use the partition $P = \left\{0, \dfrac{1}{n}, \dfrac{2}{n}, \ldots, \dfrac{n}{n}\right\}$. Draw the picture.

9. Let $f(x) = x^2$. Then $\displaystyle\int_0^1 f(x)\, dx = \frac{x^3}{3}\Big|_0^1 = 1/3$. On the interval $[0, 1]$ use the

partition $P = \{0, 1/n, \ldots, n/n\}$. Draw the picture.

10. $\displaystyle\int_1^n \log x\, dx = (x \log x - x)\Big|_1^n = n \log n - n + 1.$

The lower sum is $\log 1 + \log 2 + \cdots + \log(n-1) = \log(n-1)!$ and

$$e^{\text{lower sum}} = (n-1)!, \qquad \text{because} \qquad e^{\log u} = u.$$

On the other hand,

$$e^{n \log n - n + 1} = e^{n \log n} e^{-n} e^1 = n^n e^{-n} e.$$

Since lower sum $\leqq$ integral, we get $e^{\text{lower sum}} \leqq e^{\text{integral}}$, so

$$(n-1)! \leqq n^n e^{-n} e.$$

The upper sum is $\log 2 + \log 3 + \cdots + \log n = \log n!$ and so $e^{\text{upper sum}} = n!$. Since integral $\leq$ upper sum, we have $e^{\text{integral}} \leq e^{\text{upper sum}}$. Since the integral is $n \log n - n + 1$, we get

$$n^n e^{-n} e \leq n!.$$

X, §2, p. 325

1. x^4 **2.** $3x^5/5 - x^6/6$ **3.** $-2\cos x + 3 \sin x$ **4.** $\frac{9}{5}x^{5/3} + 5 \sin x$
5. $5e^x + \log x$ **6.** 0 **7.** 0 **8.** $e^2 - e^{-1}$ **9.** $4 \cdot 28/3$

10. In this problem, the curves intersect at $x = 0$ and $x = 1$.]

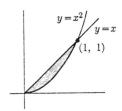

Hence the area is

$$\int_0^1 (x - x^2)\, dx = \frac{x^2}{2} - \frac{x^3}{3}\Big|_0^1 = \frac{1}{2} - \frac{1}{3}.$$

11. $\frac{1}{2}$ from -1 to 1 **12.** $\frac{1}{12}$
13. $\frac{1}{2}$ **14.** $\frac{8}{3} + \frac{5}{12}$.
15. $\sqrt{2} - 1$. In this problem the graph is as follows.

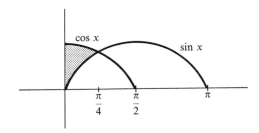

The first point of intersection is when $x = \pi/4$. Hence the area between the curves is

$$\int_0^{\pi/4} (\cos x - \sin x)\, dx.$$

16. 9/2   **17.** $\frac{1}{3} - \frac{1}{2}$ **18.** 0   **19.** $\pi^2/2 - 2$ **20.** 4   **21.** 4   **22.** 1   **23.** 4   **24.** $-\pi^2$
25. 4   **26.** (a) 14   (b) 14   (c) $2n$

X, §4, p. 334

1. Yes, $\sqrt{2}$ **2.** No
3. Yes, $\pi/2$. We have

$$\int_0^B \frac{1}{1 + x^2}\, dx = \arctan x \Big|_0^B = \arctan B - \arctan 0$$

$$= \arctan B.$$

As $B \to \infty$, $\arctan B \to \pi/2$. Remember the graph of $\arctan x$ which is as follows.

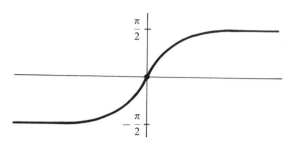

4. No. Let $0 < b < 5$. Then

$$\int_0^b \frac{1}{5 - x} \, dx = -\log(5 - x) \Big|_0^b = -[\log(5 - b) - \log 5]$$

$$= \log 5 - \log(5 - b).$$

As b approaches 5, $5 - b$ approaches 0, and $\log(5 - b)$ becomes large negative. Hence the integral from 0 to 5 does not exist.

5. Let $2 < a < 3$. Then

$$\int_a^3 \frac{1}{x - 2} \, dx = \log(x - 2) \Big|_a^3 = \log 1 - \log(a - 2)$$

$$= -\log(a - 2).$$

As $a \to 2$, $\log(a - 2) \to -\infty$, so $-\log(a - 2) \to \infty$ and the integral from 2 to 3 does not exist.

6. No.

7. Here we have $0 \leqq x < 2$. Remember that we have the indefinite integral.

$$\int \frac{1}{x} \, dx = \log(-x) \qquad \text{if} \quad x < 0.$$

Therefore, if $x < 2$ then

$$\int \frac{1}{x - 2} \, dx = \log(2 - x).$$

You can verify this by the chain rule, differentiating the right-hand side. Don't forget the -1. Note that when $x < 2$ then $2 - x > 0$. The log is not defined at negative numbers. Now you can find the definite integral for $0 < a < 2$. We have

$$\int_0^a \frac{1}{x - 2} \, dx = \log(2 - x) \Big|_0^a = \log(2 - a) - \log 2.$$

The right-hand side approaches $-\infty$ as $a \to 2$, so the integral

$$\int_0^2 \frac{1}{x-2}\, dx = \lim_{a \to 2} \int_0^a \frac{1}{x-2}\, dx$$

does not exist.

8. Improper integral does not exist. Let $2 < c < 3$. Then

$$\int_c^3 \frac{1}{(x-2)^2}\, dx = \int_c^3 (x-2)^{-2}\, dx = -(x-2)^{-1} \Big|_c^3$$

$$= -\left[1 - \frac{1}{c-2}\right].$$

As $c \to 2$, the quotient $1/(c-2)$ becomes arbitrarily large.

9. Does not exist.

10. Exists. Let $1 < c < 4$. Then

$$\int_c^4 (x-1)^{-2/3}\, dx = 3(x-1)^{1/3} \Big|_c^4 = 3[3^{1/3} - (c-1)^{1/3}].$$

As $c \to 1$, $(c-1)^{1/3} \to 0$, and so

$$\lim_{c \to 1} \int_c^4 (x-1)^{-2/3}\, dx = 3 \cdot 3^{1/3}.$$

11. Integral exists for $s < 1$. Let $0 < a < 1$. Then for $s \neq 1$, we get

$$\int_a^1 x^{-s}\, dx = \frac{x^{-s+1}}{-s+1} \Big|_a^1 = \frac{1}{1-s} - \frac{a^{1-s}}{1-s}.$$

If $s < 1$, then $1 - s > 0$ and a^{1-s} approaches 0 as a approaches 0. Hence the limit of the right-hand side as $a \to 0$ exists and is equal to $1/(1-s)$. On the other hand, if $s > 1$, then

$$a^{1-s} = \frac{1}{a^{s-1}},$$

and $s - 1 > 0$, so $a^{s-1} \to 0$ as $a \to 0$, and $1/a^{s-1} \to \infty$, so the integral from 0 to 1 does not exist. When $s = 1$,

$$\int_a^1 \frac{1}{x}\, dx = \log x \Big|_a^1 = \log 1 - \log a = -\log a.$$

When $a \to 0$, $\log a \to -\infty$ so the integral does not exist.

12. The integral $\int_1^\infty x^{-s}\, dx$ exists for $s > 1$ and does not exist for $s < 1$. Evaluate

$$\int_1^B x^{-s}\, dx = \frac{B^{-s+1}}{1-s} - \frac{1}{1-s}.$$

If $s > 1$ then we write $B^{-s+1} = 1/B^{s-1}$ and $s - 1 > 0$, so $1/B^{s-1} \to 0$ as $B \to \infty$. The limit of the integral exists and is equal to $1/(s-1)$. If $s < 1$, then $B^{-s+1} = B^{1-s}$ becomes arbitrarily large when B becomes large.

13. Yes. Let $B > 1$. Then

$$\int_1^B e^{-x}\,dx = -e^{-x}\Big|_1^B = -[e^{-B} - e^{-1}] = \frac{1}{e} - \frac{1}{e^B}.$$

As B becomes large, $1/e^B$ approaches 0, and the integral from 1 to B approaches $1/e$. It exists.

14. Does not exist.

15. $-\frac{1}{2}e^{-2B} + \frac{1}{2}e^{-4}$. Yes, $\frac{1}{2}e^{-4}$.

XI, §1, p. 339

1. $e^{x^2}/2$ **2.** $-\frac{1}{4}e^{-x^4}$ **3.** $\frac{1}{6}(1 + x^3)^2$ **4.** $(\log x)^2/2$

5. $\dfrac{(\log x)^{-n+1}}{1-n}$ if $n \neq 1$, and $\log(\log x)$ if $n = 1$. **6.** $\log(x^2 + x + 1)$

7. $x - \log(x + 1)$ **8.** $\dfrac{\sin^2 x}{2}$ **9.** $\dfrac{\sin^3 x}{3}$ **10.** 0 **11.** $\frac{2}{5}$ **12.** $-\arctan(\cos x)$

13. $\frac{1}{2}(\arctan x)^2$ **14.** 2/15. Let $u = 1 - x^2$. **15.** $-\frac{1}{4}\cos(\pi^2/2) + \frac{1}{4}$

16. (a) $-(\cos 2x)/2$ (b) $(\sin 2x)/2$ (c) $-(\cos 3x)/3$ (d) $(\sin 3x)/3$ (e) $e^{4x}/4$ (f) $e^{5x}/5$ (g) $-e^{-5x}/5$ **17.** $-\frac{1}{2}e^{-B^2} + \frac{1}{2}$. Yes, $\frac{1}{2}$ **18.** $-\frac{1}{3}e^{-B^3} + \frac{1}{3}$. Yes, $\frac{1}{3}$

XI, §1, Supplementary Exercises, p. 340

1. $\frac{1}{4}\log(x^4 + 2)$ **3.** $\dfrac{\sin^5 x}{5}$ **5.** $\sqrt{x^2 - 1}$ **7.** $\dfrac{-1}{6(3x^2 + 5)}$ **9.** $\dfrac{-1}{2\sin^2 x}$

11. $\dfrac{1}{3}\left[\dfrac{u^{17/5}}{17/5} - \dfrac{u^{12/5}}{12/5}\right]$ where $u = x^3 + 1$. **13.** $\dfrac{-\cos 3x}{3}$ **15.** $-\cos e^x$

17. $\log(\log x)$ **19.** $\log(e^x + 1)$ **21.** $\frac{1}{4}$ **23.** $\dfrac{4\sqrt{2}}{3} - \dfrac{2}{3}$ **25.** $\dfrac{\pi}{4}$ **27.** $\dfrac{\pi^2}{72}$ **29.** $e - \dfrac{1}{e}$

XI, §2, p. 344

1. $x \arcsin x + \sqrt{1 - x^2}$. Let

$$u = \arcsin x, \qquad dv = dx,$$

$$du = \frac{1}{\sqrt{1 - x^2}}\,dx, \qquad v = x.$$

Then

$$\int \arcsin x \, dx = \int u \; dv = x \arcsin x - \int \frac{x}{\sqrt{1 - x^2}} \, dx.$$

Now the integral $\int \dfrac{x}{\sqrt{1 - x^2}} \, dx$ can be done by substitution, letting $u = 1 - x^2$ and $du = -2x \, dx$, so

$$\int \frac{x}{\sqrt{1 - x^2}} \, dx = \frac{-1}{2} \int u^{-1/2} \, du.$$

2. $x \arctan x - \frac{1}{2} \log(x^2 + 1)$. Let $u = \arctan x$.

3. $\dfrac{e^{2x}}{13} (2 \sin 3x - 3 \cos 3x)$. Let $I = \displaystyle\int e^{2x} \sin 3x \, dx$. Let

$$u = e^{2x}, \qquad dv = \sin 3x \, dx,$$
$$du = 2e^{2x} \, dx, \qquad v = -\tfrac{1}{3} \cos 3x.$$

Then

$$I = -\tfrac{1}{3}e^{2x} \cos 3x + \frac{2}{3} \int e^{2x} \cos 3x \; dx.$$

Apply the same procedure to this second integral, with

$$u = e^{2x}, \qquad dv = \cos 3x \, dx,$$
$$du = 2e^{2x} \, dx, \qquad v = \tfrac{1}{3} \sin 3x.$$

Then

$$I = -\tfrac{1}{3}e^{2x} \cos 3x + \tfrac{2}{3}[\tfrac{1}{3}e^{2x} \sin 3x - \tfrac{2}{3}I]$$
$$= -\tfrac{1}{3}e^{2x} \cos 3x + \tfrac{2}{9}e^{2x} \sin 3x - \tfrac{4}{9}I.$$

Thus we see I appearing on the right-hand side with some constant factor. We can solve for I to get

$$\tfrac{13}{9}I = \tfrac{2}{9}e^{2x} \sin 3x - \tfrac{1}{3}e^{2x} \cos 3x.$$

Multiply by 9 and divide by 13 to get the answer.

4. $\tfrac{1}{10}e^{-4x} \sin 2x - \tfrac{1}{5}e^{-4x} \cos 2x$

5. $x(\log x)^2 - 2x \log x + 2x$. Let $u = (\log x)^2$, so $du = 2(\log x)\dfrac{1}{x}\, dx$. Let $dv = dx$, $v = x$. Then

$$\int (\log x)^2\, dx = x(\log x)^2 - \int \frac{x}{x} \log x \, dx.$$

Then $x/x = 1$, and you are reduced to $\int \log x\, dx = x \log x - x$.

6. $(\log x)^3 x - 3 \int (\log x)^2\, dx$ **7.** $x^2 e^x - 2x e^x + 2e^x$

8. $-x^2 e^{-x} - 2x e^{-x} - 2e^{-x}$ **9.** $-x \cos x + \sin x$ **10.** $x \sin x + \cos x$

11. $-x^2 \cos x + 2 \int x \cos x\, dx$ **12.** $x^2 \sin x - 2 \int x \sin x\, dx$

13. $\frac{1}{2}[x^2 \sin x^2 + \cos x^2]$. Write

$$I = \int x^3 \cos x^2\, dx = \int x^2 (\cos x^2)\, x dx = \frac{1}{2} \int x^2 (\cos x^2)\, 2x dx.$$

First let $u = x^2$, $du = 2x dx$ and use substitution to get

$$I = \frac{1}{2} \int u \cos u\, du.$$

Then use integration by parts.

14. $-\frac{1}{3}(1 - x^2)^{3/2} + \frac{2}{5}(1 - x^2)^{5/2} - \frac{1}{7}(1 - x^2)^{7/2}$. Let $u = 1 - x^2$:

$$\int x^5 \sqrt{1 - x^2}\, dx = \int x^4 \sqrt{1 - x^2}\, x\, dx$$

$$= -\frac{1}{2} \int (1 - u)^2 u^{1/2}\, du$$

$$= -\frac{1}{2} \int (u^{1/2} - 2u^{3/2} + u^{5/2})\, du$$

15. $\frac{1}{3}x^3 \log x - \frac{1}{9}x^3$. Let $u = \log x$ and $dv = x^2\, dx$. Then $du = (1/x)\, dx$ and $v = x^3/3$ so

$$\int x^2 \log x\, dx = \frac{1}{3}x^3 \log x - \frac{1}{3} \int x^2\, dx.$$

16. $(\log x)\dfrac{x^4}{4} - \dfrac{x^4}{16}$

17. $(\log x)^2 \dfrac{x^3}{3} - \dfrac{2}{3} \int x^2 \log x\, dx$. Repeat the procedure to get the complete answer.

18. $-\frac{1}{2}x^2 e^{-x^2} - \frac{1}{2}e^{-x^2}$. Let $u = x^2$ first.

19. $\dfrac{1}{4}\left(\dfrac{1}{1 - x^4}\right) + \frac{1}{4} \log(1 - x^4)$ **20.** -4π

21. $-Be^{-B} - e^{-B} + 1$. Yes, 1. First evaluate the indefinite integral $\int xe^{-x}\,dx$ by parts. Let

$$u = x, \qquad dv = e^{-x}\,dx,$$

$$du = dx, \qquad v = -e^{-x}.$$

Then the integral is equal to

$$\int xe^{-x}\,dx = -xe^{-x} + \int e^{-x}\,dx = -xe^{-x} - e^{-x}.$$

Now put in the limits of integration:

$$\int_0^B xe^{-x}\,dx = -xe^{-x}\Big|_0^B - e^{-x}\Big|_0^B = -Be^{-B} - [e^{-B} - 1].$$

As $B \to \infty$ the two terms with B approach 0, so you get the answer.

22. Yes, $5/e$ **23.** Yes, $16/e$

24. $-\dfrac{1}{\log B} + \dfrac{1}{\log 2}$. Yes, $1/\log 2$. Let $u = \log x$. When $x = 2$, $u = \log 2$. When $x = B$, $u = \log B$. Then

$$\int_2^B \frac{1}{x(\log x)^2}\,dx = \int_{\log 2}^{\log B} u^{-2}\,du.$$

25. Yes, $1/3(\log 3)^3$

26. $2 \log 2 - 2$. The indefinite integral is $x \log x - x$. Let $0 < a < 2$. Then

$$\int_a^2 \log x\,dx = x \log x - x\Big|_a^2 = 2 \log 2 - 2 - (a \log a - a).$$

As $a \to 0$ we know from Chapter VIII, §5, Exercise 14 that $a \log a$ approaches 0. Hence $a \log a - a$ approaches 0 as a approaches 0, which gives the answer.

XI, §2, Supplementary Exercises, p. 346

1. $\frac{1}{2}(x^2 \arctan x + \arctan x - x)$. Use $u = \arctan x$ and $dv = x\,dx$. Also use the trick

$$\int \frac{x^2}{x^2 + 1}\,dx = \int \frac{x^2 + 1 - 1}{x^2 + 1}\,dx = \int 1\,dx - \int \frac{1}{x^2 + 1}\,dx.$$

2. (a) If I is the integral, then

$$I = x\sqrt{1 - x^2} + \arcsin x - 1$$

so $2I = x\sqrt{1 - x^2} + \arcsin x$, and dividing by 2 yields the answer.

(b) Integrating by parts reduces the integral to (a): $u = \arcsin x$, $dv = x\,dx$. Use a trick as in Exercise 1.

3. $\frac{1}{4}(2x^2 \arccos x - \arccos x - x\sqrt{1 - x^2})$ 5. $1 - \dfrac{\pi}{2}$ 7. $\dfrac{\pi}{32}$ 9. $\dfrac{2}{e}$

10.

$$\int_0^1 x^3\sqrt{1 - x^2}\,dx = \int_0^1 x^2(1 - x^2)^{1/2}x\,dx.$$

Let $u = 1 - x^2$, $du = -2x\,dx$. When $x = 0$, $u = 1$ and when $x = 1$, $u = 0$. Then

$$\int_0^1 x^2(1 - x^2)^{1/2}\,dx = \frac{-1}{2}\int_1^0 (1 - u)u^{1/2}\,du = \frac{-1}{2}\int_1^0 [u^{1/2} - u^{3/2}]\,du$$

$$= \frac{-1}{2}\left[\frac{u^{3/2}}{3/2} - \frac{u^{5/2}}{5/2}\right]\Big|_1^0$$

$$= \frac{-1}{2}\left[0 - \left(\frac{2}{3} - \frac{2}{5}\right)\right] = \frac{2}{15}.$$

11. $-2e^{-\sqrt{x}}(x^{3/2} + 3x + 6x^{1/2} + 6)$. Let $x = u^2$ first. Then

$$\int xe^{-\sqrt{x}}\,dx = \int u^2 e^{-u} 2u\,du = 2\int u^3 e^{-u}\,du.$$

13. Let $u = (\log x)^n$ and $dv = dx$.
14. Let $u = x^n$ and $dv = e^x\,dx$.
15. Let $u = (\log x)^n$ and $dv = x^m\,dx$. Then

$$du = n(\log x)^{n-1}\,\frac{1}{x}\,dx \qquad \text{and} \qquad v = x^{m+1}/(m + 1).$$

16. First we find the indefinite integral by parts with $u = x^n$, $dv = e^{-x}\,dx$, so

$$\int x^n e^{-x}\,dx = -x^n e^{-x} + n\int x^{n-1}e^{-x}\,dx.$$

Then we have the definite integral

$$\int_0^B x^n e^{-x}\,dx = -B^n e^{-B} + n\int_0^B x^{n-1}e^{-x}\,dx.$$

Taking the limit as $B \to \infty$ and using that $B^n e^{-B} \to 0$, we find:

$$\int_0^\infty x^n e^{-x}\,dx = n\int_0^\infty x^{n-1}e^{-x}\,dx.$$

Let $I_n = \int_0^\infty x^n e^{-x}\,dx$. This last equality can be rewritten in the form

$$I_n = nI_{n-1}.$$

Thus we have reduced the evaluation of the integral to the next step. For instance, $I_{10} = 10I_9$; $I_9 = 9I_8$; $I_8 = 8I_7$; and so on. Continuing in this way, it takes n steps to get

$$I_n = n!I_0 = n! \int_0^\infty e^{-x}\,dx,$$

where $n! = n(n-1)(n-2)\cdots 3 . 2 . 1$ is the product of the first n integers. This final integral is easily evaluated, namely

$$\int_0^\infty e^{-x}\,dx = \lim_{B \to \infty} \int_0^B e^{-x}\,dx = \lim_{B \to \infty} -e^{-x}\Big|_0^B$$

$$= \lim_{B \to \infty} -[e^{-B} - 1] = 1.$$

XI, §3, p. 354

1. $-\frac{1}{4}\sin^3 x \cos x - \frac{3}{8}\sin x \cos x + \frac{3}{8}x$ **2.** $\frac{1}{3}\cos^2 x \sin x + \frac{2}{3}\sin x$

3. $\dfrac{\sin^3 x}{3} - \dfrac{\sin^5 x}{5}$ **4.** 3π **5.** 8π **6.** πab (if $a,\ b > 0$) **7.** πr^2

8. (a) $-2\sqrt{2}\cos \theta/2$ (b) $2\sqrt{2}\sin \theta/2$ **13.** $-\log \cos x$ **14.** $\arcsin \dfrac{x}{3}$

15. $\arcsin \dfrac{x}{\sqrt{3}}$ **16.** $\frac{1}{2}\arcsin(\sqrt{2}x)$ **17.** $\dfrac{1}{b}\arcsin \dfrac{bx}{a}$. Let $x = au/b$, $dx = (a/b)\,du$.

18. (a) $c_0 = a_n = 0$ all n, $b_n = -(2/n)\cos n\pi$.
  (b) $c_0 = \pi^2/3$, $a_n = -(4/n^2)\cos n\pi$, $b_n = 0$ all n.
  (c) $c_0 = \pi/2$, $a_n = 2(\cos n\pi - 1)/\pi n^2$, $b_n = 0$ all n.
19. (b) all a_n and $c_0 = 0$

XI, §3, Supplementary Exercises, p. 356

1. $\log \sin x - \dfrac{\sin^2 x}{2}$

2. Write $\tan^2 x = \tan^2 x + 1 - 1$ and note that $d \tan x/dx = \tan^2 x + 1$.

3. $-\cos e^x$ **4.** Let $x = 2u$, $dx = 2du$ **5.** $\dfrac{\pi}{4}$ **7.** $\dfrac{\pi}{4}$ **9.** $\dfrac{\pi}{16}$

11. $\frac{1}{8}\arcsin x - \frac{1}{8}x(1 - 2x^2)\sqrt{1 - x^2}$ **13.** $\dfrac{\pi}{2}$ **14.** $-\arcsin x - \dfrac{1}{x}\sqrt{1 - x^2}$

15. $-16u^{1/2} + \frac{1}{3}u^{3/2}$ where $u = 16 - x^2$

16. Let $u + 1 + x^2$. Then

$$\int \frac{x^3}{\sqrt{1+x^2}}\, dx = \frac{1}{2}\int \frac{x^2\, 2x}{\sqrt{1+x^2}}\, dx = \frac{1}{2}\int \frac{u-1}{u^{1/2}}\, du = \frac{1}{2}\left[\frac{u^{3/2}}{3/2} - \frac{u^{1/2}}{1/2}\right].$$

The rest of the exercises are done by letting $x = \sin\theta$ or $x = a\sin\theta$, $dx = a\cos\theta\, d\theta$. We give the answers, but work out Exercise 19 in full.

17. $\dfrac{-1}{a}\log\left[\dfrac{a+\sqrt{a^2-x^2}}{x}\right]$ **18.** $\dfrac{a^2}{2}\arcsin(x/a) - \tfrac{1}{2}x\sqrt{a^2-x^2}$

19. $\dfrac{-\sqrt{a^2-x^2}}{2a^2x^2} - \dfrac{1}{2a^3}\log\left[\dfrac{a+\sqrt{a^2-x^2}}{x}\right]$. We have a choice of whether to let $x = a\cos\theta$ or $x = a\sin\theta$. The principle is the same. Let us do as usual,

$$x = a\sin\theta, \qquad dx = a\cos\theta\, d\theta.$$

Then

$$\int \frac{1}{x^3\sqrt{a^2-x^2}}\, dx = \int \frac{1}{a^3\sin^3\theta(a\cos\theta)}\, a\cos\theta\, d\theta = \frac{1}{a^3}\int \frac{1}{\sin^3\theta}\, d\theta.$$

It's a pain, but we show how to do it. Recall that to integrate positive powers of sine, we used integration by parts. We try a similar method here. Thus let

$$I = \int \frac{1}{\sin^3\theta}\, d\theta = \int \frac{1}{\sin\theta}\frac{1}{\sin^2\theta}\, d\theta = \int \frac{1}{\sin\theta}\csc^2\theta\, d\theta.$$

In analogy with the tangent, we have

$$\frac{d\cot\theta}{d\theta} = -\csc^2\theta,$$

so we let

$$u = \frac{1}{\sin\theta}, \qquad\qquad dv = \csc^2\theta\, d\theta,$$

$$du = -\frac{1}{\sin^2\theta}\cos\theta\, d\theta, \qquad v = -\cot\theta.$$

Then

$$I = -\frac{\cot\theta}{\sin\theta} - \int \frac{\cos^2\theta}{\sin^3\theta}\, d\theta = -\frac{\cos\theta}{\sin^2\theta} - \int \frac{1-\sin^2\theta}{\sin^3\theta}\, d\theta$$

and so

$$I = -\frac{\cos\theta}{\sin^2\theta} - I + \int \frac{1}{\sin\theta}\, d\theta,$$

whence

$$I = \frac{1}{2}\left[-\frac{\cos\theta}{\sin^2\theta} - \log(\csc\theta + \cot\theta)\right].$$

You may leave the answers in terms of θ, this is usually done. But if you want the answer in terms of x, then use:

$$\sin\theta = \frac{x}{a}, \qquad \cos\theta = \sqrt{1 - \sin^2\theta} = \frac{1}{a}\sqrt{a^2 - x^2},$$

$$\csc\theta = \frac{1}{\sin\theta} = \frac{a}{x}, \qquad \cot\theta = \frac{\cos\theta}{\sin\theta} = \frac{\sqrt{a^2 - x^2}}{x}.$$

20. $-\dfrac{1}{a^2}\cot\theta$ where $x = a\sin\theta$, $dx = a\cos\theta\, d\theta$.

21. $\sqrt{1 - x^2} - \log\left(\dfrac{1 + \sqrt{1 + x^2}}{x}\right)$. The method is the same as Exercise 19.

22. Let $x = at$, $dx = a\, dt$, and reduce to Exercise 14.

23. $\dfrac{x}{\sqrt{a^2 - x^2}} - \arcsin\dfrac{x}{a}$

XI, §4, p. 370

1. $-\frac{1}{8}\log(x - 1) + \frac{17}{8}\log(x + 7)$

2. $\dfrac{-1}{2(x^2 - 3)}$. Don't use partial fractions here, use the substitution $u = x^2 - 3$ and $du = 2x\, dx$.

3. (a) $\frac{1}{5}[\log(x - 3) - \log(x + 2)]$ (b) $\log(x + 1) - \log(x + 2)$

4. $-\frac{1}{2}\log(x + 1) + 2\log(x + 2) - \frac{3}{2}\log(x + 3)$

5. $2\log x - \log(x + 1)$ **6.** $\log(x + 1) + \dfrac{1}{x + 1}$

7. $-\log(x + 1) + \log(x + 2) - \dfrac{2}{x + 2}$

8. $\log(x - 1) + \log(x - 2)$ **9.** $\dfrac{x}{2(x^2 + 1)} + \frac{1}{2}\arctan x$

10. (a) $\dfrac{1}{4}\dfrac{x}{(x^2 + 1)^2} + \dfrac{3}{8}\dfrac{x}{(x^2 + 1)} + \dfrac{3}{8}\arctan x$

11. $\dfrac{-1}{x^2 + 1} - 3\left[\dfrac{x}{2(x^2 + 1)} + \dfrac{1}{2}\arctan x\right]$

12. $\dfrac{1}{2}\dfrac{-1}{x^2 + 9} + \dfrac{1}{18}\dfrac{x}{x^2 + 9} + \dfrac{1}{54}\arctan\dfrac{x}{3}$ **13.** $\dfrac{1}{8}\dfrac{x}{x^2 + 16} + \dfrac{1}{32}\arctan\dfrac{x}{4}$

14. $\frac{1}{4}\log\dfrac{(x + 1)^2}{x^2 + 1} + \dfrac{1}{2}\arctan x$. Factorization:

$$x^3 - 1 = (x - 1)(x^2 + x + 1) \qquad \text{and} \qquad x^4 - 1 = (x + 1)(x - 1)(x^2 + 1).$$

15. $C_1 = -\frac{33}{100}$, $C_2 = -\frac{11}{100}$, $C_3 = -\frac{130}{100}$, $C_4 = -\frac{110}{100}$, $C_5 = \frac{11}{100}$

16. (a) Let $x = bt$, $dx = b\, dt$ (b) Let $x + a = bt$, $dx = b\, dt$.

17. (a) $-\frac{1}{2}\arctan x + \frac{1}{4}\log\left(\dfrac{x-1}{x+1}\right)$ (b) $\frac{1}{4}[\log(x^2-1)-\log(x^2+1)]$

18. (a) $\frac{1}{3}\log(x-1)-\frac{1}{6}\log(x^2+x+1)-\dfrac{1}{\sqrt{3}}\arctan\dfrac{2x+1}{\sqrt{3}}$

  (b) $\frac{1}{2}\log\dfrac{x^2}{x^2+x+1}-\dfrac{1}{\sqrt{3}}\arctan\dfrac{2x+1}{\sqrt{3}}$

19. $-\log(x-1)+\log(x^2+x+1)$

XI, §5, p. 377

1. $2\sqrt{1+e^x}-\log(\sqrt{1+e^x}+1)+\log(\sqrt{1+e^x}-1)$ **2.** $x-\log(1+e^x)$
3. $\arctan(e^x)$ **4.** $-\log(\sqrt{1+e^x}+1)+\log(\sqrt{1+e^x}-1)$
5. Let $y=f(x)=\frac{1}{2}(e^x-e^{-x})=\sinh x$. Then

$$f'(x)=\tfrac{1}{2}(e^x+e^{-x})=\cosh x.$$

But $\cosh x > 0$ for all x, so f is strictly increasing for all x. If x is large negative, then e^x is small, and e^{-x} is large positive, so $f(x)$ is large positive. If x is large positive, then e^x is large positive, and e^{-x} is small. Hence

$$f(x)\to\infty\qquad\text{as}\quad x\to\infty,$$

$$f(x)\to-\infty\qquad\text{as}\quad x\to-\infty.$$

By the intermediate value theorem, the values of $f(x)$ consist of all numbers. Hence the inverse function $x=g(y)$ is defined for all numbers y. We have

$$g'(y)=\dfrac{1}{f'(x)}=\dfrac{1}{\cosh x}=\dfrac{1}{\sqrt{\sinh^2 x+1}}=\dfrac{1}{\sqrt{y^2+1}}.$$

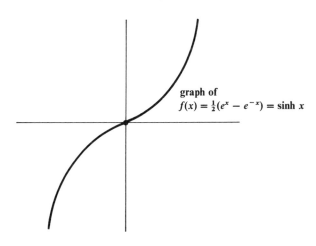

graph of
$f(x)=\frac{1}{2}(e^x-e^{-x})=\sinh x$

6. Let $y = f(x) = \frac{1}{2}(e^x + e^{-x}) = \cosh x$. Then

$$f'(x) = \frac{1}{2}(e^x - e^{-x}) = \sinh x.$$

If $x > 0$ then $e^x > 1$ and $0 < e^{-x} < 1$ so $f'(x) > 0$ for all $x > 0$. Hence f is strictly increasing, and the inverse function

$$x = g(y) = \text{arccosh } y$$

exists. We have $f(0) = 1$. As $x \to \infty$, $e^x \to \infty$ and $e^{-x} \to 0$, so $f(x) \to \infty$. Hence the values of $f(x)$ consist of all numbers ≥ 1 when $x \geq 0$. Hence the inverse function g is defined for all numbers ≥ 1. We have

$$g'(y) = \frac{1}{f'(x)} = \frac{1}{\sinh x} = \frac{1}{\sqrt{\cosh^2 x - 1}} = \frac{1}{\sqrt{y^2 - 1}}.$$

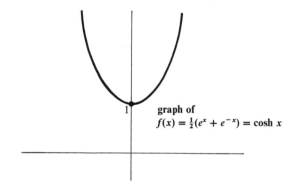

graph of
$f(x) = \frac{1}{2}(e^x + e^{-x}) = \cosh x$

Finally, let $u = e^x$. Then $y = \frac{1}{2}(u + 1/u)$. Multiply by $2u$ and solve the quadratic equation to get

$$e^x = u = y + \sqrt{y^2 - 1} \qquad \text{or} \qquad e^x = u = y - \sqrt{y^2 - 1}.$$

The graph of $f(x) = \cosh x$ for all numbers x bends as shown on the figure, and there are two possible inverse functions depending on whether we look at the interval

$$x \leq 0 \qquad \text{or} \qquad x \geq 0.$$

Taking

$$x = \log(y + \sqrt{y^2 - 1})$$

is the inverse function for $x \geq 0$ and taking

$$x = \log(y - \sqrt{y^2 - 1})$$

is the inverse functions for $x \leq 0$. Indeed suppose we take the solution with the minus sign. Then by simple algebra, you can see that

$$y - \sqrt{y^2 - 1} \leq 1.$$

[*Proof*: you have to check that $y - 1 \leq \sqrt{y^2 - 1}$. Since $y \geq 1$, it suffices to check that $(y - 1)^2 \leq y^2 - 1$, which amounts to

$$y^2 - 2y + 1 \leq y^2 - 1,$$

or $y \geq 1$, which checks.]
     Thus $y - \sqrt{y^2 - 1} \leq 1$, whence

$$\log(y - \sqrt{y^2 - 1}) \leq 0.$$

For $x \geq 0$ it follows that we have to use the solution of the quadratic equation for u in terms of y with the plus sign, that is

$$e^x = u = y + \sqrt{y^2 - 1} \quad \text{and} \quad x = \log(y + \sqrt{y^2 - 1}).$$

7. Let

$$I = \int \frac{x^2}{\sqrt{x^2 + 4}}\, dx.$$

Let $x = 2 \sinh t$, $dx = 2 \cosh t\, dt$. Then $x^2 + 4 = 4 \cosh^2 t$. Hence

$$I = \int \frac{4 \sinh^2 t}{2 \cosh t}\, 2 \cosh t\, dt$$

$$= 4 \int \sinh^2 t\, dt = \frac{4}{4} \int (e^{2t} - 2 + e^{-2t})\, dt$$

$$= \tfrac{1}{2} e^{2t} - 2t - \tfrac{1}{2} e^{-2t}.$$

8. $\log(x + \sqrt{x^2 + 1})$

9. Let

$$I = \int \frac{x^2 + 1}{x - \sqrt{x^2 + 1}}\, dx.$$

Let $x = \sinh t$, $dx = \cosh t\, dt$. Then

$$I = \int \frac{\cosh^2 t}{\sinh t - \cosh t}\, \cosh t\, dt$$

$$= \int \frac{\cosh^3 t}{\dfrac{e^t - e^{-t}}{2} - \dfrac{e^t + e^{-t}}{2}}\, dt$$

$$= \int \frac{\left(\dfrac{e^t + e^{-t}}{2}\right)^3}{-e^{-t}}\, dt$$

$$= -\frac{1}{8} \int e^t(e^{3t} + 3e^t + 3e^{-t} + e^{-3t})\, dt$$

$$= -\frac{1}{8}\left(\frac{e^{4t}}{4} + \frac{3e^{2t}}{2} + 3t + \frac{e^{-2t}}{-2}\right).$$

10. $-\frac{1}{2}\log(x + \sqrt{x^2 - 1}) + \frac{1}{2}x\sqrt{x^2 - 1}$ (see Exercise 11)

11. $-\frac{1}{2}\log(B + \sqrt{B^2 - 1}) + \frac{1}{2}B\sqrt{B^2 - 1}$. The graph of the equation $x^2 - y^2 = 1$ is a hyperbola as drawn.

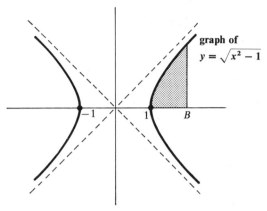

The top part of the hyperbola in the first quadrant is the graph of the function

$$y = \sqrt{x^2 - 1},$$

with the positive square root, and $x \geq 1$. Hence the area under the graph between 1 and B is

$$\text{Area} = \int_1^B \sqrt{x^2 - 1} \; dx.$$

We want to make the expression under the square into a perfect square. We use the substitution

$$x = \cosh t \qquad \text{and} \qquad dx = \sinh t \; dt.$$

Then you get into an integral consisting of powers of e^t and e^{-t} which is easy to evaluate. Change the limits of integration in the way explained in the last example of the section. Let $u = e^t$ and solve a quadratic equation for u. You will find the given answer.

12. $\log(B + \sqrt{B^2 + 1}) + B\sqrt{B^2 + 1}$

13. Let $y = a \cosh(x/a)$. Then

$$\frac{dy}{dx} = a \sinh(x/a) \cdot \frac{1}{a} = \sinh(x/a),$$

$$\frac{d^2y}{dx^2} = \cosh(x/a) \cdot \frac{1}{a} = \frac{1}{a}\cosh(x/a)$$

$$= \frac{1}{a}\sqrt{1 + \sinh^2(x/a)}$$

$$= \frac{1}{a}\sqrt{1 + (dy/dx)^2}.$$

14. Let $x = az$, $dx = adz$, and reduce to the worked-out case.

XII, §1 p. 384

1. $\frac{4}{3}\pi r^3$ **2.** π **3.** $\frac{\pi^2}{8} - \frac{\pi}{4}$ **4.** $\frac{\pi^2}{8} + \frac{\pi}{4}$ **5.** $\frac{2 \cdot 5^4 \pi}{3}$ **6.** $\pi(e - 2)$ **7.** πe^2

8. $\pi[2(\log 2)^2 - 4\log 2 + 2]$. Integrate $\int (\log x)^2 \, dx$ by parts, $u = (\log x)^2$,

$du = \frac{2 \log x}{x} \, dx, \; dv = dx$

9. 16π **10.** (a) $\frac{\pi}{2}\left(\frac{1}{e^2} - \frac{1}{e^{2B}}\right)$; yes $\frac{\pi}{2e^2}$

    (b) $\frac{\pi}{4}\left(\frac{1}{e^4} - \frac{1}{e^{4B}}\right)$; yes $\frac{\pi}{4e^4}$ (c) $\frac{\pi}{4}\left(\frac{1}{e^2} - \frac{1}{e^{2B^2}}\right)$; yes $\frac{\pi}{4e^2}$

11. Equation of the line is $y = \frac{r}{h} x$. Volume is $\frac{\pi r^2 h}{3}$ **12.** $2\pi(1 - \sqrt{a}), \; 2\pi$

13. $\frac{\pi}{24} - \frac{\pi}{3B^3}$; yes $\frac{\pi}{24}$ **14.** For all $c > 1/2$, $\pi/(2c - 1)$

15. For all $c < 1/2$, $\pi/(1 - 2c)$

XII, §1, Supplementary Exercises, p. 385

1. $f(x) = R + \sqrt{a^2 - x^2}$ and $g(x) = R - \sqrt{a^2 - x^2}$. The volume is

$$V = \pi \int_{-a}^{a} f(x)^2 \, dx - \pi \int_{-a}^{a} g(x)^2 \, dx$$

which after easy algebra comes out equal to

$$4\pi R \int_{-a}^{a} \sqrt{a^2 - x^2} \, dx = 2\pi^2 R a^2.$$

2. $\frac{32\pi}{5}$ **3.** 12π **4.** 2π **5.** $\frac{2\pi}{3}$ **6.** $\frac{5\pi}{14}$ **7.** $\frac{\pi}{3}$ **8.** $\frac{4}{3}\pi a^2 b$ **9.** $\frac{\pi}{2}(e^{-2} - e^{-10})$

10. $\pi[2(\log 2)^2 - 4\log 2 + 2]$ **11.** $\pi\left[\sqrt{3} - \frac{\pi}{3}\right]$ **12.** $\pi\left(1 - \frac{1}{B}\right)$, π as $B \to \infty$

13. $\frac{\pi}{3}\left(1 - \frac{1}{B^3}\right)$, $\frac{\pi}{3}$ as $B \to \infty$ **14.** $\pi \log B$

15. $\pi \log \frac{1}{a}$. No limit as $a \to 0$. The volume increases without bound.

16. $\pi\left(\frac{1}{a} - 1\right)$. No limit as $a \to 0$. The volume increases without bound.

17. $\pi\left[\frac{\sqrt{2}}{2} - \cos a + \log(\sqrt{2} - 1) - \log(\csc a - \cot a)\right]$

No limit as $a \to 0$. The volume increases without bound.

XII, §2, p. 390

1. 6π **2.** a^2 (using symmetry and values of θ such that $\sin 2\theta \geq 0$, the problem reduces to $\int_0^{\pi/2} a^2 \sin 2\theta \, d\theta$)

3. πa^2　**4.** $\dfrac{\pi}{12}$　**5.** $3\pi/2$　**6.** $3\pi/2$　**7.** $9\pi/2$　**8.** $\pi/3$

XII, §2, Supplementary Exercises, p. 390

1. 25π　**2.** $\dfrac{3\pi}{2}$　**3.** π　**4.** $\dfrac{9\pi}{2}$　**5.** $\dfrac{3\pi}{8}$　**6.** $\dfrac{3\pi}{2}$　**7.** $2\pi + \dfrac{3\sqrt{3}}{2}$　**8.** $\dfrac{3\pi}{2}$

9. $\dfrac{\pi}{4}$　**10.** $\dfrac{9\pi}{2}$　**11.** $10\frac{2}{3}$　**12.** $10\frac{2}{3}$　**13.** $\frac{4}{3}$　**14.** $\frac{4}{3}$　**15.** $\dfrac{5\sqrt{5}}{6}$　**16.** 10

XII, §3, p. 397

1. $\frac{8}{27}(10^{3/2} - 1)$　**2.** $\dfrac{\sqrt{5}}{2} + \log\!\left(\dfrac{4 + 2\sqrt{5}}{1 + \sqrt{5}}\right)$

3. $\sqrt{e^4 + 1} + 2 - \sqrt{2} + \log\!\left(\dfrac{1 + \sqrt{2}}{1 + \sqrt{e^4 + 1}}\right)$　**4.** $2\sqrt{17} + \log\!\left(\dfrac{\sqrt{17} + 4}{\sqrt{17} - 4}\right)^{1/4}$

5. $\sqrt{1 + e^2} + \frac{1}{2}\log\dfrac{\sqrt{1 + e^2} - 1}{\sqrt{1 + e^2} + 1} - \left(\sqrt{2} + \frac{1}{2}\log\dfrac{\sqrt{2} - 1}{\sqrt{2} + 1}\right)$

6. $\frac{1}{27}(31^{3/2} - 13^{3/2})$　**7.** $e - \dfrac{1}{e}$

8. We work out Exercise 8 in full.

$$\text{length} = \int_0^{3/4} \sqrt{1 + \left(\frac{-2x}{1 - x^2}\right)^2}\, dx$$

$$= \int_0^{3/4} \sqrt{1 + \frac{4x^2}{(1 - x^2)^2}}\, dx$$

$$= \int_0^{3/4} \frac{\sqrt{1 - 2x^2 + x^4 + 4x^2}}{1 - x^2}\, dx$$

$$= \int_0^{3/4} \frac{\sqrt{(1 + x^2)^2}}{1 - x^2}\, dx$$

$$= \int_0^{3/4} \frac{1 + x^2}{1 - x^2}\, dx$$

$$= \int_0^{3/4} \frac{2}{1 - x^2}\, dx + \int_0^{3/4} \frac{x^2 - 1}{1 - x^2}\, dx$$

$$= 2\int_0^{3/4} \frac{1}{2}\left(\frac{1}{1 - x} + \frac{1}{1 + x}\right) dx - \int_0^{3/4} dx$$

$$= -\log(1 - x) + \log(1 + x)\Big|_0^{3/4} - \frac{3}{4}$$

$$= \log\!\left(\frac{1 + 3/4}{1 - 3/4}\right) - \tfrac{3}{4} = \log 7 - 3/4$$

9. $\dfrac{1}{2}\left(e - \dfrac{1}{e}\right)$ **10.** $\log(2 + \sqrt{3})$

XII, §4, p. 407

2. $2\pi r$ **3.** $\sqrt{2}(e^2 - e)$ **4.** (a) $\frac{3}{4}$ (b) 3   **5.** $2\sqrt{5} + \log(\sqrt{5} + 2)$

6. $4\sqrt{2} + 2\log\dfrac{\sqrt{2} + 1}{\sqrt{2} - 1}$ **7.** $2\sqrt{3}$ **8.** 5   **9.** 8   **10.** $4a$ **12.** $\sqrt{2}(e^2 - e)$

13. $\sqrt{2}(e^{\theta_2} - e^{\theta_1})$ **14.** $(8^{3/2} - 5^{3/2})$ **15.** $\dfrac{\sqrt{17}}{4}(e^{-4} - e^{-8})$ **16.** $\dfrac{3\pi}{4}$

17. $\sqrt{5} - \dfrac{\sqrt{17}}{4} + \log\left[\dfrac{8 + 2\sqrt{17}}{1 + \sqrt{5}}\right]$ **18.** $4\sin\dfrac{\pi}{8} = 2\sqrt{2 - \sqrt{2}}$ **19.** 4   **20.** 2

21. 8   **22.** π **23.** $2\sqrt{3}$

XII, §5, p. 415

1. $12\pi a^2/5$ **2.** $\dfrac{\pi}{27}(10\sqrt{10} - 1)$ **3.** $\dfrac{2\pi}{3}(26\sqrt{26} - 2\sqrt{2})$ **4.** $4\pi^2 a^2$ **5.** $4\pi^2 aR$

6. $\dfrac{\pi}{6}(17\sqrt{17} - 1)$

XII, §6, p. 418

1. 5 lb/in.; 80 in.–lb   **2.** $\dfrac{10}{\sin\dfrac{\pi}{9}}$ lb/in.; $\dfrac{180}{\pi}\left[\cot\dfrac{\pi}{9} - \dfrac{1}{2}\csc\dfrac{\pi}{9}\right]$ in.–lb

3. $c\left[\dfrac{1}{r_1} - \dfrac{1}{r}\right]$ **4.** yes; $\dfrac{c}{r_1}$ **5.** $\dfrac{99c}{200}$ where c is the constant of proportionality

6. 2×10^{-6} pound-miles   **7.** $\dfrac{E}{6}$ in.–lb

8. (a) $-90\ CmM$ dyne-cm   (b) $9\ CmM$ dyne-cm   **9.** $c\left(\dfrac{1}{r_1} - \dfrac{1}{r_2}\right)$

10. 1500 log 2 in.–pounds. If A is the area of the cross section of the cylinder, and $P(x)$ is the pressure, then

$$P(x) \cdot Ax = C = \text{constant.}$$

If a is the length of the cylinder when the volume is 75 in^3, then

$$\text{Work} = \int_a^{2a} \text{Force } dx = \int_a^{2a} P(x)A\,dx = \int_a^{2a} \dfrac{C}{x}\,dx = C\log 2.$$

But $C = 20 \cdot 75 = 1{,}500$ from initial data. This gives the answer.

XII, §7, p. 423

1. $\dfrac{3}{4}\left(\dfrac{15^4 - 5^4}{15^3 - 5^3}\right)$ **2.** 10   **3.** 10/log 3

XIII, §1, p. 434

1. (a) $f^{(k)}(x) = \dfrac{(-1)^{k+1}(k-1)!}{(1+x)^k}$

(b) $f^{(k)}(0) = (-1)^{k+1}(k-1)!$

(c) Since $(k-1)!/k! = 1/k$, we get from (b)

$$\frac{f^{(k)}(0)}{k!} = \frac{(-1)^{k+1}(k-1)!}{k!} = \frac{(-1)^{k+1}}{k}$$

and

$$P_n(x) = \sum_{k=0}^{n} \frac{f^{(k)}(0)}{k!}\, x^k.$$

This proves that when $f(x) = \log(1+x)$ then the n-th Taylor polynomial is given by

$$P_n(x) = \sum_{k=0}^{n} \frac{(-1)^{k+1}}{k}\, x^k.$$

2. For $f(x) = \cos x$, $f^{(n)}(x) = f^{(n+4)}(x)$. Use this and the formula for $P_n(x)$ to derive $P_n(x)$ for the function $f(x) = \cos x$.

XIII, §3, p. 446

1. $1 - \dfrac{x^2}{2!} + \dfrac{x^4}{4!}$

2. $|f^{(n)}(c)| \le 1$ for all n and all numbers c so the estimate follows from Theorem 2.1.

3. $1 - \dfrac{0.01}{2} + R_4(0.1) = 0.995 + R_4(0.1)$ **4.** $|R_3| \le \dfrac{(0.1)^3}{3!} = \dfrac{1}{6} \times 10^{-3}$.

5. $|R_4| \le \tfrac{2}{3}10^{-4}$

6. $P_4(x) = x + \dfrac{x^3}{3}$.   Let $f(x) = \tan x$. You have to find first all derivatives $f^{(1)}(x)$, $f^{(2)}(x)$, $f^{(3)}(x)$, $f^{(4)}(x)$, and then $f^{(1)}(0),\ldots,f^{(4)}(0)$. Then use the general formula for the Taylor polynomial

$$P_4(x) = f(0) + f^{(1)}(0)x + \cdots + f^{(4)}(0)\frac{x^4}{4!}.$$

7. $|R_5| \le 10^{-4}$ by crude estimates

8. (a) $\sin\left(\dfrac{\pi}{6}+\dfrac{\pi}{180}\right)=\dfrac{1}{2}+\dfrac{\sqrt{3}}{2}\left(\dfrac{\pi}{180}\right)+E$ and $|E|<10^{-3}$

(b) $\cos\left(\dfrac{\pi}{6}+\dfrac{\pi}{180}\right)=\dfrac{\sqrt{3}}{2}-\dfrac{1}{2}\left(\dfrac{\pi}{180}\right)+E$

(c) $\sin\left(\dfrac{\pi}{4}+\dfrac{2\pi}{180}\right)=\dfrac{\sqrt{2}}{2}+\dfrac{\sqrt{2}}{2}\left(\dfrac{2\pi}{180}\right)+E$

(d) $\cos\left(\dfrac{\pi}{4}+\dfrac{2\pi}{180}\right)=\dfrac{\sqrt{2}}{2}-\dfrac{\sqrt{2}}{2}\left(\dfrac{2\pi}{180}\right)+E$

(e) $\sin\left(\dfrac{\pi}{6}+\dfrac{2\pi}{180}\right)=\dfrac{1}{2}+\dfrac{\sqrt{3}}{2}\left(\dfrac{2\pi}{180}\right)+E$

(f) $\cos\left(\dfrac{\pi}{6}+\dfrac{2\pi}{180}\right)=\dfrac{\sqrt{3}}{2}-\dfrac{1}{2}\left(\dfrac{2\pi}{180}\right)+E$

We carry out the steps for part (e) in full.

$$\sin 32° = \sin\left(\frac{\pi}{6}+\frac{2\pi}{180}\right)=\sin\left(\frac{\pi}{6}+\frac{\pi}{90}\right)=\sin(a+h)$$

$$=\sin\frac{\pi}{6}+\left(\cos\frac{\pi}{6}\right)\frac{\pi}{90}+R_2(h)$$

$$=\frac{1}{2}+\frac{\sqrt{3}}{2}\frac{\pi}{90}+R_2\left(\frac{\pi}{90}\right),$$

and

$$\left|R_2\left(\frac{\pi}{90}\right)\right|\leqq\left(\frac{\pi}{90}\right)^2\frac{1}{2}\leqq\left(\frac{3.15}{90}\right)^2\frac{1}{2}\leqq(3.5\times10^{-2})^2\frac{1}{2}\leqq10^{-3}.$$

9. $\cos\left(\dfrac{\pi}{6}+\dfrac{\pi}{180}\right)=\dfrac{\sqrt{3}}{2}-\dfrac{1}{2}\left(\dfrac{\pi}{180}\right)+E$

10. $\sin\left(\dfrac{\pi}{3}+\dfrac{\pi}{180}\right)=\dfrac{\sqrt{3}}{2}+\dfrac{1}{2}\left(\dfrac{\pi}{180}\right)+E$

11. $\cos\left(\dfrac{\pi}{3}+\dfrac{\pi}{180}\right)=\dfrac{1}{2}-\dfrac{\sqrt{3}}{2}\left(\dfrac{\pi}{180}\right)+E$

12. (a) $\dfrac{\sin x}{x}=1-\dfrac{x^2}{3!}+\dfrac{R_5(x)}{x}$ so

$$\int_0^1 \frac{\sin x}{x}\,dx=1-\frac{1}{3\cdot3!}+E\qquad\text{where}\qquad|E|\leqq\frac{1}{5\cdot5!}.$$

(b) $-\dfrac{(0.1)^2}{4}+E\qquad$ where $\qquad|E|\leqq\dfrac{10^{-4}}{4\cdot4!}$

(c) Write $\sin u = u - \dfrac{u^3}{3!} + R_5(u)$, so that for $u = x^2$ we get

$$\sin x^2 = x^2 - \frac{x^6}{3!} + R_5(x^2).$$

We have, for $u \geq 0$,

$$|R_5(u)| \leq \frac{u^5}{5!} = \frac{x^{10}}{5!}.$$

Hence

$$\int_0^1 \sin x^2 \, dx = \frac{1}{3} - \frac{1}{7 \cdot 3!} + E,$$

where

$$|E| \leq \int_0^1 \frac{x^{10}}{5!} \, dx = \frac{1}{11 \cdot 5!} \leq 10^{-3}.$$

(d) $\dfrac{1}{2} - \dfrac{1}{6 \cdot 3!} + E,$ and $|E| \leq \dfrac{1}{10 \cdot 5!}$

(e) $1 - \dfrac{1}{5 \cdot 2!} + \dfrac{1}{9 \cdot 4!} + E,$ and $|E| \leq \dfrac{1}{13 \cdot 6!}$

(f) $1 - \dfrac{1}{5 \cdot 3!} + E,$ and $|E| \leq \dfrac{1}{9 \cdot 5!}$

13.

$$\int_0^{1/2} \frac{\cos x - 1}{x} \, dx = \int_0^{1/2} \frac{1 - \dfrac{x^2}{2} + \dfrac{x^4}{4!} + R_6(x) - 1}{x} \, dx$$

$$= \int_0^{1/2} \left(-\frac{x}{2} + \frac{x^3}{4!} + \frac{R_6(x)}{x} \right) dx$$

$$= -\frac{x^2}{4} + \frac{x^4}{96} \Big|_0^{1/2} + E$$

$$= -\frac{1}{16} + \frac{1}{16 \cdot 96} + E,$$

and

$$|E| \leq \frac{1}{6!} \int_0^{1/2} x^5 \, dx = \frac{1}{6!} \frac{x^6}{6} \Big|_0^{1/2} = \frac{1}{6! \cdot 6 \cdot 2^6} < 10^{-5}.$$

XIII, §4, p. 448

1. $1 - x + \dfrac{x^2}{2!} - \dfrac{x^3}{3!} + \dfrac{x^4}{4!} - \dfrac{x^5}{5!}$ **2.** $|R_3| \leq e^{1/2} \dfrac{(\frac{1}{2})^3}{3!} \leq \dfrac{2(\frac{1}{8})}{6} = \dfrac{1}{24}$

3. $|R_4| \leq e^{10^{-2}} \dfrac{(10^{-2})^4}{4!} \leq \dfrac{2(10^{-8})}{4!} = \dfrac{10^{-8}}{12}$

4. $|R_3| \leq e^{10^{-2}} \dfrac{(10^{-2})^3}{3!} \leq \dfrac{2(10^{-6})}{6} = \dfrac{10^{-6}}{3}$

6. $|R_7| \leq e^{-1} \dfrac{|-1|^7}{7!} \leq \dfrac{1}{7!} = \dfrac{1}{5040}$

7. (a) $|R_4| \leq e^2 \dfrac{|2|^4}{4!} \leq 6$ (b) $|R_4| \leq e^3 \dfrac{|3|^4}{4!} \leq 214$

8. (a) $|R_5| \leq e^2 \dfrac{|2|^5}{5!} \leq \dfrac{64}{15}$ (b) $|R_5| \leq e^3 \dfrac{|3|^5}{5!} \leq \dfrac{648}{5}$

9. (a) $|R_{13}| \leq \dfrac{16 \cdot 2^{12}}{12!}$ using $e < 4$ (b) $|R_{16}| \leq \dfrac{16 \cdot 2^{16}}{16!}$ using $e < 4$

10. $e = 1 + \displaystyle\sum_{n=1}^{13} \dfrac{1}{n!} + E$ **11.** $e^{-2} = \displaystyle\sum_{k=0}^{12} \dfrac{(-2)^k}{k!} + E$ and $|E| \leq \dfrac{2^{13}}{13!}$

12. (a) $1 + \dfrac{1}{2 \cdot 2!} + \dfrac{1}{3 \cdot 3!} + \cdots + \dfrac{1}{7 \cdot 7!} + E$, where $|E| \leq \dfrac{e}{8 \cdot 8!}$

(b) Write $e^u = 1 + u + \cdots + \dfrac{u^4}{4!} + R_5(u)$.   For $u \leq 0$ we have $e^u \leq 1$, and

hence $|R_5(u)| \leq \dfrac{|u|^5}{5!}$.   Now put $u = -x^2$.   Then

$$e^{-x^2} = 1 - x^2 + \dfrac{x^4}{2!} - \dfrac{x^6}{3!} + \dfrac{x^8}{4!} + R_5(-x^2).$$

Integrate the first part consisting of powers of x term by term.   We get

$$\int_0^1 e^{-x^2}\, dx = 1 - \dfrac{1}{3} + \dfrac{1}{5 \cdot 2!} - \dfrac{1}{7 \cdot 3!} + \dfrac{1}{9 \cdot 4!} + E,$$

and

$$|E| \leq \int_0^1 |R_5(-x^2)|\, dx \leq \int_0^1 \dfrac{x^{10}}{5!}\, dx = \dfrac{1}{11 \cdot 5!} \leq 10^{-3}.$$

(c) $1 + \dfrac{1}{3} + \dfrac{1}{5 \cdot 2!} + \dfrac{1}{7 \cdot 3!} + \dfrac{1}{9 \cdot 4!} + \dfrac{1}{11 \cdot 5!} + E$,

where $|E| \leq \dfrac{e}{13 \cdot 6!} < \dfrac{1}{3} \times 10^{-3}$.

(d) $e^u = 1 + u + R_2(u)$, and we put $u = x^2$.   For $0 \leq x \leq 0.1$ we find $e^u \leq 2$
(generous estimate).   Hence

$$|R_2(u)| \leq \dfrac{2u^2}{2!} = u^2.$$

We have $e^{x^2} = 1 + x^2 + R_2(x^2)$, and $|R_2(x^2)| \leq x^4$.   Hence

$$\int_0^{0.1} e^{x^2}\, dx = 0.1 + \dfrac{(0.1)^3}{3} + E, \qquad \text{where} \qquad |E| \leq \int_0^{0.1} x^4\, dx \leq \dfrac{10^{-5}}{5}.$$

(e) $0.1 - (\tfrac{1}{3})10^{-3} + E$, where $|E| \leq 10^{-6}$.

XIII, §5, p. 455

1. (a) $\log 1.2 = 0.2 - \dfrac{(0.2)^2}{2} + \dfrac{(0.2)^3}{3} + R_4, \; |R_4| \leqq 4 \cdot 10^{-4}$

(b) $\log 0.9 = -\log 10/9 = -\log\left(1 + \dfrac{1}{9}\right)$

$$= -\left[\frac{1}{9} - \frac{(1/9)^2}{2}\right] - R_3(1/9) = -\frac{1}{9} + \frac{1}{2 \cdot 81} - R_3(1/9)$$

and $\left|-R_3(1/9)\right| < \frac{1}{2} \times 10^{-3}$

(c) $\log 1.05 = 0.05 - \dfrac{(0.05)^2}{2} + R_3$ and $|R_3| \leqq \dfrac{5^3}{3} \cdot 10^{-6}$

(d) $\log 9/10 = -\log 10/9$ as in (b).

(e) $\log 24/25 = -\log 25/24 = -\log\left(1 + \dfrac{1}{24}\right)$

$$= \frac{-1}{24} + \frac{1}{2(24)^2} - R_3(1/24),$$

and $|R_3(1/24)| < 10^{-3}$.

(f) $\log 26/25 = 0.04 + R_2(1/25), \; |R_2| \leqq 8 \times 10^{-4}$

2. (a) We transform the right-hand side until it is equal to the left-hand side. We have:

$$-7 \log \frac{9}{10} + 2 \log \frac{24}{25} + 3 \log \frac{81}{80} = \log\left(\frac{10}{9}\right)^7 \left(\frac{24}{25}\right)^2 \left(\frac{81}{80}\right)^3$$

$$= \log \frac{2^7 5^7 (2^3 \cdot 3)^2 (3^4)^3}{(3^2)^7 5^4 (2^4)^3 5^3}$$

$$= \log \frac{2^7 2^6 3^2 3^{12} 5^7}{2^{12} 3^{14} 5^7}$$

$$= \log 2.$$

The case of $\log 3$ is done in the same way. Note that each fraction $9/10$, $24/25$, $81/80$ is close to 1, and hence the value of the log is approximated very well by just a few terms from the Taylor formula. For instance,

$$\log \frac{9}{10} = -\log \frac{10}{9} = -\log\left(1 + \frac{1}{9}\right).$$

Then

$$\log\left(1 + \frac{1}{9}\right) = \frac{1}{9} - \frac{(1/9)^2}{2} + \frac{(1/9)^3}{3} - \frac{(1/9)^4}{4} + \frac{(1/9)^5}{5} + R_6(1/9),$$

and

$$|R_6(1/9)| \leqq \frac{(1/9)^6}{9} < \frac{1}{2} \times 10^{-6}.$$

Hence

(1) $$7 \log \frac{10}{9} = 7A_1 + E_1,$$

where $E_1 = 7R_6(1/9)$ and

$$|E_1| = 7|R_6(1/9)| < 3.5 \times 10^{-6}.$$

Observe the factor of 7 which comes in throughout at this point.
    Next, we have

$$\log \frac{25}{24} = \log\left(1 + \frac{1}{25}\right) = \frac{1}{25} - \frac{1}{25^2} + \frac{1}{25^3} + R_4(1/25)$$

$$= A_2 + R_4(1/25),$$

and

$$|R_4(1/25)| \leq \frac{(1/25)^4}{4} < 10^{-6}.$$

Hence

(2) $$-2 \log \frac{25}{24} = -2A_2 - 2R_4(1/25) = -2A_2 + E_2,$$

where

$$|E_2| = |2R_4(1/25)| < 2 \times 10^{-6}.$$

Thirdly, we have

$$\log \frac{81}{80} = \log\left(1 + \frac{1}{80}\right) = \frac{1}{80} - \frac{1}{80^2} + R_3(1/80)$$

$$= A_3 + R_3(1/80),$$

and

$$|R_3(1/80)| \leq \frac{(1/80)^3}{3} < \frac{1}{1.5} \times 10^{-6}.$$

Hence

(3) $$3 \log \frac{81}{80} = 3A_3 + E_3,$$

where

$$|E_3| = 3|R_3(1/80)| < \frac{3}{1.5} \times 10^{-6} \leq 2 \times 10^{-6}.$$

    We may now put together the computations of the three terms, and
we find:

$$\log 2 = 7A_1 - 2A_2 + 3A_3 + E, \qquad \text{where} \quad E = E_1 + E_2 + E_3$$

and

$$|E| \leq |E_1| + |E_2| + |E_3| < 3.5 \times 10^{-6} + 2 \times 10^{-6} + 2 \times 10^{-6}$$
$$< 10^{-5}.$$

This concludes the computation of $\log 2$.

The computation for $\log 3$ is similar. In each case, note that the factors $7, -2, 3$ for $\log 2$ and $11, -3, 5$ for $\log 3$ have to be taken into account, and contribute to the error term.

XIII, §6, p. 459

1.
$$2 \arctan u = \arctan u + \arctan u = \arctan \frac{u + u}{1 - u^2} = \arctan \frac{2u}{1 - u^2},$$

$$3 \arctan u = \arctan u + 2 \arctan u = \arctan u + \arctan \frac{2u}{1 - u^2}.$$

Now let $v = 2u/(1 - u^2)$ and use the formula for $\arctan u + \arctan v$.

2. (a) Let $u = 1/2$ and $v = 1/3$ in the formula for $\arctan u + \arctan v$.

(b) Let $u = 1/5$ and $v = 1/8$ in this same formula.

(c) We have to apply the addition formula repeatedly. We start with

$$2 \arctan \frac{1}{5} = \arctan \frac{1}{5} + \arctan \frac{1}{5} = \arctan \frac{1/5 + 1/5}{1 - 1/25}$$

$$= \arctan \frac{5}{12}.$$

Next,

$$\arctan \frac{5}{12} + \arctan \frac{1}{7} = \arctan \frac{5/12 + 1/7}{1 - 5/84} = \arctan \frac{47}{79}.$$

Next,

$$2 \arctan \frac{1}{8} = \arctan \frac{1}{8} + \arctan \frac{1}{8} = \arctan \frac{1/8 + 1/8}{1 - 1/64}$$

$$= \arctan \frac{16}{63}.$$

Then finally

$$2 \arctan \frac{1}{5} + \arctan \frac{1}{7} + 2 \arctan \frac{1}{8} = \arctan \frac{47}{79} + \arctan \frac{16}{63}$$

$$= \arctan \frac{47/79 + 16/63}{1 - 47 \cdot 16/79 \cdot 63}$$

$$= \arctan 1 = \pi/4.$$

3. By Taylor's formula, we have

$$\arctan\frac{1}{5} = \frac{1}{5} - \frac{1}{3}\left(\frac{1}{5}\right)^3 + R_5\left(\frac{1}{5}\right)$$

so

(1)
$$8 \cdot \arctan\frac{1}{5} = \frac{8}{5} - \frac{8}{3}\left(\frac{1}{5}\right)^3 + E_1,$$

where

$$E_1 = 8R_5\left(\frac{1}{5}\right) \qquad \text{and} \qquad |E_1| \le 8\,\frac{1}{5}\left(\frac{1}{5}\right)^5 = \frac{16}{3} \times 10^{-4}.$$

Second, we have

$$\arctan\frac{1}{7} = \frac{1}{7} - \frac{1}{3}\left(\frac{1}{7}\right)^3 + R_5\left(\frac{1}{7}\right)$$

so

(2)
$$4 \cdot \arctan\frac{1}{7} = \frac{4}{7} - \frac{4}{3}\left(\frac{1}{7}\right)^3 + E_2,$$

where

$$E_2 = 4R_5\left(\frac{1}{7}\right) \qquad \text{and} \qquad |E_2| \le 4\,\frac{1}{5}\left(\frac{1}{7}\right)^5 < \frac{2}{5} \times 10^{-4}.$$

Third, we have

$$\arctan\frac{1}{8} = \frac{1}{8} - \frac{1}{3}\left(\frac{1}{8}\right)^3 + R_5\left(\frac{1}{8}\right)$$

so

(3)
$$8 \cdot \arctan\frac{1}{8} = 1 - \frac{8}{3}\left(\frac{1}{8}\right)^3 + E_3$$

where

$$E_3 = 8R_5\left(\frac{1}{8}\right) \qquad \text{and} \qquad |E_3| \le 8\,\frac{1}{5}\left(\frac{1}{8}\right)^5 < \frac{8}{15} \times 10^{-4}.$$

Adding the three expressions, we find by Exercise 2(c):

$$\pi = 8 \arctan\frac{1}{5} + 4 \arctan\frac{1}{7} + 8 \arctan\frac{1}{8}$$

$$= \frac{8}{5} - \frac{8}{3}\left(\frac{1}{5}\right)^3 + \frac{4}{7} - \frac{4}{3}\left(\frac{1}{7}\right)^3 + 1 - \frac{8}{3}\left(\frac{1}{8}\right)^3 + E,$$

where

$$E = E_1 + E_2 + E_3 \qquad \text{and} \qquad |E| \le |E_1| + |E_2| + |E_3| < 10^{-3}.$$

4. $\arctan(1/5) + \arctan(1/5) = \arctan 5/12$ using $u = v = 1/5$,
$\arctan(1/5) + \arctan(5/12) = \arctan(37/55)$ using $u = 1/5$ and $v = 5/12$,
$\arctan(1/5) + \arctan(37/55) = \arctan(120/119)$.

This last value is equal to $4 \arctan(1/5)$. Then

$$4 \arctan(1/5) - \arctan(1/239) = \arctan(120/119) + \arctan(-1/239).$$

Let $u = 120/119$ and $v = -1/239$ and use arithmetic to get $\arctan 1$.

XIII, §7, p. 467

In the answers, we give only the approximating value, except in a couple of cases to illustrate an estimate for the error term. But you should include the estimate in your work.

1. (a) $|R_2| \leq \dfrac{3}{32} \cdot 10^{-4} < 10^{-5}$. We use $s = 1/4$ and

$$R_2(x) = \frac{1}{4} \left| -\frac{3}{4} \right| \frac{1}{2} (1 + c)^{-7/4} |x|^2.$$

Since $(1 + c)^{-7/4} \leq 1$, we get

$$|R_2(0.1)| \leq \frac{3}{32} (0.1)^2 \leq \frac{3}{32} 10^{-4} < 10^{-5}.$$

(b) $|R_2| \leq \dfrac{3}{8} \cdot 10^{-2}$ (c) $|R_2| \leq \dfrac{3}{32} \cdot 10^{-2}$

2. (a) $|R_3| \leq 5 \cdot 10^{-4}$ (b) $|R_3| \leq \dfrac{1}{16} (0.8)^{-5/2}(0.2)^3 \leq \dfrac{1}{2 \cdot 8^3} \leq 10^{-3}$

(c) $|R_3| \leq \dfrac{1}{16} \cdot 10^{-4}$

3. Estimate $R_2(x)$ for $(1 + x)^{1/3}$ and $-0.1 \leq x \leq 0.1$. The general expression for R_2 with $s = 1/3$ is

$$|R_2(x)| = \left| \frac{(1/3)(1/3 - 1)}{2} \right| (1 + c)^{1/3 - 2} |x|^2$$

so the term $(1 + c)^{-5/3} = 1/(1 + c)^{5/3}$ will be biggest when $x = 0.1$. Also $|x|^2$ is biggest when $x = 0.1$. Hence

$$|R_2(x)| \leq \frac{1}{3} \frac{2}{3} \frac{1}{2} (0.9)^{-5/3}(0.1)^2$$

$$\leq \frac{1}{9} \left(\frac{10}{9} \right)^{5/3} 10^{-2}$$

$$\leq \frac{1}{9} \left(\frac{10}{9} \right)^2 10^{-2} = \frac{1}{729} < \frac{1}{7} \times 10^{-2}.$$

4. (a) $|R_2| \leq \frac{1}{2} \cdot (0.8)^{-3/2} \cdot 10^{-2} \leq \frac{1}{2} \cdot \frac{1}{(0.8)^2} \cdot 10^{-2} \leq 10^{-2}$ (b) $|R_2| \leq \frac{1}{8} \cdot 10^{-2}$

5. (a) $5\left(+\frac{1}{3} \cdot \frac{1}{125}\right) + E$ (b) $5\left(1 + \frac{1}{3} \cdot \frac{1}{25} - \frac{1}{9} \cdot \frac{1}{625}\right) + E$

(c) $5\left(1 + \frac{1}{3} \cdot \frac{6}{125} - \frac{1}{9} \cdot \frac{6^2}{125^2}\right) + E.$

In this part we include the estimate for the error. We write

$$131 = 125 + 6 = 125\left(1 + \frac{6}{125}\right)$$

so

$$(131)^{1/3} = 5\left(1 + \frac{6}{125}\right)^{1/3}.$$

Then

$$\left|R_3\left(\frac{6}{125}\right)\right| \leq \frac{1}{3} \frac{2}{3} \frac{5}{3} \frac{1}{3!} \left(\frac{6}{125}\right)^3 \leq \frac{1}{9} \times 10^{-4}.$$

Hence

$$(131)^{1/3} = 5\left(1 + \frac{1}{3} \frac{6}{125} - \frac{1}{9} \frac{6^2}{(125)^2}\right) + E,$$

where

$$|E| = \left|5R_3\left(\frac{6}{125}\right)\right| \leq \frac{5}{9} \times 10^{-4} < 10^{-4}.$$

(d) $6\left(1 + \frac{1}{3} \cdot \frac{4}{6^3}\right) + E$

6. (a) $10\left(1 - \frac{1}{2} \cdot \frac{3}{100} - \frac{1}{8} \cdot \frac{3^2}{100^2}\right) + E$ (b) $10\left(1 + \frac{1}{2} \cdot \frac{2}{100}\right) + E$

(c) $10\left(1 + \frac{1}{2} \cdot \frac{5}{100} - \frac{1}{8} \cdot \frac{5^2}{100^2}\right) + E$

(d) $5\left(1 + \frac{1}{2} \cdot \frac{3}{25} - \frac{1}{8} \cdot \frac{3^2}{25^2} + \frac{1}{3!} \cdot \frac{3}{8} \cdot \frac{3^3}{25^3}\right) + E$

By writing $28 = 25 + 3 = 25(1 + 3/25)$ you can apply the same method as in the examples, and $E = 5R_4(3/25)$, so we have to estimate $R_4(3/25)$. We have:

$$\left|R_4\left(\frac{3}{25}\right)\right| \leq \frac{1}{2} \frac{1}{2} \frac{3}{2} \frac{5}{2} \frac{1}{4!} \left(\frac{3}{25}\right)^4 \leq \frac{1}{8} \times 10^{-4}$$

so $|E| \leq (5/8) \times 10^{-4}$ which is within the desired accuracy.

XIII, §8, p. 471

1. 0   **2.** 1/4!   **3.** 2   **4.** 0   **5.** 1   **6.** 1   **7.** 1   **8.** 1   **9.** 1   **10.** 2   **11.** $\frac{1}{2}$ **12.** 0
13. $-\frac{1}{2}$ **14.** 1   **15.** 1   **16.** 1   **17.** $-\frac{1}{2}$ **18.** -1 **19.** 2   **20.** $\frac{1}{2}$ **21.** 1   **22.** 1

23. -1 **24.** 1 **25.** 1 **26.** 0 **27.** $-\frac{1}{6}$ **28.** 0 **29.** $\frac{1}{2}$ **30.** 1 **31.** $-\frac{1}{8}$ **32.** $-\frac{1}{9}$
33. (a) 0 (b) 0 (c) 0 **34.** 1 **35.** $-\frac{1}{2}$ **36.** 0 **37.** $\dfrac{1}{5!}$ **38.** 0 **39.** -1

XIV, §2, p. 480

3. No **4.** Yes **5.** No **6.** No **7.** No **8.** Yes **9.** Yes

XIV, §3, p. 482

1. Yes **2.** Yes **3.** No **4.** Yes **5.** No **6.** Yes **7.** No **8.** Yes **9.** No
10. Yes **11.** No **12.** Yes **13.** No **14.** Yes **15.** Yes **16.** Yes **17.** Yes
18. Yes

XIV, §4, p. 485

3. Yes **4.** Yes **5.** Yes **6.** Yes **7.** Yes **8.** Yes **9.** Yes **10.** Yes

XIV, §5, p. 488

1. Yes **2.** Yes **3.** Yes **4.** Yes **5.** Yes
6. Converges, but not absolutely **7.** Yes **8.** Converges, but not absolutely
9. Converges, but not absolutely **11.** Converges, but not absolutely
12. Converges, but not absolutely
13. Does not converge; does not converge absolutely
14. Does not converge **15.** Converges, but not absolutely
17. Converges, but not absolutely **18.** Yes
19. Converges, but not absolutely **20.** Converges, but not absolutely

XIV, §6, p. 494

2. (a) $4/e^2$ (b) $2^2 5^5 e^{-4}/3^3$ **3.** (a) 0 (b) ∞ **4.** 1 **5.** 1 **6.** 1 **7.** 1 **8.** $\frac{1}{2}$
9. 2 **10.** 0 **11.** 1 **12.** 1 **13.** $\frac{1}{2}$ **14.** 1 **15.** $\frac{1}{4}$ **16.** $\dfrac{1}{e}$ **17.** 27 **18.** $\dfrac{4}{e^2}$ **19.** 0
20. 2 **21.** 2 **22.** 3 **23.** 1 **24.** ∞ **25.** 1 **26.** ∞ **27.** 1 **28.** ∞ **29.** e
30. ∞

App., §1, p. 504

1. (a) glb is 2; lub does not exist. (b) glb is 1; lub does not exist.
   (c) glb does not exist; lub does not exist.
2. (a) glb is 0; lub is $\sqrt[3]{5}$. (b) glb is 0; lub is $\sqrt[3]{5}$. (c) glb is -2; lub is 2.
   (d) glb does not exist; lub is $\frac{11}{2}$.

App., §2, p. 513

4. $f(x)$ exist for all x; $f(x) = 1$, $|x| \geq 1$; $f(x) = 0$, $|x| < 1$
5. (a) $f(1) = 0$, $f(\frac{1}{2}) = -1$, $f(2) = 1$
(b) $\lim\limits_{x \to 1} f(x)$ does no exist (c) $\lim\limits_{x \to -1} f(x)$ does not exist
6. (a) $f(1) = 0$, $f(\frac{1}{2}) = 1$, $f(2) = 1$
(b) $\lim\limits_{x \to 1} f(x) = 1$ (c) $\lim\limits_{x \to -1} f(x) = 1$
7. (a) 0 (b) 0 (c) 0 (d) 0 **8.** (a) 0 (b) 0 (c) 0

XV, §1, p. 530

	$A + B$	$A - B$	$3A$	$-2B$
1.	$(1, 0)$	$(3, -2)$	$(6, -3)$	$(2, -2)$
2.	$(-1, 7)$	$(-1, -1)$	$(-3, 9)$	$(0, -8)$
3.	$(1, 0, 6)$	$(3, -2, 4)$	$(6, -3, 15)$	$(2, -2, -2)$
4.	$(-2, 1, -1)$	$(0, -5, 7)$	$(-3, -6, 9)$	$(2, -6, 8)$
5.	$(3\pi, 0, 6)$	$(-\pi, 6, -8)$	$(3\pi, 9, -3)$	$(-4\pi, 6, -14)$
6.	$(15 + \pi, 1, 3)$	$(15 - \pi, -5, 5)$	$(45, -6, 12)$	$(-2\pi, -6, 2)$

XV, §2, p. 534

1. No **2.** Yes **3.** No **4.** Yes **5.** No **6.** Yes **7.** Yes **8.** No

XV, §3, p. 537

1. (a) 5 (b) 10 (c) 30 (d) 14 (e) $\pi^2 + 10$ (f) 245
2. (a) -3 (b) 12 (c) 2 (d) -17 (e) $2\pi^2 - 16$ (f) $15\pi - 10$
4. (b) and (d)

XV, §4, p. 551

1. (a) $\sqrt{5}$ (b) $\sqrt{10}$ (c) $\sqrt{30}$ (d) $\sqrt{14}$ (e) $\sqrt{10 + \pi^2}$ (f) $\sqrt{245}$
2. (a) $\sqrt{2}$ (b) 4 (c) $\sqrt{3}$ (d) $\sqrt{26}$ (e) $\sqrt{58 + 4\pi^2}$ (f) $\sqrt{10 + \pi^2}$
3. (a) $(\frac{3}{2}, -\frac{3}{2})$ (b) $(0, 3)$ (c) $(-\frac{2}{3}, \frac{2}{3}, \frac{2}{3})$ (d) $(\frac{17}{26}, -\frac{51}{26}, \frac{34}{13})$

   (e) $\dfrac{\pi^2 - 8}{2\pi^2 + 29}$ $(2\pi, -3, 7)$ (f) $\dfrac{15\pi - 10}{10 + \pi^2}$ $(\pi, 3, -1)$

4. (a) $(-\frac{6}{5}, \frac{3}{5})$ (b) $(-\frac{6}{5}, \frac{18}{5})$ (c) $(\frac{2}{15}, -\frac{1}{15}, \frac{1}{3})$ (d) $-\frac{17}{14}(-1, -2, 3)$

   (e) $\dfrac{2\pi^2 - 16}{\pi^2 + 10}$ $(\pi, 3, -1)$ (f) $\dfrac{3\pi - 2}{49}$ $(15, -2, 4)$

5. (a) $\dfrac{-1}{\sqrt{5}\sqrt{34}}$ (b) $\dfrac{-2}{\sqrt{5}}$ (c) $\dfrac{10}{\sqrt{14}\sqrt{35}}$ (d) $\dfrac{13}{\sqrt{21}\sqrt{11}}$ (e) $\dfrac{-1}{\sqrt{12}}$

6. (a) $\dfrac{35}{\sqrt{41\cdot35}}, \dfrac{6}{\sqrt{41\cdot6}}, 0$ (b) $\dfrac{1}{\sqrt{17\cdot26}}, \dfrac{16}{\sqrt{41\cdot17}}, \dfrac{25}{\sqrt{26\cdot41}}$

7. Let us dot the sum

$$c_1 A_1 + \cdots + c_r A_r = O$$

with A_i. We find

$$c_1 A_1 \cdot A_i + \cdots + c_i A_i \cdot A_i + \cdots + c_r A_r \cdot A_i = O \cdot A_i = 0.$$

Since $A_j \cdot A_i = 0$ if $j \neq i$ we find

$$c_i A_i \cdot A_i = 0.$$

But $A_i \cdot A_i \neq 0$ by assumption. Hence $c_i = 0$, as was to be shown.

8. (a) $\|A + B\|^2 + \|A - B\|^2 = (A + B) \cdot (A + B) + (A - B) \cdot (A - B)$

$$= A^2 + 2A \cdot B + B^2 + A^2 - 2A \cdot B + B^2$$

$$= 2A^2 + 2B^2 = 2\|A\|^2 + 2\|B\|^2$$

9. $\|A - B\|^2 = A^2 - 2A \cdot B + B^2 = \|A\|^2 - 2\|A\|\,\|B\|\cos\theta + \|B\|^2$

XV, §5, p. 556

1. (a) Let $A = P_2 - P_1 = (-5, -2, 3)$. Parametric representation of the line is
$X(t) = P_1 + tA = (1, 3, -1) + t(-5, -2, 3)$.
(b) $(-1, 5, 3) + t(-1, -1, 4)$

2. $X = (1, 1, -1) + t(3, 0, -4)$ **3.** $X = (-1, 5, 2) + t(-4, 9, 1)$

4. (a) $(-\tfrac{3}{2}, 4, \tfrac{1}{2})$ (b) $(-\tfrac{2}{3}, \tfrac{11}{3}, 0), (-\tfrac{7}{3}, \tfrac{13}{3}, 1)$ (c) $(0, \tfrac{17}{5}, -\tfrac{2}{5})$ (d) $(-1, \tfrac{19}{5}, \tfrac{1}{5})$

5. $P + \tfrac{1}{2}(Q - P) = \dfrac{P + Q}{2}$

XV, §6, p. 562

1. The normal vectors $(2, 3)$ and $(5, -5)$ are not perpendicular because their dot product $10 - 15 = -5$ is not 0.

2. The normal vectors are $(-m, 1)$ and $(-m', 1)$, and their dot product is $mm' + 1$. The vectors are perpendicular if and only if this dot product is 0, which is equivalent with $mm' = -1$.

3. $y = x + 8$ **4.** $4y = 5x - 7$ **6.** (c) and (d)

7. (a) $x - y + 3z = -1$ (b) $3x + 2y - 4z = 2\pi + 26$ (c) $x - 5z = -33$

8. (a) $2x + y + 2z = 7$ (b) $7x - 8y - 9z = -29$ (c) $y + z = 1$

9. $(3, -9, -5), (1, 5, -7)$ (Others would be constant multiples of these.)

10. $(-2, 1, 5)$ **11.** $(11, 13, -7)$

12. (a) $X = (1, 0, -1) + t(-2, 1, 5)$
    (b) $X = (-10, -13, 7) + t(11, 13, -7)$ or also $(1, 0, 0) + t(11, 13, -7)$

13. (a) $-\frac{1}{3}$ (b) $-\dfrac{2}{\sqrt{42}}$ (c) $\dfrac{4}{\sqrt{66}}$ (d) $-\dfrac{2}{\sqrt{18}}$

14. (a) $(-4, \frac{11}{2}, \frac{15}{2})$ (b) $(\frac{25}{13}, \frac{10}{13}, -\frac{9}{13})$ **15.** $(1, 3, -2)$

16. (a) $\dfrac{8}{\sqrt{35}}$ (b) $\dfrac{13}{\sqrt{21}}$

17. (a) $-2/\sqrt{40}$ (b) $(41/17, 23/17)$ **18.** (a) $x + 2y = 3$ (c) $6/\sqrt{5}$
19. $-12/7\sqrt{6}$

XVI, §1, p. 575

1. $(e^t, -\sin t, \cos t)$ **2.** $\left(2\cos 2t, \dfrac{1}{1+t}, 1\right)$ **3.** $(-\sin t, \cos t)$

4. $(-3\sin 3t, 3\cos 3t)$ **6.** B

7. $\left(\dfrac{1}{2}, \dfrac{\sqrt{3}}{2}\right) + t\left(\dfrac{1}{2}, \dfrac{\sqrt{3}}{2}\right)$, $(-1, 0) + t(-1, 0)$, or $y = \sqrt{3}x$, $y = 0$

8. (a) $ex + y + 2z = e^2 + 3$ (b) $x + y = 1$

11. $\sqrt{(X(t) - Q) \cdot (X(t) - Q)}$

If t_0 is a value of t which minimizes the distance, then it also minimizes the square of the distance, which is easier to work with because it does not involve the square root sign. Let $f(t)$ be the square of the distance, so

$$f(t) = (X(t) - Q)^2 = (X(t) - Q) \cdot (X(t) - Q).$$

At a minimum, the derivative must be 0, and the derivative is

$$f'(t) = 2(X(t) - Q) \cdot X'(t).$$

Hence at a minimum, we have $(X(t_0) - Q) \cdot X'(t_0) = 0$, and hence $X(t_0) - Q$ is perpendicular to $X'(t_0)$, i.e. is perpendicular to the curve. If $X(t) = P + tA$ is the parametric representation of a line, then $X'(t) = A$, so we find

$$(P + t_0 A - Q) \cdot A = 0.$$

Solving for t_0 yields $(P - Q) \cdot A + t_0 A \cdot A = 0$, whence

$$t_0 = \frac{(Q - P) \cdot A}{A \cdot A}.$$

13. Differentiate $X'(t)^2 = \text{constant}$ to get

$$2X'(t) \cdot X''(t) = 0.$$

14. Let $v(t) = \|X'(t)\|$. To show $v(t)$ is constant, it suffices to prove that $v(t)^2$ is constant, and $v(t)^2 = X'(t) \cdot X'(t)$. To show that a function is constant it suffices to prove that its derivative is 0, and we have

$$\frac{d}{dt}\, v(t)^2 = 2X'(t) \cdot X''(t).$$

By assumption, $X'(t)$ is perpendicular to $X''(t)$, so the right-hand side is 0, as desired.

15. Differentiate the relation $X(t) \cdot B = t$, you get

$$X'(t) \cdot B = 1,$$

so $\|X'(t)\|\, \|B\| \cos \theta = 1$. Hence $\|X'(t)\| = 1/\|B\| \cos \theta$ is constant. Hence the square $X'(t)^2$ is constant. Differentiate, you get

$$2X'(t) \cdot X''(t) = 0,$$

so $X'(t) \cdot X''(t) = 0$, and $X'(t)$ is perpendicular to $X''(t)$, as desired.

16. (a) $(0,\ 1,\ \pi/8) + t(-4,\ 0,\ 1)$ (b) $(1,\ 2,\ 1) + t(1,\ 2,\ 2)$
(c) $(e^3,\ e^{-3},\ 3\sqrt{2}) + t(3e^{-3},\ -3e^{-3},\ 3\sqrt{2})$ (d) $(1,\ 1,\ 1) + t(1,\ 3,\ 4)$

18. Let $X(t) = (e^t,\ e^{2t},\ 1 - e^{-t})$ and $Y(\theta) = (1 - \theta,\ \cos \theta,\ \sin \theta)$. Then the two curves intersect when $t = 0$ and $\theta = 0$. Also

$$X'(t) = (e^t,\ 2e^{2t},\ e^{-t}) \quad \text{and} \quad Y'(\theta) = (-1,\ -\sin \theta,\ \cos \theta)$$

so

$$X'(0) = (1,\ 2,\ 1) \quad \text{and} \quad Y'(0) = (-1,\ 0,\ 1).$$

The angle between their tangents at the point of intersection is the angle between $X'(0)$ and $Y'(0)$, which is $\pi/2$, because

$$\text{cosine of the angle} = \frac{X'(0) \cdot Y'(0)}{\|X'(0)\|\, \|Y'(0)\|} = 0.$$

19. $(18, 4, 12)$ when $t = -3$ and $(2, 0, 4)$ when $t = 1$.
By definition, a point $X(t) = (x(t), y(t), z(t))$ lies on the plane if and only if

$$3x(t) - 14y(t) + z(t) - 10 = 0.$$

In the present case, this means that

$$3(2t^2) - 14(1 - t) + (3 + t^2) - 10 = 0.$$

This is a quadratic equation for t, which you solve by the quadratic formula. You will get the two values $t = -3$ or $t = 1$, which you substitute back in the parametric curve $(2t^2,\ 1 - t,\ 3 + t^2)$ to get the two points.

20. (a) Each coordinate of $X(t)$ has derivative equal to 0, so each coordinate is constant, so $X(t) = A$ for some constant A.
(b) $X(t) = tA + B$ for constant vectors $A \neq O$ and B.

21. Let $E = (0, 0, 1)$ be the unit vector in the direction of the z-axis. Then $X'(t) = (-a \sin t, \ a \cos t, \ b)$ and

$$\cos \theta(t) = \frac{X'(t) \cdot E}{\|X'(t)\|} = \frac{b}{\sqrt{a^2 + b^2}}.$$

23. Differentiate the relation $X(t) \cdot B = e^{2t}$, you get

$$X'(t) \cdot B = 2e^{2t} = \|X'(t)\| \, \|B\| \cos \theta.$$

But $\|B\| = 1$ by assumption, so the speed is $v(t) = \|X'(t)\| = 2e^{2t}/\cos \theta$. Square this and differentiate. You find

$$X'(t) \cdot X''(t) = \frac{8e^{4t}}{\cos^2 \theta}.$$

25. (a) To say that $B(t)$ lies on the surface means that the coordinates of $B(t)$ satisfy the equation of the surface, that is

$$z(t)^2 = 1 + x(t)^2 - y(t)^2.$$

Differentiate. You get

$$2z(t)z'(t) = 2x(t)x'(t) - 2y(t)y'(t),$$

which after dividing by 2 yields

(*) $$z(t)z'(t) = x(t)x'(t) - y(t)y'(t).$$

Now

$$B(t) \cdot B'(t) = x(t)x'(t) + y(t)y'(t) + z(t)z'(t).$$

$$= 2x(t)x'(t) \quad \text{by (*)}.$$

(b) Given any point (x, y, z) the distance of this point to the yz-plane is just $|x|$. So if x is positive, the distance is x itself. We use the derivative test: if $x'(t) \geq 0$ for all t then x is increasing. We have:

$$2x(t)x'(t) = B(t) \cdot B'(t) \quad \text{by (a)}$$

$$= \|B(t)\| \, \|B'(t)\| \cos \theta(t).$$

By assumption, $\cos \theta(t)$ is positive, and the norms $\|B(t)\|, \|B'(t)\|$ are ≥ 0, so if $x(t) > 0$, dividing by $2x(t)$ shows that $x'(t) \geq 0$, whence $x(t)$ is increasing, as was to be shown.

26. (a) $(1, 1, \frac{2}{3}) + t(1, 2, 2)$ (b) $x + 2y + 2z = 1$

27. We have $C'(t) = (-e^t \sin t + e^t \cos t, \ e^t \cos t + e^t \sin t)$. Let θ be the angle between $C(t)$ and $C'(t)$ (the position vector). Then

$$\cos \theta = \frac{C(t) \cdot C'(t)}{\|C(t)\| \, \|C'(t)\|}$$

and a little algebra will show you it is independent of t.

XVI, §2, p. 579

1. $\sqrt{2}$ 2. (a) $2\sqrt{13}$ (b) $\dfrac{\pi}{8}\sqrt{17}$

3. (a) $\dfrac{3}{2}(\sqrt{41}-1)+\dfrac{5}{4}\left(\log\dfrac{6+\sqrt{41}}{5}\right)$ (b) $e-\dfrac{1}{e}$

4. (a) 8   (b) $4-2\sqrt{2}$

   The integral for the length is $L(t)=\displaystyle\int_a^b\sqrt{2-2\cos t}\,dt$. Use the formula

$$\sin^2 u=\frac{1-\cos 2u}{2},$$

with $t=2u$.

5. (a) $\sqrt{5}-\sqrt{2}+\log\dfrac{2+2\sqrt{2}}{1+\sqrt{5}}=\sqrt{5}-\sqrt{2}+\dfrac{1}{2}\log\left(\dfrac{\sqrt{5}-1}{\sqrt{5}+1}\dfrac{\sqrt{2}+1}{\sqrt{2}-1}\right)$

   The speed is $\|X'(t)\|=\sqrt{1+(1/t)^2}$ so the length is

$$L=\int_1^2\frac{1}{t}\sqrt{1+t^2}\,dt=\int_{\sqrt{2}}^{\sqrt{5}}\frac{u^2}{u^2-1}\,du$$

$$=\int_{\sqrt{2}}^{\sqrt{5}}\frac{u^2-1+1}{u^2-1}\,du=\int_{\sqrt{2}}^{\sqrt{5}}du+\int_{\sqrt{2}}^{\sqrt{5}}\frac{1}{u^2-1}\,du.$$

But

$$\frac{1}{u^2-1}=\frac{1}{2}\left(\frac{1}{u-1}-\frac{1}{u+1}\right).$$

These last integrals give you logs, with appropriate numbers in front.

(b) $\sqrt{26}-\sqrt{10}+\dfrac{1}{2}\log\left(\dfrac{\sqrt{26}-1}{\sqrt{26}+1}\cdot\dfrac{\sqrt{10}+1}{\sqrt{10}-1}\right)=\sqrt{26}-\sqrt{10}+\log\dfrac{5}{3}\left(\dfrac{1+\sqrt{10}}{1+\sqrt{26}}\right)$

6. $\log(\sqrt{2}+1)$ 7. 5/3   8. 8

XVII, §1, p. 586

1.   2.

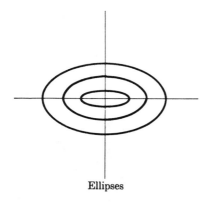

Ellipses

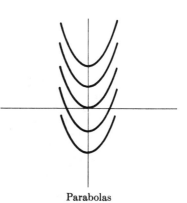

Parabolas

4.

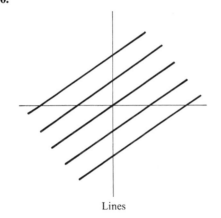

Parabolas

6.

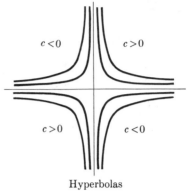

Hyperbolas

10.

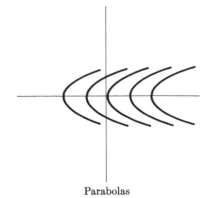

Lines

11.

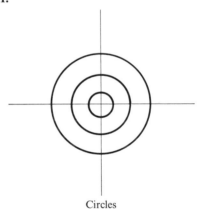

Circles

12.

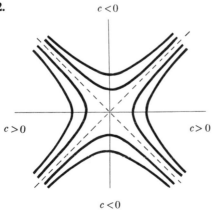

Hyperbolas

XVII, §2, p. 592

	$\partial f/\partial x$	$\partial f/\partial y$	$\partial f/\partial z$
1.	y	x	1
2.	$2xy^5$	$5x^2y^4$	0
3.	$y\cos(xy)$	$x\cos(xy)$	$-\sin(z)$
4.	$-y\sin(xy)$	$-x\sin(xy)$	0
5.	$yz\cos(xyz)$	$xz\cos(xyz)$	$xy\cos(xyz)$
6.	yze^{xyz}	xze^{xyz}	xye^{xyz}
7.	$2x\sin(yz)$	$x^2z\cos(yz)$	$x^2y\cos(yz)$
8.	yz	xz	xy
9.	$z+y$	$z+x$	$x+y$
10.	$\cos(y-3z)+\dfrac{y}{\sqrt{1-x^2y^2}}$	$-x\sin(y-3z)+\dfrac{x}{\sqrt{1-x^2y^2}}$	$3x\sin(y-3z)$

11. (1) $(2, 1, 1)$ (2) $(64, 80, 0)$ (6) $(6e^6, 3e^6, 2e^6)$ (8) $(6, 3, 2)$ (9) $(5, 4, 3)$

12. (4) $(0, 0, 0)$ (5) $(\pi^2\cos\pi^2, \pi\cos\pi^2, \pi\cos\pi^2)$
(7) $(2\sin\pi^2, \pi\cos\pi^2, \pi\cos\pi^2)$

13. $(-1, -2, 1)$ **14.** $\dfrac{\partial x^y}{\partial x}=yx^{y-1}$ $\dfrac{\partial x^y}{\partial y}=x^y\log x$

15. $(-2e^{-2}\cos\pi^2, -\pi e^{-2}\sin\pi^2, -\pi e^{-2}\sin\pi^2)$ **16.** $\left(\frac{3}{2}, \frac{1}{2}, -\dfrac{5\sqrt{3}}{2}\right)$

XVII, §3, p. 598

1. $2, -3$ **2.** a, b **3.** a, b, c
5. Select first $H=(h,0)=hE_1$. Then $A\cdot H=ha_1$ if $A=(a_1, a_2)$. Divide both sides of the relation

$$f(X+H)-f(X)=a_1h+|h|g(H)$$

by $h\neq 0$ and take the limit to see that $\alpha_1 = D_1f(X)$. Similarly use $H=(0,h)=hE_2$ to see that $\alpha_2 = D_2f(x,y)$. Similar argument for three variables.

XVIII, §1, p. 603

1. $\dfrac{d}{dt}(P+tA)=A$, so this follows directly from the chain rule.
2. 5. Indeed, $C'(t)=(2t, -3t^{-4}, 1)$ and $C'(1)=(2, -3, 1)$. Dot this with given grad $f(1, 1, 1)$ to find 5.

3. $C'(0) = (0, 1)$

Let $C'(0) = (a, b)$. Now $\operatorname{grad} f(C(0)) = (9, 2)$ and $\operatorname{grad} g(C(0)) = (4, 1)$, so using the chain rule on the functions f and g, respectively, we obtain

$$2 = \frac{d}{dt} f(C(t))\Big|_{t=0} = (9, 2) \cdot (a, b) = 9a + 2b,$$

$$1 = \frac{d}{dt} g(C(t))\Big|_{t=0} = (4, 1) \cdot (a, b) = 4a + b.$$

Solving for the above simultaneous equations yields $C'(0) = (0, 1)$.

4. (a) $\operatorname{grad} f(tP) \cdot P$.

(b) Use 4(a) and let $t = 0$.

5. Viewing x, y as constant, put $P = (x, y)$ and use Exercise 4(a). Then put $t = 1$. If you expand out, you will find the stated answer.

7. (a) $\partial f/\partial x = x/r$ and $\partial f/\partial y = y/r$ if $r = \sqrt{x^2 + y^2}$.

(b) $\dfrac{\partial f}{\partial x} = \dfrac{x}{(x^2 + y^2 + z^2)^{1/2}}$, $\dfrac{\partial f}{\partial y} = \dfrac{y}{(x^2 + y^2 + z^2)^{1/2}}$, $\dfrac{\partial f}{\partial z} = $ guess what?

8. $\dfrac{\partial r}{\partial x_i} = \dfrac{x_i}{r}$

9. (a) $\partial f/\partial x = (3x^2 y + 4x) \cos(x^3 y + 2x^2)$

$\partial f/\partial y = x^3 \cos(x^3 y + 2x^2)$

(b) $\partial f/\partial x = -(6xy - 4) \sin(3x^2 y - 4x)$

$\partial f/\partial y = -3x^2 \sin(3x^2 y - 4x)$

(c) $\partial f/\partial x = \dfrac{2xy}{(x^2 y + 5y)}$, $\dfrac{\partial f}{\partial y} = \dfrac{x^2 + 5}{x^2 y + 5y} = \dfrac{1}{y}$

(d) $\partial f/\partial x = \frac{1}{2}(2xy + 4)(x^2 y + 4x)^{-1/2}$

$\partial f/\partial y = \frac{1}{2}x^2 (x^2 y + 4x)^{-1/2}$

XVIII, §2, p. 609

1.

	Plane	Line
(a)	$6x + 2y + 3z = 49$	$X = (6, 2, 3) + t(12, 4, 6)$
(b)	$x + y + 2z = 2$	$X = (1, 1, 0) + t(1, 1, 2)$
(c)	$13x + 15y + z = -15$	$X = (2, -3, 4) + t(13, 15, 1)$
(d)	$6x - 2y + 15z = 22$	$X = (1, 7, 2) + t(-6, 2, -15)$
(e)	$4x + y + z = 13$	$X = (2, 1, 4) + t(8, 2, 2)$
(f)	$z = 0$	$X = (1, \pi/2, 0) + t(0, 0, \pi/2 + 1)$

2. (a) $(3, 0, 1)$ (b) $X - \left(\log 3, \dfrac{3\pi}{2}, -3\right) + t(3, 0, 1)$

(c) $3x + z = 3 \log 3 - 3$

3. (a) $X = (3, 2, -6) + t(2, -3, 0)$ (b) $X = (2, 1, -2) + t(-5, 4, -3)$

(c) $X = (3, 2, 2) + t(2, 3, 0)$

4. $\|C(t) - Q\|$ and see Exercise 11 of Chapter II, §1.
5. (a) $6x + 8y - z = 25$ (b) $16x + 12y - 125z = -75$
   (c) $\pi x + y + z = 2\pi$ **6.** $x - 2y + z = 1$
7. (b) $x + y + 2z = 2$ **8.** $3x - y + 6z = 14$
9. $(\cos 3)x + (\cos 3)y - z = 3\cos 3 - \sin 3.$

10. $3x + 5y + 4z = 18$ **11.** (a) $\dfrac{1}{\sqrt{27}}$ (5, 1, 1)   (b) $5x + y + z - 6 = 0$

12. $\dfrac{-10}{3\sqrt{12}}$ **13.** (a) 0   (b) 6   **14.** $4ex + 4ey + 4ez = 12e$

XVIII, §3, p. 614

1. (a) $\frac{5}{3}$ (b) max $= \sqrt{10}$, min $= -\sqrt{10}$

2. (a) $\dfrac{3}{2\sqrt{5}}$ (b) $\frac{48}{13}$ (c) $2\sqrt{145}$

3. Increasing $\left(-\dfrac{9\sqrt{3}}{2}, -\dfrac{3\sqrt{3}}{2}\right)$, decreasing $\left(\dfrac{9\sqrt{32}}{2}, \dfrac{3\sqrt{3}}{2}\right)$

4. (a) $\left(\dfrac{9}{2\cdot 6^{7/4}}, \dfrac{3}{2\cdot 6^{7/4}}, -\dfrac{6}{2\cdot 6^{7/4}}\right)$ (b) $(1, 2, -1, 1)$

5. (a) $-2/\sqrt{5}$ (b) $\sqrt{116}$ **6.** $\dfrac{1}{\sqrt{54}}$ (5, 2, 5), $6\sqrt{6}$ **7.** $\left(\dfrac{-1}{\sqrt{2}}, \dfrac{1}{\sqrt{2}}\right)$, $\sqrt{2}$

8. $\dfrac{1}{\sqrt{3}}(2e - 5)$ **9.** (a) 0   (b) $-\sqrt{1 + 2\pi^2}$

10. For any unit vector A, the function of t given by $f(P + tA)$ has a maximum at $t = 0$ (for small values of t), and hence its derivative is 0 at $t = 0$. But its derivative is grad $f(P + tA) \cdot A$, which at $t = 0$ is grad $f(P) \cdot A$. This is true for all A, whence grad $f(P) = O$. (For instance, let A be any one of the standard unit vectors in the directions of the coordinate axes.)

Although the above argument is the one which will work in Problem 11, there is a basically easier way to see the assertion. Fix all but one variable, and say x_1 is the variable. Let

$$g(x) = f(x, a_2, \ldots, a_n), \quad \text{where} \quad P = (a_1, \ldots, a_n).$$

Then g is a function of one variable, which has a maximum at $x = a_1$. Hence $g'(a_1) = 0$ by last year's calculus. But

$$g'(a_1) = D_1 f(a_1, \ldots, a_n).$$

Similarly $D_i f(P) = 0$ for all i, as asserted.

XVIII, §4, p. 619

1. $\dfrac{\partial f}{\partial x} = \dfrac{dg}{dr}\dfrac{\partial r}{\partial x} = \dfrac{dg}{dr}\dfrac{x}{r}$. Replace x by y and z. Square each term and add. You can factor

$$\frac{x^2}{r^2} + \frac{y^2}{r^2} + \frac{z^2}{r^2} = 1.$$

2. (a) $-X/r^3$ (b) $2X$ (c) $-3X/r^5$ (d) $-2e^{-r^2}X$ (e) $-X/r^2$
(f) $-4mX/r^{m+2}$ (g) $-(\sin r)X/r$

3. $F(t)^2 = (\cos t)^2 A^2 + 2(\cos t)(\sin t) A \cdot B + (\sin t)^2 B^2 = 1$,
because $A^2 = B^2 = 1$ since A, B are unit vectors and $A \cdot B = 0$ by assumption. Hence $\|F(t)\| = 1$, so $F(t)$ lies on the sphere of radius 1.

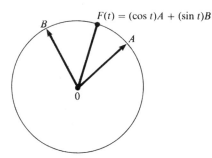

$F(t) = (\cos t)A + (\sin t)B$

4. Note that $L(t) = (1 - t)P + tQ$. If $L(t) = 0$ for some value of t, then

$$(1 - t)P = -tQ$$

Square both sides, use $P^2 = Q^2 = 1$ to get $(1 - t)^2 = t^2$. It follows that $t = 1/2$, so $\dfrac{1}{2}P = \dfrac{-1}{2} Q$, whence $P = -Q$.

5. By Exercise 4, $L(t) \neq 0$ if $0 \le t \le 1$. Then $L(t)/\|L(t)\|$ is a unit vector, and this expression is composed of differentiable expressions so is differentiable. Furthermore, we have

$$L(0) = P \qquad \text{and} \qquad L(1) = Q.$$

Thus if we put $C(t) = L(t)/\|L(t)\|$, then $\|C(t)\| = 1$ for all t, and the curve $C(t)$ lies on the sphere. Also

$$C(0) = P \qquad \text{and} \qquad C(1) = Q.$$

Hence $C(t)$ is a curve on the sphere which joins P and Q.
The picture looks as follows.

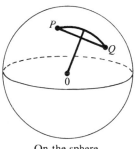

On the sphere

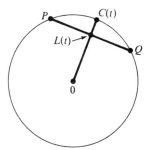

cross section

Note that $C(t)$ is the unit vector in the direction of $L(t)$.

6. Suppose P, Q are two points on the sphere, but $P = -Q$. In this case we cannot apply Exercise 5, but we can apply Exercise 3. We let

$$C(t) = (\cos t)P + (\sin t)A,$$

where A is a unit vector perpendicular to P. Then $C(t)^2 = 1$, so $C(t)$ lies on the sphere, and we have

$$C(0) = P, \qquad C(\pi) = -P.$$

Thus $C(t)$ is a curve on the sphere joining P and $-P$.

7. Let $x = a \cos t$ and $y = b \sin t$.

9. Let P, Q be two points on the sphere of radius a. It suffices to prove that $f(P) = f(Q)$. By Exercises 5 and 6, there exists a curve $C(t)$ on the sphere which joins P and Q, that is $C(t)$ is defined on an interval, and there are two numbers t_1 and t_2 such that $C(t_1) = P$ and $C(t_2) = Q$. In those exercises, we did it only for the sphere or radius 1, but you can do it for a sphere of arbitrary radius a by considering $aC(t)$ instead of the $C(t)$ in Exercises 5 or 6. Now, it suffices to prove that the function $f(C(t))$ is constant (as function of t). Take its derivative, get by the chain rule

$$\frac{d}{dt} f(C(t)) = \operatorname{grad} f(C(t)) \cdot C'(t) = h(C(t))C(t) \cdot C'(t).$$

But $C(t)^2 = a^2$ because $C(t)$ is on the sphere of radius a. Differentiating this with respect to t yields $2C(t) \cdot C'(t) = 0$, so $C(t) \cdot C'(t) = 0$, which you plug in above to see that the derivative of $f(C(t)) = 0$. Hence $f(C(t_1)) = f(C(t_2))$ so $f(P) = f(Q)$.

10. $\operatorname{grad} f(X) = \left(g'(r)\dfrac{x}{r}, g'(r)\dfrac{y}{r}, g'(r)\dfrac{z}{r} \right) = \dfrac{g'(r)}{r} X$ (say in three variables), and $g'(r)/r$ is a scalar factor of X, so $\operatorname{grad} f(X)$ and X are parallel.

XVIII, §5, p. 623

1. $k \log \|X\|$ **2.** $-\dfrac{k}{2r^2}$ **3.** $\begin{cases} \log r, & k = 2 \\[2mm] \dfrac{1}{(2-k)r^{k-2}}, & k \neq 2 \end{cases}$

Exercises 1 and 2 are special cases of 3. Let.

$$F(X) = \frac{1}{r^k} X.$$

We have to find a function $g(r)$ such that if we put $f(X) = g(r)$ then $F(X) = \operatorname{grad} f(X)$. This means we must solve the equation

$$\frac{1}{r^k} X = \frac{g'(r)}{r} X,$$

or in other words

$$g'(r) = r^{1-k}.$$

Then

$$g(r) = \int r^{1-k}\, dr,$$

which is an integral in one variable. You should know how to find it.

Index

65. $\int \dfrac{\cos ax}{\sin ax}\,dx = \dfrac{1}{a}\ln|\sin ax| + C$

66. $\int \cos^n ax \sin ax\,dx = -\dfrac{\cos^{n+1} ax}{(n+1)a} + C, \qquad n \neq -1$

67. $\int \dfrac{\sin ax}{\cos ax}\,dx = -\dfrac{1}{a}\ln|\cos ax| + C$

68. $\int \sin^n ax \cos^m ax\,dx = -\dfrac{\sin^{n-1} ax \cos^{m+1} ax}{a(m+n)} + \dfrac{n-1}{m+n}\int \sin^{n-2} ax \cos^m ax\,dx,$
$$n \neq -m \qquad \text{(If } n = -m, \text{ use No. 86.)}$$

69. $\int \sin^n ax \cos^m ax\,dx = \dfrac{\sin^{n+1} ax \cos^{m-1} ax}{a(m+n)} + \dfrac{m-1}{m+n}\int \sin^n ax \cos^{m-2} ax\,dx,$
$$m \neq -n \qquad \text{(If } m = -n, \text{ use No. 87.)}$$

70. $\int \dfrac{dx}{b + c\sin ax} = \dfrac{-2}{a\sqrt{b^2 - c^2}}\tan^{-1}\left[\sqrt{\dfrac{b-c}{b+c}}\tan\left(\dfrac{\pi}{4} - \dfrac{ax}{2}\right)\right] + C, \qquad b^2 > c^2$

71. $\int \dfrac{dx}{b + c\sin ax} = \dfrac{-1}{a\sqrt{c^2 - b^2}}\ln\left|\dfrac{c + b\sin ax + \sqrt{c^2 - b^2}\cos ax}{b + c\sin ax}\right| + C, \qquad b^2 < c^2$

72. $\int \dfrac{dx}{1 + \sin ax} = -\dfrac{1}{a}\tan\left(\dfrac{\pi}{4} - \dfrac{ax}{2}\right) + C$

73. $\int \dfrac{dx}{1 - \sin ax} = \dfrac{1}{a}\tan\left(\dfrac{\pi}{4} + \dfrac{ax}{2}\right) + C$

74. $\int \dfrac{dx}{b + c\cos ax} = \dfrac{2}{a\sqrt{b^2 - c^2}}\tan^{-1}\left[\sqrt{\dfrac{b-c}{b+c}}\tan\dfrac{ax}{2}\right] + C, \qquad b^2 > c^2$

75. $\int \dfrac{dx}{b + c\cos ax} = \dfrac{1}{a\sqrt{c^2 - b^2}}\ln\left|\dfrac{c + b\cos ax + \sqrt{c^2 - b^2}\sin ax}{b + c\cos ax}\right| + C, \qquad b^2 < c^2$

76. $\int \dfrac{dx}{1 + \cos ax} = \dfrac{1}{a}\tan\dfrac{ax}{2} + C$ 77. $\int \dfrac{dx}{1 - \cos ax} = -\dfrac{1}{a}\cot\dfrac{ax}{2} + C$

78. $\int x\sin ax\,dx = \dfrac{1}{a^2}\sin ax - \dfrac{x}{a}\cos ax + C$ 79. $\int x\cos ax\,dx = \dfrac{1}{a^2}\cos ax + \dfrac{x}{a}\sin ax + C$

80. $\int x^n \sin ax\,dx = -\dfrac{x^n}{a}\cos ax + \dfrac{n}{a}\int x^{n-1}\cos ax\,dx$

81. $\int x^n \cos ax\,dx = \dfrac{x^n}{a}\sin ax - \dfrac{n}{a}\int x^{n-1}\sin ax\,dx$

82. $\int \tan ax\,dx = -\dfrac{1}{a}\ln|\cos ax| + C$ 83. $\int \cot ax\,dx = \dfrac{1}{a}\ln|\sin ax| + C$

84. $\int \tan^2 ax\,dx = \dfrac{1}{a}\tan ax - x + C$ 85. $\int \cot^2 ax\,dx = -\dfrac{1}{a}\cot ax - x + C$

86. $\int \tan^n ax\,dx = \dfrac{\tan^{n-1} ax}{a(n-1)} - \int \tan^{n-2} ax\,dx, \qquad n \neq 1$

87. $\int \cot^n ax\,dx = -\dfrac{\cot^{n-1} ax}{a(n-1)} - \int \cot^{n-2} ax\,dx, \qquad n \neq 1$

88. $\int \sec ax\,dx = \dfrac{1}{a}\ln|\sec ax + \tan ax| + C$ 89. $\int \csc ax\,dx = -\dfrac{1}{a}\ln|\csc ax + \cot ax| + C$

Continued overleaf.

90. $\int \sec^2 ax \, dx = \dfrac{1}{a} \tan ax + C$

91. $\int \csc^2 ax \, dx = -\dfrac{1}{a} \cot ax + C$

92. $\int \sec^n ax \, dx = \dfrac{\sec^{n-2} ax \tan ax}{a(n-1)} + \dfrac{n-2}{n-1} \int \sec^{n-2} ax \, dx, \qquad n \neq 1$

93. $\int \csc^n ax \, dx = -\dfrac{\csc^{n-2} ax \cot ax}{a(n-1)} + \dfrac{n-2}{n-1} \int \csc^{n-2} ax \, dx, \qquad n \neq 1$

94. $\int \sec^n ax \tan ax \, dx = \dfrac{\sec^n ax}{na} + C, \qquad n \neq 0$

95. $\int \csc^n ax \cot ax \, dx = -\dfrac{\csc^n ax}{na} + C, \qquad n \neq 0$

96. $\int \sin^{-1} ax \, dx = x \sin^{-1} ax + \dfrac{1}{a}\sqrt{1 - a^2 x^2} + C$

97. $\int \cos^{-1} ax \, dx = x \cos^{-1} ax - \dfrac{1}{a}\sqrt{1 - a^2 x^2} + C$

98. $\int \tan^{-1} ax \, dx = x \tan^{-1} ax - \dfrac{1}{2a}\ln\left(1 + a^2 x^2\right) + C$

99. $\int x^n \sin^{-1} ax \, dx = \dfrac{x^{n+1}}{n+1} \sin^{-1} ax - \dfrac{a}{n+1}\int \dfrac{x^{n+1}\, dx}{\sqrt{1 - a^2 x^2}}, \qquad n \neq -1$

100. $\int x^n \cos^{-1} ax \, dx = \dfrac{x^{n+1}}{n+1} \cos^{-1} ax + \dfrac{a}{n+1}\int \dfrac{x^{n+1}\, dx}{\sqrt{1 - a^2 x^2}}, \qquad n \neq -1$

101. $\int x^n \tan^{-1} ax \, dx = \dfrac{x^{n+1}}{n+1} \tan^{-1} ax - \dfrac{a}{n+1}\int \dfrac{x^{n+1}\, dx}{1 + a^2 x^2}, \qquad n \neq -1$

102. $\int e^{ax} \, dx = \dfrac{1}{a} e^{ax} + C$

103. $\int b^{ax} \, dx = \dfrac{1}{a}\dfrac{b^{ax}}{\ln b} + C, \qquad b > 0, \ b \neq 1$

104. $\int x e^{ax} \, dx = \dfrac{e^{ax}}{a^2}(ax - 1) + C$

105. $\int x^n e^{ax} \, dx = \dfrac{1}{a} x^n e^{ax} - \dfrac{n}{a}\int x^{n-1} e^{ax} \, dx$

106. $\int x^n b^{ax} \, dx = \dfrac{x^n b^{ax}}{a \ln b} - \dfrac{n}{a \ln b}\int x^{n-1} b^{ax} \, dx, \qquad b > 0, \ b \neq 1$

107. $\int e^{ax} \sin bx \, dx = \dfrac{e^{ax}}{a^2 + b^2}(a \sin bx - b \cos bx) + C$

108. $\int e^{ax} \cos bx \, dx = \dfrac{e^{ax}}{a^2 + b^2}(a \cos bx + b \sin bx) + C$

109. $\int \ln ax \, dx = x \ln ax - x + C$

110. $\int x^n \ln ax \, dx = \dfrac{x^{n+1}}{n+1} \ln ax - \dfrac{x^{n+1}}{(n+1)^2} + C, \qquad n \neq -1$

111. $\int x^{-1} \ln ax \, dx = \dfrac{1}{2}(\ln ax)^2 + C$

112. $\int \dfrac{dx}{x \ln ax} = \ln |\ln ax| + C$

·113. $\int \sinh ax \, dx = \dfrac{1}{a} \cosh ax + C$

114. $\int \cosh ax \, dx = \dfrac{1}{a} \sinh ax + C$

115. $\int \sinh^2 ax \, dx = \dfrac{\sinh 2ax}{4a} - \dfrac{x}{2} + C$

116. $\int \cosh^2 ax \, dx = \dfrac{\sinh 2ax}{4a} + \dfrac{x}{2} + C$

117. $\int \sinh^n ax \, dx = \dfrac{\sinh^{n-1} ax \cosh ax}{na} - \dfrac{n-1}{n}\int \sinh^{n-2} ax \, dx, \qquad n \neq 0$